DIGITAL ELECTRONICS
An Integrated Laboratory Approach

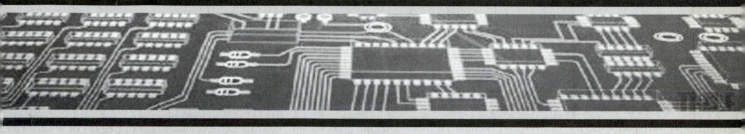

DIGITAL ELECTRONICS
An Integrated Laboratory Approach

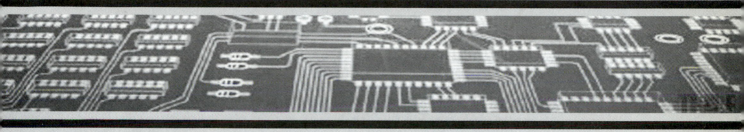

Terry L. M. Bartelt

Fox Valley Technical College

Upper Saddle River, New Jersey
Columbus, Ohio

Library of Congress Cataloging-in-Publication Data

Bartelt, Terry L. M.
    Digital electronics : an integrated laboratory approach / Terry L.M. Bartelt.
        p. cm.
    Includes index.
    ISBN 0-13-093102-0
        1. Digital electronics, I. Title.

TK7868.D5 B436 2002
621.381–dc21   2001040920

Editor in Chief: Stephen Helba
Product Manager: Scott J. Sambucci
Production Editor: Rex Davidson
Design Coordinator: Karrie Converse-Jones
Cover Designer: Jason Moore
Cover Photo: PhotoDisc
Illustrations: Jane Lopez and Steve Botts
Production Manager: Pat Tonneman
Project Management: Holly Henjum, Clarinda Publication Services

This book was set in Times Roman by The Clarinda Company and was printed and bound by The Banta Book Group. The cover was printed by The Lehigh Press, Inc.

Electronics Workbench® and MultiSim® are registered trademarks of Electronics Workbench.

Pearson Education Ltd., *London*
Pearson Education Australia Pty. Limited, *Sydney*
Pearson Education Singapore Pte. Ltd.
Pearson Education North Asia Ltd., *Hong Kong*
Pearson Education Canada, Ltd., *Toronto*
Pearson Educación de Mexico, S.A. de C.V.
Pearson Education—Japan, *Tokyo*
Pearson Education Malaysia Pte. Ltd.
Pearson Education, *Upper Saddle River, New Jersey*

10 9 8 7 6 5 4 3 2 1
ISBN: 0-13-093102-0

To my parents,
from a grateful son.

PREFACE

This book is an introductory text for students new to the electronics field. Before using the text, students should have had a course in DC circuits; and while using it, students should be concurrently enrolled in a basic solid-state electronics class.

This text can be the basis for a one- or two-semester course in digital electronics at the vocational school, technical school, or community college level, and in industrial education at the university level. It is also appropriate for students enrolled in continuing education evening courses and for advanced high school students studying digital electronics. With a background in individualized instruction, the author has taken great care to describe the concepts completely and with clarity, making the book also appropriate for book club members, for military schools, and as a reference in self-study courses.

Because each chapter of the book builds on information presented in the preceding chapter, it is important that the reader comprehend the initial material before advancing to the more difficult material covered in later chapters. The exception is chapter 4 on Boolean algebra. This chapter is designed to be studied either in the numerical sequence where it is placed or at the end of the book, or it may be deleted.

Each chapter begins with a list of objectives and concludes with a summary and end-of-chapter problems. Illustrative examples and review questions are placed throughout each chapter for consistent reinforcement and the figures are designed to enhance the explanation of the concepts. Explanations about the use of components and circuits in practical digital devices are provided to give the reader a better insight and understanding.

Many of the digital components or circuits in this book also exist in integrated circuit form. These ICs are included and their operation is described in detail. To reinforce the theory of digital components and circuits, there are 31 experiments included in the book. All of these lab activities have been class tested and can be performed on protoboards.

A test file that contains exams for each chapter is provided with an answer key in an instructor's guide. The guide also has answers to all the review questions and chapter problems in the text, and solutions for all the laboratory experiments. The author has constructed some of the circuits for the laboratory experiments on the simulation software product produced by Electronics Workbench (MultiSim). These circuits are provided on a CD-ROM that accompanies the instructor's guide.

An appendix on troubleshooting procedures and test equipment used for defective digital circuitry is included. The last section of all but two chapters describes troubleshooting principles that apply to the types of circuitry covered in each chapter. Chapter problems include exercises to help the reader envision and experience the troubleshooting process. The last chapter explains how circuits described throughout the book can be used in the construction of functional digital equipment.

I wish to thank the reviewers of the manuscript: Terry Collett, Lake Michigan College; Richard Skelton; and Ray Williams, Tennessee State University.

I would like to express my appreciation to my colleagues and students who have given me assistance and feedback during the development of the text. I would also like to acknowledge my editor, Scott Sambucci, and Holly Henjum from Clarinda Publication Services. Special thanks to my wife Carol for her patience and support during this long project.

Terry Bartelt

CONTENTS

DIGITAL ELECTRONICS
An Integrated Laboratory Approach

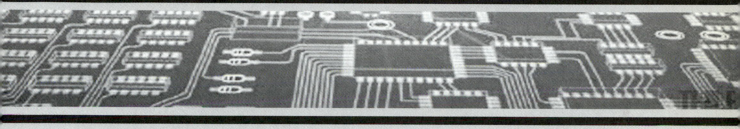

CHAPTER 1

INTRODUCTION TO DIGITAL ELECTRONICS

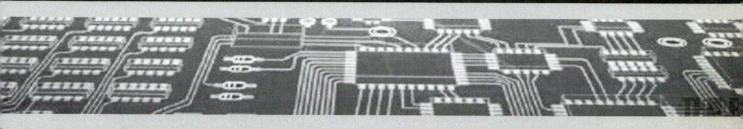

When you complete this chapter, you will be able to:

1. Categorize variables as being either analog or digital in nature.
2. Describe the difference between analog and digital signals.
3. List the events that have influenced the evolution of digital devices.
4. List and describe the difference between memory and decision-making devices.
5. List and describe the advantages of digital techniques over analog.
6. Describe the fundamental operation of various devices that perform complex logic and storage functions.

1.1 INTRODUCTION

Over the past century, electronic equipment has evolved from rather primitive creations such as the telegraph to the sophisticated devices used today, that is, programmable radio receivers, talking cash registers, etc. Perhaps the first large-scale application of electronic equipment was the telephone. Within 20 years, information was transmitted great distances by radio. Today, television instantly brings us news from anywhere in the world through satellite communications.

Applications of electronic equipment have not been limited to the area of communications. For decades, electronic devices have been used in the home, in business, in transportation, for military equipment, and for entertainment.

One element shared by most of these types of equipment is that their operation depends on analog electrical signals. Much of the equipment being developed today still operates by using analog voltage and current values. However, many applications requiring equipment that uses digital signals are being developed at an even faster rate.

1.2 COMPARING ANALOG AND DIGITAL SIGNALS

There are basically two types of electrical-electronic signals, analog and digital.

Analog Signals

Analog signal:

A voltage or current value that is proportional to the quantity it represents.

The word *analog* is derived from the word *analogous,* which means "similar to." Therefore, an **analog signal** is defined as one whose size is proportional to the quantity it represents. Because this signal is continuous, it has an infinite set of possible values.

A representation of how an analog signal is both developed and reacted to is illustrated in Fig. 1-1. A simple series circuit with an ammeter and a light bulb is shown. Both the amount of current through the meter and the intensity of the bulb change gradually as the rheostat is varied. In the real world, most quantities, such as sound, velocity, length, weight, temperature, distance, pressure, and light intensity, are analog in nature. For this reason, analog signals, which often represent these values are the most common type monitored, processed, or controlled by electronic circuitry.

FIGURE 1.1
Representation of an analog signal.

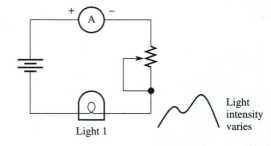

Light 1

Light intensity varies

Digital Signals

Digital signal:

A voltage or current value that abruptly alternates between two different levels.

A **digital signal** is one that alternates between two discrete levels of voltage. These changes are very abrupt. A typical digital signal is shown in Fig. 1-2. It illustrates that the most positive fixed voltage represents what is called a 1 state. Likewise, the most negative fixed voltage represents a 0 state.

A representation of how a digital signal is both developed and reacted to is illustrated in Fig. 1-3. A series circuit is shown with a light bulb and ammeter. When the switch is closed, the meter movement and intensity of the light bulb increase abruptly, representing a 1 state. When the switch is opened, the bulb intensity and meter movement abruptly vanish, which represents a 0 state.

Analog signals represent quantities that are proportional; digital signals represent quantities that often mean *yes* or *no, true* or *false, on* or *off, high* or *low,* and *1* or *0.*

For most applications, a 1 or a 0 by itself does not provide enough information to be useful. However, when grouped together in some type of orderly format, 1s and 0s are capable of representing an unlimited variety of information.

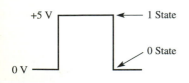

FIGURE 1.2
A digital signal.

FIGURE 1.3
Representation of a digital signal.

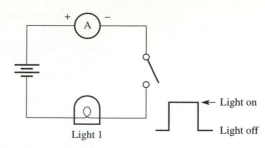

Light 1

Light on

Light off

DIGITAL CIRCUIT DEVICES

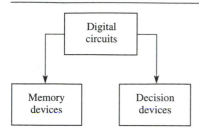

FIGURE 1.4
Digital circuit categories.

Memory device:

A device capable of storing both types of digital signals.

Flip-flop:

A memory device that is capable of storing one digital level until it is made to store the opposite digital signal level.

Gate:

A device with at least two inputs and one output that is capable of making logic decisions.

Digital circuits process and manipulate information in ways similar to a human brain. The brain's thinking function requires remembering facts and making decisions. Digital circuits are divided into two different categories that fulfill the same functions. These are memory devices and decision devices; see Fig. 1-4.

Memory Devices

A **memory device** is capable of remembering, or storing, one digital signal level until it is directed to change to the opposite digital signal level. Such a device is called a **flip-flop;** see Fig. 1-5. The name is derived from the way it operates. A flip-flop has the ability to hold one signal until a control signal causes it to flip or flop to the opposite state. When several flip-flops are connected together, they are capable of storing groups of digital data.

Decision Devices

The basic decision process is performed by a device called a **gate,** which has at least two inputs and one output; see Fig. 1-6. Depending on what combinations of inputs are applied, a particular output results. Therefore, a decision has been made. Gates are often referred to as logic devices because they are capable of making logic decisions.

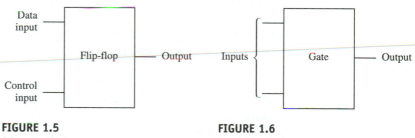

FIGURE 1.5
Memory device.

FIGURE 1.6
Decision device.

When gates and flip-flops are connected into various configurations, they are capable of performing an unlimited variety of applications, many of which are described throughout this book.

■ REVIEW QUESTIONS

1. Which of the following are analog in nature?
   (a) A light switching on and off.
   (b) Light intensity.
   (c) Temperatures.
   (d) Velocity.

2. How many distinct voltage levels does a digital signal have?
  (a) One.
  (b) Two.
  (c) Ten.
  (d) Infinity.

3. Flip-flops are used to do which of the following?
  (a) Generate analog signals.
  (b) Make logic decisions.
  (c) Store digital data.
  (d) Store analog signals.

4. Gates are used to do which of the following?
  (a) Generate analog signals.
  (b) Store digital data.
  (c) Make logic decisions.
  (d) Convert analog signals to digital signals.

1.4   EVOLUTION OF DIGITAL CIRCUITS

Perhaps the greatest application of digital techniques today is in computers. Actually, the design of the modern computer was devised in 1834 by the English mathematician Charles Babbage. He invented a machine called the analytical engine, which consisted of all the basic sections that are used by modern digital computers. The machine used punched cards for the input and output of data. However, the machine was not practical. The technology of the time had not advanced sufficiently to produce the required logic components that would function within the necessary tolerances. It was not until the development of the relay, a descendant of the original electrically controlled telephone switch, that the computer became a reality. This occurred in 1944 when the first general-purpose computer called the Mark I was designed. It used gears, relays, and punched cards. Its development was the result of a joint venture between IBM and Harvard University. A year later, the University of Pennsylvania developed the first all-electronic computer, which replaced the relay with electron tubes. Named ENIAC (Electronic Numerical Integrator and Calculator), it consisted of over 18,000 electron tubes as the primary logic component along with hundreds of thousands of resistors, capacitors, and inductors. There were several major problems associated with ENIAC. It took up 15,000 square feet of floor space, consumed 200 kilowatts of power and experienced an incredible number of tube failures (19,000 in 1952). To perform a desired function, ENIAC had to be programmed with a set of instructions by being wired a certain way. Unlike today's computers that require a keyboard for reprogramming, ENIAC had to be rewired. During the late 1940s and early 1950s, commercially available computers made from relays and vacuum tubes were programmed this way. Therefore, they were used only for single-job operations. They were labeled first-generation computers.

  With the advent of the transistor in 1948, the cumbersome volume of vacuum tubes and relays in computers was replaced. Thus, the second-generation general-purpose computers were smaller, faster, less expensive, and more reliable than the ones they replaced. See Fig. 1-7.

  The most recent and one of the most significant advances in digital technology occurred in the early 1960s. Advances in semiconductor technology permitted an entire circuit to be located on a small wafer of silicon. This device, called an **integrated circuit (IC),** is shown in Fig. 1-8. The cutaway view of one type of IC package shows a chip mounted inside a plastic case with connections to pins that are used for inputs and outputs. The chip contains miniature components, such as diodes, transistors, capacitors, and resistors, and conductor paths to which they are connected. Also called *chips,* ICs have the advantage of being small, inexpensive, low-power consumers, and very reliable. Recognizing these features when the IC was developed, electronic engineers began using them in the equipment that they designed. They are presently used almost exclusively in today's

Integrated circuit (IC):

A miniature chip of silicon or germanium on which an entire electronic circuit consisting of resistors, capacitors, diodes, and transistors is built.

FIGURE 1.7
Comparison of the logic devices used during the evolution of digital equipment.

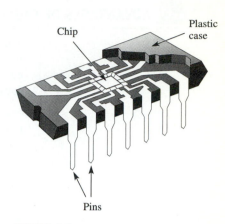

FIGURE 1.8
Cutaway view of an IC package.

Microprocessor:

The central processing unit of a computer, which is in the form of an integrated circuit.

state-of-the-art equipment. Although digital techniques had been known for years, they were not practical until the development of the IC. Integrated circuits were responsible for the development of the third-generation computer. As semiconductor technology advanced even further, a fourth-generation computer was developed that is really a component. It is called a **microprocessor** and is contained on a single integrated chip; see Fig. 1-9.

Digital circuitry enables the computer to process and store tremendous amounts of information. This is especially evident in the fields of data processing and information systems. Advantages of digital methods are also being realized in the fields of industrial

FIGURE 1.9
A microprocessor integrated circuit chip. (Reprinted by permission of Texas Instruments.)

controls, communications, consumer electronics, and instrumentation and in many other areas as well.

ADVANTAGES OF DIGITAL ELECTRONICS

The primary reason for the broad use of digital techniques is the constant advancement in IC technology. In recent years, significant improvements have been made to electronic equipment as a result of the IC. However, digital devices have also become popular because of all the advantages that they provide, some of which are listed next.

Convenience

Digital instruments and equipment are easier to use, primarily because the direct display of data is convenient to read. A digital multimeter displaying a voltage is shown in Fig. 1-10.

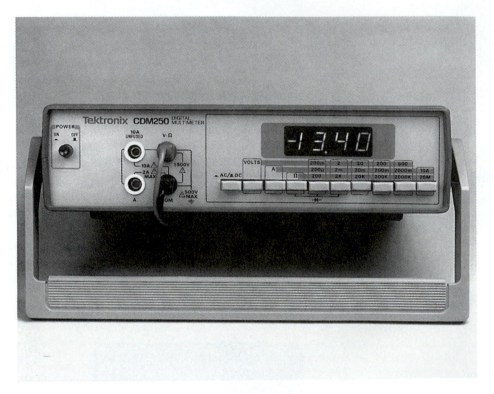

FIGURE 1.10
A numerical voltage value displayed by a digital multimeter. (Courtesy of Tektronix, Inc.)

Versatility

Digital circuits can perform most of the functions that were once considered strictly analog. Figure 1-11 demonstrates how a digital circuit can measure and control an analog value. It is a block diagram of an industrial process control system that measures and controls the temperature of a furnace. The temperature sensor develops an analog voltage that is proportional to the temperature. The voltage is then changed to an equivalent digital quantity by an analog-to-digital (A/D) converter. The digital quantity is fed into the central processing unit (CPU). The CPU determines which action should be taken if the temperature is too high, too low, or at the desired value. The digital output of the central processing unit is then changed back into an equivalent analog voltage by a digital-to-analog (D/A)

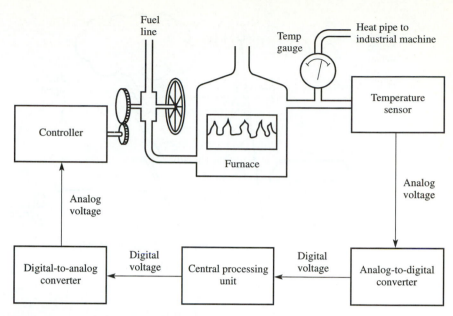

FIGURE 1.11
Temperature-control system.

converter. The analog output is applied to the controller unit. This device opens or closes the fuel line so that the temperature can be adjusted to the desired level.

Greater Stability

Analog circuits are susceptible to undesirable instability problems. Component-tolerance problems can develop from environmental factors such as extreme temperatures and high humidity. Resistor, capacitor, inductor, and transistor values can change drastically under these conditions. Component aging can also create similar problems. Analog circuitry that depends on precise voltages cannot tolerate such conditions. Digital circuit components encounter the same problems. However, the circuits continue to operate properly even if distorted signals resemble 1- or 0-state wave shapes and voltage levels.

Fewer Transmission Problems

Noise:

An electromagnetic field produced by various environmental effects and artificial devices that can induce an unwanted signal into a circuit conductor path.

When electrical signals are transmitted from one location to another over wires or by radio waves, they are exposed to different types of interference called **noise.** For example, when data are sent to a satellite, environmental factors such as radiation from lightning storms or sun spots can distort the signal. Because analog circuitry often depends on precise voltages, such deviations in the original signal can create problems. If the data are sent in digital form instead, no problems would arise so long as the distorted signals resembled the 1 and 0 states. See Fig. 1-12.

Greater Accuracy

Digital signals are more accurate than analog. The reason for this is that digital circuits are more precise when representing quantities. For example, the addition function can be performed by an analog computer using a two-input amplifier, as shown in Fig. 1-13(a). The circuit shows a 3-volt analog signal applied to one input and a 2-volt analog signal applied to the other input. If 1 volt represents a decimal digit, the output of the amplifier should generate a sum of 5 volts. However, the voltage may be 4.85 or 5.13 volts instead, depending on the accuracy of the amplifier.

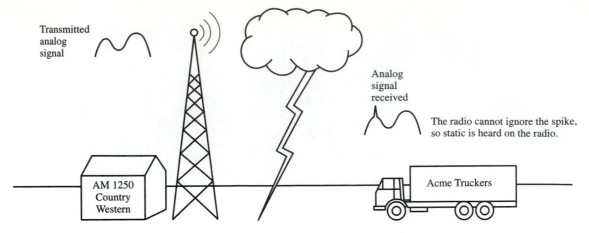

Lightning bolt causing static on AM radio

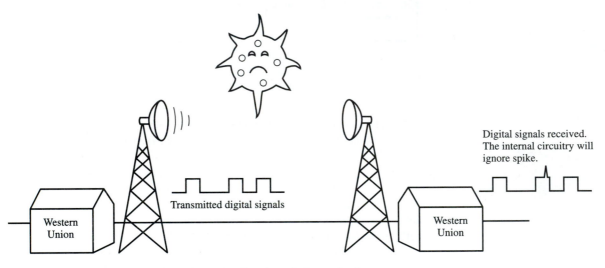

Sun spots causing distortions of teletype signals

FIGURE 1.12
Unwanted induced radiation of transmitted signals.

If binary digits equivalent to the two decimal numbers 2 and 3 were added by the calculator shown in Fig. 1-13(b), the answer would be exactly 5. This is because the discrete signals manipulated by the calculator are precise values.

Memory Capability

One of the major advantages of digital equipment over analog equipment is the ability to receive, hold, and retrieve information. For example, a pocket calculator receives information resulting from a keyboard entry, holds the information as 1s and 0s when the *Memory/Store* button is pressed, and retrieves the information when the *Memory Recall* button is pressed. Data is also stored in digital circuitry when a VCR is programmed, an alarm on a digital clock is set, or the channels on a television are pressed.

Easier Circuit Modifications

To modify the circuit operation of a linear circuit, it is either necessary to change the value of a variable component or replace a component with one of a different value. With digital circuits, simply changing the pattern of the data bits stored can alter the operation. The

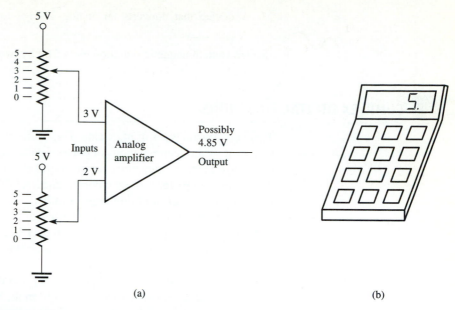

(a)

(b)

FIGURE 1.13
(A) Analog adder. (B) Digital adder.

modification is often made by changing the contents of stored data with a programming device. Figure 1-14 shows an example of a down counter that decrements its count by 1 each time an input pulse is applied to its input. A pulse arrives each second. In part (A), a number 6 is stored before the first clock pulse is applied to its input. The counter will cause an alarm to activate for six seconds until the number decrements to zero. The circuit operation can be modified very easily without changing components so that it drives an alarm for twice the duration when a 12 is loaded into the down counter.

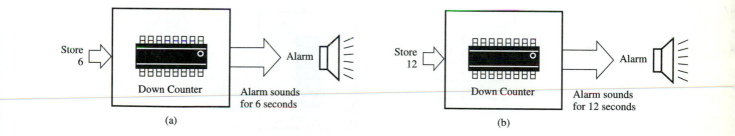

(a)

(b)

FIGURE 1.14
Programming a digital circuit.

■ REVIEW QUESTIONS

5. Which digital device has had the most significant effect on making digital techniques practical?
   (a) Relay.
   (b) Vacuum tube.
   (c) Transistor.
   (d) Integrated circuit.

6. Which of the following is not an advantage of a digital device?
   (a) Accuracy.
   (b) Fewer transmission problems.
   (c) Faster response.
   (d) Greater stability.
   (e) Memory capability.

7. A device that converts an analog voltage to an equivalent digital quantity is an _____.

8. Unwanted magnetic radiation picked up by wires is called _____.

1.6 COMPLEX DIGITAL FUNCTIONS

There are several types of logic gate. They differ in the way in which they produce an output when various combinations of binary data are applied to their inputs. A single gate can only perform simple functions. Several types of gates connected together into various configurations can perform more complex and useful logic operations. These gated arrangements are called **combination logic circuits.**

There are also several types of basic storage devices. They are referred to as *latches* and *flip-flops*. A single storage device can only store one bit of data. By connecting several storage devices together into various configurations, they can perform many complex and useful operations.

This section of the chapter provides a general overview of several important combination and storage circuits. These devices form the building blocks of digital systems, such as computers. They will be covered in detail in later chapters.

COMPLEX LOGIC FUNCTIONS

The Encoding Operation

The encoding operation is performed by a circuit called an **encoder.** This device converts a number or alphabetic character into some coded form. An example of an encoder is shown in Fig. 1-15. It converts each decimal digit, 0 through 9, to a binary code.

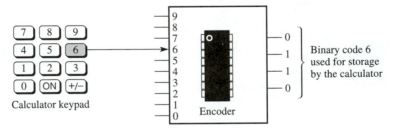

FIGURE 1.15
An encoder operation.

The Decoding Operation

The decoding operation is performed by a circuit called a **decoder.** This device converts a coded value into an encoded form. An example of a decoder is shown in Fig. 1-16. It converts a binary code representing 6 into a signal that activates a display device.

FIGURE 1.16
A decoder operation.

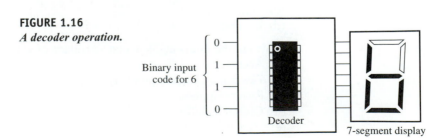

The Addition Operation

The addition operation is performed by a circuit called an **adder.** This device adds two binary numbers and produces their sum. An example of this circuit adding 2 plus 4 is shown in Fig. 1-17. The adder circuit also performs subtraction, multiplication, and division.

FIGURE 1.17
An adder operation.

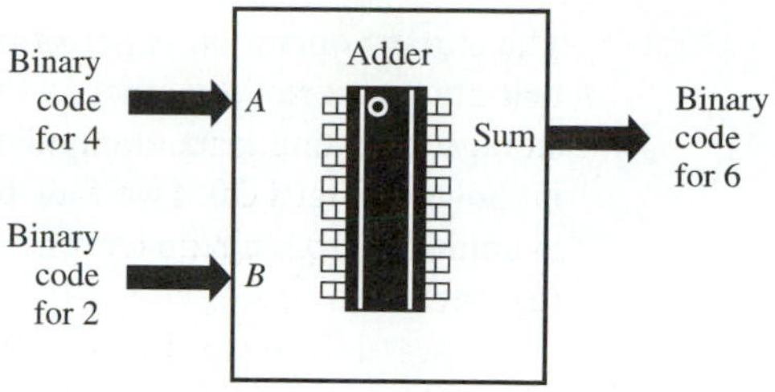

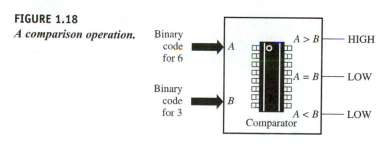

The Comparison Operation

The comparison operation is performed by a circuit called a **magnitude comparator.** This device compares two binary numbers and produces one of three outputs to indicate if they are equal, or which one is of a greater value. An example of this circuit comparing 6 and 3 is shown in Fig. 1-18. The output produced indicates that the number applied to input A is greater than the number applied to input B.

FIGURE 1.18
A comparison operation.

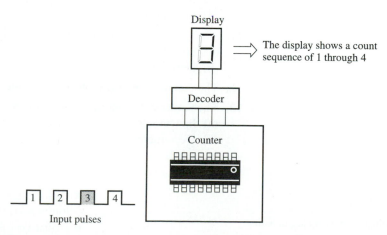

COMPLEX STORAGE FUNCTIONS

The Counting Operation

The counting operation is performed by a circuit called a **counter.** It contains several flip-flops cascaded together. Its basic function is to count or produce a particular sequence of counts in a binary code. This device stores a number and increments when activated by a pulse. An example of a counter is shown in Fig. 1-19. Each time a pulse is applied to its

FIGURE 1.19
A counter operation.

input, the binary code increments one count. The output of the counter is connected to a decoder. It converts the code into a signal that drives a module that displays the corresponding number. There are also digital counters that decrement.

The Storage Operation

The storage operation is performed by a circuit called a **register.** It uses several flip-flops that are connected together in various configurations to perform one of four operations: storage, counting, data manipulation, and arithmetic. An example of how a register is used is shown in Fig. 1-20. Two four-bit numbers are applied to an adder. Each of these numbers is unloaded from a register. The sum produced by the adder is loaded and stored in a third register.

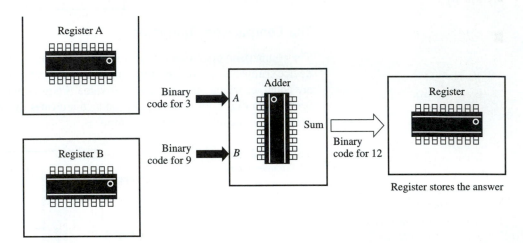

FIGURE 1.20
An adding operation that use registers for storage.

■ REVIEW QUESTIONS

9. Registers and counters are constructed from a combination of _____.
   (a) Gates.
   (b) Flip-flops.

10. Encoders are constructed from a combination of _____.
    (a) Gates.
    (b) Flip-flops.

11. A(n) _____ converts a number or alphabetic character into some coded form.
    (a) Encoder.
    (b) Decoder.

12. The comparison operation compares the magnitude of how many binary numbers?

13. The numerical count in a digital counter _____.
    (a) Increments.
    (b) Decrements.
    (c) a or b.

■ SUMMARY

■ Most information is analog in nature.

■ Analog signal voltage levels vary gradually, whereas digital signal voltage levels change abruptly.

■ Digital signals represent quantities that often mean *yes* or *no, true* or *false, on* or *off, high* or *low,* and *1* or *0.*

■ Digital devices are divided into two categories. The memory device uses flip-flops to store digital signals, and the decision device makes logic decisions about digital data.

- Digital circuitry has utilized relays, vacuum tubes, transistors, and integrated circuits to perform memory and decision-making operations.
- Advantages of digital techniques over analog are convenience, versatility, greater accuracy, fewer data transmission problems, memory capability, and easier circuit modification.
- Logic gates connected together into various configurations are called combination logic circuits.
- Digital storage devices are constructed from a combination of latches and flip-flops.
- Encoding, decoding, binary adding, and the comparison of binary numbers are complex logic functions.
- Counting and storage operations are complex storage functions.

■ PROBLEMS

1. Analog signals change _____, whereas digital signals change _____. (1-2)

2. Electronic signals developed directly from sound, temperature, pressure, light, and weight are (analog or digital) in nature. (1-2)

3. Analog signals represent quantities that are _____. (1-2)

4. Draw two different series circuits, one to represent an analog device and the other to represent a digital device. (1-2)

5. What are the different ways that digital quantities are often represented? (1-2)

6. A flip-flop is referred to as a _____ device, whereas a gate is referred to as a _____ device. (1-3)

7. What are the four electronic devices from which digital circuitry has been made? (1-4)

8. Which of the following is the most significant reason for the increased use of digital techniques? (1-4)
   (a) The development of consumer electronics.
   (b) The space program.
   (c) Automatic telephone switching systems.
   (d) Development of IC technology.

9. A fourth-generation computer utilizes the _____. (1-4)

10. Unwanted interference, called noise, that distorts electrical signals is the result of what factors? (1-5)

11. What are three advantages of digital techniques over analog? (1-5)

12. A(n) _____ converts a binary code into a signal that activates a display device. (1-6)
    (a) Encoder.
    (b) Decoder.

13. An adder circuit performs which of the following arithmetic functions? (1-6)
    (a) Addition.
    (b) Subtraction.
    (c) Division.
    (d) Multiplication.
    (e) All of the above.

14. List the three outputs produced by the magnitude comparator. (1-6)

15. The contents stored in a digital counter are in the form of _____. (1-6)
    (a) A binary code.
    (b) Decimal values.
    (c) Alphanumeric values.

16. Digital registers perform which of the following functions? (1-6)
    (a) Storage.
    (b) Counting.
    (c) Data manipulation.
    (d) Arithmetic.
    (e) All of the above.

■ ANSWERS TO REVIEW QUESTIONS

1. (b), (c), and (d)　　2. (b)　　3. (c)　　4. (c)　　5. (d)
6. (c)　　7. A/D converter　　8. noise　　9. (b)　　10. (a)
11. (a)　　12. two　　13. (c)

CHAPTER **2**

DIGITAL NUMBER SYSTEMS

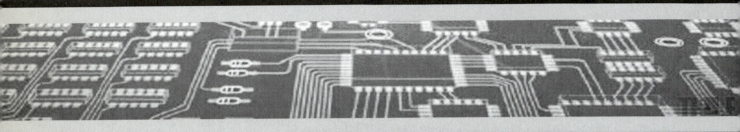

When you complete this chapter, you will be able to:

1. List practical applications of decimal, binary, octal, BCD (binary-coded decimal), and hexadecimal number systems.
2. Count in binary, octal, BCD, and hexadecimal.
3. Convert from decimal to binary and binary to decimal, decimal to octal and octal to decimal, decimal to BCD and BCD to decimal, and from decimal to hexadecimal and hexadecimal to decimal.
4. Add binary numbers.
5. Subtract binary numbers.
6. Use a table to convert the ASCII code into other number systems.

2.1 INTRODUCTION

Digital circuits use several number codes and systems to perform their functions. These number systems represent various types of information. This information can be either numbers, letters of the alphabet, punctuation, or special characters in their simplest form. When combined into more complex arrangements, they represent instructions for software programming, memory addresses, or data to be processed by computers. To understand the operation of the digital circuits that use number codes, it is also necessary to understand what the codes represent.

This chapter first describes the properties of the decimal number system because many of the concepts associated with it are already familiar. The decimal concepts are then used as a reference when the other number systems are explained. The chapter is divided into four parts:

1. Understanding Number Systems
2. Number Conversions (converting quantities from one number system to another)
3. Binary Arithmetic
4. ASCII Code

UNDERSTANDING NUMBER SYSTEMS

2.2   DECIMAL NUMBERS

Decimal number system:

Also referred to as the base 10 number system, it contains the 10 characters 0 to 9.

The **decimal number system** was first used about 2000 years ago by Hindu mathematicians. The word *decimal* is derived from the word *deci*, meaning 10. This number system has evolved naturally because humans have 10 fingers. The word *digit* is the Latin word for finger.

Using the decimal number system, early mathematicians discovered three basic concepts that simplify the operations needed to manipulate numbers. They are as follows:

1. Assigning a number value based on the type of symbol used.
2. Assigning a number value based on the position of the symbol with respect to a fixed reference (such as a decimal point).
3. The use of zero.

Type of Symbol

The type of symbol (or character) determines the value of a number. The decimal number system uses the following symbols:

0—zero

1—one

2—two

3—three

4—four

5—five

6—six

7—seven

8—eight

9—nine

Radix:

The sum of the different symbols used in a given number system.

Every number system has a **radix,** or base. The radix is equal to the sum of the different symbols used. For example, the decimal number system with a radix of 10 has 10 different digits, 0 through 9. A number system with a radix of 6 has 6 different digits, 0–5. Also, the value of the radix is always one unit greater than the value of the largest symbol used. For example, 7 is the largest digit of a number system with a radix of 8.

Symbol Position

The position of a symbol with respect to a reference point also affects its value. The further symbol digits extend to the left of the reference point, the higher their value will be.

Figure 2-1 shows the decimal positional values of several columns where digits are placed to the left of the reference (or decimal) point. At the bottom of each column, a radix of 10 with a superscript number beside it is shown. This superscript number is called an **exponent.** The exponents begin with 0 in the column next to the decimal point and increment by 1 for every column they extend to the left.

Exponent:

A symbol written above and to the right of a larger number to indicate how many times the number is multiplied by itself.

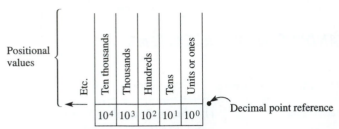

FIGURE 2.1
Positional values of each column for the decimal number system.

Whenever a radix has an exponent of zero, the positional value of that column is always 1. Notice that the positional value of each column is 10 times greater than the value in the column to its immediate right. Each positional value is determined by multiplying 10 by itself the number of times indicated by the exponent. This procedure is shown in Table 2-1.

TABLE 2.1
Procedure for determining the positional value of each decimal column

RADIX	*EXPONENT*	
10	0	$= 10^0 = 1$ (Units)
10	1	$= 10^1 = 10$ (Tens)
10	2	$= 10^2 = 10 \times 10 = 100$ (Hundreds)
10	3	$= 10^3 = 10 \times 10 \times 10 = 1000$ (Thousands)
10	4	$= 10^4 = 10 \times 10 \times 10 \times 10 = 10,000$ (Ten Thousands)
	etc.	

EXAMPLE 2.1

For the decimal number system, determine the positional value of the fourth column to the left of the reference point.

Solution

Step 1. Using Fig. 2-1, find the exponent value for the fourth column to the left of the reference point. The exponent is 3 (10^3).

Step 2. Multiply the base of 10 by itself the number of times indicated by the exponent.

$$10 \times 10 \times 10 = 1000$$

To determine the quantitative value of each decimal character, it is necessary to multiply its symbol value by the positional value of the column where it is placed.

EXAMPLE 2.2

Determine what quantity the number 6 represents in the multidigit number 632.

Solution

Step 1. Determine the column where 6 is placed. The third column from the decimal point is 10^2.

Step 2. Multiply 10 by itself the number of times indicated by the exponent. 10×10 provides the positional value of 100.

Step 3. Multiply 6 by the positional value of 100.

$$6 \times 100 = 600$$

When the digit 4 is used in the following three multidigit numbers, 124, 241, and 420, it has three different values. The digit 4 in the number 124 represents its basic unit, or absolute value of 4. In the second number, 241, the digit 4 has the value of 40. In the last number, 420, the digit 4 represents 400.

Use of the Zero

The use of the zero is also important. For example, sometimes a position in a number does not have a value between 1 and 9. Yet this position cannot be left out, otherwise there would be no way to differentiate between the quantities 306 and 36. Therefore, the zero (0) represents an absence of a value at a given position. Thus, the number 306 indicates that there are three hundreds, no tens, and six units.

The use of these three concepts makes counting and the representation of quantities quite easy to understand. These same concepts apply to other numbering systems that are described in subsequent sections of this chapter.

Decimal Application

The decimal number system is often used in digital electronics by peripheral devices. These include keyboards that are used as data inputs and display readouts that are used as data outputs.

■ REVIEW QUESTIONS

1. Another word for radix is _____.
2. The value of a symbol is based on what two factors?
3. The value of the radix is always one unit greater than the value of the _____ symbol being used.
4. The exponent assigned to the radix located to the immediate left of the reference point is always _____. The decimal radix with this exponent has a positional value equal to _____.
5. For the decimal number system, the positional value of each column is _____ times greater than the column to its immediate right.
6. The _____ is used to represent the absence of a value at a given position.

2.3   BINARY NUMBERS

Binary number system:

Also called the base 2 system, it contains the two functional characters 0 and 1.

Base 2 numbers, known as the **binary number system,** comprise the simplest number system. Like decimal numbers, their values are also determined by the symbol used and the position of the symbol. Instead of using digits 0 to 9, binary numbers use bits (derived from <u>bi</u>nary di<u>gits</u>) 0 and 1. These bits can represent any decimal value. For example, the binary number 101101 represents the decimal value of 45.

When a decimal digit-by-digit count is made and uses the last available digit (9), the count begins again at 0 and a 1 is placed in the immediate left column. When a binary bit-by-bit count is made, it also starts at zero and increases to 1. However, this is the last possible bit, so in the next count, a 0 is placed in the column and a 1 is placed in the immediate column to the left. Table 2-2 illustrates this concept and compares incrementing binary numbers to decimal numbers.

TABLE 2.2
Comparing binary numbers to decimal numbers

QUANTITY	DECIMAL (0, 1, 2, 3, 4, 5, 6, 7, 8, 9)	BINARY (0.1)
Zero	0	0
One	1	1
Two	2	10
Three	3	11
Four	4	100
Five	5	101
Six	6	110
Seven	7	111
Eight	8	1000
Nine	9	1001
Ten	10	1010
Eleven	11	1011
Twelve	12	1100
Thirteen	13	1101

Like decimal numbers, the position of a binary symbol is a factor in determining the value of the number. However, the value of each column is different for binary numbers as compared to decimal numbers.

Figure 2-2 illustrates what the decimal-equivalent positional value of each binary number represents. Each quantity is determined by multiplying the radix by itself the number of times indicated by the exponent located in the column. Notice that the value of each column is twice the amount of the column to the immediate right.

FIGURE 2.2
The decimal-equivalent positional values of each column for the binary number system.

The value of a binary 1 is determined by multiplying 1 by the positional value of the column where it is located.

Any binary 0 placed in a column does not represent a value because 0 times its positional value equals zero.

Digital system work with binary values grouped together into a specific number of digits. These groups are often referred to as **words.** Common word lengths are 8, 16, 32, and 64.

Reading and writing decimal numbers can become very cumbersome when the value is very large, and it is easy to make a mistake. To help reduce errors, decimal digits are separated by commas into groups of three. Working with binary numbers is even more diffi-

EXPERIMENT

Basic Gates and Integrated Circuits

Objectives

- To wire and operate logic gates.
- To use test equipment to detect the operation of IC logic gates.
- To use data sheets.

Materials

1—+5-V DC Power Supply	1—7432 Integrated Circuit
1—7402 Integrated Circuit	1—14 Pin IC Socket
1—7404 Integrated Circuit	4—Logic Switches or SPDT Switches
1—7411 Integrated Circuit	1—Voltmeter or Logic Probe
1—7420 Integrated Circuit	1—TTL Logic Circuit IC Data Manual

Introduction

In this experiment, you will develop truth tables for several different gate circuits by applying inputs to the gate while monitoring the output. By examining the truth table, you will identify which type of logic gates are contained in each IC package. By using a logic circuit data book, you will verify your answers.

Each IC listed contains two to six gates with one to four inputs. All gates within each IC used in this experiment are of the same TTL IC family. Be aware that the 7400 TTL series of ICs includes many more types than the ones you will be exposed to in this experiment. Many of the 7400 integrated circuits contain much more complex circuitry than simple gates.

Background Information

- All of the ICs in this experiment use pin 14 as the V_{CC} supply voltage (+5 volts) and pin 7 as ground. The experiment will not function properly if these connections are not made.
- Any inputs that are left disconnected are recognized as a logic 1 state by a TTL IC.
- Be sure to shut the power off before an IC is inserted into or removed from the socket.

Procedure Information

You will be given information that identifies the pin numbers of the inputs and output for each IC used.

Procedure

Step 1. Set up the input circuit shown in Fig. 3-23. Connect the switch to the +5-volt supply position if a logic 1 state is required at an input lead and to the ground position if a logic 0 state is required.

FIGURE 3.23

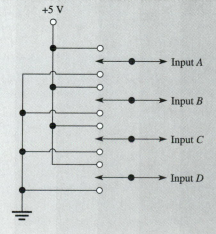

Step 2.   Using a 7402 IC, connect the circuit using the following pin numbers:

	Input		Output
	A	*B*	*Y*
Pins	2	3	1

Step 3.   Fill in the truth table for the 7402 IC and identify the gate type.

TRUTH TABLE

A	*B*	*Y*

7402 Gate Type

_____

Step 4.   Remove the 7402 IC from the socket and replace it with a 7404 IC. Connect the circuit according to the following pin numbers:

	Input	Output
	A	*Y*
Pins	5	6

Step 5.   Fill in the truth table for the 7404 IC and identify the gate type.

TRUTH TABLE

A	*Y*

7404 Gate Type

_____

Step 6.   Remove the 7404 IC from the socket and replace it with a 7420 IC. Connect the circuit according to the following pin numbers:

	Input				Output
	A	*B*	*C*	*D*	*Y*
Pins	1	2	4	5	6

Step 7. Fill in the truth table for the 7420 IC and identify the gate type.

TRUTH TABLE   7420 Gate Type

A	B	C	D	Y

_____

Step 8. Remove the 7420 IC from the socket and replace it with a 7432 IC. Connect the circuit according to the following pin numbers:

	Input		Output
	A	B	Y
Pins	1	2	3

Step 9. Complete a truth table for the 7432 IC and identify the gate type.

TRUTH TABLE   7432 Gate Type

A	B	Y

_____

Step 10. Remove the 7432 IC from the socket and replace it with a 7411 IC. Connect the circuit according to the following pin numbers:

	Input			Output
	A	B	C	Y
Pins	1	2	13	12

Step 11. Fill in the truth table for the 7411 IC and identify the gate type.

Do not disassemble the circuit at this time.

TRUTH TABLE   7411 Gate Type

A	B	C	Y

_____

Step 12. Ask the instructor for a logic data manual that contains a listing of the 7400 TTL series of integrated circuits. Verify that the gate types you identified in this experiment are correct.

Step 13. Disconnect the power supply connections to pins 7 and 14 of the 7411 IC. Apply logic 1 states to each of the inputs of the gate and record the logic state present at the output.

Output _____

Procedure Question 1

What should be the logic state produced at the output in Step 14?

Procedure Question 2

Why didn't the circuit operate properly?

Step 14. Reconnect the power supply connections to pins 7 and 14 of the 7411 IC. Leave each of the input pins of the AND gate open and record the logic state present at its output.

_____ Logic State

Procedure Question 3

What logic state is observed at each input lead of the AND gate?

Step 15. Dismantle the circuit.

■ EXPERIMENT QUESTIONS

1. Will the logic gate in a TTL IC package operate properly if a +5-V V_{CC} and GND are not connected to the device?
2. Any input of a TTL IC logic gate that is not connected is recognized as a logic _____ (0,1) state.
3. What rule should be observed when inserting or removing an IC from a socket?

3.9 EXCLUSIVE-OR GATE

Exclusive-OR gate:

A basic logic device that produces a high at its output when the logic states applied to its inputs are different.

The OR gate described in Section 3-5 is also known as an inclusive-OR gate because it produces a high output when either one or all inputs are high. An **exclusive-OR gate** operates differently in that the output is high when any input is high, but not when all inputs are high. Because the logic function of this gate excludes the "all inputs = 1" condition from producing a 1 output, the gate is termed exclusive-OR. The standard logic symbol for an exclusive-OR gate is shown in Fig. 3-24(a). Unlike other types of gates that have two or more inputs, the exclusive-OR gate always has two inputs. The truth table for the exclusive-OR gate is shown in Fig. 3-24(b).

The operation of an exclusive-OR gate is shown by the simple circuit of Fig. 3-24(c). The light representing the output turns on only when both input switches are set at opposite 1 and 0 state positions.

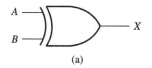

(a)

Inputs		Output
B	A	X
0	0	0
0	1	1
1	0	1
1	1	0

(b)

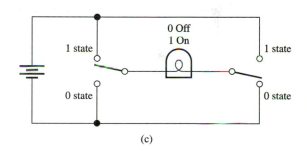

(c)

FIGURE 3.24
(a) Exclusive-OR gate symbol. (b) Exclusive-OR gate truth table. (c) Equivalent circuit.

EXAMPLE 3.4

If two waveforms, *A* and *B* of Fig. 3-25, are applied to the exclusive-OR gate, what would be the resulting waveform produced at the output?

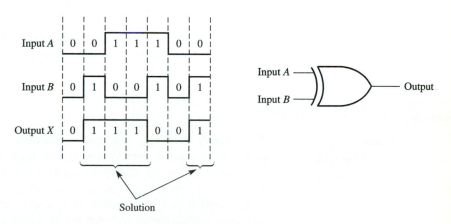

FIGURE 3.25
Input and output waveforms for the exclusive-OR gate of Example 3-4.

Solution The output goes high only when both inputs are at opposite states, as shown in the figure.

The Boolean expression for the exclusive-OR gate is

$$X = A\overline{B} + \overline{A}B$$

This expression is read as "X equals A and not B or not A and B." This simply means that the output of the gate is 1 when A is 1 and B is *not* 1, or when A is *not* 1 and B is 1. The exclusive-OR expression is sometimes represented using the inclusive-OR expression and placing a circle around the "+" symbol. Therefore, $A\overline{B} + \overline{A}B = A \oplus B$.

The pin diagram of a popular exclusive-OR gate IC package is shown in Fig. 3-26.

FIGURE 3.26
Exclusive-OR gate IC pin diagram.

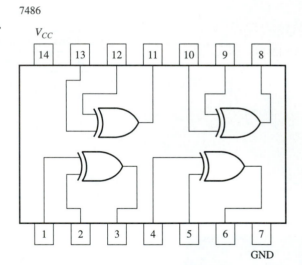

The exclusive-OR gate has a wide variety of applications that require two input conditions to be compared to determine if their states are alike or not.

3.10 EXCLUSIVE-NOR GATE

Exclusive-NOR gate:

A basic logic device that produces a high at its output when the logic states applied to its inputs are the same.

An **exclusive-NOR gate** is a basic circuit with two inputs and one output. At the output, a logic 1 is produced when both inputs are logic 1 or when both inputs are logic 0. A 0 is produced at the output when both inputs are of opposite states. The operation is exactly opposite to the operation of an exclusive-OR gate. Therefore, the standard symbol for an exclusive-NOR gate has a bubble at its output terminal, as shown in Figure 3-27(a). The truth table for this circuit is shown in Fig. 3-27(b).

The operation of an exclusive-NOR gate is shown by the simple circuit of Fig. 3-27(c). Note that the light (output) turns on only when both switches are in the same 0 or 1 state position. It is off when both switches are in opposite positions.

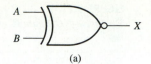

(a)

Inputs		Output
B	A	X
0	0	1
0	1	0
1	0	0
1	1	1

(b)

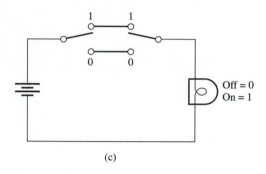

(c)

FIGURE 3.27
(a) Exclusive-NOR gate symbol. (b) Exclusive-NOR gate truth table. (c) Equivalent circuit.

EXAMPLE 3.5

If two waveforms, *A* and *B* of Fig. 3-28, are applied to the exclusive-NOR gate, what would be the resulting waveform produced at the output?

Solution The output goes high only when both inputs are at the same state, as shown in the figure.

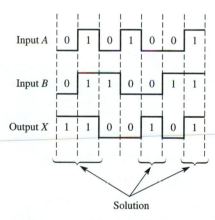

Solution

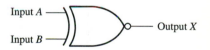

FIGURE 3.28
Input and output waveforms for the exclusive-NOR gate of Example 3-5.

The Boolean expression for the exclusive-NOR gate is

$$X = \overline{A}\,\overline{B} + AB$$

This expression is read as "*X* is equal to not *A* and not *B* or *A* and *B*." This simply means that the output of the gate is 1 when *A* is *not* 1 and B is *not* 1, or *A* is 1 and *B* is 1. The exclusive-NOR expression is sometimes represented by using the inclusive-NOR expression and placing a circle around the "+" symbol. Therefore, $\overline{A}\,\overline{B} + AB = \overline{A \oplus B}$.

The pin diagram of a popular exclusive-NOR gate IC package is shown in Fig. 3-29.

FIGURE 3.29
Exclusive-NOR gate IC pin diagram.

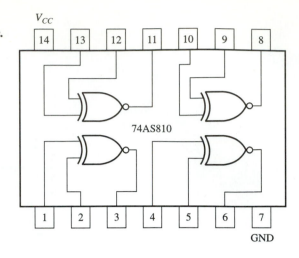

■ REVIEW QUESTIONS

15. An _____ gate produces a high on its output only when all of its inputs are at the same level.

16. An _____ gate produces a high on its output only when its inputs are at different levels.

17. The exclusive-OR and exclusive-NOR gates have no more than _____ inputs.

18. The Boolean expression for an exclusive-NOR gate with inputs *A* and *B* is _____.

EXPERIMENT

Exclusive-OR and NOR Gates

Objectives

- To wire and operate the exclusive-OR gate.
- To complete the truth table for an exclusive-OR gate.
- To learn the various functions that exclusive-OR and exclusive-NOR gates are used for.

Materials

1—+5-V DC Power Supply

5—Logic Switches or SPDT Switches

1—7486 Integrated Circuit

4—LEDs

4—100-Ohm Resistors

Introduction

The exclusive-OR (XOR) gate has two inputs. Its output goes high when its inputs are at opposite states. The XOR gate is available in IC form. The IC used in this experiment contains four gates.

The exclusive-OR gate is used for a variety of applications, including the following:

1. Acting as a programmable inverter.
2. Performing binary arithmetic functions.
3. Comparing binary data.

Procedure

Step 1. Using the 7486 XOR IC chip, construct the circuit shown in Fig. 3-30.

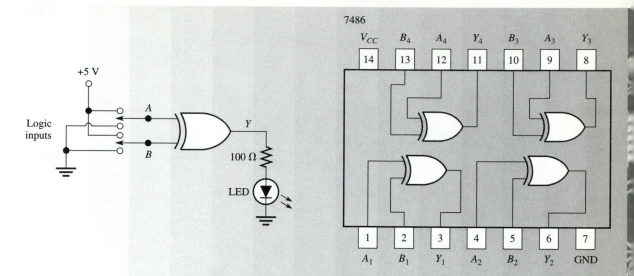

FIGURE 3.30

Step 2. Set the logic switches according to the input column of Table 3-1.
Step 3. Observe the output to determine the operation. Place the correct responses in the output column.

TABLE 3.1

B	A	Y
0	0	
0	1	
1	0	
1	1	

Background Information

Programmable Inverter

An important characteristic of the XOR gate is that it can be programmed to operate as either a straight wire or an inverter. Observe Fig. 3-31. Binary data is applied to the top input lead of the gate. The bottom lead is connected to a logic switch. When the switch is low, any binary numbers applied to the data input are passed straight through to the output. When the switch is high, binary bits at the data input are inverted as they pass through to the output.

FIGURE 3.31

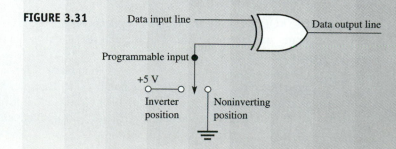

Procedure

Step 4. Construct the circuit shown in Fig. 3-32.

Step 5. Set the data switches D_1 to D_4 and the programmable switch (P) according to the input column of Table 3-2.

Step 6. Observe the outputs to determine the operation. Place the correct responses in the output column.

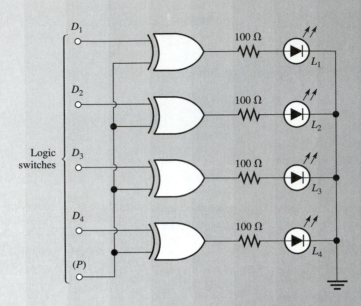

FIGURE 3.32

TABLE 3.2

Data Input Switches					Output Displays			
D_1	D_2	D_3	D_4	(P)	L_1	L_2	L_3	L_4
0	1	1	0	0				
0	0	1	0	1				
1	1	0	0	1				
1	0	1	1	0				
1	0	1	0	1				
1	1	1	1	0				

Background Information

Comparing Binary Data

The XOR gate is often called a *comparing device* because it is capable of determining if its inputs are *alike* or *unlike.* By cascading XOR gates, the number of bits that can be compared increases.

One type of compare function is to indicate if the number of 1s or 0s applied to the input of a circuit is an even or odd number. For example, a circuit called an *odd detector* is used to determine if the number of 1s or 0s at its input leads is an odd number. It produces a 1 at its output if an odd number of 1s or 0s is applied, and a 0 if an even number of 1s or

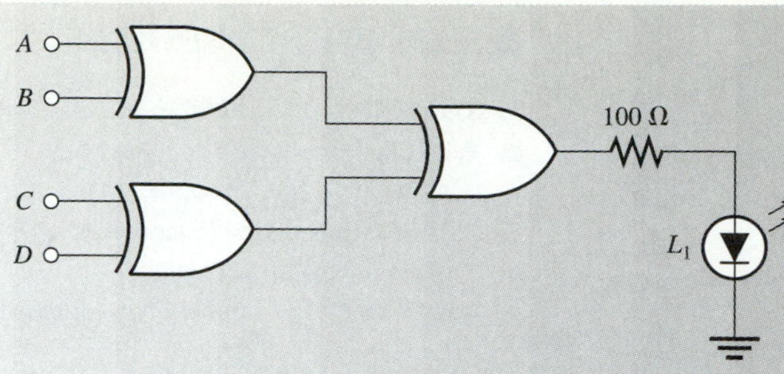

FIGURE 3.33

TABLE 3.3

	Data Switches			Display
A	**B**	**C**	**D**	**L₁**
0	0	0	0	
0	0	0	1	
0	0	1	0	
0	0	1	1	
0	1	0	0	
0	1	0	1	
0	1	1	0	
0	1	1	1	
1	0	0	0	
1	0	0	1	
1	0	1	0	
1	0	1	1	
1	1	0	0	
1	1	0	1	
1	1	1	0	
1	1	1	1	

0s is applied. An application of this device is found in a circuit called a *parity checker*, which detects whether binary data is transmitted properly.

Procedure

Step 7. Construct the circuit shown in Fig. 3-33.

Step 8. Set the logic switches *A–D* according to the input portion of Table 3-3 labeled *Data Switches*.

Step 9. Observe the output to determine the operation. Place the correct responses in the output column labeled *Display*.

Procedure Question 1

Does the light turn on when the number of 1s is even or odd?

Background Information

Exclusive-NOR gates operate exactly opposite of XOR gates. Using XNOR gates configured like the circuit in Fig. 3-33, if the number of 1s (or 0s) applied to the inputs is even, the output lead produces a logic 1 state.

■ EXPERIMENT QUESTIONS

1. The XOR gate output is __________ (low, high) when both inputs are high.
2. When the programmable input of the Programmable Inverter in Fig. 3-31 is connected to a high, what happens to a bit applied to the data input lead?
3. If 1001 is applied to the inputs of Fig. 3-33, the output will be __________ (low, high).
4. The circuit in Fig. 3-33 using exclusive- __________ (OR, NOR) gates produces a logic 1 at the output lead if the number of 1s (or 0s) applied to the inputs is an even number.

3.11 MEMORIZING TRUTH TABLES

It is important to memorize the truth tables of each one of the six gates. As complex circuits made up of these basic gates are studied, it will be much easier to analyze them if basic logic gate truth tables are committed to memory.

The process of memorizing logic gate truth tables can be simplified by utilizing a few basic principles. For example, Fig. 3-34(a) shows a NAND gate and an equivalent circuit consisting of an AND gate with an inverter connected to its output lead. The comparison illustrates that the output of an AND gate is opposite that of a NAND gate. Therefore, if the AND gate truth table is memorized, the NAND gate truth table with the opposite output is memorized. The truth table in Fig. 3-34(b) further proves this concept.

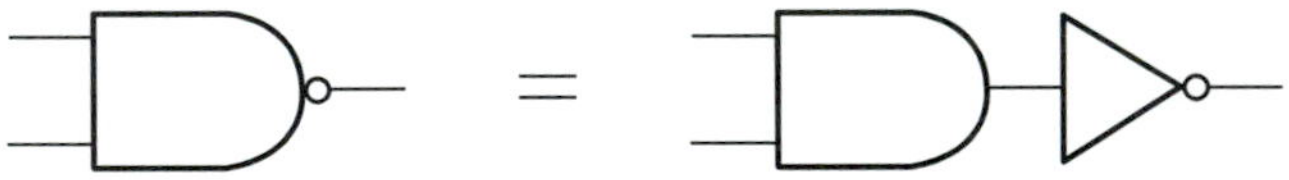

(a)

Inputs		AND output	NAND output
B	*A*		
0	0	0	1
0	1	0	1
1	0	0	1
1	1	1	0

(b)

FIGURE 3.34
(a) Equivalent NAND operations. (b) Truth table showing AND and NAND operations opposite.

The same idea applies to the other types of gates. The outputs of the OR and NOR gates are exactly opposite, as are the outputs of the exclusive-OR and exclusive-NOR gates. Therefore, by learning the truth tables for only three gates, the AND, OR, and exclusive-OR, the memorization process is simplified.

Because truth tables summarize the complete operation of logic circuits so well, they are used extensively when analyzing both simple and complex circuitry. Therefore, it is often necessary for a technician to develop a truth table to effectively analyze a circuit.

When developing a truth table, the first step is to determine the number of inputs to the circuit. By using the number as an exponent for the binary number 2, the number of possible input combinations of 1s and 0s applied to the truth table is determined. For example:

		NUMBER OF TRUTH TABLE INPUT COMBINATIONS
Two-Input Circuit	$2^2 =$	4
Three-Input Circuit	$2^3 =$	8
Four-Input Circuit	$2^4 =$	16

A truth table with eight input combinations for a three-input circuit is shown in Table 3-4. The input combinations of 1s and 0s should be set up in a binary counting sequence. This ensures that all possible combinations are covered.

TABLE 3.4
Truth table showing eight input combinations for a three-input gate[a]

	BINARY		
DECIMAL	4	2	1
0	0	0	0
1	0	0	1
2	0	1	0
3	0	1	1
4	1	0	0
5	1	0	1
6	1	1	0
7	1	1	1

[a] Decimal numbers 0 through 7 and their binary equivalents.

3.12 GATE SUBSTITUTIONS

It can be useful for electronics personnel to know that certain logic gates connected in various configurations can perform the same functions as other logic gates. The NAND gate is the most versatile of the devices that have this capability. It is referred to as the *universal gate* because it can be wired to operate as an inverter and as each of the six different logic gates, as shown in Fig. 3-35.

It is possible to achieve most gate functions from any other types of gates. This is accomplished by using inverters on the inputs, output, or a combination of both. The ability to substitute one gate for another is very useful, especially when only one type is available but another gate is needed.

There are three different types of gate inversion procedures: output inversion, input and output inversion, and input inversion. Figure 3-36 shows all of the gate inversions possible.

Output Inversion

To perform the output inversion procedure, simply add a NOT gate to the output lead. This converts one type of gate function to its complemented gate function. It allows an AND gate to operate as a NAND gate, a NAND gate as an AND, an OR as a NOR, and a NOR as an OR. The top section of Fig. 3-36 shows which conversions are made using the output inversion procedure.

Input and Output Inversion

To perform the input and output inversion procedure, simply add inverters to the inputs and output. This converts back and forth from AND to OR and from NAND to NOR. The

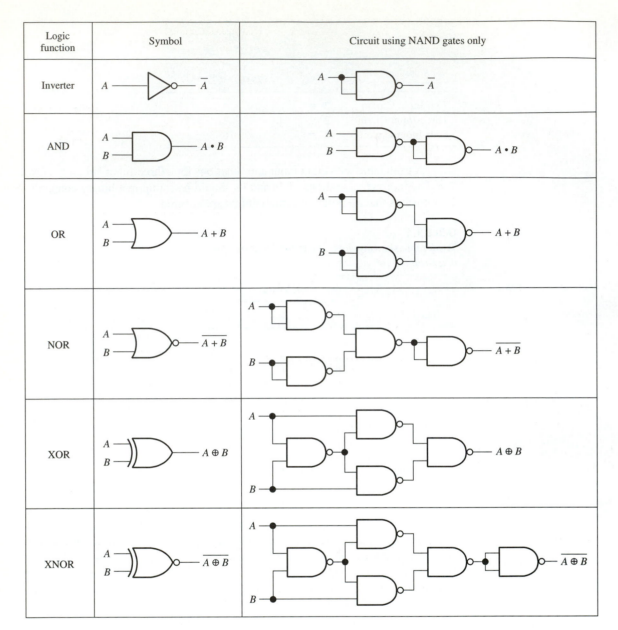

FIGURE 3.35
Substituting NAND gates.

middle section of Fig. 3-36 summarizes which conversions are made by using the input and output inversion procedure.

Input Inversion

To perform the input inversion procedure, simply add inverters to the inputs. This converts back and forth from NAND to OR and NOR to AND. The bottom section of Fig. 3-36 summarizes which conversions are made by using the input inversion procedure.

A reference chart for showing which type of gate inversion procedure to use when substituting one type of gate for another is shown in Figure 3-37.

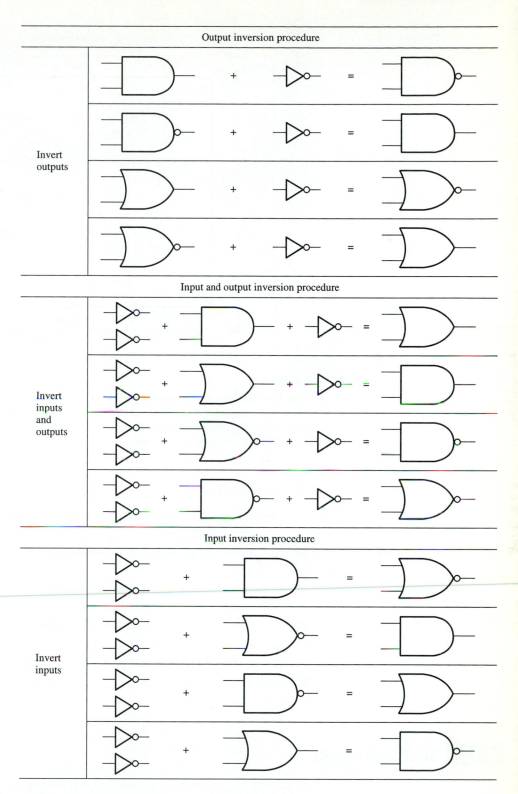

FIGURE 3.36
Gate conversion procedures using inverters.

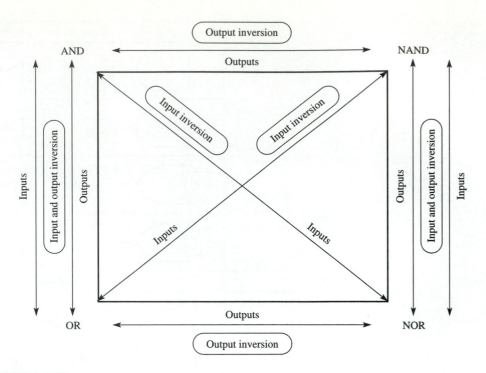

FIGURE 3.37
Gate inversion chart. The arrows indicate the following inversions: (1) The output is inverted when replacing gates that are horizontal to each other. (2) The inputs and outputs are inverted when replacing gates that are vertical to each other. (3) The inputs are inverted when replacing gates that are diagonal to each other.

■ REVIEW QUESTIONS

19. The operation of all six logic gates is more easily remembered by memorizing only _____ logic gates.

20. The number of possible input combinations of 1s and 0s for a four-input gate is _____.

21. To ensure that all input combinations of 1s and 0s are placed in the input section of the truth table, it should be set up by using the _____ counting sequence.

22. To perform _____ inversion, only invert the output of the gate.

23. Input and output inversion allows an _____ to operate as an OR, an _____ to operate as an AND, a NOR as a _____, and a NAND as a _____.

24. Input inversion allows an _____ to operate as an OR, and a NOR as a _____.

3.13 STATE INDICATORS

In addition to the standard logic symbols that have been described in this chapter, there are other types called *state indicator symbols*. In most schematic diagrams, state indicator symbols are not used. Many circuit diagrams still use standard gates exclusively. However, they often appear in manufacturer manuals. These manuals illustrate the internal circuitry of integrated circuits, especially combination logic circuits, which are covered in Chapter 5. State indicator symbols provide two types of information. One type is to show a logic function that is performed. The other type is to indicate which logic state should be present at an input terminal or an output terminal during the operation of the circuit. The advantage

EXAMPLE 3.6

Using the gate inversion chart, determine how to substitute an AND gate for a NOR gate.

Solution   The diagonal line from the top left AND corner of the chart to the bottom right corner indicates that the inverters be placed at the AND gate inputs. See Fig. 3-38.

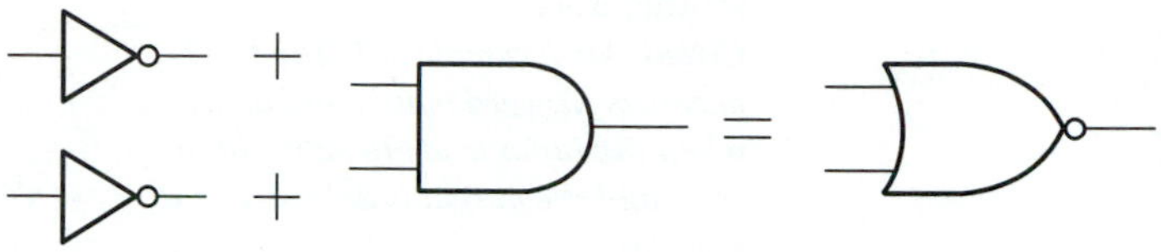

FIGURE 3.38
Substituting an AND gate for a NOR gate in Example 3-6.

of using state indicator symbols in schematic diagrams is that the information they provide is useful for troubleshooting.

An example of state indicator symbology is shown in Fig. 3-39. Two symbols are used for the NOT gate. Both operate identically by inverting the signals applied to their inputs. The difference between the symbols is the placement of the small circle called a *state indicator*. Its location indicates the way in which the inverter is being used during the activated condition, that being where it is causing something to happen (e.g., turning on a device). A lead with a circle indicates that it must be low to activate a device that it controls. Likewise, the absence of a circle indicates that the lead must be high to activate a device that it controls.

FIGURE 3.39
Two inverter symbols.

Figure 3-40 shows a NOT gate with a state indicator at the input. The **light-emitting diode (LED)** is the device being controlled by the circuit. See Fig. 3-41. The diagram shows the circuit in its inactive (resting) condition. With the push button open, a high is applied to the inverter input. Therefore, a low is generated at the output. Because the LED is not forward biased, it does not turn on. When the push button is pressed, the circuit goes into the *active* condition. A low is applied to the inverter input lead with the circle, and a high is produced at the inverter output without the circle. The LED turns on (becomes activated) because it is forward biased. This circuit has what is called an *active low* input and an *active high* output.

Light-emitting diode (LED):

A device that allows current to flow through it in one direction and gives off light when current passes through it.

FIGURE 3.40
State indicator at input. The LED becomes forward biased and turns on when the input is at the active low state and the output is activated to a high state.

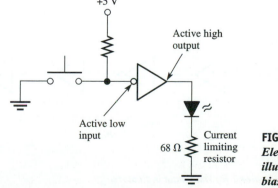

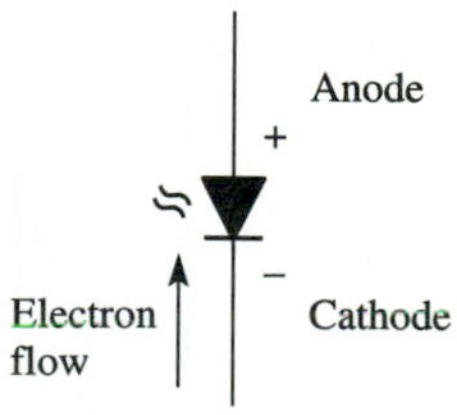

FIGURE 3.41
Electron-flow diagram for an illuminated LED when forward biased.

EXAMPLE 3.7

Draw a circuit similar to the one in Fig. 3-40, showing how an inverter with an *active high* input and an *active low* output turns on a LED during an activated condition.

Solution See Fig. 3-42.

FIGURE 3.42
Circuit for Example 3-7. The LED becomes forward biased and turns on when the input is at the active high state and the output is activated to a low state.

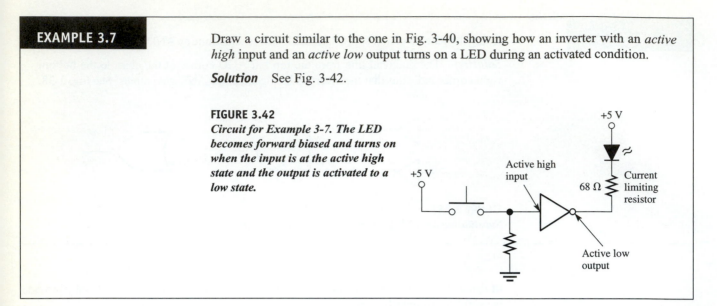

Applications of State Indicators

The use of state indicators helps make circuit operation much easier to follow. They show which logic states input and output leads are in during their active and inactive circuit conditions. This information is especially useful when troubleshooting a circuit with many gates.

Figure 3-43(a) shows a schematic diagram of six inverters connected in series to delay a signal 70 nanoseconds. Suppose it is suspected that one of the inverters is faulty. To find the source of the problem, it is necessary to trace logic states from the input circuitry to the output circuitry. As many as seven readings may be required before the fault is found. By using the schematic diagram with state indicators in Fig. 3-43(b), signals do not have to be traced throughout the circuitry. Instead, this information is provided by the small circles. For example, the bubbles on the leads between inverters 5 and 6 indicate that the signal should be low during the active condition. The leads between inverters 2 and 3 without the bubbles should be high during the active condition.

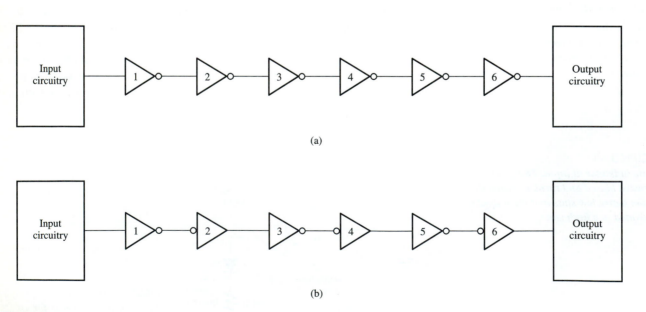

FIGURE 3.43
Delay circuit: (a) Without state indicators. (b) With state indicators.

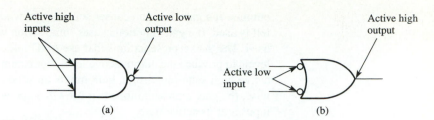

FIGURE 3.44
Logic gate state indicators. (a) Output goes low when all inputs are high. (b) Output goes high when any input is low.

The logic state of each lead during the circuit's inactive condition is always opposite to that shown by the state indicators. For example, the lead between inverters 4 and 5 without bubbles present should be low during the resting condition. Likewise, a high should be observed on the leads with the bubbles, such as between inverters 1 and 2.

The inverter is not the only logic symbol that uses state indicators. They are also used at the inputs and outputs of logic gates. Figure 3-44 shows two ways that the NAND function is identified. These include a standard NAND gate and an equivalent symbol consisting of an OR gate with small circles at the inputs. Although these two gate representations are equivalent, they are interpreted differently. The decision as to which representation should be used depends on which logic state is present at the output lead when it activates the device it is controlling. For example, suppose that the output of a NAND operating device normally rests in the high state and goes to a low state when activated. The standard NAND gate with the circle at the output is used in Fig. 3-44(a). Likewise, when the output of a NAND operating device is normally resting at a low and is activated to a high, the equivalent NAND device in Fig. 3-44(b) that has an output lead without the circle is used.

It should be mentioned that in an actual wired circuit, only the standard NAND gate is used to perform the NAND operation. Equivalent symbols in the form of state indicators do not exist. Standard symbols and equivalent state indicator symbols are only used to provide information about the normal resting and active states of logic gate devices.

Figure 3-45 shows the equivalent logic representation for five logic devices. The representations on the left have active low outputs and those on the right have active high

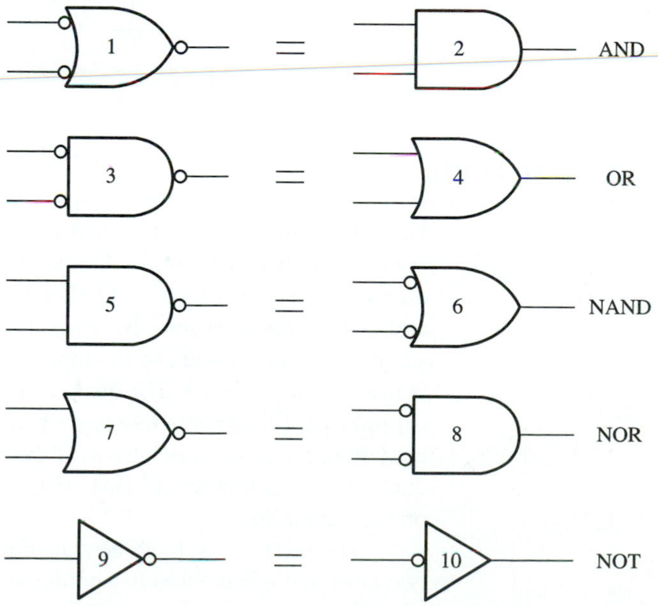

FIGURE 3.45
Equivalent logic representations.

outputs. If a gate function causes some action when its output is 0, one of the gates on the left is used. If a gate function causes some action when its output is 1, a gate on the right is used. The gate representations that use AND logic require that all inputs be at their active levels to produce the output active level. For example, when the output of gate 8 goes to an active high state (no circle), both inputs must be at their active low states (circles). Likewise, the gate representations that use OR logic will produce an active output level if any input is at its active state.

State indicators are not used in all schematic diagrams. Many circuit diagrams still use standard gate symbols exclusively. However, many circuits do use state indicators, so it is essential that they be understood. The ability to understand digital electronics depends largely on how well this concept is mastered.

EXAMPLE 3.8

Illustrate how a LED connected to a state indicator logic device is turned on by a low output state.

Solution The cathode of the LED is tied to the gate output and the anode is connected to a +5-volt source; see Fig. 3-46. When either one or both inputs of the gate are low, the output is driven low, which forward biases the LED to turn it on.

FIGURE 3.46
Circuit for Example 3-8.

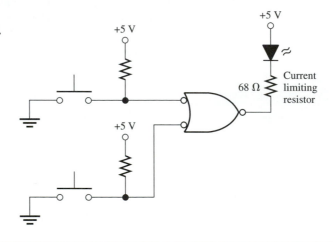

3.14 THREE-STATE LOGIC

Two-state logic:

Logic devices that produce either a 0 or 1 at their output.

All of the logic devices described so far produce outputs that are either a 0 or 1 state. This is known as **two-state logic.** Another type of logic device, called a *three-state buffer,* produces three states. Figure 3-47 shows a three-state buffer. It consists of an input, output, and control (enable) line. Because it does not have a bubble at its output terminal, the same logic state applied to the input is developed at the output only when the enable line is in an active low state. The third output condition, called a *high-impedance (high-Z) state,* is produced if the enable line is not activated because a logic high is applied. In the high-*Z* condition, the device operates as if the output is electronically disconnected from the conductor to which it is wired. This circuit action can be compared to opening a switch located on the output line.

The purpose of the three-state device is to make it possible to connect the output of two or more logic devices to a common conductor called a *bus.* Normally, several standard logic device outputs are not connected to the same line. The reason is that if any logic device produces a high, it will be pulled low by any of the other logic devices producing a 0 state. However, several three-state devices can be connected to a common bus as long as the outputs of all but one logic device is in the high-*Z* state.

FIGURE 3.47
(a) A simple three-state buffer can be represented by a switch. (b) Three-state buffer logic symbol. (c) Three-state buffer truth table.

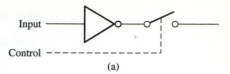

Input ————
Control —————

(a)

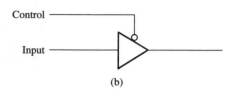

Control ————
Input ————

(b)

Control	Input	Output
1	0	HIGH Z
1	1	HIGH Z
0	0	0
0	1	1

(c)

The basic inverter is available as a three-state device. A bubble is placed on its output lead. Many other types of logic gates and digital circuits that use the three-state function are also available. Three-state logic devices are also called *tristate buffers*.

■ REVIEW QUESTIONS

25. When used as a state indicator, the presence of a circle indicates that a lead must be in a _____ state during the time the gate is activated. The absence of a circle indicates that a lead must be in a _____ state during the time the gate is deactivated.

26. Equivalent logic gate devices consisting of a gate and state indicators (do or do not) exist.

27. The equivalent gate representation using OR logic produces an active output state if (any or all) inputs are in their active state.

28. The equivalent gate representation using AND logic produces an active output state if (any or all) inputs are in their active state.

29. A three-state logic device has three lines called _____, _____, and _____.

EXPERIMENT

Tristate Logic Devices

Objectives

■ To wire and operate a tristate logic device.
■ To wire and operate several logic devices that share the same bidirectional bus line.
■ To troubleshoot a defective tristate logic device.

Materials

1—+5-V DC Power Supply   2—LEDs

1—Logic Pulse Input Switch   1—68-Ohm Resistor

1—Clock Pulse Input Generator (5–10 Hz)   4—Logic Switches

1—74125 Integrated Circuit   1—SPDT Switch

Introduction

Similar to standard logic gates, the *tristate* buffer produces two logic signals defined as a low and a high. The third state produced by a tristate buffer is a high-impedance (high-Z) state. When this condition exists, the internal circuitry of the device can be considered as disconnected from the output terminal.

Each type of tristate logic device has an *enable* input. When it is activated, the tristate device will produce either a logic 1 or 0 state. When it is deactivated, the tristate device will operate in the high-impedance state. The tristate logic device is available in IC form. The operation of the 74125 IC, which consists of four tristate logic devices, will be examined in this experiment.

Part A

Procedure

Step 1. Assemble the circuit shown in Fig. 3-48.

FIGURE 3.48

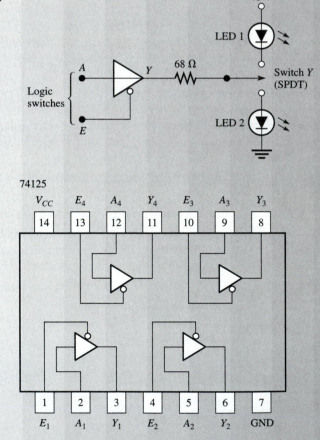

Background Information

The tristate logic device used in this experiment is also called a *tristate buffer.* When the Enable input is activated by a low, the device acts like a closed switch; the logic state applied to its input is passed on straight through to its output terminal. When the Enable input is deactivated by a high, no input data can reach the output. Therefore, the tristate buffer is effectively an open switch.

Step 2. Set switches *A* and *E* according to the input section of Table 3-5. For each switch setting, record the output states in the *Y* column. Place switch *X* in both positions to observe one of the following results:

<u>Output State</u>

LED 1 On = Logic 0

LED 2 On = Logic 1

Both LEDs Off = High-*Z*

TABLE 3.5
Tristate buffer truth table

A	E	Y
0	0	
0	1	
1	0	
1	1	

Part B

Step 3. Assemble the circuit in Fig. 3-49.

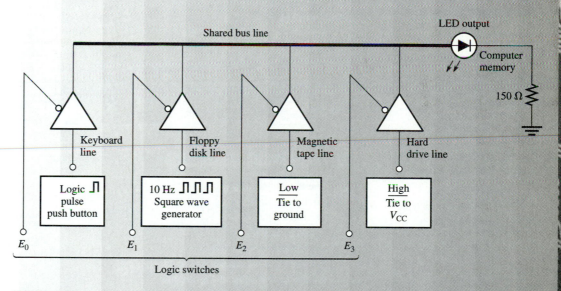

FIGURE 3.49

Background Information

Many types of digital equipment such as computers have the outputs of several logic devices connected to one bus. A bus is an electronic path over which digital information is transmitted. The use of one bus is possible only if one device uses the bus at any one time. Tristate buffers are used to control which device is operated. The tristate buffer that is enabled passes the data from its input to the bus line. All other tristate devices are disabled and in their high-impedance state.

Procedure Information

Four different signals are used to simulate the data transmission from four input sources to a computer memory.

Step 4. Fill in Table 3-6 by using the following procedure:
   a. For each observation, set the enable switches E_0 through E_3 according to the input portion of the truth table.
   b. For each observation, apply these signals to the tristate buffer input lines and determine which signal is passed on to the bus line:

   Input Data Sources   Simulated Signals

   Keyboard input   = Press the logic pulse push button once

   Floppy disk input   = Apply a 10-Hz square wave

   Magnetic tape input = Connect a logic low

   Hard drive input   = Connect a logic high

   c. Observe the LED to determine which type of input signal is passed to the bus line. Record your findings in the column of the table labeled *LED Output* by using *Pulse, Square Wave, Low* or *High* to describe the signal.
   d. Fill in the right column of the table to indicate which input source data is being sent to the bus line.

TABLE 3.6

Observation Number	Enable Input Lines				LED Output: Type of Signal	Input Data Source
	E_0	E_1	E_2	E_3		
1	0	1	1	1		
2	1	1	0	1		
3	1	1	1	0		
4	1	0	1	1		

■ EXPERIMENT QUESTIONS

1. The third logic state in tristate logic is a
   (a) Low logic state.
   (b) High logic state.
   (c) Low-impedance state.
   (d) High-impedance state.
2. A tristate buffer has
   (a) An input and an output.
   (b) An input, an output, and an enable input.
   (c) An enable input and an output.
3. A 74125 tristate buffer IC that is enabled operates like a(n)
   (a) Open switch.
   (b) Closed switch.
   (c) Inverter.
4. A 74125 tristate buffer that is disabled operates like a(n)
   (a) Open switch.
   (b) Closed switch.
   (c) Inverter.
5. A(n) _____ is an electronic path over which digital information is transmitted.
6. What are tristate buffers primarily used for?

The final section of this and each remaining chapter provides troubleshooting examples of at least one digital circuit described in the chapter. Appendix A provides information about techniques and equipment used to troubleshoot digital circuitry and should be reviewed by the reader who chooses to read any of the troubleshooting sections.

3.15 TROUBLESHOOTING

When troubleshooting logic gates, the technician must first know the output of each type of gate for a set of given inputs. When a gate output is tested, the input signals may be provided by the circuitry of the equipment under test or injected artificially by the troubleshooter.

This section describes some of the common failures that occur in digital logic gates and basic troubleshooting techniques that are used for finding them.

Open Input

If an open develops at the input of a logic gate, only that gate will be affected. Perform a static test using a logic pulser and probe to trace test signals throughout the circuit being examined, as shown in Fig. 3-50. The pulser will cause the logic state being read by the probe to change if the gates or conductor is functioning properly.

FIGURE 3.50
Using a logic pulser and probe to trace test signals.

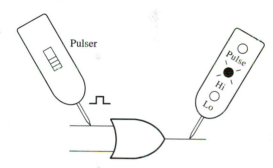

Step 1. See Fig. 3-51. Connect the pulser to pin 1 of ICZ and apply a train of pulses. Observe the signal at its output, pin 3. *Symptom:* Signal present.

Step 2. Connect the pulser to pin 2 of ICZ and observe the signal at its output, pin 3. *Symptom:* Signal present.

Step 3. Connect the pulser to pin 3 of ICX and the probe to pin 1 of ICZ and observe the waveform. *Symptom:* Signal present.

Step 4. Connect the pulser to pin 3 of ICY and the probe to pin 2 of ICZ and observe the waveform, as shown in Fig. 3-51(b). *Symptom:* Signal not present.

Step 5. Connect the pulser to pin 3 of ICY and the probe to test point *A* and observe the waveform. *Symptom:* Signal present.

The symptoms indicate that an open is present somewhere between junction X and input pin 2 of ICZ. Any inputs that follow an open will usually pull themselves to about 1.5 volts. Connecting a scope or voltmeter to pin 2 of ICZ verifies the open wire. Even though this potential is in the invalid logic voltage range, gate ICZ may operate as if a permanent 1 state is present at pin 2.

Short-Circuit Failures

Node:

A circuit junction point that is common to two or more gates or other elements.

A short-circuit failure affects all of the circuits that are connected to the **node** or gate that is shorted. A short can develop to either ground or V_{CC}. If a short-circuit ground exists, all inputs and outputs common to the short will be permanently low. If a short to V_{CC} exists, all inputs and outputs common to the short will be permanently high. Therefore, the troubleshooting procedure used to find internal and external shorts are the same.

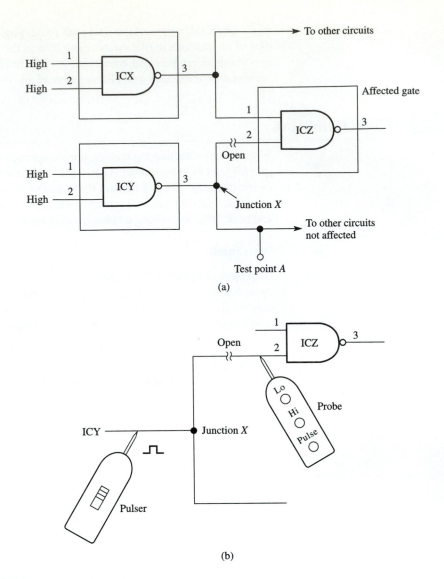

(a)

(b)

FIGURE 3.51
External open input.

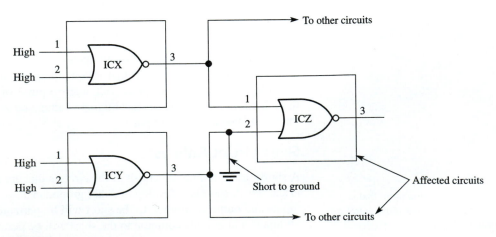

FIGURE 3.52
Short-circuit ground.

Troubleshooting Procedure for Short-Circuit Ground

Perform a static test using a logic pulser and probe to trace test signals throughout the circuit being examined.

Step 1. See Fig. 3-52. Connect the pulser to pin 1 of ICZ and observe the signal at output pin 3. *Symptom:* Signal present.

Step 2. Connect the pulser to pin 2 of ICZ and observe the signal at output pin 3. *Symptom:* Signal not present.

Note: A pulser output cannot change the logic state of a grounded node. Therefore, the test node remains at a 0-volt low.

The symptom indicates that input 1 and the output function properly. Therefore, a problem exists at input 2 of ICZ or the node connected to it.

Step 3. Connect a voltmeter or scope to pin 2. *Symptom:* 0 volts.

The symptom indicates that there is a short to ground. To find the origin of a short to ground, use a pulser and current tracer.

Using a Current Tracer to Find a Short

When a circuit node is shorted to ground, as shown in Fig. 3-53(a), the tracer and a pulser can be used to find the location of the short. Place the pulser on the circuit path midway between the two gates, and place the tracer between the pulser and ICY. If the indicator

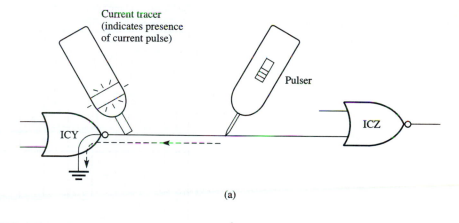

(a)

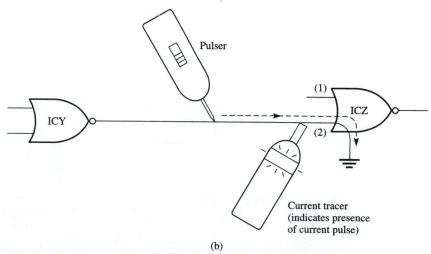

(b)

FIGURE 3.53

Using a logic pulser and current tracer to locate a shorted ground.

on the tracer lights with each pulse from the pulser, this indicates that current is flowing and the output of ICY is shorted to ground. If no current is detected, move the tracer between the pulser and input pin 2 of ICZ, as shown in Fig. 3-53(b). If the tracer shows that current is flowing, input 2 of ICZ is shorted to ground.

By moving the pulser and probe, the same procedure can be used to find a short to ground that is located on the conductor path between ICY and ICZ in Fig. 3-52.

Short Between Nodes

A short sometimes develops across both inputs of a gate. This short may be located between two nodes external to the IC or internal to the IC. Whenever the shorted input pins are high simultaneously or low simultaneously, the gate responds properly. However, if the inputs are different, they will either oppose each other so that the inputs go to a bad voltage level, as shown in Fig. 3-54(a), or the low input will pull down the high input voltage to its low-voltage level. Figure 3-54(b) shows a short between two IC inputs.

Troubleshooting Procedure for a Short Between Nodes

Step 1. Pulse input 1 with a probe located at input 2 of ICZ. If there is a short between the pulsed and the probed inputs, the probe will detect the short by indicating a bad voltage level or a momentary opposite state.

Step 2. To verify the short, switch the pulser to pin 2 and the probe to pin 1 of ICZ and check again.

FIGURE 3.54
Short between nodes. (a) Bad logic level shown on the screen of an oscilloscope. (b) A gate with shorted inputs due to a solder splash.

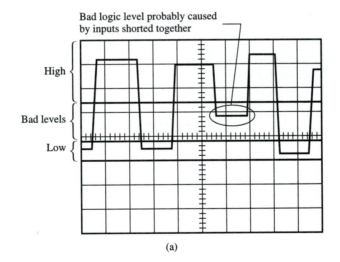

(a)

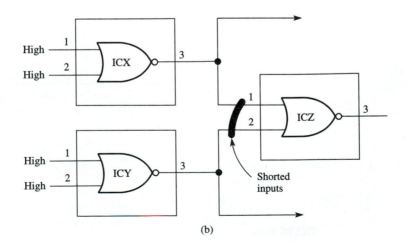

(b)

Step 3. To further check, a scope can be used to observe the signal shown in Fig. 3-54(a), or an ohmmeter can be placed between both pins to detect the short when the power in the circuit is off.

The most common short is a solder bridge between two nodes. If it cannot be visibly found, the most likely problem is a defective IC.

■ REVIEW QUESTIONS

30. A gate often recognizes an open input lead as a (1 or 0) state.

31. A _____ is a circuit junction point that is common to two or more gates or other elements.

32. If a short-circuit ground exists, all inputs and outputs common to the short will read:
(a) $V_{CC} = 5$ volts.
(b) 1.5 volts.
(c) 2.5 volts.
(d) 0 volts ground.

33. If a short to V_{CC} exists, all inputs and outputs common to the short will read:
(a) $V_{CC} = 5$ volts.
(b) 1.5 volts.
(c) 2.5 volts.
(d) 0 volts ground.

■ SUMMARY

- The only time the output of an AND gate goes high is when all inputs are high.
- The output of an OR gate goes high when at least one input is high.
- The output of an inverter (NOT gate) is always the opposite logic state of its inputs.
- The output of a NAND gate is high when any of its inputs are low.
- The output of a NOR gate is high when all inputs are low.
- The output of an exclusive-OR gate will be high only if its inputs are at different logic states.
- The output of an exclusive-NOR gate will be high only if its inputs are the same logic levels.
- The operation of a basic logic device can be described by a symbol, truth table, timing diagram, or Boolean algebra equation.
- By inverting input leads, output leads, or a combination of both, most gate functions can be achieved by using other gates.
- State indicators show the logic levels needed at the inputs and output of a logic gate or inverter to activate a device it is controlling.
- Three-state buffers, which have three output states (high, low, and high-Z), allow the outputs of more than one of these devices to be connected to a common bus line.
- Logic probes and pulsers are low-cost testers designed specifically to troubleshoot digital logic circuits.

■ PROBLEMS

1. Fill in the missing areas of Fig. 3-55. (3-4 to 3-10)

2. A logic circuit has five inputs. How many of the following possible input combinations can it have? (3-11)
(a) 2   (d) 16
(b) 4   (e) 32
(c) 8

3. Write a Boolean algebra expression to represent each of the following basic logic functions. (3-4 to 3-10)
(a) NOR   (c) OR
(b) Exclusive-OR   (f) AND
(c) Inverter   (g) Exclusive-NOR
(d) NAND

4. Figure 3-56 shows the input waveforms for three different types of logic gates. Draw the output waveforms for each. (3-4 to 3-10)

5. Figure 3-57 shows 7409 and 7427 integrated circuit packages. Connect the inputs and output of each gate to the correct pins. (3-4 to 3-10)

6. To what are the pins of an IC consisting of basic logic gates connected? (3-3)

7. What are the standard logic gates that operate exactly opposite each other? (3-11)

8. A logic gate has _____ or more inputs and _____ output(s). (3-1)

9. What logic device is also called a NOT gate and why? (3-6)

| Gate rule | Truth table | Switch analogy | Symbol |

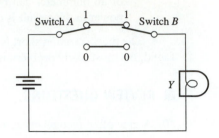

	Inputs		Output
	Switch A	Switch B	Y
Output = 1 if inputs are the same	0	0	1
	0	1	0
	1	0	0
	1	1	1

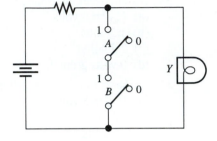
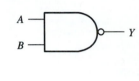

A	B	Y
0	0	
0	1	
1	0	
1	1	

Output = 0 if and only if all inputs = 1

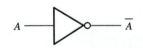

A	$\overline{A}$
0	1
1	0

Output = 1 if input = 0 and vice versa

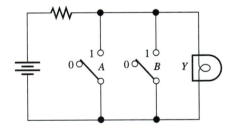
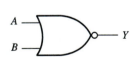

A	B	Y
0	0	1
0	1	0
1	0	0
1	1	0

FIGURE 3.55
Diagram for Problem 1.

10. An IC can function without being connected to an external +5-V and ground potentials. True or false? (3-3)

11. When viewed from the top, the pins of a dual in-line IC package are numbered (clockwise or counterclockwise) from the reference identification marker. (3-3)

12. What are three forms of notation that serve to precisely describe the operation of logic gates? (3-2)

13. Some of the basic gates are available in three-state buffer versions. True or false? (3-14)

14. Given a NAND gate, draw how it can be configured to operate as an inverter. (3-7)

15. Given 2 two-input AND gates, how would you connect them to produce a three-input AND gate? (3-4)

16. What is the maximum number of inputs an exclusive-OR gate can have? (3-9)

17. If the output of a two-input exclusive-OR gate is high, which of the following would be the inputs? (3-9)
   (a) Both low.   (c) The same.
   (b) Both high.   (d) Opposite.

18. If the output of a two-input exclusive-NOR gate is high, which of the following would be the inputs? (3-10)
   (a) Both low.   (c) The same.
   (b) Both high.   (d) Opposite.

19. The format of the timing diagram shows the logic level on the (vertical or horizontal) axis and time on the (vertical or hori-

FIGURE 3.56
Waveforms for Problem 4.

OR gate

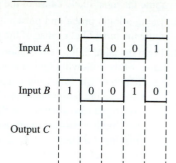

NAND gate

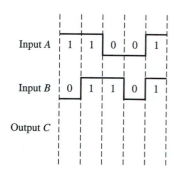

Exclusive-NOR gate

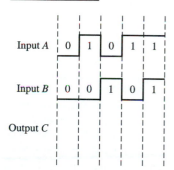

zontal) axis. The diagram is read from (left or right) to (left or right). (3-2)

20. Draw the waveform at the output of the circuit of Fig. 3-58 when the switch is (a) in position *A* and (b) in position *B*. (3-4)

21. Using a battery, switches (resistor, if necessary), and light bulb, draw a series, parallel, or combination circuit that represents the operation of the following gates. (3-4, 3-6, 3-8, 3-9)
    (a) AND   (c) Inverter
    (b) NOR   (d) Exclusive-OR

22. Several standard logic gates cannot be connected to the same output line. Why? (3-14)

23. Given a NAND gate and inverters, connect them to produce the following functions. (3-12)
    (a) AND   (b) OR

24. The input inversion procedure allows a/an _____ to operate as an OR, a NAND to operate as a/an _____, a NOR to operate as a/an _____, and an AND to operate as a/an _____. (3-12)

25. To perform the output inversion procedure, invert the _____ of the gate, which allows a/an _____ to operate as an OR, a/an _____ as an AND, a NOR as a/an _____, and a NAND as a/an _____. (3-12)

26. Explain the use of state indicators. (3-13)

27. Place 1s and 0s at the input and output leads of each gate of Fig. 3-59 to show their active states. (3-13)

28. Place 1s and 0s at the input and output leads of each gate of Fig. 3-60 to show their inactive, or resting, states. (3-13)

29. By observing the inputs of Fig. 3-61 applied to the three-state inverter with an active-high control input line; label the output waveform as 0, 1, or high-Z. (3-14)

30. Which of the following will be the reading for an IC input lead that follows an open? (3-15)
    (a) 0 volts.   (c) 2.5 volts.
    (b) 1.5 volts.   (d) 5 volts.

Troubleshooting

31. What should be the logic level at test point *A* of the circuit of Fig. 3-62 when the output device is activated? (3-15)

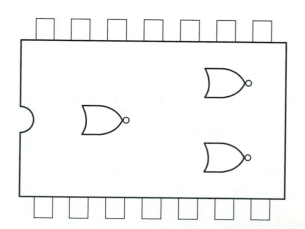

FIGURE 3.57
IC packages for Problem 5.

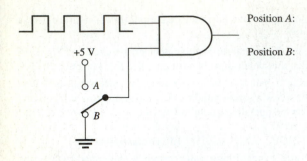

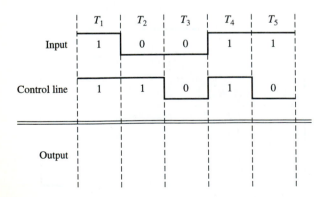

FIGURE 3.58
Circuit for Problem 20.

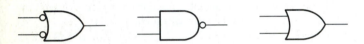

FIGURE 3.59
Gates for Problem 27.

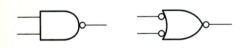

FIGURE 3.60
Gates for Problem 28.

Position *A*:

Position *B*:

32. Using a logic probe and pulser, as shown in Fig. 3-63, determine the most likely cause of the gate failure. (3-15)

33. Using a logic probe and pulser, as shown in Fig. 3-64, determine the most likely cause of the gate failure. (3-15)

34. From the table in Fig. 3-65, determine which is the most likely cause of the following OR gate malfunctions. (3-15)
    (a) Which of the following is the most likely internal condition causing fault condition 1?
        (i) Input pins shorted together.
        (ii) Input *A* shorted to ground.
        (iii) Input *B* shorted to ground.
        (iv) All of the above.
    (b) Which of the following is the most likely internal condition causing fault 2?
        (i) Input *A* shorted to ground.
        (ii) Input *A* shorted to V_{CC}.
        (iii) Input *B* shorted to ground.
        (iv) Input *B* shorted to V_{CC}.

35. From the table in Fig. 3-66, determine which is the most likely cause of the following AND gate malfunctions. (3-15)
    (a) What is the most likely condition causing fault 1 to occur?
        (i) Output pin lead is open.
        (ii) Input of the next succeeding gate is shorted to ground.
        (iii) Output pin is shorted to ground.
        (iv) Both (ii) and (iii).
    (b) What is the most likely internal condition causing fault 2?
        (i) Input *A* is shorted to ground.
        (ii) Input *A* is shorted to V_{CC}.
        (iii) Input *B* is shorted to ground.
        (iv) Input *B* is shorted to V_{CC}.

36. From the table in Fig. 3-67, determine which is the most likely cause of the following inverter malfunctions. (3-15)
    (a) What is the most likely internal condition causing fault 1?
        (i) Input is shorted to V_{CC}.
        (ii) Input is shorted to the output.
        (iii) Output is shorted to ground.
        (iv) Both (i) and (iii).
    (b) What is the most likely internal condition causing fault 2?
        (i) Input is shorted to ground.
        (ii) Input is shorted to the output.
        (iii) Output is shorted to V_{CC}.
        (iv) Both (i) and (iii).

FIGURE 3.61
Waveforms for Problem 29.

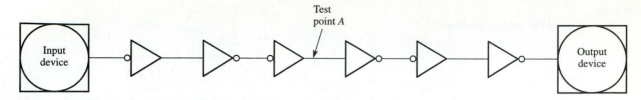

FIGURE 3.62
Circuit for Problem 31.

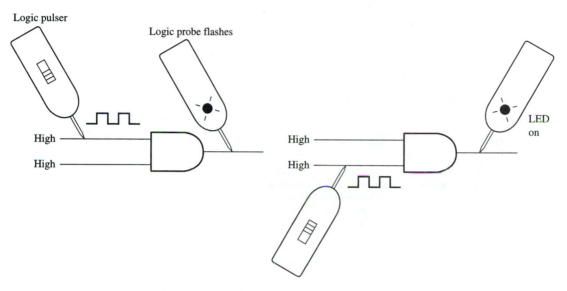

FIGURE 3.63
Logic probe and pulser positions for Problem 32.

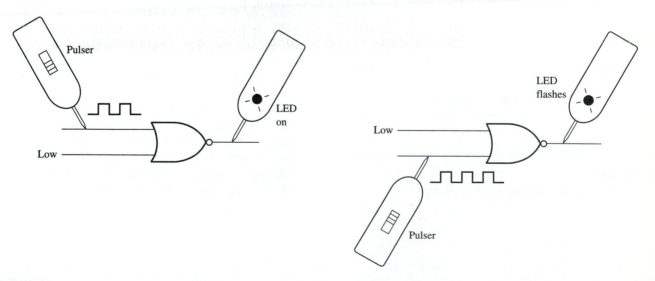

FIGURE 3.64
Logic probe and pulser positions for Problem 33.

FIGURE 3.65
Input/output table and OR gate for Problem 34.

Input (volts)		Output (volts)		
B	A	Normal condition	Fault condition 1	Fault condition 2
0	0	0	0	0
0	5	5	0	5
5	0	5	0	0
5	5	5	5	5

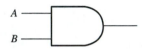

Input (volts)		Output (volts)		
B	A	Normal condition	Fault condition 1	Fault condition 2
0	0	0	0	0
0	5	0	0	5
5	0	0	0	0
5	5	5	0	5

FIGURE 3.66
Input/output table and AND gate for Problem 35.

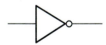

Input (volts)	Output (volts)		
	Normal condition	Fault condition 1	Fault condition 2
0	5	0	5
5	0	0	5

FIGURE 3.67
Input/output table and inverter for Problem 36.

■ ANSWERS TO REVIEW QUESTIONS

1. Gates, inverters   **2.** one, two   **3.** symbols   **4.** Truth tables, waveform diagrams, and Boolean algebra equations   **5.** counterclockwise   **6.** (c)   **7.** (b)   **8.** The name NOT gate is derived from an inverter producing a NOT A ($\overline{A}$) when an A is applied to its input.   **9.** OR, AND   **10.** $A \cdot B$ **11.** (c)   **12.** (a)   **13.** NAND, NOR   **14.** $\overline{A + B}$ **15.** exclusive-NOR   **16.** exclusive-OR   **17.** two   **18.** $\overline{A \oplus B}$ **19.** three   **20.** 16   **21.** binary   **22.** output   **23.** AND, OR, NAND, NOR   **24.** NAND, AND   **25.** low, low   **26.** do not   **27.** any   **28.** all   **29.** input, output, control (enable)   **30.** 1   **31.** node   **32.** (d)   **33.** (a)

CHAPTER 4

BOOLEAN ALGEBRA

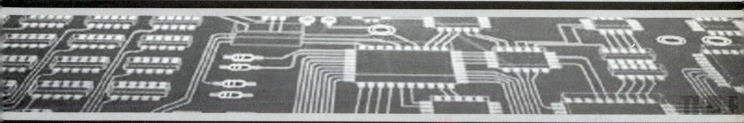

When you complete this chapter, you will be able to:

1. Define Boolean algebra.
2. Describe the difference between conventional algebra and Boolean algebra.
3. Define the meaning of letters and operation symbols of a Boolean expression.
4. List some applications of Boolean algebra.
5. Write the Boolean expression that corresponds to a given logic circuit.
6. Draw a logic circuit that corresponds to a given Boolean expression.
7. Describe the operation of a circuit through the use of sum-of-products and product-of-sums equations.
8. Simplify a given logic expression using Boolean algebra.

4.1   INTRODUCTION

The Greek philosopher Aristotle, who lived from 384 to 322 B.C., devised a formal logic thinking system that was based on two conclusions: *true* or *false*. Aristotle wrote six famous works on the subject that were studied for many centuries and that helped to influence the development of organized reasoning. Many unsuccessful attempts were made by mathematicians to solve Aristotle's logic theories by using conventional algebra. However, it was not until 1854 that English philosopher and mathematician George Boole successfully developed a mathematical system of logic that solved Aristotle's logic problems. For decades, Boole's works remained as a form of pure theoretical mathematics, considered interesting but useless for any practical application then known.

In 1938, Dr. Claude E. Shannon of MIT wrote a paper entitled "A Symbolic Analysis of Relay Switching Circuits." Its purpose was to provide a solution to the telephone company's problems in meeting the large volume of relay circuit requirements needed to satisfy the rapidly growing telephone usage during that period. For nearly 50 years prior to this time, telephone company engineers relied almost exclusively on intuition and experience to design these systems. As a result, it was difficult to know whether an optimum design had been achieved. Thus, this technique could no longer be relied on. Dr. Shannon realized that Boolean mathematical equations could be written to describe logic circuit functions that operated in either the *on* or *off* condition. He proposed that the laws of Boolean algebra could then be used to simplify the equation while still retaining the same circuit functions, thus reducing the number of switching circuits required for implementation.

Applications for Boolean algebra have expanded beyond telephone switching circuits. It is presently used for designing digital and computer circuits that also operate at two voltage levels. Whenever used properly in the design process, the applications of Boolean algebra usually result in the most effective performance of a logic circuit. By using fewer components, the size, power consumption, and costs are reduced, whereas the operating speed and reliability are increased. Boolean algebra is also used for analyzing digital circuits and for effectively troubleshooting computers.

4.2   BOOLEAN ALGEBRA CHARACTERISTICS

There are two major differences between conventional algebra and Boolean algebra:

1. Unlike conventional algebra, which uses digits to represent arithmetic quantities (such as decimal integers 0–9), Boolean algebra uses a 1 or a 0 to represent only two conditions.
2. Conventional algebra uses the operations of addition, subtraction, multiplication, and division. Boolean algebra uses three operations: Boolean addition, Boolean multiplication, and Boolean complementation.

4.3   SYMBOLOGY

As mentioned in the introduction, mathematical equations (also called *expressions*) can be used to describe the operation of logic circuit functions. Capital letters of the alphabet are used as variables in the equations. These variables represent the 1 and 0 state inputs applied to the logic circuit.

Three symbols, called *operators,* are used in a Boolean equation. They indicate which one of the three Boolean operations is being performed.

($\cdot$) A center dot or no symbol located between letter variables indicates Boolean multiplication.

($+$) A plus sign located between letter variables indicates Boolean addition.

($\overline{\text{X}}$) A bar over a letter variable (called an *overbar*) indicates the complementary operation.

The *three* Boolean operations are performed by only *three* basic logic devices:

1. The *AND* gate performs Boolean multiplication.
2. The *OR* gate performs Boolean addition.
3. The *inverter* performs the complementary operation.

AND

The multiplication operation of a two-input AND gate can be expressed in equation form as

$$A \cdot B$$

Boolean multiplication:

Occurs when binary numbers are multiplied by an AND function.

The basic rules for **Boolean multiplication** with two input variables are as follows:

Multiplication Operation

$$0 \cdot 0 = 0$$
$$0 \cdot 1 = 0$$
$$1 \cdot 0 = 0$$
$$1 \cdot 1 = 1$$

The numbers that are multiplied represent the binary numbers applied to the inputs of an AND gate. The product (answer) is the binary number generated by the AND gate.

Notice that the number arrangement is the same as for a two-input AND gate truth table.

OR

The addition operation of a two-input OR gate can be expressed in equation form as

$$A + B$$

Boolean addition:

Occurs when binary numbers are added by an OR function.

The basic rules for **Boolean addition** with two input variables are as follows:

Addition Operation

$$0 + 0 = 0$$
$$0 + 1 = 1$$
$$1 + 0 = 1$$
$$1 + 1 = 1^{\dagger}$$

The numbers that are added represent the binary numbers applied to the inputs of an OR gate. The sum is the binary number generated by the OR gate.

Notice that the number arrangement is the same as for a two-input OR gate truth table.

Inverter

Complementary operation:

Occurs when a binary number is changed to the opposite binary number by an inverter function.

The **complementary operation** of an inverter can be expressed in equation form as

$$\overline{A}$$

The complementary operation of an inverter can be examined by using the equation variables in the following truth table:

Complementary Operation

INPUT		OUTPUT
A: 0	=	$\overline{A}$: 1
A: 1	=	$\overline{A}$: 0

Even though they are not Boolean functions, the NAND and NOR gate operations can also be performed by these three basic logic operators. This is accomplished by simply connecting an inverter to the output of an AND gate or an OR gate.

An overbar is placed over the entire output expression of the gate after it is inverted. Therefore, $\overline{AB}$ is the output expression of a NAND gate, and $\overline{A + B}$ is the expression that represents a NOR gate.

---

† Notice the difference between decimal and Boolean addition when two 1s are added. Therefore, Boolean algebra differs from binary arithmetic.

Figure 4-1 illustrates the symbology concepts by showing a Boolean equation with its equivalent multigate logic circuit representation.

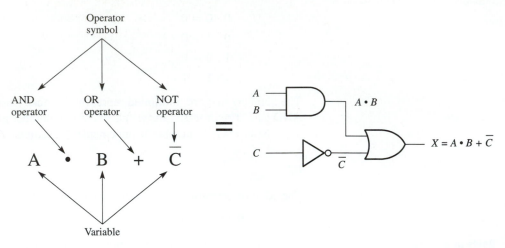

FIGURE 4.1
A Boolean equation and its equivalent logic circuit.

■ REVIEW QUESTIONS

1. Boolean algebra uses which of the following to represent arithmetic quantities?
   (a) Decimal digits.
   (b) Exponents.
   (c) Binary bits.
   (d) Fractions.
   (e) All of the above.

2. Which of the following operations are used by Boolean algebra?
   (a) Boolean addition.
   (b) Boolean multiplication.
   (c) Boolean complementation.
   (d) All of the above.

3. Capital letters are used as _____ in Boolean equations. They represent the 1 and 0 states _____ applied to the logic circuit.

4. To indicate which of the three Boolean operations are being performed, symbols called _____ are used in a Boolean equation.

5. What are the logic devices that perform the following Boolean operations?
   (a) Multiplication.
   (b) Addition.
   (c) Complementation.

6. The Boolean expression for addition of inputs *A* and *B* is _____.

BOOLEAN ALGEBRA APPLICATIONS

There are several applications of Boolean algebra that can be used when working with logic circuits:

1. Boolean equations provide a shorthand notation of a logic circuit configuration.
2. Boolean equations describe the operation of a logic circuit.
3. Boolean expressions can be used to design logic circuits (circuit simplification).
4. Boolean expressions can be used in troubleshooting.

4.4 BOOLEAN EQUATIONS AS A SHORTHAND FOR LOGIC CIRCUIT CONFIGURATION

Boolean algebra provides a shorthand method to describe a logic circuit configuration by a mathematical equation. A Boolean expression indicates how many logic gates are used, what types of gates are needed, and how they are connected. When working with Boolean algebra, it is important to understand how to draw logic diagrams for a given expression and write an expression for a given logic diagram when the inputs are known.

Determining Output Expressions

When deriving logic expressions from a circuit, a rule of thumb is to work from the inputs toward the output. For example, Fig. 4-2 shows two gates with their given inputs.

FIGURE 4.2
Deriving an expression from a logic circuit.

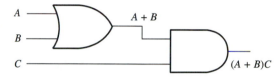

To develop the eventual output expression, it is necessary first to write the output expression of the OR gate, which is $A + B$. Next, the output of the OR is ANDed with input C and generates an output of $(A + B)C$. The parentheses indicate that $A + B$ is not the same input as C. Parentheses are sometimes used to artificially group parts of an expression for the purpose of eliminating mistakes. Each group originates from a separate input. The examples in Fig. 4-3 illustrate how the outputs are grouped for an AND and an OR gate to indicate which parts of the expression originated from different inputs.

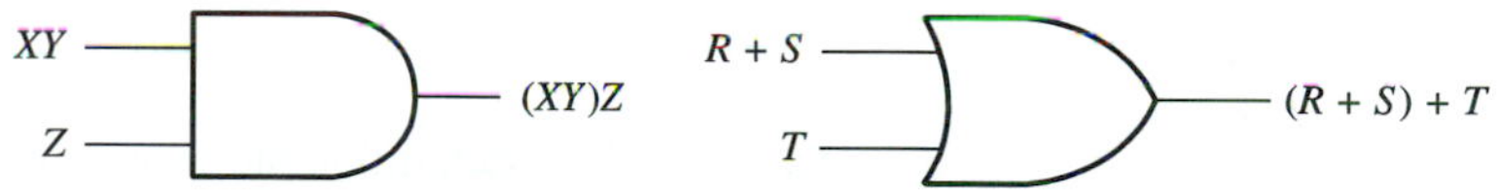

XY ──┐ $(XY)Z$ $R + S$ ──┐ $(R + S) + T$
Z ──┘ T ──┘

FIGURE 4.3
Parentheses are used to artificially group parts of an expression.

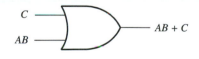

FIGURE 4.4
An expression representing an ANDed input to an OR gate without parentheses.

In some situations, it is not necessary to artificially group expressions because natural groups exist. For example, any ANDed input to an OR logic gate does not need parentheses because the AND represents a natural group, which is shown in Fig. 4-4.

If an input already has parentheses, brackets [ ] are used to provide for additional grouping of an output expression. The examples in Fig. 4-5 illustrate how they are used.

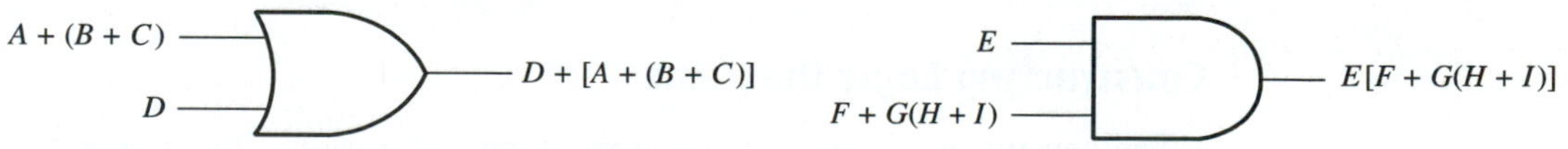

$A + (B + C)$ ──┐ $D + [A + (B + C)]$ E ──┐ $E[F + G(H + I)]$
D ──┘ $F + G(H + I)$ ──┘

FIGURE 4.5
Brackets are used for additional grouping.

The examples in Fig. 4-6 show the steps that are required to derive the output expression from a circuit consisting of several logic gates. The procedure begins with the

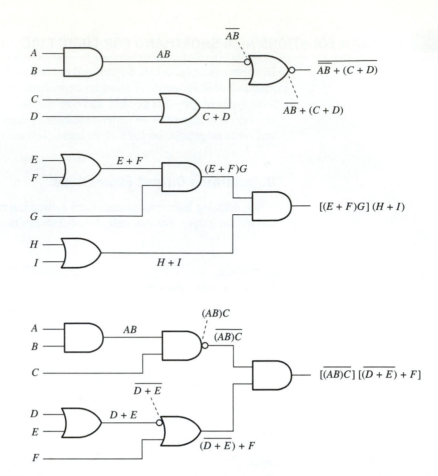

FIGURE 4.6
Expressions from circuits consisting of several logic gates.

inputs at the left. To determine the expression, move to the right, using the output of each preceding gate as the input to the next gate.

■ REVIEW QUESTIONS

7. What are four applications of Boolean algebra?
8. When deriving logic expressions from a circuit, work from the (inputs or output) toward the (inputs or output).
9. Any _____ input to an _____ gate does not need parentheses.
10. Write the output expressions for the gates in Fig. 4-7.

Constructing Logic Diagrams

When constructing logic diagrams from output expressions, it is necessary to start at the output and work toward the input side of the diagram. The expression is dissected into groups that become gates. The groups are then dissected until they become individual letters that represent the inputs of the diagram.

The first step in developing a logic circuit from an expression is to look for natural groups. These are letters or groups of letters that are ANDed together. Then separate natural groups from other ORed groups or letters before splitting the natural groups themselves

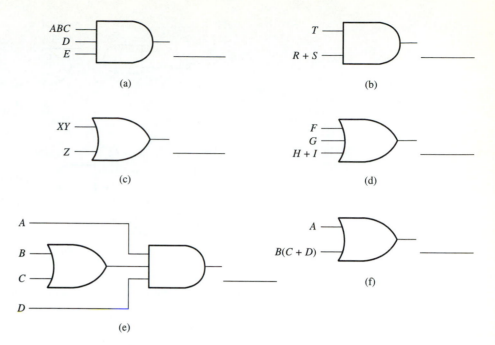

FIGURE 4.7
Gates for Review Question 10.

where they are ANDed. For example, the natural group AB is first split from C in the expression $AB + C$. This indicates that an OR gate develops the final output. Next, the AB input to the OR gate is split and becomes an AND gate. Fig. 4-8 illustrates how the circuit was developed to eventually show individual letters as the inputs.

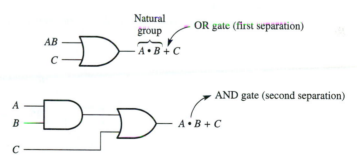

FIGURE 4.8
Constructing a logic diagram from an output expression with a natural group.

In some instances, artificial groups separated by parentheses indicate which gate should be drawn first. For example, to develop a logic circuit from the expression $A(B + C)$, the parentheses are used to establish the first group. Therefore, an AND gate is developed, as shown in Fig. 4-9(a). Next, $B + C$ indicates that an OR gate be drawn, as shown in Fig. 4-9(b).

Figure 4-10 shows how a diagram with both natural and artificial groups is converted to a logic circuit.

The overbar is also used to determine groupings when it appears over more than one letter. For example, Fig. 4-11 shows the circuit for the expression $\overline{ABC} + D$.

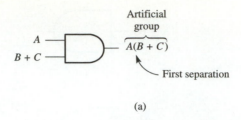

(a)

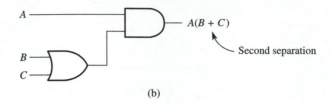

(b)

FIGURE 4.9
An expression with an artificial group converted to a logic circuit.

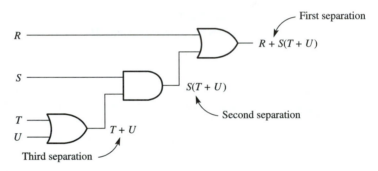

FIGURE 4.10
An expression with artificial and natural groups converted to a logic circuit.

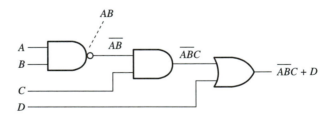

FIGURE 4.11
Overbars extending over more than one variable indicate groupings.

■ REVIEW QUESTIONS

11. When constructing a logic diagram from an expression, start at the (inputs or output) and work toward the (inputs or output).

12. Develop logic diagrams from the following expressions:
   (a) $B + (C + D + E)$
   (b) $XY + Z$

(c) $[E(F + G)](HI + JK)$

(d) $\overline{(\overline{A} + BC) + \overline{\overline{D(EF + \overline{G})}}}$

(e) $\overline{(\overline{EFG} + H + I)[(J + \overline{K})\overline{LM}]}$

4.5   BOOLEAN EQUATIONS THAT DESCRIBE LOGIC CIRCUIT OPERATIONS

Sum-of-products equation:

A logic expression that represents inputs that feed AND gates and outputs that feed an OR gate.

Product-of-sums equation:

A logic expression that represents inputs that feed OR gates and outputs that feed an AND gate.

There are two forms of Boolean expressions that are more commonly used than others. They are called the **sum-of-products** and the **product-of-sums** equations, which provide enough information to describe the basic operation of the logic circuit they represent. The term *product* refers to the multiplication of variables performed by the AND function ($A \ and \ B = AB$). The term *sum* refers to the addition of variables performed by the OR function ($A \ or \ B = A + B$). The sum-of-products and product-of-sums expressions combine the AND and OR functions.

Product-of-Sums Expression

$(A + B + C)(D + E + F)$

The distinguishing feature of this circuit is that the inputs feed OR gates (that produce the sum) and the output is derived from an AND gate (that produces the product).

Sum-of-Products Expression

$ABC + DE$

The feature of this circuit is that the inputs feed AND gates (that produce the product) and the output is derived from an OR gate (that produces the sum). Laws associated with Boolean algebra allow most expressions to be translated into one of these forms.

Expressions are capable of describing the operation of the circuit by identifying what input conditions produce a 1 state output. For example, the expression for an exclusive-OR gate is $A\overline{B} + \overline{A}B$. A letter without an overbar represents a 1 state, and a letter with an overbar represents a 0 state. Therefore, the expression states that when input $A = 1$ and $B = 0$ or $A = 0$ and $B = 1$, the output is 1. All other input conditions produce a 0. This can be verified by the truth table in Fig. 4-12.

FIGURE 4.12
(a) Exclusive-OR gate logic symbol. (b) Exclusive-OR gate truth table.

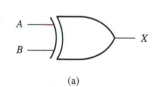

(a)

Inputs		Output
B	A	
0	0	0
0	1	1
1	0	1
1	1	0

(b)

EXAMPLE 4.1

Develop a truth table from the following expression: $\overline{A}\overline{B} + AB$. Try to determine what type of gate it represents.

Solution
Step 1. Develop a truth table for a two-input gate. See Fig. 4-13.

FIGURE 4.13
Example 4-1: (a) Exclusive-NOR gate logic symbol. (b) Exclusive-NOR gate truth table.

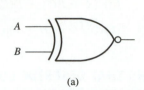

(a)

Inputs		
B	A	Output
0	0	1
0	1	0
1	0	0
1	1	1

(b)

Step 2. Assume that a letter without an overbar is a 1 and a letter with an overbar is a 0. Insert a 1 at the outputs that correspond to the conditions specified by the expression. A 1 is generated when both inputs A and B are 0 *or* when both inputs are 1.

Step 3. Both the expression and the truth table indicate that the equation represents an exclusive-NOR gate.

The group of letters inside the parentheses of a product-of-sums expression is called a *minterm*. In a sum-of-products expression, letters separated by OR operators are also called *minterms*.

Product-of-sums: $(A + B + C)(D + F)$

Minterm

Sum-of-products: $\overline{ABC} + \overline{DE}$

4.6 DESIGNING LOGIC CIRCUITS USING BOOLEAN EXPRESSIONS

A gate network is designed from a conditional statement that includes the variable and the state required for each variable to produce a desired output. The process involves several steps:

Step 1. The first step in designing the gate network is to define the desired function of a circuit. ***Example:*** Purchase a cup of coffee for 25 cents from a vending machine.

Step 2. The second step is to develop a statement that describes the necessary conditions. ***Example:*** A cup of coffee can be purchased only if one quarter, *or* a nickel *and* a nickel *and* a nickel *and* a nickel *and* a nickel, *or* a dime *and* a nickel *and* a nickel *and* a nickel, *or* a dime *and* a dime *and* a nickel are inserted.

Step 3. The third step is to identify each of the input conditions in terms of letter symbols.

Q_1 = quarter

N_1 = nickel

N_2 = nickel

N_3 = nickel

N_4 = nickel

N_5 = nickel

D_1 = dime

D_2 = dime

Step 4. The fourth step is to restate the statement problem using the letter symbols. ***Example:*** Q_1, or (N_1 and N_2 and N_3 and N_4 and N_5), or (D_1 and N_1 and N_2 and N_3), or (D_1 and D_2 and N_1) = output high.

Step 5. The fifth step is to establish a Boolean expression by replacing the words and/or with operator symbols +/·. ***Example:*** $Q_1 + (N_1 \cdot N_2 \cdot N_3 \cdot N_4 \cdot N_5) + (D_1 \cdot N_1 \cdot N_2 \cdot N_3) + (D_1 \cdot D_2 \cdot N_1)$.

Step 6. Simplify the Boolean expression if possible through the use of Boolean laws. ***Example:*** This equation cannot be reduced any further. How equations are simplified using this method is explained in the remainder of this chapter.

Step 7. Construct the logic circuit from the expression. The circuit is shown in Fig. 4-14.

FIGURE 4.14
Designing a logic circuit using a Boolean expression.

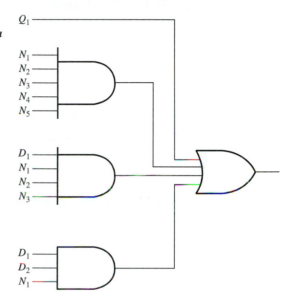

4.7 BOOLEAN ALGEBRA SIMPLIFICATION

The circuit of Fig. 4-15(a) shows the logic gate configuration used to perform certain specified operations. A close examination using deductive reasoning reveals that it contains a number of unnecessary gates. The following simplification steps show why the single gate of Fig. 4-15(b) is capable of performing the same function as the circuit of Fig. 4-15(a).

Solution by Using Deductive Reasoning

Step 1. A close examination of A_1 and N_1 reveals that both share the same inputs. Because the operations of an AND gate and a NAND gate are opposite, the outputs of these gates always complement each other.

Step 2. Because A_1 and N_1 complement each other, there is always a 1 at the output of either one gate or the other. Therefore, there is always a 1 applied to the OR gate.

Step 3. Because there is always a 1 applied to one of the inputs of the OR gate, there is always a 1 at its output lead.

Step 4. AND gate A_2 has two input Bs. Whether there are one B or two Bs, a 1 or 0 state applied to B affects the AND gate the same way. Therefore, one of the two B inputs can be eliminated.

Step 5. A 1 state from 01 is always applied to the top input of A_3. Therefore, the output of A_3 is ultimately determined by which state arrives from A_2 at its bottom input. As a result, A_3 can be replaced with a straight wire.

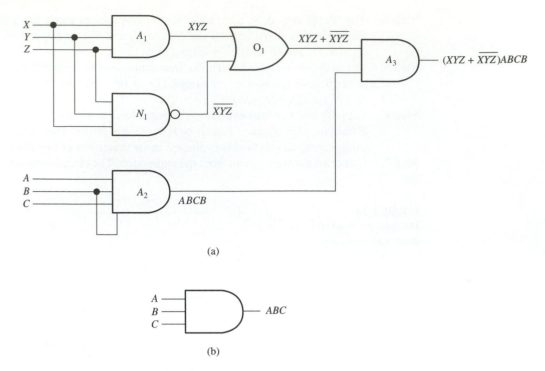

FIGURE 4.15
Simplifying a logic circuit using deductive reasoning. (a) Original expression. (b) Reduced expression.

Conclusion: Because the ultimate output of the circuit depends on the output of AND gate A_2, the circuit consisting of five gates is reduced to a one-gate circuit.

BOOLEAN ALGEBRA AND ASSOCIATED LAWS

A more scientific method of simplification than deductive reasoning is the use of Boolean algebra and the laws associated with it. In this section, the mathematical laws used to reduce logic circuits to their simplest form are explained. The laws fall into two categories:

1. Conventional algebra laws
2. Boolean algebra laws

Each law performs at least one of two primary functions:

1. *Rearrangement:* Two or more expressions may be equal to each other even though they are written differently. For example, by rearranging *BAC* into alphabetical order it becomes *ABC*. Therefore, both expressions equal each other.
2. *Reduction:* The use of mathematical laws sometimes allows equations to be reduced without changing their original representation.

Tables 4-1 and 4-2 summarize the mathematical laws that fall into each category, equations that apply to them, and the functions that they perform.

Boolean algebra laws apply to situations where the output of a logic gate is determined by a single input. Some of the laws allow the replacement of logic gates with the wire connected between the influential input terminal and the output. These laws are explained in detail throughout the remainder of this section. However, for clarity, they are presented in a different order than they were in the tables.

TABLE 4.1
Conventional Algebra Laws*

LAWS	*EQUATION*	*FUNCTION*
Commutative law	$AB = BA, A + B = B + A$	Rearrangement
Associative law	$A(BC) = (AB)C = ABC$	Reduction
	$A + (B + C) = (A + B) + C = A + B + C$	

*These two laws are also commonly used in conventional algebra.

TABLE 4.2
Boolean Algebra Laws*

LAWS	*BOOLEAN MULTIPLICATION*	*BOOLEAN ADDITION*	*BOOLEAN COMPLEMENTATION*
Idempotent	$A \cdot A = A$	$A + A = A$	
Complementary	$A \cdot \overline{A} = 0$	$A + \overline{A} = 1$	
Intersection	$A \cdot 1 = A$		
	$A \cdot 0 = 0$		
Union		$A + 1 = 1$	
		$A + 0 = A$	
Double negation			$\overline{\overline{A}} = A$

*Boolean algebra laws are all used to perform the *reduction* functions. Five of them are listed and the equation that represents each one is placed in the column that shows which one of the three operations it performs.

Commutative law:

When two variables are ANDed or ORed, they will yield the same result.

Commutative Law

$AB = BA$ and $A + B = B + A$ *Function:* Rearrangement

This law states that it does not matter what order variables are ANDed or ORed; they still yield the same results. Therefore, letters in an ANDed or ORed expression can be written in any order.

How the Law Affects Logic Circuits

The order in which the input variables are applied does not affect the binary value of the output. See Fig. 4-16.

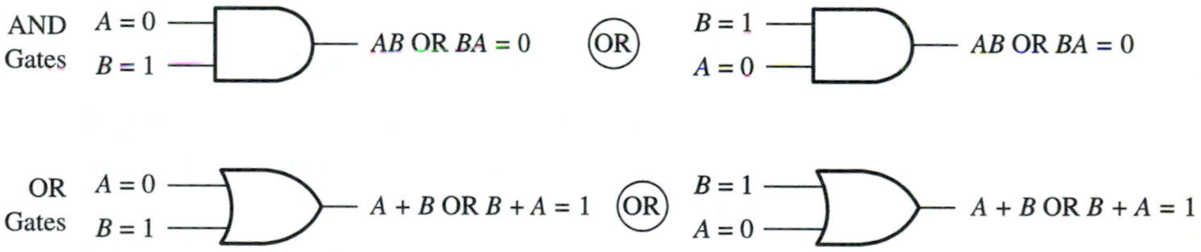

FIGURE 4.16
Commutative law diagrams.

■ REVIEW QUESTIONS

Which expressions are true according to the commutative law?

13. $A(B + C)$ and $(A + B)C$.

14. $RS + T + V$ and $V + T + RS$.

15. $UV + W$ and $VW + U$.

16. $E(F + \overline{G} + HI)$ and $(\overline{G} + F + IH)E$.

Associative law:

Allows parentheses to be removed when an enclosed expression is ANDed and that expression is also ANDed with an outside expression; the same rule applies to an ORed expression ORed to an external expression.

Associative Law

$$A(BC) = ABC \qquad A + (B + C) = A + B + C \qquad \textit{Function: } \text{Reduction}$$

This law allows the removal of parentheses from an expression and applies to AND gates and OR gates.

How the Law Affects AND Gates

(This is a Boolean multiplication operation using AND gates). This law applies to an enclosed ANDed expression that is ANDed with an outside expression. For example,

$$(AB)C = ABC$$
$$A(BC) = ABC$$

Through the Boolean multiplication operation, two AND gates are replaced by one AND gate.

How the Law Affects Logic Circuits

It does not matter which variables are ANDed first, second, third, and so forth; they are all ANDed together, and the result is one AND gate. Therefore, 2 two-input AND gates are reduced to 1 three-input AND gate. See Fig. 4-17.

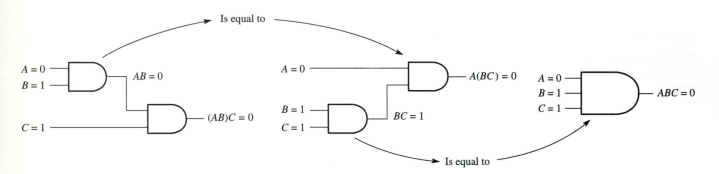

FIGURE 4.17
Associative (multiplication) law diagrams.

How the Law Affects OR Gates

(This is a Boolean addition operation using OR gates.) This law allows the removal of parentheses from an enclosed ORed expression that is ORed with an outside expression. For example,

$$(A + B) + C = A + B + C$$
$$A + (B + C) = A + B + C$$

Through the Boolean addition operation, 2 two-input OR gates are replaced by 1 three-input OR gate.

How the Law Affects Logic Circuits

It does not matter what variables are ORed first, second, third, and so forth; they are all ORed together; the result is one OR gate. Therefore, 2 two-input OR gates are reduced to 1 three-input OR gate. See Fig. 4-18.

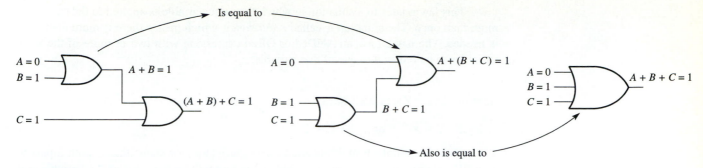

FIGURE 4.18
Associative (addition) law diagrams.

Double-negation law:

A mathematical method that states that if a binary bit is complemented twice, the result is the binary bit itself.

Double-Negation Law

$$\overline{\overline{A}} = A \qquad \textit{Function: Reduction}$$

(This is a Boolean complementation operation.) When two overbars of equal length appear over a variable or expression, they may be removed. The following examples show how two overbars are removed from expressions using this law:

$$\overline{\overline{A}}\,\overline{\overline{B}}\,\overline{\overline{C}} = ABC \qquad \overline{\overline{\overline{\overline{X}}}} = \overline{X} \qquad \overline{\overline{S}}\,\overline{\overline{T}} + U = \overline{S}\overline{T} + U$$

$$\overline{\overline{\overline{E}\,\overline{\overline{F}}G}} = \overline{E}FG \qquad \overline{\overline{(R + \overline{S})}}T = (R + \overline{S})T$$

How the Law Affects Logic Circuits

Figure 4-19 shows that when a variable is inverted twice, it becomes the variable itself. For example, if the variable is a 1 and it is inverted, it becomes a 0; when inverted again, it becomes a 1. Therefore, the two inverters can be replaced by a straight wire.

FIGURE 4.19
Double-negation diagrams.

■ REVIEW QUESTIONS

Simplify the following expressions that apply to the associative and double-negation laws.

17. $RST + (UV)W$

18. $AB + (C + D) + E$

19. $\overline{\overline{\overline{ABC}}} + \overline{\overline{DE}}$

20. $\overline{\overline{\overline{EFG}}\,\overline{\overline{H}} + \overline{\overline{I}}}$

21. $G(HI + J)(KL)(M + N)$

Idempotent law:

A mathematical method that states if a variable is ANDed or ORed with itself, it will give the variable itself as an output.

Idempotent Law

$$X \cdot X = X \qquad X + X = X \qquad \textit{Function: Reduction}$$

This law relates to a situation in which the same variable is applied to the same gate more than once. The situation is called *redundancy,* which means there is more than what is needed. The law allows an ANDed or ORed expression with two or more of the same variables to be reduced to one of that variable.

How the Law Affects AND Gates

$X \cdot X = X$ or $XX = X$

(This is a Boolean multiplication operation.) This law states that if both inputs to an AND gate represent the same variable, the output will be the same as the variable. For example,

$$ABCABC = ABC$$
$$(VW)(VW) = VW$$
$$(GH)EF(GH) = EF(GH)$$ (This can be reduced to *EFGH* by the associative law.)

How the Law Affects Logic Circuits

Figure 4-20 illustrates why this law is true. Therefore, the gate can be replaced with a wire located between where the input variables are connected and the output line.

X = 0
0
0
XX OR X = 0
Assume: X = 0
X = 1
1
1
XX OR X = 1
Assume: X = 1

FIGURE 4.20
Idempotent (multiplication) law diagrams.

How the Law Affects OR Gates

$X + X = X$

(This is a Boolean addition operation.) This law states that if both inputs to an OR gate represent the same variable, the output will be the same as the variable. For example,

$$R + R = R$$
$$X + Y + Z + X + Y + Z = X + Y + Z$$
$$\overline{E}FG + FG\overline{E} = \overline{E}FG$$

How the Law Affects Logic Circuits

Figure 4-21 illustrates why this law is true. Therefore, the gate can be replaced with a wire located between where the input variables are connected and the output line.

X = 0
0
0
X + X OR X = 0
Assume: X = 0
X = 1
1
1
X + X OR X = 1
Assume: X = 1

FIGURE 4.21
Idempotent (addition) law diagrams.

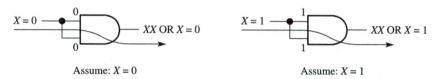

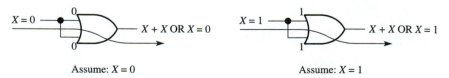

■ REVIEW QUESTIONS

Commutative:	$AB = BA, A + B = B + A$
Associative:	$A(BC) = ABC, A + (B + C) = A + B + C$
Double Negation:	$\overline{\overline{A}} = A$
Idempotent:	$AA = A, A + A = A$

Using these four basic laws, simplify the following expressions to their simplest possible form. List the laws used for each step and their purpose in the reduction process.

22. $TTU + ABC + CAB$

23. $AB + (CD + BA)$

24. $F(G + H) + IJ + \overline{KL} + IJ + \overline{LK} + (H + G)F$

25. $(\overline{U + V})(\overline{\overline{\overline{V + U}}})WXY$

26. $(LMN + OP) + \overline{\overline{MLN}}$

Complementary law (complementation law):

If 0 and 1 are ANDed together, the result is 0; if 0 and 1 are ORed together, the result is 1.

Complementary Law

$$A\overline{A} = 0 \qquad A + \overline{A} = 1 \qquad \textit{Function: Reduction}$$

This law relates to a situation in which a variable and its complement are applied as inputs to a gate. The law allows complementing variables in an ANDed expression to be replaced by a 0, and complementing variables in an ORed expression to be replaced by a 1.

How the Law Affects AND Gates

$$X\overline{X} = 0$$

(This is a Boolean multiplication operation.) This law states that if an AND gate has complementary signals applied to its inputs, the output will always be 0 because both inputs are never 1 at the same time. For example,

$$T\overline{T} = 0$$
$$AB\overline{AB} = 0$$
$$(\overline{LM + N})(LM + N) = 0$$

How the Law Effects Logic Circuits

Figure 4-22 illustrates why this law is true. Because the output is always 0, the gate can be replaced with a wire located between a ground lead and the output.

Assume: $X = 0$ Assume: $X = 1$

FIGURE 4.22
Complementary (multiplication) law diagrams.

How the Law Affects OR Gates

$$X + \overline{X} = 1$$

(This is a Boolean addition operation.) This law states that if an OR gate has complementary signals applied to its inputs, the output will always be 1 because at least one input is 1 at all times. For example,

$$XYZ + \overline{XYZ} = 1$$
$$T + \overline{T} = 1$$
$$\overline{AB + C} + AB + C = 1$$

How the Law Affects Logic Circuits

Figure 4-23 illustrates why this law is true. Because the output is always 1, the gate can be replaced with a wire connected between the power source and the output.

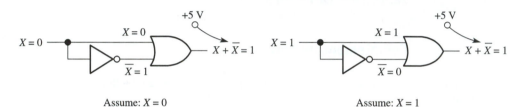

FIGURE 4.23
Complementary (addition) law diagrams.

■ REVIEW QUESTIONS

Commutative:	$AB = BA, A + B = B + A$
Associative:	$A(BC) = ABC, A + (B + C) = A + B + C$
Double Negation:	$\overline{\overline{A}} = A$
Idempotent:	$AA = A, A + A = A$
Complementary:	$X\overline{X} = 0, X + \overline{X} = 1$

Using these basic laws, simplify the following expressions to their simplest possible form. List the laws used for each step and their purpose in the reduction process.

27. $(A + B) + \overline{\overline{C}} + (C + \overline{\overline{B + A}})$

28. $\overline{RS} + \overline{\overline{T}} + U + \overline{\overline{\overline{SR}}} + \overline{\overline{T}}$

29. $\overline{E\overline{E\overline{E}}}$

30. $J(\overline{\overline{J}} + J)\overline{J}$

31. $F(\overline{GH}) + \overline{\overline{GH}}F$

32. $[\overline{\overline{(\overline{WX})\overline{Y}} + Z}](\overline{\overline{Z}} + \overline{WX\overline{Y}})$

Law of intersection:

A mathematical method which states that if a 0 is ANDed with a 1, the result is 0; or if a 1 is ANDed with a 1, the result is 1.

Law of Intersection

$$X \cdot 0 = 0 \qquad X \cdot 1 = X \qquad \textit{Function:} \text{ Reduction}$$

This law pertains to the AND gate as it performs the multiplication operation. When a 0 state is applied to an input, the AND gate is disabled; when a 1 state is applied to an input, the AND gate is enabled. Any ANDed expression with a variable that is 0 replaces all other variables with the 0. Any ANDed expression with a variable that is 1 is removed, leaving the remaining variable.

Disabled AND Gate

$$X \cdot 0 = 0$$

If one input to an AND gate is 0, the output will always be 0.

$$RT \cdot 0 = 0$$
$$(P + E) \cdot 0 = 0$$
$$(\overline{R} + ST) \cdot 0 = 0$$

How the Law Affects Logic Circuits

Figure 4-24 illustrates why this law is true. Despite the condition applied to the X input, the output is always 0. Therefore, the gate can be replaced with a wire placed between the low input and the output.

Assume: $X = 0$ Assume: $X = 1$

FIGURE 4.24
Law of intersection with a disabling input.

Enabled AND Gate

$$X \cdot 1 = X$$

If one input to an AND gate is 1, the output will always be equal to the X input. For example,

$$AC \cdot 1 = AC$$
$$(E + F) \cdot 1 = E + F$$
$$(YZ)(YZ) \cdot 1 = (YZ)(YZ) = YZ \quad \text{(through the idempotent law)}$$

How the Law Affects Logic Circuits

Figure 4-25 illustrates why this law is true. The output is controlled by the condition of input X. Therefore, the gate can be replaced by a wire placed between the X input and the output.

Assume: $X = 0$ Assume: $X = 1$

FIGURE 4.25
Law of intersection with an enabling input.

Law of union:

A mathematical method which states that if an ORed input is 0, the output depends on the other inputs; if an input is 1, the output is 1.

Law of Union

$$A + 0 = A \qquad A + 1 = 1 \qquad \textit{Function: Reduction}$$

This law pertains to the OR gate as it performs the addition operation. When an ORed input is 0, the output depends on the condition of the other inputs. When an ORed input is 1, the output is 1. Any ORed expression with a variable that is 0 is removed, leaving the remaining variables. Any ORed expression with a variable that is 1 replaces all other variables with 1.

$A + 0 = A$

If one input to an OR gate is 0, the output will always be equal to the A input. For example,

$$RT + 0 = RT$$
$$ABC + 0 = ABC$$
$$(XY)Z + 0 = (XY)Z \text{ (associative law)} = XYZ$$

How the Law Affects Logic Circuits

The gate can be replaced with a wire connected between the X input and the output. See Fig. 4-26.

Assume: $X = 0$ Assume: $X = 1$

FIGURE 4.26
Law of union with at least one input at 0.

$X + 1 = 1$

If one input to an OR gate is in a 1 state, the output will always be 1. For example,

$$OP + 1 = 1$$
$$X + \bar{Y} + Z + 1 = 1$$
$$1 + UV + W + 0 = 1$$

How the Law Affects Logic Circuits

The gate can be replaced with a wire connected between the 1 state input lead and the output. See Fig. 4-27.

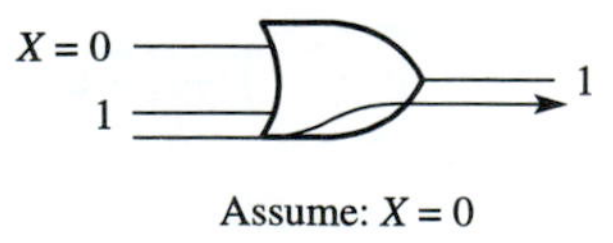

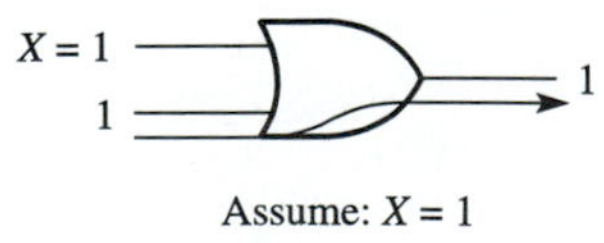

Assume: $X = 0$ Assume: $X = 1$

FIGURE 4.27
Law of union with at least one input at 1.

■ REVIEW QUESTIONS

Commutative:	$AB = BA, A + B = B + A$
Associative:	$A(BC) = ABC, A + (B + C) = A + B + C$
Double Negation:	$\bar{\bar{A}} = A$
Idempotent:	$AA = A, A + A = A$
Complementary:	$X\bar{X} = 0, X + \bar{X} = 1$
Intersection:	$X \cdot 0 = 0, X \cdot 1 = X$
Union:	$A + 0 = A, A + 1 = 1$

Using these basic laws, simplify the following expressions to their simplest possible form. List the laws used for each step and their purpose in the reduction process.

33. $\bar{A}BA$

34. $EFGF\bar{H}IHI$

35. $T(\bar{R}R + \bar{\bar{S}}S)$

36. $(U\bar{V} + X\bar{Y})(Z + \bar{Z})$

37. $(\overline{DE + GH})(F + \bar{F})(\overline{\overline{HG + ED}})$

38. $(AB)(\overline{CD}) + \overline{A(BC)D}$

The laws of Boolean algebra are summarized in Fig. 4-28. The figure shows how they affect logic gates and the expressions that represent them. Note that in each situation, the output is determined by a single input. Therefore, no logic gates are required to satisfy any of the expressions that represent each Boolean law. Mathematically, when any Boolean law expression appears, it can be replaced with any of the appropriate symbols (X, 1, or 0).

The circuit in Figure 4-15, which was reduced to one gate by using deductive reasoning, can also be simplified using the conventional and Boolean algebra laws.

$(XYZ + \overline{XYZ})ABCB$	Original Output Expression
$(XYZ + \overline{XYZ})ABBC$	Commutative Law

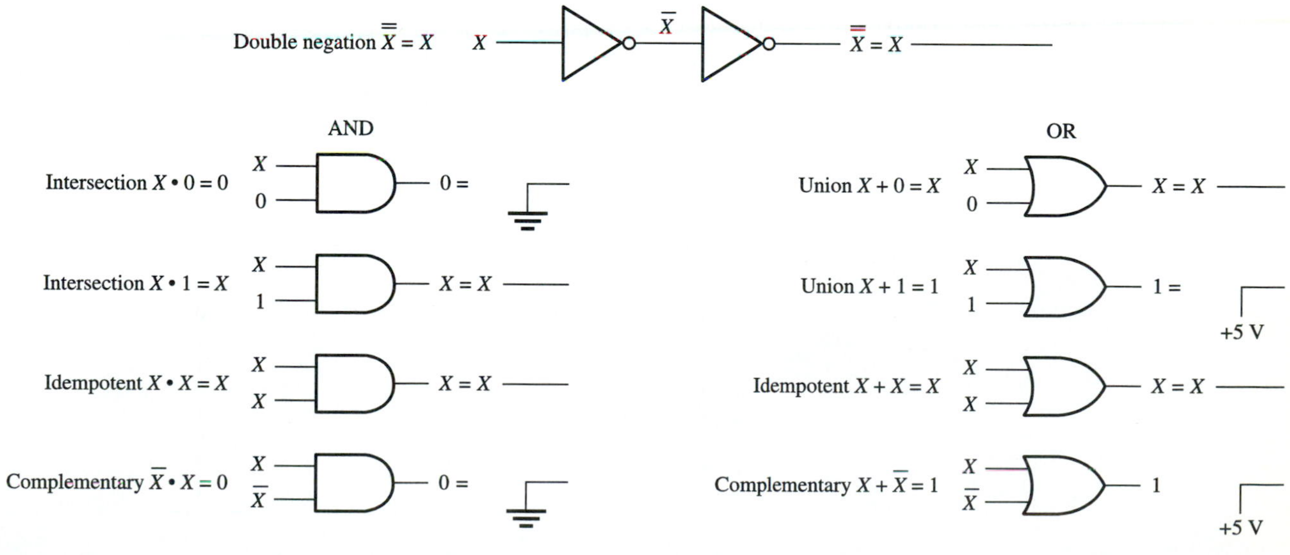

Double negation $\bar{\bar{X}} = X$

AND

Intersection $X \cdot 0 = 0$

Intersection $X \cdot 1 = X$ $X = X$

Idempotent $X \cdot X = X$ $X = X$

Complementary $\bar{X} \cdot X = 0$

OR

Union $X + 0 = X$ $X = X$

Union $X + 1 = 1$

Idempotent $X + X = X$ $X = X$

Complementary $X + \bar{X} = 1$

FIGURE 4.28

Summary of the Boolean algebra laws. The Boolean laws make it possible to simplify circuits because the output is determined by a single input.

$$(XYZ + \overline{XYZ})ABC \qquad \text{Idempotent Law}$$
$$(1)ABC \qquad \text{Complementary Law}$$
$$ABC \qquad \text{Intersention Law}$$

To make it possible to further simplify some equations, they can sometimes be manipulated by mathematical techniques such as factoring, removing parentheses, expanding, or DeMorganizing. These techniques include mathematical laws such as DeMorgan's theorem, the distributive law, and derived expressions. When used, they increase the reduction capabilities of Boolean algebra. Another method of simplifying expressions is called Karnaugh mapping, which uses a graphical technique to reduce Boolean logic equations. For those readers who want to learn more, a supplement for this textbook is available through Prentice-Hall that describes the various laws associated with Boolean algebra.

4.8 TROUBLESHOOTING

One function of a Boolean equation is to describe the circuit configuration of a combination of logic gates. A description of how any of these circuits operates is often provided by a truth table. Boolean equations are also capable of providing the same information as a truth table, but in a more concise form. This information is sometimes useful in troubleshooting.

Figure 4-29 shows a binary full adder that is capable of adding binary numbers one column at a time. The circuit includes a Boolean equation at both the sum and carry-out output lines. Each letter represents an input of the full adder. A letter with an overbar specifies when a binary 0 is applied to the input it represents. Likewise, a letter without an overbar specifies when a binary 1 is applied to the input. Each equation describes what state each input line must be in to cause the output to go high.

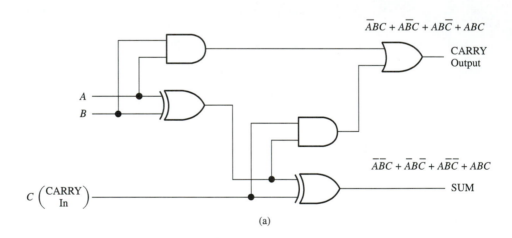

(a)

A	B	C (CARRY In)	CARRY Out	SUM
$0-\overline{A}$	$0-\overline{B}$	$0-\overline{C}$	0	0
$0-\overline{A}$	$1-B$	$0-\overline{C}$	0	1
$1-A$	$0-\overline{B}$	$0-\overline{C}$	0	1
$1-A$	$1-B$	$0-\overline{C}$	1	0
$0-\overline{A}$	$0-\overline{B}$	$1-C$	0	1
$0-\overline{A}$	$1-B$	$1-C$	1	0
$1-A$	$0-\overline{B}$	$1-C$	1	0
$1-A$	$1-B$	$1-C$	1	1

(b)

FIGURE 4.29
Full-Adder: (a) Logic circuit. (b) Numerical and Boolean truth table.

The Boolean algebra equation for the sum output is

$$S = \overline{A}\,\overline{B}C + \overline{A}B\overline{C} + A\overline{B}\,\overline{C} + ABC$$

This means that output S (sum) goes high if

A and B are low and C is high, or

B is high and A and C are low, or

A is high and B and C are low, or

A and B and C are high.

The equation also indirectly specifies that any other input combination applied to inputs A, B, and C causes the sum output to be 0.

The Boolean algebra equation for the carry output is

$$C = \overline{A}BC + A\overline{B}C + AB\overline{C} + ABC$$

This means that output C (carry output) goes high if

A is low and B and C are high, or

A and C are high and B is low, or

A and B are high and C is low, or

A and B and C are high.

Any other input combination causes the carry output to be 0. The truth table in part (B) can be used to verify these statements.

Figure 4-29 shows how the troubleshooter uses Boolean equations to troubleshoot. A logic probe is placed at all of the inputs to determine the logic state condition. Assume that $A = 1$, $B = 0$, and $C = 1$; the Boolean equation for this input combination is $A\overline{B}C$. The troubleshooter examines each output to determine if $A\overline{B}C$ is listed. Because it is located at the carry output, the logic probe is used to detect if a high is present. Because $A\overline{B}C$ is not located at the sum output, the logic probe is used to determine if a low is present.

By not being required to refer to a truth table, or not having to test the inputs and output of each gate, the troubleshooter saves time.

■ SUMMARY

- Boolean algebra uses only a 1 or a 0 to represent two conditions.
- Boolean algebra performs three operations: addition, multiplication, and complementation.
- Boolean expressions are made up of letters that represent input variables and operators that describe the addition, multiplication, and complementation functions.
- AND gates perform Boolean multiplication.
- OR gates perform Boolean addition.
- An inverter performs Boolean complementation.
- AND, OR, and NOT gates are fundamental operators. All digital circuitry uses some combination of these basic functions.
- Four primary applications of Boolean algebra are as follows:
  1. Boolean equations describe logic circuit configurations.
  2. Certain types of Boolean equations describe logic circuit operations.
  3. Boolean algebra is used in the reduction of logic circuits.
  4. Boolean algebra is used for logic circuit troubleshooting.

■ PROBLEMS

1. Identify the relationship among the symbols, operators, and functions associated with Boolean algebra. (4-2, 3)

SYMBOLS	OPERATORS	FUNCTIONS
⊐D-		Boolean multiplication
⊐D-	$(+)$	
	$\overline{X}$	Complementation

2. Write the Boolean output expression for the logic gates of Fig. 4-30. (4-4)

3. When deriving logic expressions from a circuit, a rule of thumb is to work from the _____ toward the _____ . (4-4)

4. When constructing logic diagrams from output expressions, it is necessary to start at the _____ and work toward the _____ side of the diagram. (4-4)

FIGURE 4.30
Logic gates for Problem 2.

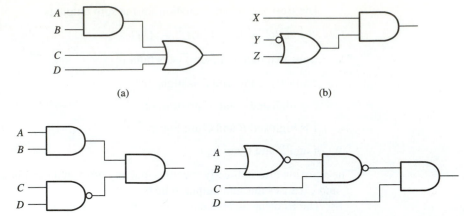

(a)

(b)

(c)

(d)

5. (a) The expression $(A + B)(C + D)$ is an example of the _____ form. (4-5)

(b) The expression $AB + CD$ is an example of the _____ form. (4-5)

6. To the left of each of the following steps, place a number to represent the order in which they should be followed in the logic design process. (4-6)

_____ Establish a Boolean expression by replacing the words "and/or" with operator symbols.

_____ Identify each input condition with a letter symbol.

_____ Define the desired function of the circuit.

_____ Develop a statement that describes the necessary conditions.

_____ Construct the logic circuit from the simplified expression.

_____ Restate the statement problem using the letter symbols.

_____ Simplify the Boolean expression.

7. Apply the indicated law to the following expressions.

(a) $BA =$ (commutative)

(b) $A(BC) =$ (associative)

(c) $A + (B + C) =$ (associative)

8. Complete the following basic Boolean equations and list the laws associated with them. (4-7)

(a) $A + 0 =$

(b) $\bar{\bar{A}} =$

(c) $A + A =$

(d) $A \cdot 1 =$

(e) $A + 1 =$

(f) $A \cdot A =$

(g) $A \cdot \bar{A} =$

(h) $A \cdot 0 =$

(i) $A + \bar{A} =$

9. Boolean laws make it possible to simplify circuits because the output of a gate is determined by a single _____ . (4-7)

TROUBLESHOOTING

10. Product-of-sums and sum-of-products equations define the output function of a logic circuit by describing when a _____ is produced at the circuit output it represents. (4-5)

11. Explain what a Boolean equation such as $S = \bar{A}\,\bar{B}C + \bar{A}B\bar{C} + AB\,\bar{C} + ABC$ provides at a test point in a circuit. (4-8)

■ ANSWERS TO REVIEW QUESTIONS

1. (c) **2.** (d) **3.** variables, inputs **4.** operators

5. (a) AND gate, (b) OR gate, (c) inverter **6.** $A + B$

7. (a) To provide a shorthand notation of a logic circuit. (b) To describe the operation of a logic circuit. (c) To minimize a logic circuit. (d) To perform troubleshooting. **8.** inputs, output

9. ANDed, OR **10.** (a) $(ABC)DE$ (b) $(R + S)T$, (c) $XY + Z$, (d) $F + G + (H + I)$, (e) $AD(B + C)$ or $A(B + C)D$ or $(B + C)AD$, (f) $A + B(C + D)$ **11.** output, inputs **12.** See Fig. 4-31.

13. False **14.** True **15.** False **16.** True

17. $\overline{RST + UVW}$ **18.** $AB + C + D + E$ **19.** $\overline{\overline{ABC} + \overline{DE}}$

20. $\overline{\overline{EFGH} + I}$ **21.** $GKL(HI + J)(M + N)$

22.

$TTU + ABC + CAB$	Original expression	
$TTU + ABC + ABC$	Commutative	Rearrangement
$TU + ABC$	Idempotent	Reduction

23.

$AB + (CD + BA)$	Original expression	
$AB + (CD + AB)$	Commutative	Rearrangement
$AB + CD + AB$	Associative	Reduction
$AB + CD$	Idempotent	Reduction

FIGURE 4.31
Logic diagram answers for Review Question 12.

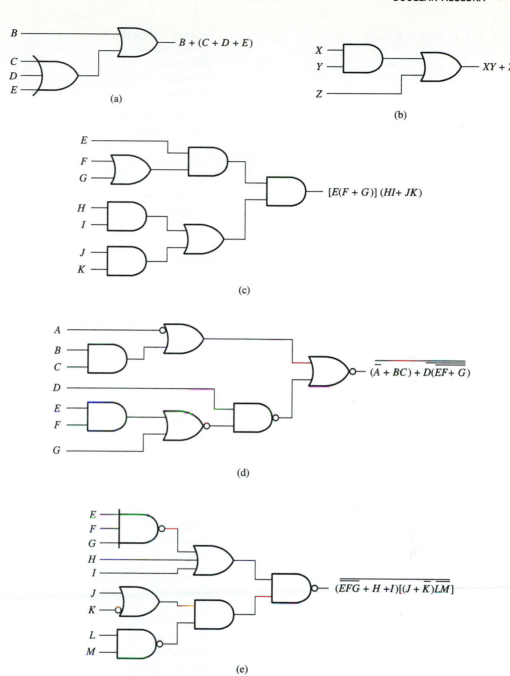

24. $F(G + H) + IJ + \overline{KL} + IJ + \overline{LK} + (H + G)F$ Original expression
     $F(G + H) + IJ + \overline{KL} + IJ + \overline{KL} + F(G + H)$ Commutative   Rearrangement
     $F(G + H) + IJ + \overline{KL}$ Idempotent   Reduction

25. $\overline{(\overline{U + V})}\overline{\overline{\overline{(\overline{V + U})}}}WXY$ Original expression
     $\overline{(\overline{U + V})}\overline{(\overline{V + U})}WXY$ Double negation   Reduction
     $\overline{(\overline{U + V})}\overline{(\overline{U + V})}WXY$ Commutative   Rearrangement
     $\overline{(\overline{U + V})}WXY$ Idempotent   Reduction

26. $(LMN + OP) + \overline{\overline{MLN}}$ Original expression
     $(LMN + OP) + MLN$ Double negation   Reduction
     $(LMN + OP) + LMN$ Commutative   Rearrangement
     $LMN + OP + LMN$ Associative   Reduction
     $LMN + OP$ Idempotent   Reduction

27. $(A + B) + \overline{\overline{C}} + (C + \overline{\overline{B} + \overline{A}})$ Original expression

    $(A + B) + C + (C + B + A)$ Double negation   Reduction

    $A + B + C + C + B + A$ Associative   Reduction

    $A + A + B + B + C + C$ Commutative   Rearrangement

    $A + B + C$ Idempotent   Reduction

28. $\overline{RS} + \overline{\overline{T}} + U + \overline{\overline{\overline{SR}}} + \overline{\overline{T}}$ Original expression

    $\overline{RS} + T + U + \overline{SR} + T$ Double negation   Reduction

    $\overline{RS} + \overline{RS} + T + T + U$ Commutative   Rearrangement

    $\overline{RS} + T + U$ Idempotent   Reduction

29. $\overline{E}E\overline{\overline{E}}$ Original expression

    $\overline{E}EE$ Double negation   Reduction

    $\overline{E}E$ Idempotent   Reduction

    0 Complementary   Reduction

30. $J(\overline{\overline{J}} + J)\overline{J}$ Original expression

    $J(\overline{J} + J)\overline{J}$ Double negation   Reduction

    $J\overline{J}(\overline{J} + J)$ Commutative   Rearrangement

    $J\overline{J}(1)$ Complementary   Reduction

    $0(1)$ Complementary   Reduction

    $0 \cdot 1 = 0$ Therefore 1 ANDed with $0 = 0$

31. $F(\overline{GH}) + \overline{\overline{GH}}F$ Original expression

    $F(\overline{GH}) + \overline{FGH}$ Commutative   Rearrangement

    $F\overline{GH} + \overline{FGH}$ Associative   Reduction

    1 Complementary   Reduction

32. $[(\overline{\overline{WX}})\overline{Y} + Z](\overline{\overline{Z} + \overline{WXY}})$ Original expression

    $[(\overline{WX})\overline{Y} + Z](\overline{Z} + \overline{WXY})$ Double negation   Reduction

    $(\overline{WX}\,\overline{Y} + Z)(\overline{Z} + \overline{WXY})$ Associative   Reduction

    $(\overline{WX}\,\overline{Y} + Z)(\overline{WX\overline{Y}} + Z)$ Commutative   Rearrangement

    0 Complementary   Reduction

33. $\overline{A}BA$ Original expression

    $A\overline{A}B$ Commutative   Rearrangement

    $0 \cdot B$ Complementary   Reduction

    0 Intersection   Reduction

34. $EFGF\overline{H}IHI$ Original expression

    $EFFG\overline{H}HII$ Commutative   Rearrangement

    $EFG\overline{H}HI$ Idempotent   Reduction

    $H\overline{H} = 0$ Complementary   Reduction

    $EFGI \cdot 0 = 0$ Intersection   Reduction

35. $T(\overline{R}R + \overline{\overline{S}}S)$ Original expression

    $T(\overline{R}R + SS)$ Double negation   Reduction

    $T(\overline{R}R + S)$ Idempotent   Reduction

    $T(0 + S)$ Complementary   Reduction

    $T(S)$ Union   Reduction

    TS Associative   Reduction

36. $(U\overline{V} + X\overline{Y})(Z + \overline{Z})$ Original expression

    $(U\overline{V} + X\overline{Y})1$ Complementary   Reduction

    $U\overline{V} + X\overline{Y}$ Intersection   Reduction

37. $(\overline{DE + GH})(F + \overline{F})(\overline{\overline{HG + ED}})$ Original expression

    $(\overline{DE + GH})(F + \overline{F})(HG + ED)$ Double negation   Reduction

    $(\overline{DE + GH})(DE + GH)(F + \overline{F})$ Commutative   Rearrangement

    $0(F + \overline{F})$ Complementary   Reduction

    0 Intersection   Reduction

38. $(AB)(\overline{\overline{CD}}) + \overline{A(BC)D}$ Original expression

    $(AB)(CD) + \overline{A(BC)D}$ Double negation   Reduction

    $ABCD + \overline{ABCD}$ Associative   Reduction

    1 Complementary   Reduction

CHAPTER 5

COMBINATION CIRCUITS

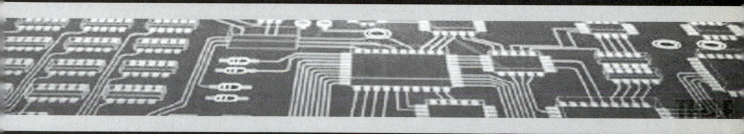

When you complete this chapter, you will be able to:

1. Explain the operation and use of encoders.
2. Explain the operation and use of decoders.
3. Explain the operation of seven-segment LED and LCD displays.
4. Explain the operation and use of multiplexers.
5. Explain the operation and use of demultiplexers.
6. Explain the operation and use of parity circuits.
7. Explain the operation and use of binary adders.
8. Explain the operation and use of comparator circuits.
9. Use and understand IC pin diagrams, logic circuits, waveform diagrams, and truth tables of combination circuits.
10. Perform basic troubleshooting of combination circuits.

5.1 INTRODUCTION

Combination circuits are digital logic circuits made up of gates and inverters. When they are tied together, they can perform a countless variety of logic functions. Some of the resulting logic systems contain only a handful of gates; others may use dozens or even hundreds of gates. The outputs of these circuits are determined by the status of the inputs, the types of gates used, and how they are configured. The combination circuits are also characterized by their fast operation because their outputs instantaneously reflect the patterns of 0s and 1s applied to the inputs.

Certain logic gate combinations are widely used in digital equipment. These circuits are called **functional logic circuits** and include encoders, decoders, multiplexers, demultiplexers, parity circuits, adders, and comparators. Each one of these standard devices is integrated onto a single chip and are available at a low cost from most IC manufacturers.

To configure nonstandard combination circuits, it is necessary to use IC circuits made up of individual gates and inverters and interconnect them, using the external pins of the IC package, into the desired pattern. Some types of nonstandard combination circuits are examined in Chapter 4, whereas the more common standard combination circuits are examined in this chapter.

There are three categories of standard combination circuits: *conversion, data transfer,* and *data processing circuits.* Each category differs from the others by the way in which it handles binary data.

CONVERSION CIRCUITS

Combination circuits:

Digital logic circuits made up of a combination of gates and inverters.

Chapter 2, "Digital Number Systems," explains how to use several different number systems and how to convert one to another. Certain conversion devices, called **encoders** and **decoders,** translate a value of one number system into its equivalent value in another number system. It is these devices especially that help make it possible for a human operator to effectively communicate with digital circuits that internally use the base-2 number system.

Figure 5-1 illustrates some of the common conversions that take place between digital circuits and input and output devices.

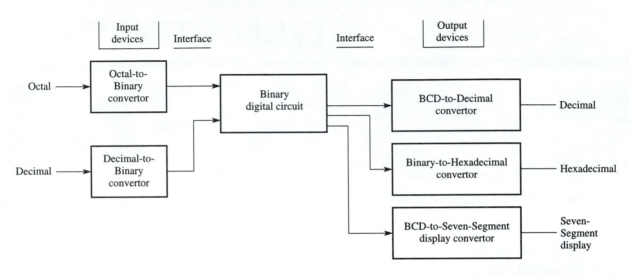

FIGURE 5.1
Common number system conversions.

Functional logic circuits:

Standard combination logic circuits such as encoders, decoders, multiplexers, demultiplexers, parity circuits, adders, and comparators.

Encoders

Encoders are usually devices that convert data into a form that can be interpreted by a digital circuit. The most common type of encoder is one that converts a nonbinary number system into an equivalent binary system. These encoders receive their inputs from various devices, including switches, punched paper tape, floppy disks, keyboards, magnetic tape, and keypads similar to those used by a basic calculator.

5.2   OCTAL-TO-BINARY ENCODER

Encoder:

Any logic device that converts a nonbinary number code into an equivalent binary code.

Decoder:

Any logic device that converts a binary code into an equivalent nonbinary code.

Octal-to-binary encoder:

A type of encoder that converts octal inputs into equivalent binary output values.

An **octal-to-binary encoder** is a device that converts octal numbers to binary. As shown in Fig. 5-2, it has eight input lines and three output lines, all of which are normally at low states. Each input line represents one of the octal digits, and they are labeled 0–7. The three output lines represent three bits of the binary number system. Together, they produce eight different 3-bit binary code values at the output, 000 to 111, that correspond to the input that is activated. Each output line is labeled with a capital letter. A letter represents one of the binary weighted positional values. The 1s column (LSB, or least significant bit) value is present at output A, the 2s column value is present at output B, and the 4s column (MSB, or most significant bit) value is present at output C.

The circuit allows only one of the input lines to be activated at any given time, so there are only eight possible input conditions. If more than one input is made high at one time, the output results will be erroneous. When line 5 is activated by a high, for example, the binary octal code $101_2 (5_8)$ is generated at the output. If input line 0 is high, the outputs will read $000_2 (0_8)$. The truth table in Fig. 5-2(c) illustrates the full operation of this circuit as it functions under normal conditions.

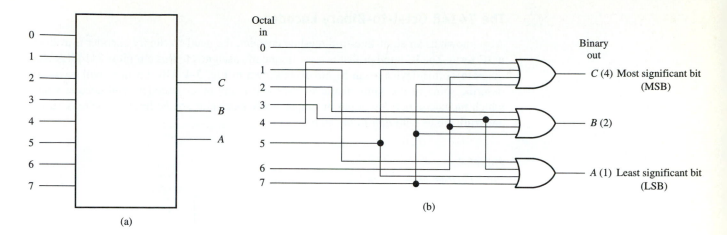

(a)

(b)

Inputs								Outputs		
0	1	2	3	4	5	6	7	C	B	A
1	0	0	0	0	0	0	0	0	0	0
0	1	0	0	0	0	0	0	0	0	1
0	0	1	0	0	0	0	0	0	1	0
0	0	0	1	0	0	0	0	0	1	1
0	0	0	0	1	0	0	0	1	0	0
0	0	0	0	0	1	0	0	1	0	1
0	0	0	0	0	0	1	0	1	1	0
0	0	0	0	0	0	0	1	1	1	1

(c)

FIGURE 5.2
Octal-to-binary encoder: (a) Block diagram. (b) Logic circuitry. (c) Function table.

EXAMPLE 5.1

What would be the output of the octal-to-binary encoder if input 6 of the circuit of Fig. 5-3 were made high?

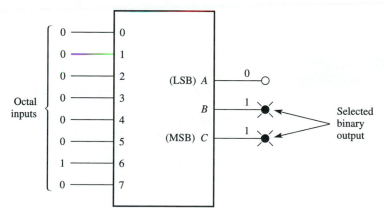

FIGURE 5.3
Circuit for Example 5.1.

Solution Output *B*, which represents a binary 2, and output *C*, which represents a binary 4, will go high and output *A*, which represents a binary 1, will remain low, as shown in the figure.

The 74148 Octal-to-Binary Encoder

Also known as an eight-line-to-three-line encoder, the octal-to-binary encoder is available in IC form. The IC number that is most frequently assigned to this circuit is 74148. A block diagram of this device as an IC circuit is shown in Fig. 5-4. This encoder will be used to describe some of the same characteristics that many other encoders and decoders have, which include *active low* input or output leads, *enable* or *strobe* lines, *priority functions*, and truth tables with *don't-care* states.

FIGURE 5.4
Block diagram of the 74148 IC.

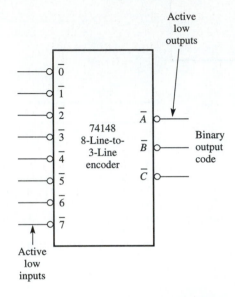

Active Low Inputs and Outputs

Figure 5-4 shows the 74148 encoder with active low inputs and outputs. When one of the input lines to the encoder is driven to 0, a binary code corresponding to the input is generated by the appropriate output lines going low. All active low inputs and outputs are labeled with an overbar over the identification number or letter (for example, $\overline{A}$). Any active high inputs and outputs are labeled without an overbar (for example, 3). Many IC circuits use active low outputs because they can provide more output current with a low state than with a high state.

EXAMPLE 5.2

When a logic 0 is applied to input line 6 of the circuit in Fig. 5-5, what are the logic states at each output line of the encoder?

FIGURE 5.5
Circuit for Example 5.2.

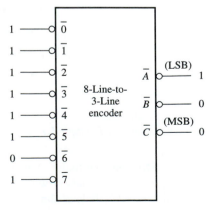

Solution The outputs are

$$(\text{LSB})\ \overline{A} = 1$$

$$\overline{B} = 0$$

$$(\text{MSB})\ \overline{C} = 0$$

Output $\overline{B}$, which represents the weighted value of 2, and line $\overline{C}$, which represents the weighted value of 4, are activated. The sum of these two values is 6, which equals the numerical value of the input that is activated.

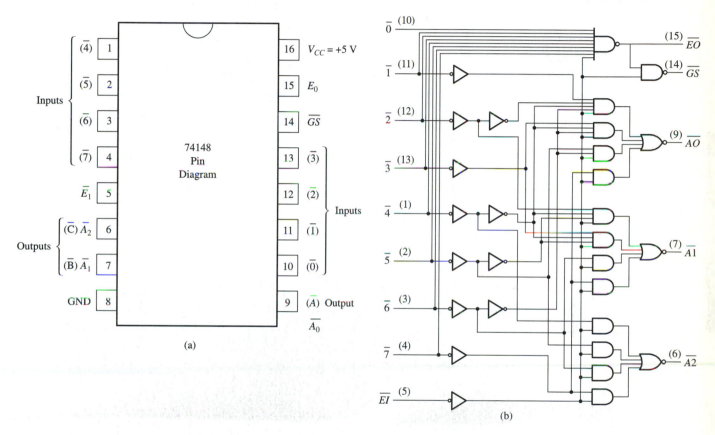

	Inputs								Outputs				
$\overline{EI}$	$\overline{0}$	$\overline{1}$	$\overline{2}$	$\overline{3}$	$\overline{4}$	$\overline{5}$	$\overline{6}$	$\overline{7}$	$\overline{A_2}$	$\overline{A_1}$	$\overline{A_0}$	$\overline{GS}$	$\overline{EO}$
H	X	X	X	X	X	X	X	X	H	H	H	H	H
L	H	H	H	H	H	H	H	H	H	H	H	H	L
L	X	X	X	X	X	X	X	L	L	L	L	L	H
L	X	X	X	X	X	X	L	H	L	L	H	L	H
L	X	X	X	X	X	L	H	H	L	H	L	L	H
L	X	X	X	X	L	H	H	H	L	H	H	L	H
L	X	X	X	L	H	H	H	H	H	L	L	L	H
L	X	X	L	H	H	H	H	H	H	L	H	L	H
L	X	L	H	H	H	H	H	H	H	H	L	L	H
L	L	H	H	H	H	H	H	H	H	H	H	L	H

(c)

FIGURE 5.6

The 74148 IC: (a) Pin diagram. (b) Logic circuit diagram. (c) Function table. Reproduced by permission of Texas Instruments. Copyright © Texas Instruments.

The pin diagram, logic circuit diagram, and truth table for this device are shown in Fig. 5-6. Output $\overline{A2}$ represents the binary weighted value of 4, output $\overline{A1}$ represents the binary weighted value of 2, and output $\overline{A0}$ represents the binary weighted value of 1.

An additional input pin, labeled $\overline{E1}$, is called the enable line. Only when it is at a low will it enable the circuit to operate. If a logic 1 is applied to this lead, the circuit will not produce an output that corresponds to the applied input.

To allow for the expansion of the 74148 IC, output pins E_0 and $\overline{GS}$ have been provided, so that by cascading the ICs, the device can encode larger numbers. Instructions on how to use these pins are provided in IC data manuals.

EXAMPLE 5.3

Draw the output waveforms of the 74148 for each combination of octal and enable inputs shown in the timing diagram of Fig. 5-7.

Solution   See Fig. 5-7.

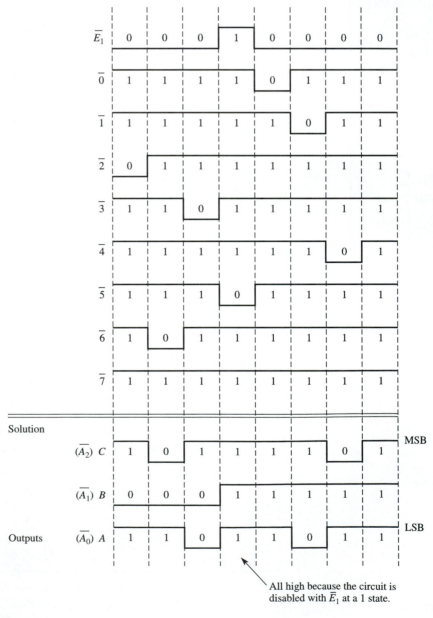

FIGURE 5.7
Timing diagram for Example 5.3.

Time Period	
1	A logic low at the enable line $\overline{E}$ allows the circuit to operate. A logic low at input $\overline{2}$ causes the encoder to produce a logic low at output $\overline{A_1}$, which represents the binary weighted value of 2.
2	A logic low at line $\overline{E_1}$ enables the circuit to operate. A logic low at input $\overline{6}$ causes the encoder to produce a logic low at output $\overline{A_1}$, which represents the binary weighted value of 2, and at output $\overline{A_2}$, which represents the binary weighted value of 4. Together, the sum of the output values are equal to the input line that is activated.
3	A logic low at line $\overline{E_1}$ enables the circuit to operate. A logic low at input $\overline{3}$ causes the encoder to produce a logic low at output $\overline{A_0}$, which represents the binary weighted value of 1, and at output $\overline{A_1}$, which represents the binary weighted value of 2. Together, their sum equals the input line that is activated.
4	A logic high at line $\overline{E_1}$ disables the circuit. This condition does not allow any of the output lines to be activated.
5	A logic low at line $\overline{E_1}$ enables the circuit to operate. None of the input lines are activated. Therefore, all of the output lines remain at a logic high because none of them is activated.
6	A logic low at line $\overline{E_1}$ enables the circuit. A logic low at input 1 causes output line A_1 to be activated by a logic low.
7	A logic low at line $\overline{E_1}$ enables the circuit. A logic low at input $\overline{4}$ causes output $\overline{A_2}$ to be activated.
8	A low at $\overline{E_1}$ enables the circuit. None of the output lines is activated because none of the input lines is activated.

Priority Function

Some encoders use push buttons to apply input signals. What would happen if more than one of the push buttons were pressed at the same time? For some encoders, such as the one described in Fig. 5-2, the outputs would be garbled. Other encoders, such as the 74148 IC, have the capability of generating the binary-coded output that corresponds to the highest-numbered input that is activated by a low. All other activated inputs are overridden.

Don't Care

Don't-care conditions:

Xs used in truth tables to specify that the inputs where they are placed are overridden by another input condition.

Many truth tables of standard ICs have Xs listed in the input section of the table, which represent **don't-care conditions.** These indicators specify that these inputs do not influence the output of the circuit because they are overridden by another input condition. As a result, the circuit "does not care" if a high or low is applied to that input. The operation of the 74148 IC priority encoder can be used to provide an example of why don't-care conditions sometimes exist. The 3rd horizontal row from the top of the truth table in Fig. 5-6(c) shows what happens when input 7 is activated. Xs are placed at each of the lower-valued inputs. This specifies that any lower-valued input can be at either state without affecting the circuit.

5.3 DECIMAL-TO-BINARY ENCODER

Decimal-to-binary encoder:

A type of encoding device that converts decimal inputs into equivalent binary values at the output leads.

The conversion of decimal to binary follows the same general concepts as octal to binary. The block diagram of a **decimal-to-binary encoder** is shown in Fig. 5-8(a). Each of the 10 inputs represents one of the decimal digits. The four output lines correspond to the BCD (binary-coded-decimal) code. Both the input and output lines are activated by logic high states. When any input is activated, the circuit produces a BCD equivalent output. The operation of the circuit is summarized by the truth table in Fig. 5-8(b). This circuit does not perform the priority function. If more than one input is made high at a given time, the output results will be garbled.

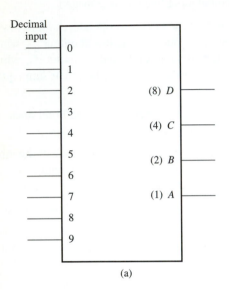

(a)

				Inputs								Outputs		
										(8)	(4)	(2)	(1)
0	1	2	3	4	5	6	7	8	9	D	C	B	A
0	0	0	0	0	0	0	0	0	1	1	0	0	1
0	0	0	0	0	0	0	0	1	0	1	0	0	0
0	0	0	0	0	0	0	1	0	0	0	1	1	1
0	0	0	0	0	0	1	0	0	0	0	1	1	0
0	0	0	0	0	1	0	0	0	0	0	1	0	1
0	0	0	0	1	0	0	0	0	0	0	1	0	0
0	0	0	1	0	0	0	0	0	0	0	0	1	1
0	0	1	0	0	0	0	0	0	0	0	0	1	0
0	1	0	0	0	0	0	0	0	0	0	0	0	1
1	0	0	0	0	0	0	0	0	0	0	0	0	0

(b)

FIGURE 5.8
Decimal-to-binary encoder: (A) Block diagram. (B) Function table.

EXAMPLE 5.4

What will be the output of the decimal-to-binary encoder if input 5 of the circuit of Fig. 5-9 is made high?

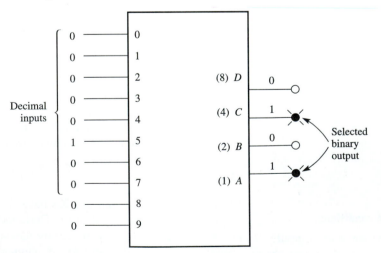

FIGURE 5.9
Circuit for Example 5.4.

Solution   Outputs A and C go high.

The 74147 Decimal-to-Binary Priority Encoder

A commonly used decimal-to-binary IC encoder is the 74147, which also has priority characteristics. As shown in Fig. 5-10, it has nine active low input lines, which are activated by push buttons, and four BCD (8-4-2-1) active low output lines. Whenever an input is activated, the circuitry converts the numerical value of the line that was selected to its equivalent output. The encoder provides priority decoding of the input data to ensure that only the highest-order line is encoded whenever multiple keys are pressed.

Each input line is tied to +5-V potentials through a pull-up resistor and is activated by pressing a key that pulls the normal high input to a low-state ground potential. Each output line is also at a normal high state, and when activated, it goes to a low.

It is not necessary to have a 0 key connected to the encoder because an inverted BCD output of 1111 (0000 BCD) is generated when none of the 1–9 output lines is activated.

The pin diagram and truth table for the 74147 encoder are shown in Fig. 5-11(a) and (b), respectively.

■ REVIEW QUESTIONS

1. _____ are usually used for entering data into digital equipment.
2. _____ _____ conditions are listed in the input sections of a truth table to indicate that other input conditions override them.
3. What are two ways to indicate that inputs and outputs are activated by a low state?
4. Only when activated, an _____ line allows an IC circuit to operate.
5. A _____ encoder is a circuit that only allows the highest input to activate the output when two or more inputs are applied simultaneously.
6. Place 1s and 0s at the output lines of the decimal-to-BCD encoder of Fig. 5-12 to show how it responds to the applied input.

EXPERIMENT

Priority Encoder

Objectives

- To demonstrate how an encoder converts signals representing a decimal number into an equivalent BCD signal.
- To show how the encoder gives priority to the highest input signal.
- To complete a truth table to show the operation of the priority encoder.
- To demonstrate active low inputs and active low outputs.

Materials

1—74147 Priority Encoder IC

4—LEDs

4—330-Ohm Resistors

9—Logic Switches

1—+5-V DC Power Supply

1—Voltmeter

Introduction

The pocket calculator uses a keypad to enter the decimal digits 0–9 for calculations. As each key is pressed, an encoder converts the signal that represents a decimal digit into an equivalent binary-coded-decimal (BCD) signal, which is used by the internal circuitry of

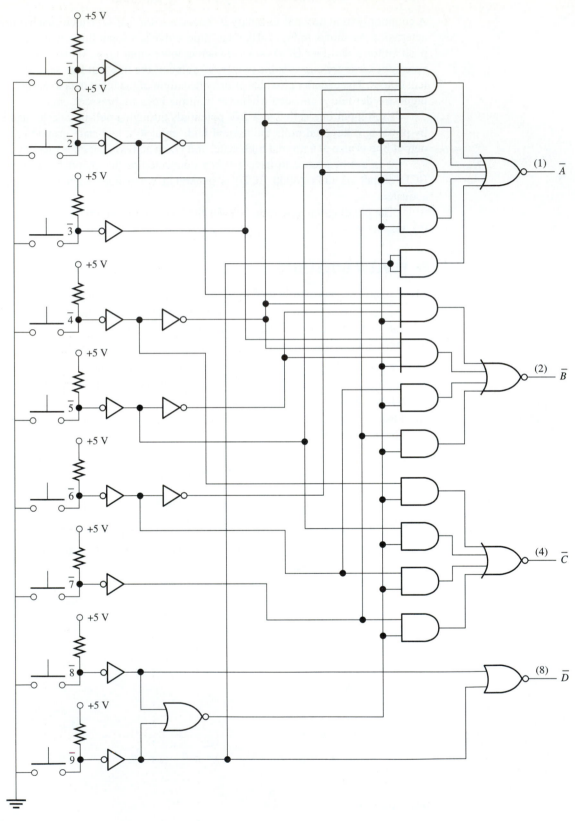

FIGURE 5.10
The 74147 decimal-to-binary encoder.

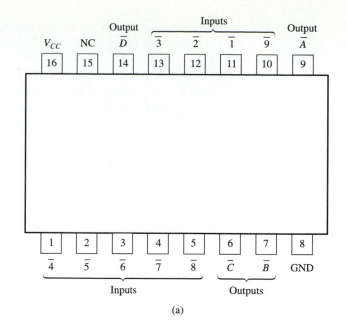

| | | Output | | Inputs | | | Output |
| | V_{CC} | NC | $\overline{D}$ | $\overline{3}$ | $\overline{2}$ | $\overline{1}$ | $\overline{9}$ | $\overline{A}$ |

(a)

				Input							Output			
Pin	10	5	4	3	2	1	13	12	11		Voltage level			
Digit	$\overline{9}$	$\overline{8}$	$\overline{7}$	$\overline{6}$	$\overline{5}$	$\overline{4}$	$\overline{3}$	$\overline{2}$	$\overline{1}$	$\overline{D}$	$\overline{C}$	$\overline{B}$	$\overline{A}$
	H	H	H	H	H	H	H	H	H	H	H	H	H
	L	X	X	X	X	X	X	X	X	L	H	H	L
	H	L	X	X	X	X	X	X	X	L	H	H	H
	H	H	L	X	X	X	X	X	X	H	L	L	L
	H	H	H	L	X	X	X	X	X	H	L	L	H
	H	H	H	H	L	X	X	X	X	H	L	H	L
	H	H	H	H	H	L	X	X	X	H	L	H	H
	H	H	H	H	H	H	L	X	X	H	H	L	L
	H	H	H	H	H	H	H	L	X	H	H	L	H
	H	H	H	H	H	H	H	H	L	H	H	H	L

(b)

FIGURE 5.11
The 74147 IC: (a) Pin diagram. (b) Function table.

FIGURE 5.12
Decimal-to-BCD encoder for Review Question 6.

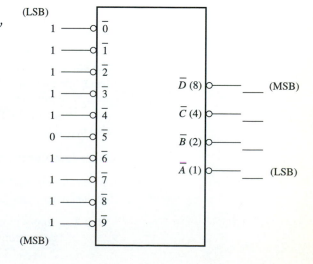

the calculator. The 74147 priority encoder IC performs an operation similar to this on decimal numbers entered into digital equipment by a keypad.

Procedure

Step 1. Assemble the circuit shown in Fig. 5-13. Use the pin diagram in Fig. 5-11. *Note:* Each output is labeled with a capital letter. Each letter represents one of the binary weighted positional values. They are as follows:

$$\overline{D} = 8$$
$$\overline{C} = 4$$
$$\overline{B} = 2$$
$$\overline{A} = 1$$

Figure 5.13

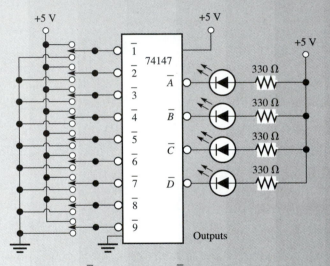

Note: Output $\overline{A}$ is the LSB; output $\overline{D}$ is the MSB

PART A—ACTIVE LOW INPUTS AND OUTPUTS

When there are logic 1 state signals applied to the decimal inputs of the encoder, the input leads are in their *inactive resting states*. When a 0 state is applied to a decimal input, it is in its *active state*. Suppose that a 0-V low is applied to decimal input 4 and +5-V logic highs are applied to the rest of the inputs. A BCD (*binary-coded-decimal*) 4 will be produced by the encoder. Therefore, the inputs are activated by an *active low* signal. Active low state leads are identified by small circles called *bubbles,* and lines called *overbars* are placed over each number (or letter) that identifies the lead.

When the BCD output leads are in their resting states, they are high. Any output lead that is activated must go low to turn on the LED. Therefore, the output leads are active low.

Procedure Question 1

When the decimal number 6 is applied to the encoder, a _____ (low, high) will be applied to pin 3 and _____ (lows, highs) will be generated at pins 6 and 7 of the IC.

Step 2. Apply power to the circuit and verify your answer for Procedure Question 1 by connecting a voltmeter to the outputs.

Step 3. Apply all of the decimal input signals that are in Table 5-1. Fill in the table with your results and compare them to the expected values.

TABLE 5.1

Activated IC Input Pin		Your Results				Expected Results			
		$\overline{D}$	$\overline{C}$	$\overline{B}$	$\overline{A}$	$\overline{D}$	$\overline{C}$	$\overline{B}$	$\overline{A}$
None						H	H	H	H
10						L	H	H	L
5						L	H	H	H
4						H	L	L	L
3						H	L	L	H
2						H	L	H	L
1						H	L	H	H
13						H	H	L	L
12						H	H	L	H
11						H	H	H	L

PART B—PRIORITY FUNCTION

The 74147 IC not only converts the selected decimal input line to its BCD equivalent value but it also gives highest priority to the largest numerical input applied. For example, if switches 1 and 2 are low at the same time, the output will produce a BCD value that represents 2.

Step 4. Connect decimal switch 3 to the ground position and verify whether the proper BCD outputs are activated. A BCD value equivalent to 3 should be produced.

Procedure Question 2

With input 3 still activated, will the LEDs display a binary 1010_2 (decimal 10) if decimal input switch 7 is also activated? (No, Yes)

Step 5. Verify your answer for Procedure Question 2 by activating input switches 3 and 7 at the same time. The LEDs that represent a BCD 7 should have illuminated.

PART C—DON'T-CARE INDICATORS

Table 5-2 shows part of the truth table for a 74147 priority encoder. The bottom portion shows that when digit 1 is activated, a BCD 1 is produced at the active low output line. Likewise, when input 2 is activated, a BCD 2 is generated at the output. However, notice that an X is placed at input 1. The X indicates a *don't-care* condition. Because input 2 is given priority over the smaller-valued input, it does not matter what logic state is applied to input 1. Therefore, an X is placed at the lower-numerical-valued input of the truth table.

Step 6. Fill in the remainder of Table 5-2 by placing Hs, Ls, or Xs at the appropriate locations.
Step 7. Place switches 5, 4, 3, 2, and 1 to their low state positions simultaneously.

Procedure Question 3

Does the output obtained in Step 7 identify the highest activated input value?

TABLE 5.2

Input		Input Digit									Output			
		$\overline{9}$	$\overline{8}$	$\overline{7}$	$\overline{6}$	$\overline{5}$	$\overline{4}$	$\overline{3}$	$\overline{2}$	$\overline{1}$	$\overline{D}$	$\overline{C}$	$\overline{B}$	$\overline{A}$
5											H	L	H	L
4											H	L	H	H
3											H	H	L	L
2		H	H	H	H	H	H	H	L	X	H	H	L	H
1		H	H	H	H	H	H	H	H	L	H	H	H	L

Procedure Question 4

Where should you have placed the Xs in Table 5-2 when input 5 was activated? Why?

■ EXPERIMENT QUESTIONS

1. If the decimal inputs 1 and 6 of the 74147 priority encoder are activated simultaneously, _____ output/s will activate.
   (a) the *A*.
   (b) the *B* and *C*.
   (c) the *A, B,* and *C*.
   (d) No.

2. A bubble is used at a terminal to show that it is activated when a _____ (low, high) logic state is present.

3. The identification number or letter of an active high lead _____ an overbar.
   (a) Uses.
   (b) Does not use.

4. True/False. An X listed in a truth table indicates that the input at which it is located is invalid when activated.

5. Place an H (high) or an L (low) next to each 74147 output to indicate the active low voltage levels that are produced when a decimal 7 is encoded.

   $\overline{A} =$

   $\overline{B} =$

   $\overline{C} =$

   $\overline{D} =$

6. A faulty 74147 encoder circuit develops the following symptoms:

Activated Input	Output Logic Levels			
	D	*C*	*B*	*A*
6	H	L	L	H
2	H	L	H	H
7	H	L	L	L
3	H	L	H	H
4	H	L	H	H
5	H	L	H	L

The most likely fault is:
(a) The circuit is open at input 4.
(b) Output C is shorted to GND.
(c) Input 4 is shorted to GND.
(d) All of the above.

Decoder

The decoder is one of the most frequently used combination circuit devices. A decoder converts a binary code (pure binary, BCD, octal, hexadecimal) into an output representing its numerical value. The output may be a singular lead that is activated or multiple leads that produce a nonbinary code. Decoders are often located at the output terminals of digital equipment and use readout devices, such as indicator lights, and LED or LCD displays, to indicate an alphanumeric value.

Decoders are also used within a digital system to execute a software instruction, set up a specific operation, recognize a unique binary address, or enable a memory chip used in microprocessors, computers, or programmable controllers to store or retrieve data.

5.4   BCD-TO-DECIMAL DECODER

BCD-to-decimal decoder:

A type of decoder that converts binary-coded decimal numbers into equivalent decimal values.

A decoder that converts binary-coded-decimal (BCD) numbers into decimal equivalents is the **BCD-to-decimal decoder.** The operation of such a circuit is shown in Fig. 5-14, which demonstrates what happens when a BCD number of 0100 (decimal 4) is applied to the input. The series of 1s and 0s show how each gate responds so that the eventual output generates a high at the output representing decimal 4. All other outputs remain low. The number in the parentheses at each output line is the input value required to activate it.

The 7442 BCD-to-Decimal Decoder

The BCD-to-decimal conversion function is performed by the 7442 IC. A block diagram of this chip is shown in Fig. 5-15. It has active high input lines and active low output lines.

EXAMPLE 5.5

How would the output of the 7442 IC respond when the BCD number 6 is applied to the inputs?

Solution   A 0110_{BCD} is applied to the input, which would cause the output pin $\overline{6}$ to go low, whereas the other nine would remain high.

The pin diagram, schematic diagram, and truth table for the 7442 IC are shown in Figure 5-16. Any binary number greater than 9 applied to the inputs is invalid. The IC will produce logic highs at all of its output lines if an invalid input is present.

EXPERIMENT

BCD–Decimal Decoder

Objectives

■ To assemble a 7442 IC.
■ To operate a decoder that converts BCD numbers to decimal values.
■ To troubleshoot a defective decoder.

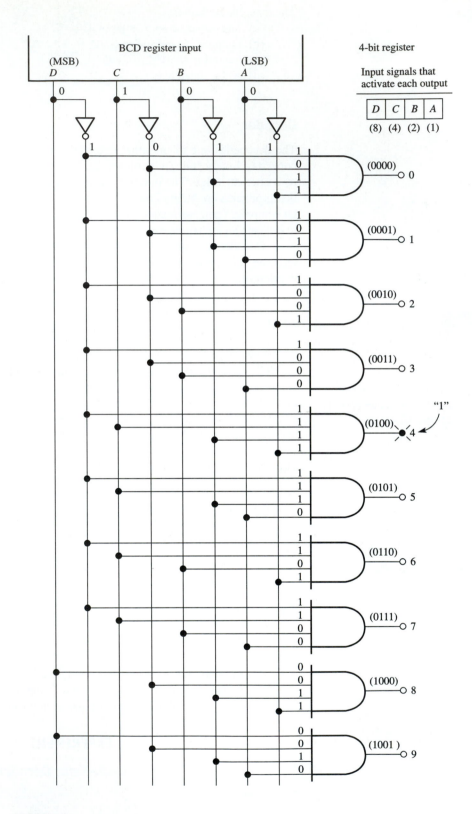

FIGURE 5.14
BCD-to-decimal decoder.

FIGURE 5.15
*Block diagram of the 7442
BCD-to-decimal decoder IC.*

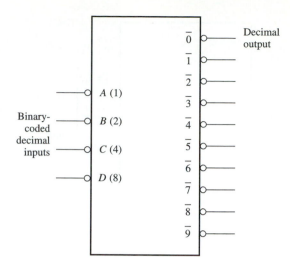

Materials

1—+5-V DC Power Supply

1—7442 IC

10—LEDs

1—150-Ohm Resistor

4—Logic Switches

1—Voltmeter

Introduction

The 7442 IC consists of a decoder that converts 4-bit binary-coded decimal numbers into equivalent decimal output values. It performs this operation by applying 1s and 0s to four input lines in a pattern that represents a binary-coded decimal count ranging from 0000 to 1001. The decoder responds by activating one of ten different output lines. Each output line represents one of the decimal values 0 to 9. The output line that is activated is determined by the BCD values applied to the input. For example, if 0011_{BCD} (3_{10}) is applied to the input, LED 3 will be activated.

Procedure

Step 1. Assemble the circuit shown in Fig. 5–17. Use the pin diagram in Fig. 5–16 for reference.

Note: **The four inputs to the IC are active high. Each input is labeled with a capital letter. Each letter represents one of the binary weighted positional values. They are as follows:**

$$D = 8$$
$$C = 4$$
$$B = 2$$
$$A = 1$$

Step 2. Set the switches so that a BCD 0011 (3_{10}) is applied to the inputs.

Step 3. Use a voltmeter to determine which logic state is developed at the IC output pin that is connected to the LED that turns on.

IC output pin number _____ equals a logic _____ (0,1) state.

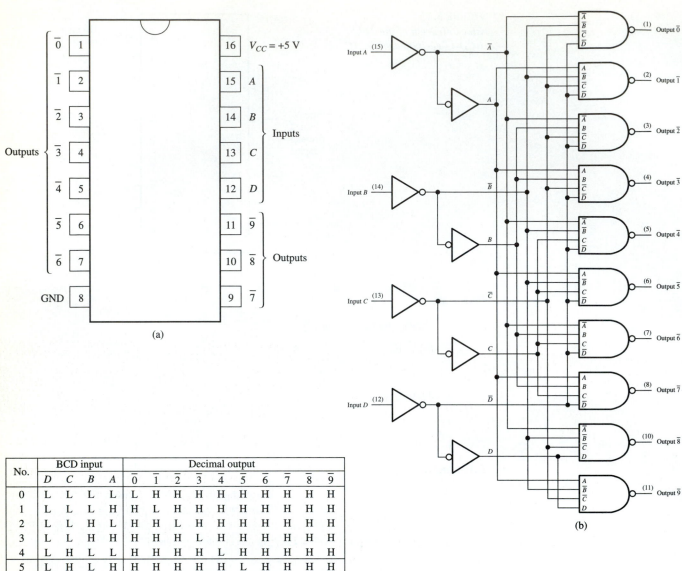

No.	BCD input				Decimal output									
	D	C	B	A	$\overline{0}$	$\overline{1}$	$\overline{2}$	$\overline{3}$	$\overline{4}$	$\overline{5}$	$\overline{6}$	$\overline{7}$	$\overline{8}$	$\overline{9}$
0	L	L	L	L	L	H	H	H	H	H	H	H	H	H
1	L	L	L	H	H	L	H	H	H	H	H	H	H	H
2	L	L	H	L	H	H	L	H	H	H	H	H	H	H
3	L	L	H	H	H	H	H	L	H	H	H	H	H	H
4	L	H	L	L	H	H	H	H	L	H	H	H	H	H
5	L	H	L	H	H	H	H	H	H	L	H	H	H	H
6	L	H	H	L	H	H	H	H	H	H	L	H	H	H
7	L	H	H	H	H	H	H	H	H	H	H	L	H	H
8	H	L	L	L	H	H	H	H	H	H	H	H	L	H
9	H	L	L	H	H	H	H	H	H	H	H	H	H	L
Invalid	H	L	H	L	H	H	H	H	H	H	H	H	H	H
	H	L	H	H	H	H	H	H	H	H	H	H	H	H
	H	H	L	L	H	H	H	H	H	H	H	H	H	H
	H	H	L	H	H	H	H	H	H	H	H	H	H	H
	H	H	H	L	H	H	H	H	H	H	H	H	H	H
	H	H	H	H	H	H	H	H	H	H	H	H	H	H

(c)

FIGURE 5.16
The 7442 BCD-to-decimal decoder IC: (a) Pin diagram. (b) Logic diagram. (c) Function table. Reproduced by permission of Texas Instruments. Copyright © Texas Instruments.

FIGURE 5.17

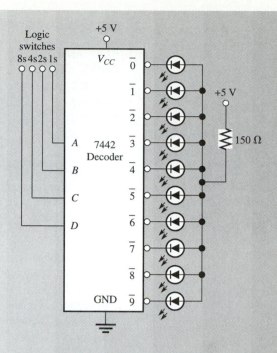

Step 4. Connect a voltmeter to one of the LEDs that does not turn on and determine which logic state develops.

IC output pin number _____ equals a logic _____ (0, 1) state.

When a BCD input number is applied to the 7442 decoder, an output line equivalent to the input value goes low and causes the LED to which it is connected to turn on. This type of output line is called active low because it is activated with a 0 state instead of a 1 state. The active low output is identified by a bubble located at the terminal and a bar located over the identification number of the output pin.

Step 5. Using the input section of Table 5-3, place the input switches at the settings to represent each BCD count of 0–9. Record the results in the boxes on the right side of the table. Place an X in a box to indicate which output is activated, and 0s at the outputs that are not activated.

TABLE 5.3

Inputs 0 = GND 1 = +5 V				Decimal Output Lit = X Not Lit = 0									
D	**C**	**B**	**A**	$\bar{0}$	$\bar{1}$	$\bar{2}$	$\bar{3}$	$\bar{4}$	$\bar{5}$	$\bar{6}$	$\bar{7}$	$\bar{8}$	$\bar{9}$
0	0	0	0										
0	0	0	1										
0	0	1	0										
0	0	1	1										
0	1	0	0										
0	1	0	1										
0	1	1	0										
0	1	1	1										
1	0	0	0										
1	0	0	1										

■ EXPERIMENT QUESTIONS

1. True/False: A 7442 IC is capable of converting the number 11 to an equivalent decimal output.

2. The BCD inputs of a 7442 decoder are active _____ (low, high) and the outputs are active _____ (low, high).

3. The following input leads are labeled with capital letters. What binary weighted value does each represent?

   $A =$ _____ $B =$ _____ $C =$ _____ $D =$ _____

4. To convert a BCD 4 to an equivalent decimal value, what logic states are at the following 7442 IC inputs and outputs?

   $A =$ ____ $B =$ ____ $C =$ ____ $D =$ ____ $\bar{0} =$ ____ $\bar{1} =$ ____ $\bar{2} =$ ____
   $\bar{3} =$ ____ $\bar{4} =$ ____ $\bar{5} =$ ____ $\bar{6} =$ ____ $\bar{7} =$ ____ $\bar{8} =$ ____ $\bar{9} =$ ____

5. What are two ways in which the active low outputs are identified?

6. Table 5-4 shows the readings taken from the BCD-to-decimal decoder shown in Fig. 5–17. What is the most likely problem?
   (a) BCD input wires A and B are crossed.
   (b) Input A is open.
   (c) Input A is shorted to ground.
   (d) Input B is open.
   (e) Input B is shorted to ground.

TABLE 5.4

Inputs 0 = GND 1 = +5 V				Decimal Output Lit = 1					Not Lit = 0				
D	C	B	A	$\bar{0}$	$\bar{1}$	$\bar{2}$	$\bar{3}$	$\bar{4}$	$\bar{5}$	$\bar{6}$	$\bar{7}$	$\bar{8}$	$\bar{9}$
0	0	0	0	0	1	0	0	0	0	0	0	0	0
0	0	0	1	0	1	0	0	0	0	0	0	0	0
0	0	1	0	0	0	0	1	0	0	0	0	0	0
0	0	1	1	0	0	0	1	0	0	0	0	0	0
0	1	0	0	0	0	0	0	0	1	0	0	0	0
0	1	0	1	0	0	0	0	0	1	0	0	0	0
0	1	1	0	0	0	0	0	0	0	0	1	0	0
0	1	1	1	0	0	0	0	0	0	0	1	0	0
1	0	0	0	0	0	0	0	0	0	0	0	0	1
1	0	0	1	0	0	0	0	0	0	0	0	0	1

5.5   BCD-TO-SEVEN-SEGMENT DECODER/DRIVER

BCD-to-seven segment decoder/driver:

A type of decoder that lights up an electronic readout to display one of the same numbers, 0–9, as the BCD value applied to its input.

The **BCD-to-seven-segment decoder/driver** is a special type of decoder circuit. This device is a logic network that allows BCD numbers to light up the segments of electronic readouts to form the decimal digits 0 through 9. Figure 5-18(a) illustrates that these readouts have seven separate segments that can be illuminated in mosaic fashion to form the decimal digits 0 to 9.

These segments, which are made up of different light-producing devices, such as incandescent filaments, light-emitting diodes (LEDs), or liquid crystal diodes (LCDs), are designated by the letters *a* through *g*. If all of the segments are illuminated at one time, the number 8 would display. The number 7 would display if segments *a*, *b*, and *c* were lighted.

FIGURE 5.18
(a) The seven-segment display. (b) Numerical designation for a seven-segment display.

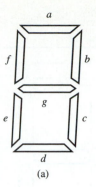

(a)

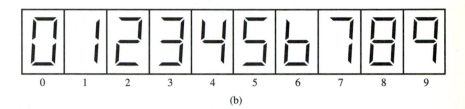

| 0 | 1 | 2 | 3 | 4 | 5 | 6 | 7 | 8 | 9 |

(b)

Figure 5-18(b) illustrates how the lighting of selected segments can display the numbers 0 to 9. Such readouts are used for digital watches, clocks, thermometers, and pocket calculators.

The Seven-Segment LED Display

The pin diagram of the MAN-72 seven-segment display is shown in Fig. 5-19. It has eight active low inputs, seven for each segment and one for a decimal point. These inputs are connected to the cathode of each LED segment. It also has three separate lines that are connected to a +5-V supply. Internally, each one of these lines has common connections at the anodes of various LED segments. For example, the anodes of segments *a* and *b* share pin 14; the anodes of segments *c* and *d* share pin 9; and segments *e, f, g* and the decimal point share pin 3. A given LED segment (*a*) lights up when its input pin is driven low. This forward biases the LED because the cathode is low and the anode is high, as shown in Fig. 5-20(a). A current-limiting resistor should be placed between each cathode pin and the device that is driving it low. Figure 5-20(b) shows how the common-anode connections are configured inside the display module for segments *e, f,* and *g*.

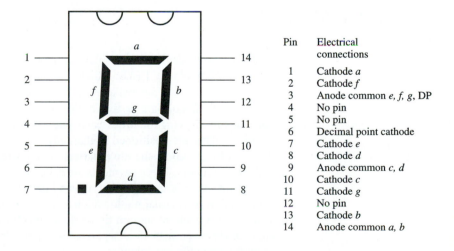

Pin	Electrical connections
1	Cathode *a*
2	Cathode *f*
3	Anode common *e, f, g*, DP
4	No pin
5	No pin
6	Decimal point cathode
7	Cathode *e*
8	Cathode *d*
9	Anode common *c, d*
10	Cathode *c*
11	Cathode *g*
12	No pin
13	Cathode *b*
14	Anode common *a, b*

FIGURE 5.19
Pin diagram of the MAN-72 LED display.

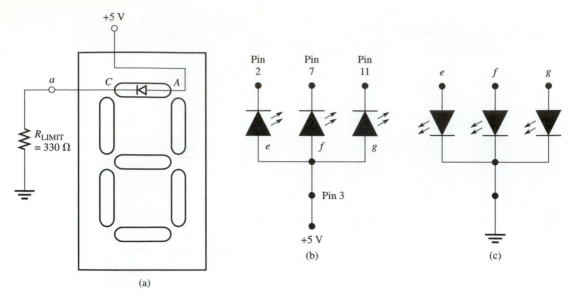

FIGURE 5.20
(a) Illuminating the a segment. (b) Common anode (active low). (c) Common cathode (active high).

Seven-segment LED displays are also configured as common-cathode devices. The cathode pins of various segments share a common terminal that is connected to an external ground. Figure 5-20(c) shows how the common-cathode connections are configured inside the display module for segments *e, f,* and *g*. A segment illuminates when a logic high is applied to its anode lead. The two types of displays allow the use of either active low or active high circuits to drive the LEDs. This option gives the designer some flexibility.

LED displays are relatively bright and can be used in a variety of lighting conditions. Digits as large as 4 feet are available, allowing the display to be read from a distance. The disadvantage of LED displays is that they require a relatively high current to turn on a segment. Therefore, they are seldom used for battery-powered applications. In general, seven-segment displays are available in red, high efficiency red, green, or yellow.

LIQUID CRYSTAL DISPLAYS

Another type of device that displays various characters and symbols is the liquid crystal display (LCD). Unlike LEDs, which illuminate figures against a dark background, LCDs produce black figures against a bright background.

The LCD consists of five basic elements, which are layered. The cross-sectional view in Fig. 5-21(a) identifies each one. Two layers of polarizers, which are made of polarized glass, are positioned so that only vertically polarized light passes through them. A layer of glass under the top polarizer contains the segments that form numbers, letters, and other symbols. These segments are called *electrodes* and are made from a thin layer of indium tin oxide, a material that allows light to pass through. Each of these segments is connected to an external pin to which a voltage is applied. A second sheet of glass has the same pattern made of indium tin oxide to which the second lead of the voltage source is applied. However, instead of separated segments, the electrodes are all connected together to form a common conductor.

A fluid that contains nematic liquid crystals is placed between the two conductive layers. These crystals contain molecules that are normally aligned in a specific pattern.

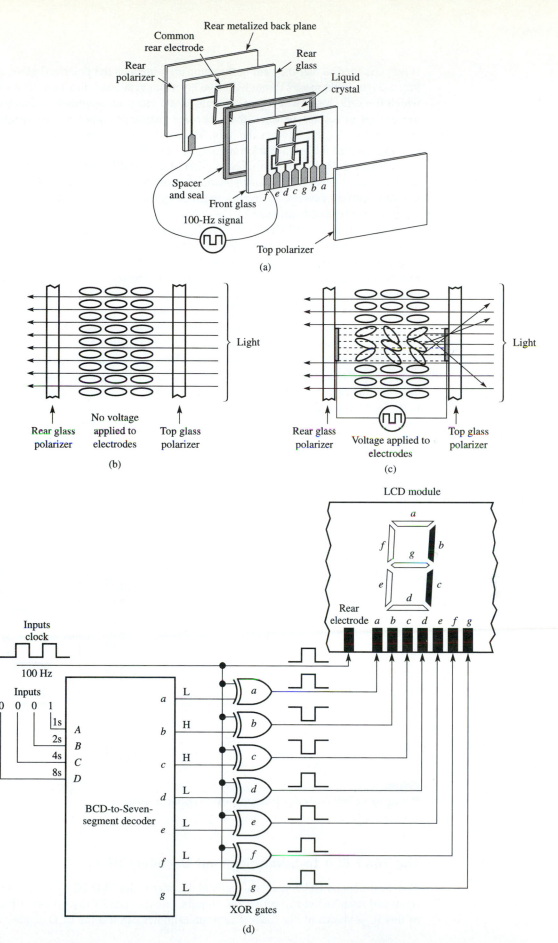

FIGURE 5.21

Liquid crystal display. (a) Cross-sectional view of layers. (b) Liquid crystal alignment with no voltage applied to electrodes. (c) Liquid crystal alignment when a voltage is applied to electrodes. (d) An LCD circuit that displays the numerical digit 1.

When arranged so that they are in the same alignment as the polarized glass, as shown in Fig. 5-21(b), light passes through all five layers and reverses direction off the back plane, which is a reflector. When any of the transparent electrode segments is energized, the liquid crystal molecules between them become rearranged until a random alignment, as shown in Fig. 5-21(c), scatters the light. The result is that an opaque, or black, appearance develops against the bright background of the reflector.

Figure 5-21(d) shows the construction of a typical liquid crystal display for one numerical digit. Unlike LED displays, which require a DC voltage to energize, LCDs are alternating current devices. If direct current is applied, they deteriorate. A 100-Hz square wave is usually used to drive an LCD.

One advantage of LCD displays is that it is easy to form the electrodes into a desired pattern, for example, the abbreviations RAD (radian) or DEG (degrees) displayed on calculators. A second advantage is that the LCD draws very little power. This characteristic makes them an ideal choice for displays that are battery powered, such as watches or calculators. Because they use very little voltage, they must be driven by a low-powered decoder/driver, such as the 74HC4543 CMOS chip.

A wiring diagram of the decoder/driver connected to an LCD device is shown in Fig. 5-22. The BCD input 0011 (decimal 3) is applied to the decoder-driver. A 100-Hz square wave is applied to both the LCD common electrode and the Ph (phase) input of the decoder. The internal circuitry of the decoder energizes the segments that form the digit 3 by inverting the square wave at outputs a, b, c, d, and g. The reason these LCDs are energized is that a potential difference forms between the two electrode planes because the square waves applied across them are out-of-phase. Because the square waves applied to the e and f electrodes are in phase, no potential forms and they remain de-energized.

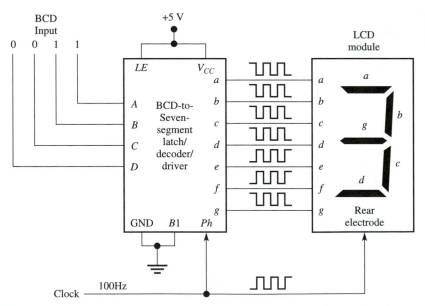

FIGURE 5.22
Wiring of a CMOS decoder/driver to a liquid crystal display.

The 7447 BCD-to-Seven-Segment Decoder/Driver

The most popular BCD-to-seven-segment decoder is the 7447 IC. It has a 4-bit active high input and seven individual active low outputs, one for each LED segment. A block diagram of this IC is shown in Fig. 5-23. The troubleshooter can test the LED display module that

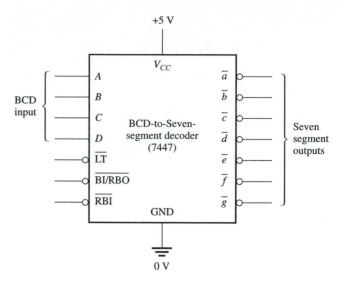

FIGURE 5.23
Block diagram of the 7447 BCD-to-seven-segment decoder/driver IC.

the 7447 chip is driving by applying a high to the $\overline{BI/RBO}$ node and a low to the $\overline{LT}$ (lamp-test) node. This test causes all LED segments to turn on simultaneously.

The $\overline{RBI}$ (ripple-blanking input) node is capable of controlling the light intensity of the segments when a square wave is applied to it. A symmetrical square wave, such as that shown in Fig. 5-24(d), will cause the LED segments to be off 50% of the time. A nonsymmetrical square wave, such as that shown in Fig. 5-24(e), will cause the LEDs to be brighter because they are off only 10% of the time. The pin diagram, schematic diagram, and truth table for the 7447 IC are shown in Fig. 5-24.

▪ REVIEW QUESTIONS

7. Decoders are usually at the _____ terminals of digital equipment.
8. _____ input lines are connected to most encoder and decoder circuits and either allow or prevent the operation of the circuit.
9. By examining the output states of the 7447 decoder in Fig. 5-25, place the appropriate 1s and 0s at the input leads that cause the circuit to operate this way.
10. What LED segments illuminate when the 7447 decoder produces the output states shown in Fig. 5-25?
11. The common-anode pins 3, 9, and 14 of the MAN-72 LED display are connected to (ground or +5 V).

DATA-TRANSFER CIRCUITS

Another useful type of combination logic device is the data-transfer circuit. Some types of data-transfer devices are used to route data to and from several transmitting lines over one wire. Circuits called *multiplexers* and *demultiplexers* perform this function. Another data-transfer device monitors logic signals and is capable of detecting a faulty digital signal during its transmission. This device is called a *parity circuit.*

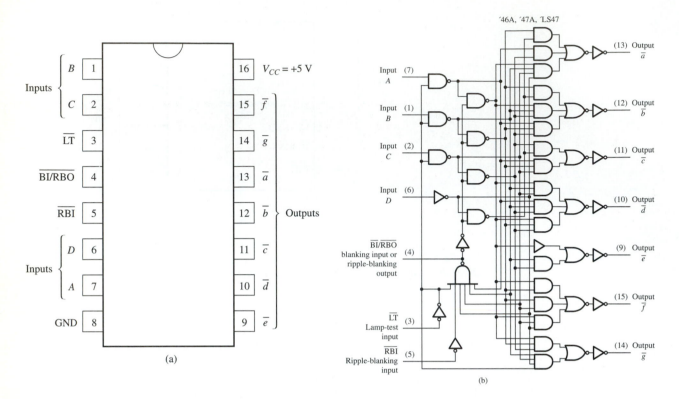

Decimal or function	Inputs						$\overline{BI}/RBO$	Outputs						
	$\overline{LT}$	$\overline{RBI}$	D	C	B	A		$\overline{a}$	$\overline{b}$	$\overline{c}$	$\overline{d}$	$\overline{e}$	$\overline{f}$	$\overline{g}$
0	H	H	L	L	L	L	H	ON	ON	ON	ON	ON	ON	OFF
1	H	X	L	L	L	H	H	OFF	ON	ON	OFF	OFF	OFF	OFF
2	H	X	L	L	H	L	H	ON	ON	OFF	ON	ON	OFF	ON
3	H	X	L	L	H	H	H	ON	ON	ON	ON	OFF	OFF	ON
4	H	X	L	H	L	L	H	OFF	ON	ON	OFF	OFF	ON	ON
5	H	X	L	H	L	H	H	ON	OFF	ON	ON	OFF	ON	ON
6	H	X	L	H	H	L	H	OFF	OFF	ON	ON	ON	ON	ON
7	H	X	L	H	H	H	H	ON	ON	ON	OFF	OFF	OFF	OFF
8	H	X	H	L	L	L	H	ON	ON	ON	ON	ON	ON	ON
9	H	X	H	L	L	H	H	ON	ON	ON	OFF	OFF	ON	ON
10	H	X	H	L	H	L	H	OFF	OFF	OFF	ON	ON	OFF	ON
11	H	X	H	L	H	H	H	OFF	OFF	ON	ON	OFF	OFF	ON
12	H	X	H	H	L	L	H	OFF	ON	OFF	OFF	OFF	ON	ON
13	H	X	H	H	L	H	H	ON	OFF	OFF	ON	OFF	ON	ON
14	H	X	H	H	H	L	H	OFF	OFF	OFF	ON	ON	ON	ON
15	H	X	H	H	H	H	H	OFF	OFF	OFF	OFF	OFF	OFF	OFF
BI	X	X	X	X	X	X	L	OFF	OFF	OFF	OFF	OFF	OFF	OFF
RBI	H	L	L	L	L	L	L	OFF	OFF	OFF	OFF	OFF	OFF	OFF
LT	L	X	X	X	X	X	H	ON	ON	ON	ON	ON	ON	ON

(c)

FIGURE 5.24

The BCD-to-seven-segment decoder/driver IC. (a) Pin diagram. (b) Logic diagram. (c) Function table. (d) Symmetrical square wave. (e) Nonsymmetrical square wave. (f) Numerical designations and resultant displays. Reproduced by permission of Texas Instruments.

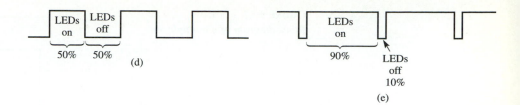

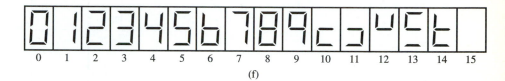

(f)

FIGURE 5.24
(Continued)

FIGURE 5.25
The 7447 decoder for Review Question 10.

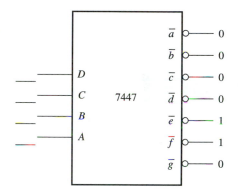

5.6 MULTIPLEXER

Multiplexer:

A logic circuit that directs data from one of several inputs to a single output.

A **multiplexer,** or *data selector,* is a switching device that accepts several data inputs and allows only one of them at a time to pass through to the output. In its simplest form, a single-pole multiposition rotary switch can be thought of as a multiplexer. Figure 5-26 illustrates an eight-input multiplexer using an eight-position rotary switch. In this example, any one of eight inputs can be selected to be connected to the output line simply by rotating the switch to the desired position. Because many operations require a multiplexer to operate at high speeds and be automatically selectable, logic gates are used instead of mechanical switches. Figure 5-27 shows a block diagram of a multiplexer. In addition to the inputs and the output, multiplexers also contain data-select (control) lines and usually *strobe* (enable) lines. The purpose of the data-select lines is to determine which one of the input lines is to be connected to the output. For example, if binary number 011_2 (3_{10}) is applied to the data-select lines, data from line 3 will be routed to the output.

FIGURE 5.26
Rotary switch illustrating the operation of a multiplexer.

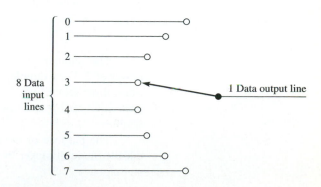

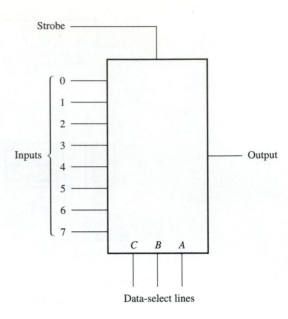

FIGURE 5.27
Block diagram of a multiplexer.

EXAMPLE 5.6

What is the binary number applied to the data-select lines that causes the data to follow the input/output path shown in Fig. 5-28?

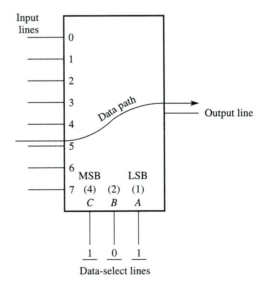

FIGURE 5.28
Circuit diagram for Example 5-6.

Solution   The number 101_2 (5_{10}) causes the data to follow the input/output path shown in the figure.

The number of data-select lines needed to select one of the inputs is always a power of 2. For example, two data-select lines ($2^2 = 4$) can select among four input data lines, three select lines ($2^3 = 8$) can select among eight input data lines, and so on. For each additional select line, the number of inputs from which the multiplexer can select doubles.

The purpose of the strobe line is to allow the multiplexer to work normally or to disable the multiplexer by not allowing the data from any of the inputs to be funneled to an output.

The 74150 Multiplexer

A variety of multiplexer ICs exist, so there is little need to construct them from individual gates. The 74150 IC shown in Fig. 5-29, for example, has 16 data lines, E_0 through E_{15}, and four data-select lines, labeled A, B, C, and D. The 4-bit binary code applied to the data select lines counts from 0000_2 (0_{10}) to 1111_2 (15_{10}) and designates which of the 16 data lines are to be used. A *strobe-enable* line is connected to all 16 AND gates. When the strobe line is low, the entire circuit is enabled. When the strobe line is high, the circuit becomes disabled. The signals from the data inputs are complemented before passing through to the output line.

Multiplexer Application

The multiplexer in Fig. 5-30 can be used to control a home security system. Eight sensors are placed at windows and doors throughout the house to monitor if they are broken into. The sensor's output representing a secured window or door produces a logic 0 state. When a sensor detects a break-in, it produces a logic 1 state.

Each sensor is connected to an input of the multiplexer. A high-speed counter applies an incrementing count of 0 to 7. Each time it reaches the count of 7, it recycles to 0 and begins incrementing again. The alarm connected to the output line is energized when a logic 1 is applied. When all of the sensors produce a logic 0, the alarm is off. If a sensor is activated, a logic 1 is passed through to the output of the multiplexer each time the count on the data-select lines is the same number as its input line. Even though the output is a logic high for one-eighth of the time, the alarm produces an audible sound because it is being strobed.

EXPERIMENT

Multiplexer/Data Selector

Objectives

■ To assemble and operate a data selector with 16 input lines.

Materials

 1—74150 IC

 5—Logic Switches

 1—100-Ohm Resistor

 1—LED

 1—+5-V DC Power Supply

 1—Square Wave Generator (3 Hz)

Introduction

Multiplexer

A multiplexer with 16 inputs and one output is shown in Fig. 5-31. It is available in IC form as a TTL 74150 chip. In addition to the 16 data Inputs, it has 4 data-select inputs. These are used to control which data input line is connected to the output. For example, if a 0101_2 (5_{10}) is applied to the data-select inputs, signals from data input 5 will be transferred to the output. The 74150 IC also has an active low strobe (enable) input line. It is

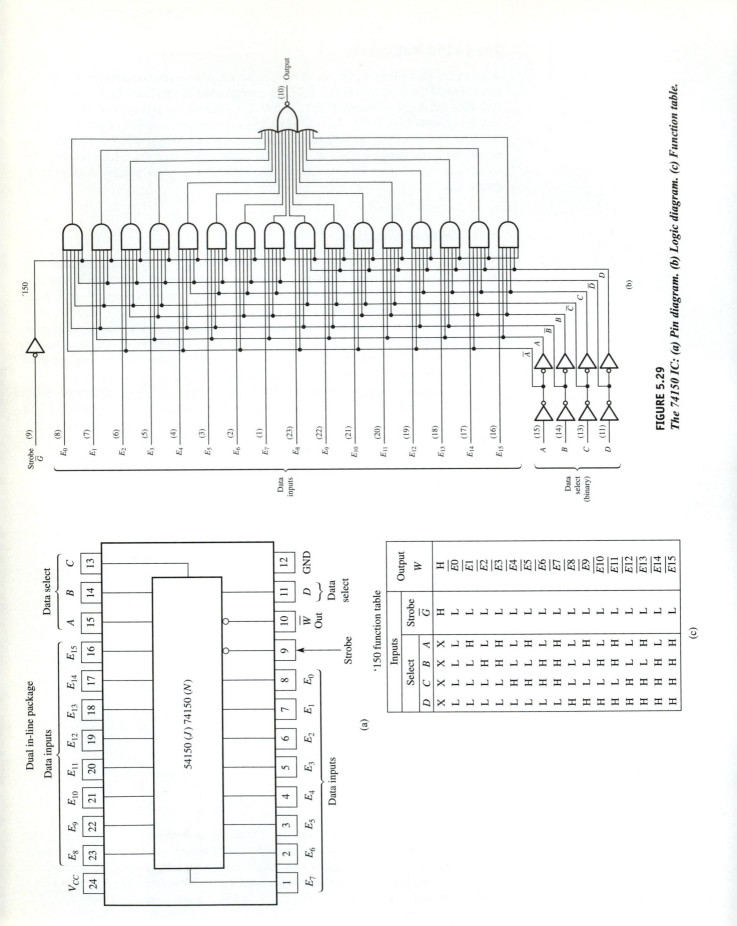

FIGURE 5.29
The 74150 IC: (a) Pin diagram. (b) Logic diagram. (c) Function table.

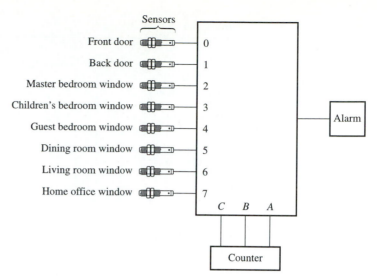

FIGURE 5.30
Multiplexer application.

used to control when data is transferred from the input to the output. If it is low, it allows the circuit to operate. If it is high, the circuit is disabled.

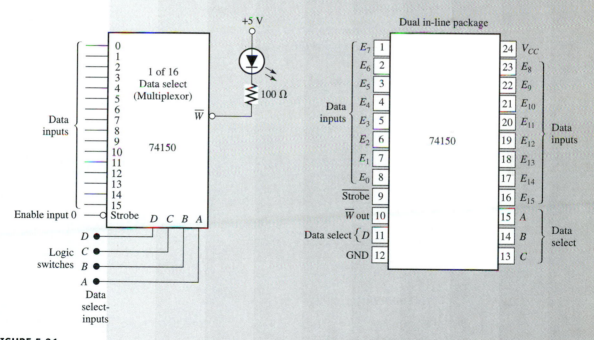

FIGURE 5.31

Procedure

Step 1. Assemble the circuit shown in Fig. 5-31.

Step 2. Set the data-select (D, C, B, and A) and enable (strobe) switches as shown on each line of Table 5-5. As each group of data select input settings is made, apply a 3-Hz square wave signal from the signal generator to each of the 16 data inputs. Fill in the boxes that represent the data-input terminals. If the LED blinks when the square wave signal is applied to the data-input line, place an X in the box that it represents. Place 0s in all the rest of the data-input boxes.

TABLE 5.5

					Inputs															
Data Select Switches																				
D	C	B	A	Enable	0	1	2	3	4	5	6	7	8	9	10	11	12	13	14	15
0	0	0	0	0																
0	0	1	0	0																
0	0	1	1	1																
0	1	0	1	0																
0	1	1	1	0																
1	0	1	0	0																
1	0	1	1	1																
1	1	0	1	0																
1	1	1	0	0																
1	1	1	1	0																

Procedure Question 1

If pulses are observed at the 74150 output lead when 1011_2 is applied to the data-select lines, what data input is the signal transferred from?

Procedure Question 2

Why are some horizontal columns in Table 5-5 without an X?

Step 3. Set the data-select inputs at $D = 0$, $C = 1$, $B = 1$, and $A = 1$, and set the enable input low.

Step 4. Connect a low to data-input pin E_7, and observe the logic state at the output terminal using the voltmeter.

Output $\overline{W}$ = _____ (low, high).

Step 5. Connect a high to data-input pin E_7, and observe the logic state at the output terminal using the voltmeter.

Output $\overline{W}$ = _____ (low, high).

Procedure Question 3

How do the logic signals applied to the inputs compare with the logic signals at the output?

■ EXPERIMENT QUESTIONS

1. A data selector is also called a _____.
2. The 74150 multiplexer sends data from _____ (1, 16) input/s to _____ (1, 16) output/s.
3. A high at the enable line makes the 74150 multiplexer _____ (enabled, disabled).

5.7 DEMULTIPLEXER

Demultiplexer:

Circuit capable of transmitting data from an input line to one of several output terminals.

A **demultiplexer,** or *data distributor,* is the opposite of a multiplexer. Instead of having several inputs and one output, the demultiplexer has a single input and several outputs. This device transmits data from the input line to one of several output ports, as demonstrated by an eight-position rotary switch placed in the opposite direction as the one used to show how a multiplexer operates (Fig. 5-26). The rotary switch and a block diagram representing a demultiplexer are shown in Fig. 5-32. Besides the input and output lines, demultiplexers also contain data-select lines. The purpose of the data-select lines is to determine which output line is to be connected to the input line. For example, if binary number 111_2 (7_{10}) is applied to the select input lines, data from the input port will be routed to data output line number 7.

FIGURE 5.32
(a) Rotary switch illustrating the operation of a demultiplexer.
(b) Block diagram of a demultiplexer.

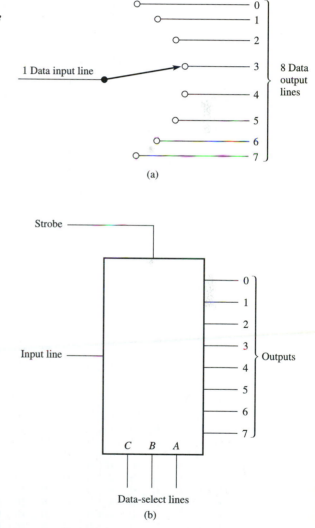

The 74154 Decoder/Demultiplexer

The 74154 IC shown in Fig. 5-33 has 16 data-output lines and performs two functions. When both strobe inputs, $\overline{E_0}$ and $\overline{E_1}$, are low, the circuit operates as a decoder as it decodes 4-bit binary-coded inputs into one of 16 mutually exclusive outputs.

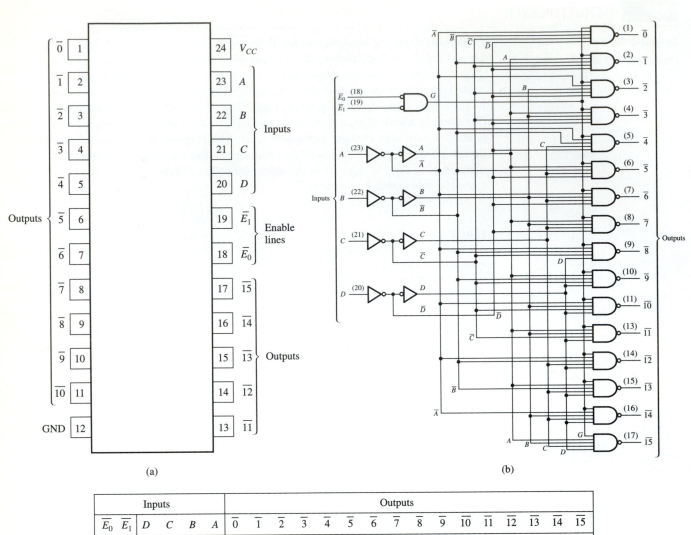

FIGURE 5.33
The 74154 decoder/demultiplexer IC: (a) Pin diagram. (b) Logic diagram. (c) Function table. Reproduced by permission of Texas Instruments. Copyright © Texas Instruments.

The circuit also operates as a demultiplexer. It performs the demultiplexing function by distributing data from one input line to any one of 16 outputs. One of the two enable lines ($\overline{E_0}$ or $\overline{E_1}$) is used as the data input when the other enable input is low. The four input lines, A, B, C, and D, are used as the data-select lines. Suppose that $\overline{E_0}$ is the data-input line, $\overline{E_1}$ is the enable line, and 0110_2 (6_{10}) is applied to the data-select lines. When a high is at $\overline{E_0}$, the addressed output line 6 (and all other output lines) is high because the demultiplexer is disabled. When a low is at $\overline{E_1}$, the demultiplexer is enabled, and data is passed from $\overline{E_0}$ straight to addressed output line 6.

Demultiplexer Applications

Multiplexers are often used with demultiplexers to help reduce wiring connections when information from many sources is transmitted over long distances. This is because it is less expensive to multiplex data onto a single wire. For example, this process was originally used with analog signals. During the early years of telephony, the demand for long-distance telephone service increased at a rapid rate. At that time, cables were used, consisting of approximately 50 wires that were capable of handling 50 phone conversations. These cables were very expensive, especially when they ran between cities hundreds of miles apart. As a result, it became necessary for telephone company engineers to devise a method that would enable several telephone conversations to be transferred over the same wire at the same time through the use of multiplexers and demultiplexers. Figure 5-34 uses switches to illustrate how four telephone conversations share the same transmission line. To operate properly, switch 1 (SW1) and switch 2 (SW2) must be first connected simultaneously to conversation 1 over a brief period of time. Then both switches increment to conversation 2, then 3, and finally 4 before they return to conversation 1 to repeat the cycle. Several separate phone conversations share the same path because each signal is chopped up considerably. However, at sufficiently high speeds, enough of a sample of the original signal is present so that it can be reconstructed from the composite at the receiving end with negligible loss in fidelity.

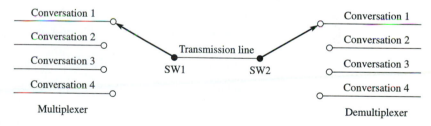

FIGURE 5.34
Rotary switches illustrating how four telephone connections can be made using one wire.

Digital equipment, especially computers, also use multiplexers and demultiplexers. Computers process and transfer tremendous amounts of data. To make separate straight-wire connections for the transfer of all these signals would be cost-prohibitive.

Figure 5-35 shows how the 74150 multiplexer and the 74154 demultiplexer work together. Their data-select lines are connected together so that the same 4-bit binary number is applied to each chip simultaneously. This configuration enables data applied to a particular multiplexer input data line to be transferred to the demultiplexer output data line with the same number. For example, if a binary 0110 is applied to the data-select line, data from input line 6 of the multiplexer is transmitted to output data line 6 of the demultiplexer.

Incidentally, because the data-select inputs address any desired data line, the pattern of bits applied to the input is often called an **address.** This is especially true in computer applications.

Address:

A numerical value that designates a specific location in a memory device or a circuit terminal to be enabled.

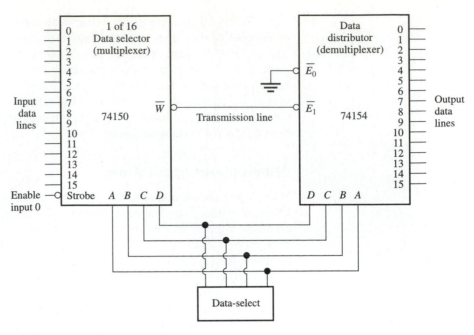

FIGURE 5.35
How a multiplexer and demultiplexer work together.

EXPERIMENT

Demultiplexer/Data-Transfer Circuit

Objectives

- To assemble and operate a data distributor with 16 output lines.
- To assemble and operate a multiplexer and demultiplexer connected to operate as a data-transfer circuit.
- To troubleshoot a defective data-transfer circuit.

Materials

       1—74150 IC

       1—74154 IC

       5—Logic Switches

       1—100-Ohm Resistor

       1—LED

       1—+5-V DC Power Supply

       1—Square Wave Generator (3 Hz)

       1—Logic Probe

Introduction

A demultiplexer with 1 input and 16 output pins is available in IC form as a TTL 74154 chip. Which of its 16 output lines is connected to the data input is determined by the 4-bit binary number applied to the data-select input pins D, C, B, and A. Data enters the demultiplexer through one of two inputs ($\overline{E_0}$ or $\overline{E_1}$), while the other E input is low.

Data-Transfer Circuit

Figure 5-36 shows how the 74150 and 74154 ICs can be connected to form a data-transfer circuit. Their data-select lines are connected so that the same 4-bit binary number is applied to each chip at the same time. This enables data applied to a multiplexer input data line to be passed on to the demultiplexer output data line with the same number. For example, if the data-select inputs are set to a binary 0110, data from input line 6 of the multiplexer is passed on to output data line 6 of the demultiplexer.

Step 1. Assemble the circuit shown in Fig. 5-36. Use the pin diagrams in Figs. 5-29 and 5-33 as a reference.

Step 2. Set the data select (D, C, B, and A) and enable (strobe) switches as shown in Table 5-6. As each 4-bit binary number is applied by the data-select logic switches, apply a 3-Hz square wave to the multiplexer input line that represents the same number.

For further directions on how to fill in the table, read notes a and b below Table 5-6.

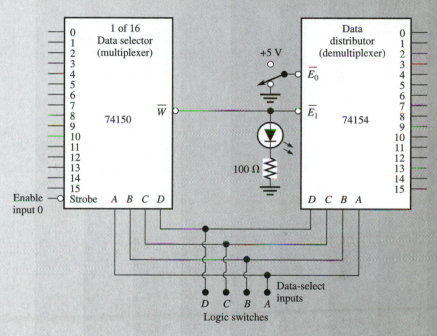

FIGURE 5.36

Procedure Information

Step 2 is performed to determine if a square wave representing data is transferred from the multiplexer to the demultiplexer input. It demonstrates whether the data is routed from the demultiplexer input to the output line that has the same numerical value as the data-select inputs. It also shows if the demultiplexer is disabled when the enable line $\overline{E_1}$ is a logic 1.

■ EXPERIMENT QUESTIONS

1. A data-transfer circuit is also called a _____.
2. In a demultiplexer, the _____ _____ inputs control to which output line the signals from the data input line are sent.
3. A high at the enable line makes the 74150 multiplexer _____ (enabled, disabled).
4. When the switches connected to the data-select lines of a demultiplexer are in the following positions, $D = 1$, $C = 0$, $B = 0$, $A = 0$, to what output is data transferred?

TABLE 5.6

74150 Multiplexer

Enable	Data Inputs — Л = Pulse Applied 0 = No Pulse															
	0	1	2	3	4	5	6	7	8	9	10	11	12	13	14	15
	Л	0	0	0	0	0	0	0	0	0	0	0	0	0	0	0
	0	0	Л	0	0	0	0	0	0	0	0	0	0	0	0	0
	0	0	0	0	0	Л	0	0	0	0	0	0	0	0	0	0
	0	0	0	0	0	0	0	Л	0	0	0	0	0	0	0	0
	0	0	0	0	0	0	0	0	0	Л	0	0	0	0	0	0
	0	0	0	0	0	0	0	0	0	0	Л	0	0	0	0	0
	0	0	0	0	0	0	0	0	0	0	0	Л	0	0	0	0
	0	0	0	0	0	0	0	0	0	0	0	0	Л	0	0	0
	0	0	0	0	0	0	0	0	0	0	0	0	0	0	Л	0
	0	0	0	0	0	0	0	0	0	0	0	0	0	0	0	Л

Data-Select Switches

D	C	B	A
0	0	0	0
0	0	1	0
0	1	0	1
0	1	1	1
1	0	0	1
1	0	1	0
1	0	1	1
1	1	0	0
1	1	1	0
1	1	1	1

74154 Demultiplexer

INPUTS		OUTPUTS — Data Outputs X = Pulses 0 = No Pulses															
Enable $\overline{E_0}$	Data $\overline{E_1}$	0	1	2	3	4	5	6	7	8	9	10	11	12	13	14	15
0																	
0																	
1																	
0																	
0																	
0																	
0																	
0																	
1																	
0																	

a. Connect the anode of a LED to input line $\overline{E_1}$ of the demultiplexer, and the cathode to ground through a 100-Ω resistor. For each data-select switch setting, place an X in the box of the $\overline{E_1}$ column if a signal is present, and a 0 if the signal is not present.

b. Connect a logic probe to each of the demultiplexer outputs for each data-select switch setting. In the output section of the table, place Xs in the appropriate boxes to show which output line receives a blinking square wave signal during each count applied to the data-select lines. Place 0s in all the rest of the data output boxes.

5. To send a square wave from input $\overline{E_1}$ of the 74154 to output 11 in Fig. 5-36, list what logic states will be at the following 74150 leads:

$\overline{\text{Strobe}} =$
$A =$
$B =$
$C =$
$D =$

6. Suppose that after the circuit in Fig. 5-36 is assembled, the readings in Table 5-7 are taken. What is the most probable reason the multiplexer is operating incorrectly?
   (a) Data-select input A is open.
   (b) Data-select input B is shorted to ground.
   (c) Wires at data selector inputs A and B are crossed.
   (d) All of the above.

5.8 PARITY CIRCUITS

Transients:

An unwanted electromagnetic signal picked up on a transmission line that carries digital signals.

Word:

Binary bits that are divided into organized groups.

Parity circuit:

A device which detects an unwanted change in a logic signal while it's being transmitted from one location to another.

Digital devices can be very complex and they operate at high speeds. Thousands of operations are performed and sent through a large number of circuit paths in a short period of time. As a result, these machines are very susceptible to picking up unwanted signals called noise, or **transients.** Such a signal may come from the magnetic field of a nearby motor. The same problem can occur when pulses of magnetic energy from sun spots are absorbed by long-distance transmission lines.

Digital data come in groups of bits called a **word.** One type of word, which consists of four bits, is called a *nibble.* Table 5-8 illustrates a sample of six nibbles. The four bits of each nibble are identified as B_1 to B_4. If an unwanted signal were picked up by a wire carrying one of these bits, a 1 state could change to a 0 state, or vice versa. Serious problems could occur if this happened.

A device called a **parity circuit** is used for detecting such an error the moment it occurs. The term *parity* refers to error detection. There are two such types of circuits, *even-parity* circuits and *odd-parity* circuits. An example of an even-parity circuit is shown in Fig. 5-37. Note that one part of the circuit is called a *parity generator* and the other is called a *parity checker.* These two parts operate the same way. A nibble enters the inputs of the parity generator and checker simultaneously. When the total number of 1 state bits in a nibble is odd, the generator and checker both generate a 1 output at the parity lines. Likewise, if the count of 1s in the word is even, the generator and checker both generate a 0 output. Again, from Table 5-8, whenever the parity bit is combined with the data bits, the total number of 1s is always even. That is why this circuit is called an even-parity circuit. The parity output lines of the generator and checker are then applied to a two-input exclusive-OR gate called a *detector.* When words are transmitted without any interference, the detector circuit remains at a 0 state because the parity-line inputs to the detector are always the same.

Assume that the data output circuit produces the following nibble: 1110 ($B_1 = 1$, $B_2 = 1$, $B_3 = 1$, and $B_4 = 0$). This is the same bit pattern that is listed in the third row of Table 5-8. If the circuits are operating properly without any interference, the parity generator and checker should both receive bit pattern 1110. The table indicates that both output lines should produce a 1 and cause the exclusive-OR gate parity detector to produce a 0.

Suppose that unwanted noise picked up on transmission line B_2 changes the 1 state to a 0 state output at parity line C. Nibble 1110 applied to the parity generator is unaffected by the noise and generates a 1 state at parity line G. Since parity lines C and G are at opposite states, the detector goes to a 1 state and an alarm light turns on.

An odd-parity circuit also exists and is used to perform the same function. It operates exactly the opposite of an even-parity circuit. Whenever an equal number of 1s are present in a data word, the parity generator produces a 1 so that an odd number of 1s are transmitted. If an odd number of 1s are present in a data word, the parity generator produces a 0.

TABLE 5.7

74150 Multiplexer

	INPUTS															
	Data Inputs ⊓ = Pulses Applied, 0 = No Pulses															
Enable	0	1	2	3	4	5	6	7	8	9	10	11	12	13	14	15
0	⊓	0	0	0	0	0	0	0	0	0	0	0	0	0	0	0
0	0	⊓	0	0	0	0	0	0	0	0	0	0	0	0	0	0
0	0	0	⊓	0	0	0	0	0	0	0	0	0	0	0	0	0
0	0	0	0	⊓	0	0	0	0	0	0	0	0	0	0	0	0
0	0	0	0	0	⊓	0	0	0	0	0	0	0	0	0	0	0
0	0	0	0	0	0	⊓	0	0	0	0	0	0	0	0	0	0
0	0	0	0	0	0	0	⊓	0	0	0	0	0	0	0	0	0
0	0	0	0	0	0	0	0	⊓	0	0	0	0	0	0	0	0
0	0	0	0	0	0	0	0	0	⊓	0	0	0	0	0	0	0
0	0	0	0	0	0	0	0	0	0	⊓	0	0	0	0	0	0
0	0	0	0	0	0	0	0	0	0	0	⊓	0	0	0	0	0
0	0	0	0	0	0	0	0	0	0	0	0	⊓	0	0	0	0
0	0	0	0	0	0	0	0	0	0	0	0	0	⊓	0	0	0
0	0	0	0	0	0	0	0	0	0	0	0	0	0	⊓	0	0
0	0	0	0	0	0	0	0	0	0	0	0	0	0	0	⊓	0
0	0	0	0	0	0	0	0	0	0	0	0	0	0	0	0	⊓

Data-Select Switches

D	C	B	A
0	0	0	0
0	0	0	1
0	0	1	0
0	0	1	1
0	1	0	0
0	1	0	1
0	1	1	0
0	1	1	1
1	0	0	0
1	0	0	1
1	0	1	0
1	0	1	1
1	1	0	0
1	1	0	1
1	1	1	0
1	1	1	1

74154 Demultiplexer

OUTPUTS																
Data Outputs X = Pulses, 0 = No Pulses																
0	1	2	3	4	5	6	7	8	9	10	11	12	13	14	15
X	0	0	0	0	0	0	0	0	0	0	0	0	0	0	0
0	X	0	0	0	0	0	0	0	0	0	0	0	0	0	0
0	0	X	0	0	0	0	0	0	0	0	0	0	0	0	0
0	0	0	X	0	0	0	0	0	0	0	0	0	0	0	0
0	0	0	0	X	0	0	0	0	0	0	0	0	0	0	0
0	0	0	0	0	X	0	0	0	0	0	0	0	0	0	0
0	0	0	0	0	0	X	0	0	0	0	0	0	0	0	0
0	0	0	0	0	0	0	X	0	0	0	0	0	0	0	0
0	0	0	0	0	0	0	0	X	0	0	0	0	0	0	0
0	0	0	0	0	0	0	0	0	X	0	0	0	0	0	0
0	0	0	0	0	0	0	0	0	0	X	0	0	0	0	0
0	0	0	0	0	0	0	0	0	0	0	X	0	0	0	0
0	0	0	0	0	0	0	0	0	0	0	0	X	0	0	0
0	0	0	0	0	0	0	0	0	0	0	0	0	X	0	0
0	0	0	0	0	0	0	0	0	0	0	0	0	0	X	0
0	0	0	0	0	0	0	0	0	0	0	0	0	0	0	X

TABLE 5-8
Truth table for an even-parity circuit

NIBBLE				PARITY LINES G AND C
B_1	B_2	B_3	B_4	
0	0	0	0	0
1	0	1	0	0
1	1	1	0	1
0	0	1	0	1
0	1	0	1	0
1	0	0	0	1

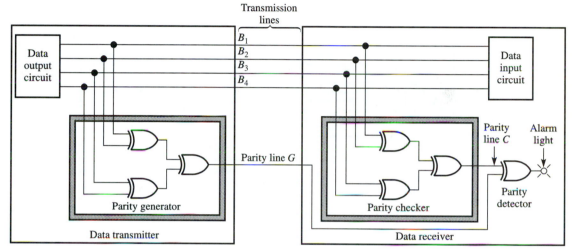

FIGURE 5.37
Even-parity generator and checker circuit.

The 74280 IC Parity Circuit

An integrated circuit that performs both the parity-generator and the parity-checker functions is the TTL 74280 IC. It is capable of determining even or odd parity for 8-bit words. A block diagram of this IC is shown in Fig. 5-38(a). It has nine input lines, labeled I_0–I_8 and two parity-bit output lines, labeled Σ_E and Σ_O. The symbol Σ at the output refers to the sum of 1s present at the input lines. If the number of 1s are even, the output with the subscript E, (for **E**ven) will produce a logic 1. If the number of 1s are odd, the output with the subscript O (for **O**dd) will produce a logic 1.

When the IC is used as an even-parity generator, the first eight inputs are used for the data. The ninth bit is permanently connected to a logic 0, and the parity-bit output line is Σ_O. Line Σ_O is connected to input line I_8 of the second IC that is used as the even-parity checker. Output Σ_O of the checker output is the error indicator. It produces a 1 if an error exists, and a 0 when there is not an error. Figure 5-39 shows an 8-bit even-parity error detection system using the two integrated circuits.

When the IC is used as an odd-parity generator, the first eight inputs are used for the data. The ninth bit is permanently connected to a logic 0, and the parity bit output line is Σ_E. Line Σ_E is connected to input line I_8 of the second IC that is used as the odd parity

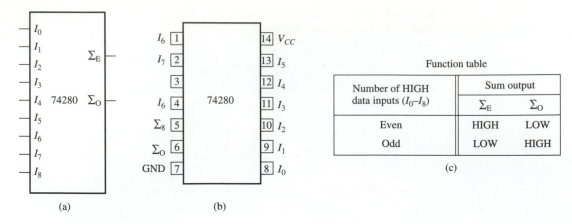

FIGURE 5.38
A 74280 parity integrated circuit. (a) Block diagram. (b) Pin diagram. (c) Function table.

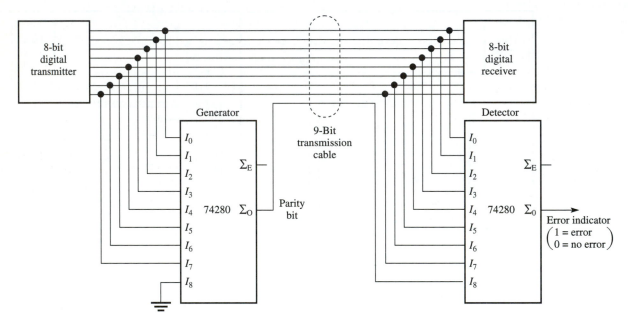

FIGURE 5.39
An 8-bit parity error-detection system.

checker. Output Σ_E of the checker's output is the error indicator. It produces a 1 if an error exists, and a 0 when there is not an error.

The pin diagram and function table of the 74280 parity IC are shown in parts (b) and (c) of Figure 5-38.

▪ REVIEW QUESTIONS

12. A _____ is a switching device that accepts several data inputs and allows only one at a time to get through to the output.

13. A _____ is a switching device that transmits data from one input line to one of several outputs.

14. The purpose of the _____ select line is to control which output line of a demultiplexer will be connected to the input line.

15. The number of select lines needed to select one of the inputs is always a power of _____. Therefore, if a demultiplexer has four data select input lines, it has _____ output lines.

16. The purpose of the data selectors and data distributors is to reduce the number of _____ _____ when information from many sources is transmitted over long distances.

17. Another name for the pattern of bits applied to the data-select input lines of the 74154 decoder/demultiplexer is the _____.

18. A device called a(n) _____ circuit is used for detecting digital error signals that occur during the transmission of data.

DATA PROCESSING CIRCUITS

Binary adder:

A logic circuit that is capable of adding two bits and a carry and that produces a sum and carry-out.

Magnitude comparator:

A circuit that compares two binary numbers and indicates whether one number is larger than, less than, or equal to the other.

A very important function performed by some types of digital equipment is the processing of binary data. One type of data processing is the arithmetic function that is performed by a **binary adder** circuit. Decision making is also a data processing function. One type of decision function is the comparison operation that compares two binary numbers. A decision is made about which number is greater, or if they are equal, as a result of the comparison. One type of circuit that performs this operation is the **magnitude comparator.**

The basic concepts of how these two types of data processing functions are performed by combination logic circuitry are explained in the following section.

5.9 BINARY ADDER CIRCUIT

The binary adder circuit is capable of adding two multibit binary numbers one column at a time. It also has the capability of adding a carry that results from the addition performed in the previous column. As shown in Fig. 5-40(a), the binary adder (also known as a *full adder*) has three inputs, A, B, and C_{IN} (carry in), and two outputs, S (sum) and C_{out} (carry out). Figure 5-40(b) contains the truth table for this circuit. The truth table states that the sum output will be 1 if only one input is 1, the C_{out} output will be 1 if any two inputs are 1, and both S and C_{out} outputs will be 1 if all three inputs are 1.

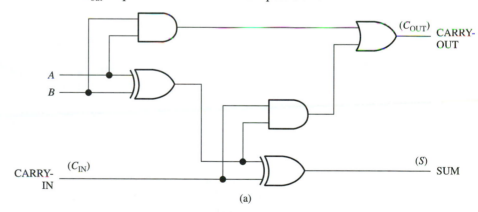

(a)

A	B	CARRY-IN (C_{IN})	CARRY-OUT (C_{OUT})	SUM (S)
0	0	0	0	0
0	1	0	0	1
1	0	0	0	1
1	1	0	1	0
0	0	1	0	1
0	1	1	1	0
1	0	1	1	0
1	1	1	1	1

(b)

FIGURE 5.40
The full adder: (a) Logic diagram. (b) Function table.

EXAMPLE 5.7

From the binary bits applied to the three inputs of the binary adder, place the appropriate 1s or 0s at each output line of the circuit of Fig. 5-41.

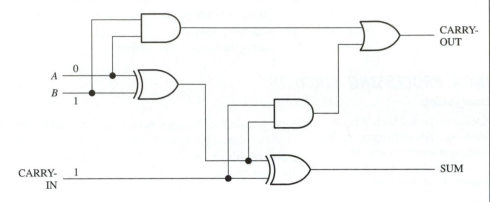

FIGURE 5.41
Circuit for Example 5.7.

Solution   A $0 + 1 + 1$ causes the full adder to generate an output of 10_2 (a sum of 0 and a carry-out of 1).

Large adders that are capable of adding several columns simultaneously are explained in Chapter 10.

Binary adders are cascaded to add two multibit binary numbers. One full adder is required for each column of the numbers. So for 2-bit numbers, two adders are used; for

EXPERIMENT

Binary Full Adder

Objective

■ To assemble and operate a full-adder circuit using logic gates.

Materials

> 1—7486 Exclusive-OR Gate IC
>
> 1—7408 AND Gate IC
>
> 1—7432 OR Gate IC
>
> 2—LEDs
>
> 2—150-Ω Resistors
>
> 3—Logic Switches
>
> 1—+5-V DC Power Supply

Introduction

One of the primary functions of digital equipment is adding binary numbers. For example, calculators and computers perform this operation quite often. The type of digital circuit that makes it possible is a binary adder circuit. By manipulating binary numbers in various ways before they enter the binary adder, the circuit is also capable of subtracting, multi plying, and dividing. This experiment will demonstrate only how the addition function is performed.

Procedure

PART B-FULL ADDER OPERATION

Step 1. Assemble the circuit shown in Fig. 5-42.

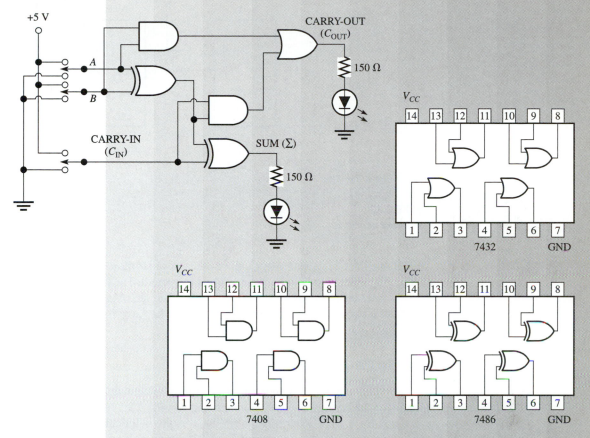

FIGURE 5.42

Step 2. Using the input section of Table 5-9, apply the eight input combinations to the full adder. Fill in the output section with your results.

TABLE 5.9

A	B	C_{IN}	Sum (Σ)	C_O
0	0	0		
0	0	1		
0	1	0		
0	1	1		
1	0	0		
1	0	1		
1	1	0		
1	1	1		

■ EXPERIMENT QUESTION

1. To use a full adder to add the LSB column of two binary numbers, what must be done with the carry-in input?

3-bit numbers, three adders are required; and so on. The carry-out terminal of each adder is connected to the carry-in lead of the next higher-order adder, as shown in the block diagram for Fig. 5-43 for a 2-bit adder.

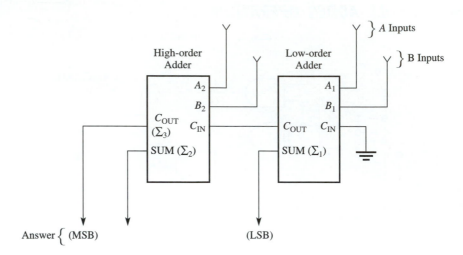

FIGURE 5.43
Cascading adders.

Since there is never a carry input to the least significant bit position, the carry input of the LSB adder is made 0 by connecting it to ground. The least significant bits (LSB) of the two numbers are represented by A_1 and B_1. The next-higher-order bits are labeled A_2 and B_2. The two sum bits are labeled Σ_1 and Σ_2. The carry-out lead of the MSB adder becomes the highest-order sum output, Σ_3.

The 7483 4-Bit Adder

An integrated circuit with four cascaded binary adders is the 7483 chip. As the logic diagram of the internal circuitry shows in Fig. 5-44(A), the two 4-bit inputs are labeled A_1 to A_4, and B_1 to B_4. The terminal labeled C_{IN} is the input carry to the least significant bit adder. The sum outputs are labeled Σ_1 (LSB) through Σ_4 (MSB). Each of the carry-out leads of the first three adders is internally connected to the carry-in leads of the next-higher-order adder. The C_4 terminal is the carry-out of the most significant bit adder.

Terminal C_{IN} should be grounded to a logic low when it adds two 4-bit binary numbers. 7483 ICs can be cascaded to add multiple 4-bit numbers. To perform this function, the C_{IN} of the low-order adder should be grounded. The C_{OUT} output of each adder should be connected to the C_{IN} input of the next-higher-order adder.

The 7483 adder is also capable of adding only one column of two binary bits. This function can be performed by using A_4 and B_4 as the input terminals, A_3 as the carry-in lead, E_4 as the sum output, and C_{OUT} as the carry-out. The terminals of C_{IN}, A_1, B_1, A_2, B_2, and A_2 leads must all be connected to ground. The pin diagram and truth table for the 7483 IC is shown in parts (b) and (c) of Fig. 5-44.

Binary Adder Applications

A binary adder is the most versatile digital circuit for performing arithmetic operations. Adders work in conjunction with temporary storage devices called *registers,* which transfer binary bits into the input terminals and out of the output terminals of the adder one column at a time. The way in which the data are manipulated before being added determines which of the four arithmetic functions (addition, subtraction, multiplication, or division) is being performed. A detailed description of how these four functions are processed is given in Chapter 10.

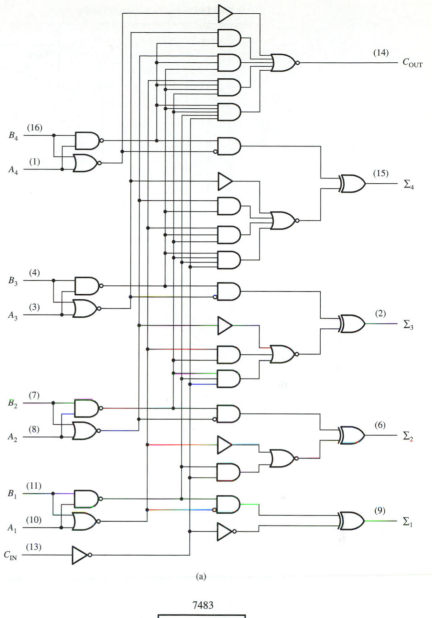

(a)

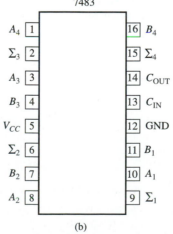

(b)

FIGURE 5.44
The 7483 4-bit binary adder IC. (a) Logic diagram. (b) Pin diagram. (c) 4-Bit function table.
(d) 1-Bit function table.

(continues)

Truth table for a 4-bit parallel adder

Inputs				When $C_0 = 0$		When $C_2 = 0$	When $C_0 = 1$		When $C_2 = 1$
A_1 / A_3	B_1 / B_3	A_2 / A_4	B_2 / B_4	Σ_1 / Σ_3	Σ_2 / Σ_4	C_2 / C_4	Σ_1 / Σ_3	Σ_2 / Σ_4	C_2 / C_4
0	0	0	0	0	0	0	1	0	0
1	0	0	0	1	0	0	0	1	0
0	1	0	0	1	0	0	0	1	0
1	1	0	0	0	1	0	1	1	0
0	0	1	0	0	1	0	1	1	0
1	0	1	0	1	1	0	0	0	1
0	1	1	0	1	1	0	0	0	1
1	1	1	0	0	0	1	1	0	1
0	0	0	1	0	1	0	1	1	0
1	0	0	1	1	1	0	0	0	1
0	1	0	1	1	1	0	0	0	1
1	1	0	1	0	0	1	1	0	1
0	0	1	1	0	0	1	1	0	1
1	0	1	1	1	0	1	0	1	1
0	1	1	1	1	0	1	0	1	1
1	1	1	1	0	1	1	1	1	1

(c)

1-Bit truth table

Inputs			Outputs	
A_4	B_4	A_3	Σ_4	C_{OUT}	← Pin indentification
A	B	C_{IN}	SUM	CARRY	← Function
0	0	0	0	0
1	0	0	1	0
0	1	0	1	0
1	1	0	0	1
0	0	1	1	0
1	0	1	0	1
0	1	1	0	1
1	1	1	1	1

(d)

FIGURE 5.44
(Continued)

5.10 MAGNITUDE COMPARATOR

The magnitude comparator circuit compares two binary numbers and indicates whether one number is larger than, less than, or equal to the other. Figures 5-45 shows the block diagram of a 4-bit magnitude comparator. It has four lines for input A, four lines for input B, and three outputs that indicate whether A is larger than B, A is less than B, or A is equal to B. The magnitude comparator has three additional inputs called the *cascade* (also referred to as *cascading, cascaded,* or *expansion*) input lines, which are labeled $I_A < B$, $I_A = B$, and $I_A > B$. Under normal operation, the $I_A = B$ input should be wired to a high and the $I_A < B$ and $I_A > B$ inputs should be wired low.

FIGURE 5.45
Block diagram of a magnitude comparator.

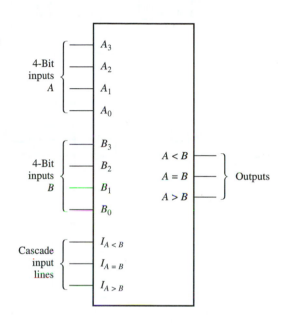

EXAMPLE 5.8

From the binary numbers applied to both inputs of the circuit of Fig. 5-46 determine which lamp is activated.

FIGURE 5.46
Circuit for Example 5.8.

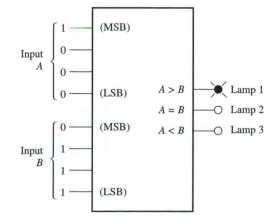

Solution Because binary 1000_2 (8_{10}) applied to input A is greater than binary number 0111_2 (7_{10}) applied to input B, output $A > B$ activates Lamp 1.

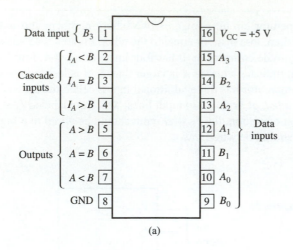

(a)

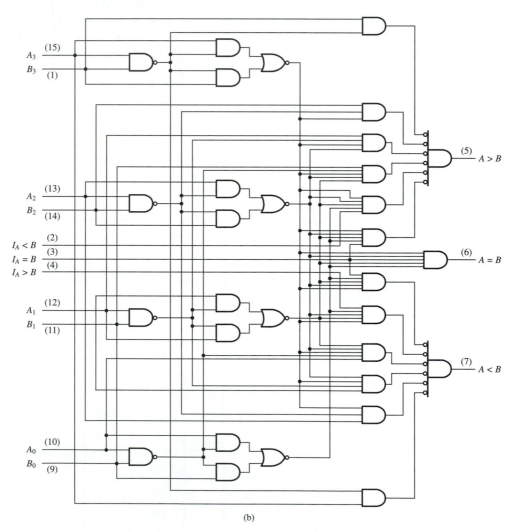

(b)

FIGURE 5.47

The 7485 magnitude comparator IC: (a) Pin diagram. (b) Logic diagram. (c) Function table. Reproduced by permission of Texas Instruments. Copyright © Texas Instruments.

FIGURE 5.47
(Continued)

Function Table

(MSB)	Comparison Input	(LSB)		Output		
A_3, B_3	A_2, B_2	A_1, B_1	A_0, B_0	$A > B$	$A = B$	$A < B$
$A_3 > B_3$	X*	X	X	H	L	L
$A_3 < B_3$	X	X	X	L	L	H
$A_3 = B_3$	$A_2 > B_2$	X	X	H	L	L
$A_3 = B_3$	$A_2 < B_2$	X	X	L	L	H
$A_3 = B_3$	$A_2 = B_2$	$A_1 > B_1$	X	H	L	L
$A_3 = B_3$	$A_2 = B_2$	$A_1 < B_1$	X	L	L	H
$A_3 = B_3$	$A_2 = B_2$	$A_1 = B_1$	$A_0 > B_0$	H	L	L
$A_3 = B_3$	$A_2 = B_2$	$A_1 = B_1$	$A_0 < B_0$	L	L	H
$A_3 = B_3$	$A_2 = B_2$	$A_1 = B_1$	$A_0 = B_0$	L	H	L

* = Don't Care   (c)

The 7485 Magnitude Comparator

The operation of the magnitude comparator is performed by the TTL 7485 integrated circuit. Figure 5-47 shows the pin diagram, logic diagram, and the truth table of the 7485 IC. The table shows that the MSBs of each 4-bit input are compared first. If those bits are not equal, the appropriate output line goes high to show which input is greater. If the MSBs are equal, the same comparison process is made with the lesser-significant bits in the next column. This comparison process continues with each column until the LSBs are reached.

If the 7485 IC compares fewer than 4 bits, the same number of input terminals are used as the number of bits that are applied, starting with the least-significant terminals. For example, if two 3-bit numbers are being compared, they should be applied to inputs A_0–A_2 and inputs B_0–B_2. Any unused input terminal (A_3 and B_3 in this example) should be connected to ground.

EXPERIMENT

Magnitude Comparator

Objectives

■ To assemble and operate the 7485 integrated circuit magnitude comparator.

Materials

> 1—7485 IC
>
> 3—LEDs
>
> 1—150-Ohm Resistor
>
> 8—Logic Switches
>
> 1—+5-V DC Power supply

Procedure Information

■ The pin diagram of a 7485 IC is shown in Fig. 5-47(a). When it is comparing only two 4-bit words, expansion inputs $I_A < B$ and $I_A > B$ must be connected to a low, and expansion input $I_A = B$ must be tied to a high.
■ The logic switch settings connected to inputs A and B are capable of applying any combination of binary numbers ranging from 0000 to 1111.
■ The LED that lights at one of the three output leads indicates how the two 4-bit numbers compare to each other.

Step 1. Construct the circuit shown in Fig. 5-48.

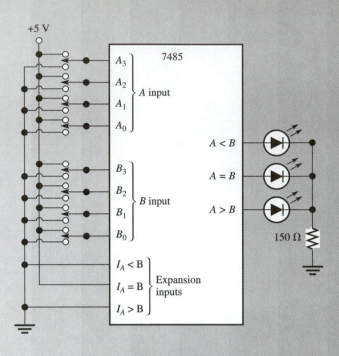

FIGURE 5.48

Step 2. Using the logic switches, apply the binary bits listed in Table 5-10 to the comparator. Fill in the table to record how the comparator responds to each set of applied inputs. Compare your results with the function table in Fig. 5-47.

TABLE 5.10

Input B				Input A				Outputs		
B_3	B_2	B_1	B_0	A_3	A_2	A_1	A_0	$A < B$	$A = B$	$A > B$
0	0	0	0	0	0	0	0			
0	1	0	0	0	0	0	1			
1	0	0	1	1	0	0	0			
0	0	1	1	0	1	0	0			
0	0	0	1	1	0	0	1			

■ EXPERIMENT QUESTIONS

1. The highest number a 4-bit magnitude comparator can compare is _____.
2. Can the 7485 4-bit comparator compare 3-bit numbers?
3. Suppose that one 7485 magnitude comparator is used to compare two 4-bit binary numbers. If the expansion inputs are incorrectly connected the following way:

$I_A < B = 1$
$I_A = B = 0$
$I_A > B = 0$

how will the circuit operate?

Cascading Comparators

When binary numbers larger than 4 bits have to be compared, several 7485 ICs can be connected together to perform this function. The block diagram of Fig. 5-49 shows how two 7485 ICs are cascaded to make an 8-bit comparator. The four least significant bits of each 8-bit word are connected to inputs A_0–A_3 and B_0–B_3 of the comparator on the left, and the four most significant bits of each 8-bit word are connected to inputs A_0–A_3 and B_0–B_3 of the comparator on the right. The $A > B$, $A = B$, and $A < B$ outputs of the least-significant comparator are connected to the expansion inputs of the most-significant comparator. The expansion lines of the least-significant comparator should be wired as if it were comparing only two 4-bit words. The comparison results of the two 8-bit words are generated at the three output lines of the most-significant comparator.

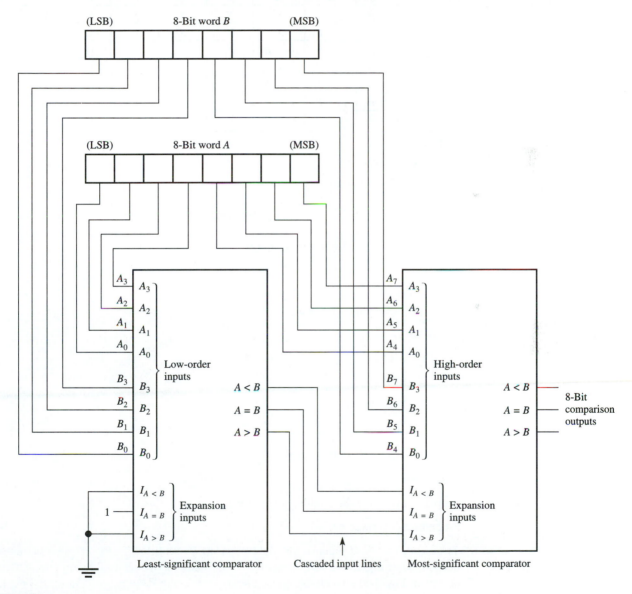

FIGURE 5.49
Eight-bit magnitude comparator.

The most significant 4 bits of inputs A and B are applied to the high-order 7485 IC. These include bits 4 through 7. If these two numbers differ, the output of the high-order comparator indicates the result. Regardless of what numbers are applied to the low-order comparator, this condition will exist. If bits 4 through 7 of inputs A and B are the same, the low-order 7485 determines the output. It does this by comparing the least-significant 4 bits of inputs A and B. The results of the low-order comparator are then fed into the cascaded inputs of the high-order comparator to produce the output. The function table of 7485 ICs that are cascaded is shown in Fig. 5-50.

(MSB)	Comparison Input		(LSB)	Cascading Inputs			Output		
A_3, B_3	A_2, B_2	A_1, B_1	A_0, B_0	$I_{A>B}$	$I_{A=B}$	$I_{A>C}$	$A>B$	$A=B$	$A<B$
$A_3 > B_3$	X*	X	X	X	X	X	H	L	L
$A_3 < B_3$	X	X	X	X	X	X	L	L	H
$A_3 = B_3$	$A_2 > B_2$	X	X	X	X	X	H	L	L
$A_3 = B_3$	$A_2 < B_2$	X	X	X	X	X	L	L	H
$A_3 = B_3$	$A_2 = B_2$	$A_1 > B_1$	X	X	X	X	H	L	L
$A_3 = B_3$	$A_2 = B_2$	$A_1 < B_1$	X	X	X	X	L	L	H
$A_3 = B_3$	$A_2 = B_2$	$A_1 = B_1$	$A_0 > B_0$	X	X	X	H	L	L
$A_3 = B_3$	$A_2 = B_2$	$A_1 = B_1$	$A_0 < B_0$	X	X	X	L	L	H
$A_3 = B_3$	$A_2 = B_2$	$A_1 = B_1$	$A_0 = B_0$	H	L	L	H	L	L
$A_3 = B_3$	$A_2 = B_2$	$A_1 = B_1$	$A_0 = B_0$	L	H	L	L	H	L
$A_3 = B_3$	$A_2 = B_2$	$A_1 = B_1$	$A_0 = B_0$	L	L	H	L	L	H

* = Don't Care

FIGURE 5.50
Function table for cascaded magnitude comparators.

EXPERIMENT

Cascading Magnitude Comparators

Objectives

■ To simulate the operation of two magnitude comparators that can be cascaded to compare two 8-bit binary numbers.

Materials

1—7485 IC

3—LEDs

1—150-Ohm Resistor

11—Logic Switches

1—+5-V DC Power Supply

Procedure Steps

Step 1. Construct the circuit shown in Fig. 5-51.

Procedure Information

To cascade two 7485 comparators for comparing two 8-bit numbers, 16 logic switches are needed. Most digital trainers do not have that many logic switches. As an alternative, the circuit in Fig. 5-51, which uses only one 7485 IC, will demonstrate the same operation. The 8 logic switches connected to its A and B inputs will represent the two 4-bit high-order bits. Three additional logic switches will be connected to the expansion inputs to simulate the operation of a low-order comparator. These switches act as the three output pins of the low-order comparator. The expansion switch that is set high indicates how the 4 least significant bits compare to each other.

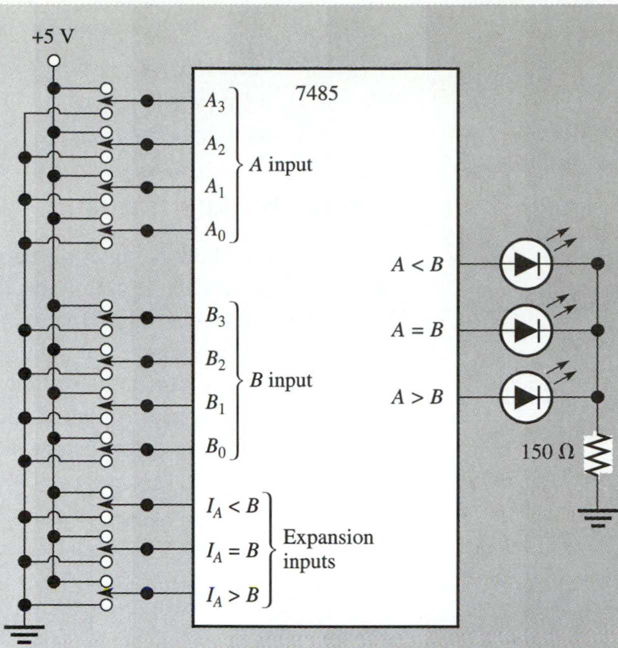

FIGURE 5.51

Step 2. Apply the input signals listed in Table 5-11 to the comparator. Fill in the table to show how it responds. Compare your results with the table in Fig. 5-50.

TABLE 5.11

Input B				Input A				Expansion Inputs			Outputs		
B_3	B_2	B_1	B_0	A_3	A_2	A_1	A_0	$I_{A>B}$	$I_{A=B}$	$I_{A>B}$	$A<B$	$A=B$	$A>B$
0	0	0	0	1	1	1	1	1	0	0			
0	0	0	1	0	0	0	1	0	0	1			
0	1	1	0	0	1	1	0	0	1	0			
1	1	1	0	1	1	0	1	0	0	1			
0	1	0	1	0	1	1	0	0	1	0			

Procedure Question 1

If the low-order bits are always overridden by the high-order bits, why use the expansion inputs?

■ EXPERIMENT QUESTIONS

1. The highest number a 4-bit magnitude comparator can compare is _____.
2. Assume that two comparators are cascaded to make the following comparisons:

   Input A = 10100010   Input B = 10110011

   List the logic state found at each expansion input of the high-order comparator:

   $I_{A<B}$ =
   $I_{A=B}$ =
   $I_{A>B}$ =

3. How many 7485 magnitude comparators are needed to compare the following numbers:

$A = 0110110101100101$
$B = 1011110100101101$

Comparator Application

Figure 5-52(A) shows a materials handling machine that sorts boxes from a main conveyor belt according to size. The A boxes are directed to conveyor 1, the B boxes go to conveyor 2, and the C boxes are diverted to conveyor 3. Figure 5-52(B) shows the three different packages traveling down the main conveyor belt toward the reader. As a box passes through a series of optical sensors, called a *light curtain,* its size is detected according to the number of light beams that it blocks. The size A boxes block two beams, size B boxes block three beams, and size C boxes block four beams.

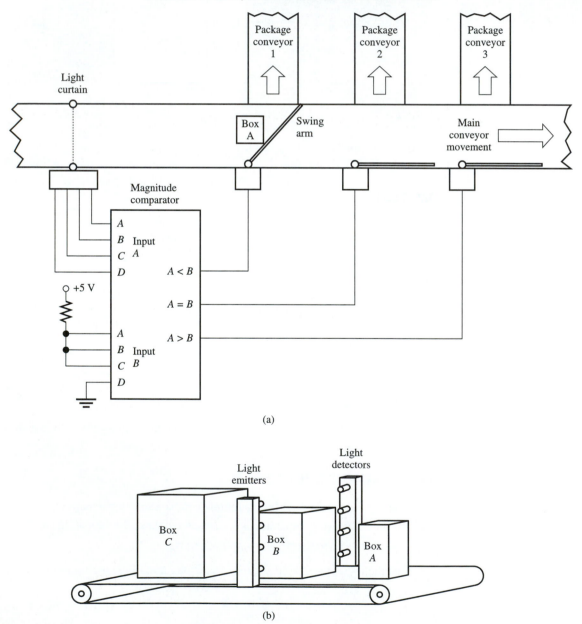

FIGURE 5.52
Comparator application. (a) Materials handling machine. (b) Detection of three different box sizes.

The magnitude comparator is used as the control circuit for this system. The comparator reads a permanent 7_{10} at its input B terminals. Input A receives its inputs from the light curtain. Any beam that is blocked produces a logic 1 at its corresponding detector. When an A box passes through the light curtain and blocks two beams, a 3_{10} is applied to input A, causing the $A < B$ output to go high. This temporary high activates a motor that swings an arm over the main conveyor for 10 seconds and deflects the box onto conveyor 1. When a B box blocks three beams, a 7_{10} is applied to input A, causing the $A = B$ output to go high. Motor 2 swings an arm over the conveyor for 10 seconds to deflect the package onto conveyor 2. When a C box is detected, four beams are blocked and a 15_{10} is applied to input A, causing the $A > B$ comparator output to go high. Motor 3 swings an arm over the conveyor for 10 seconds and deflects the package onto conveyor 3.

More detailed circuits using a magnitude comparator to control motor positioning and the division of binary numbers are described in Chapters 10 and 13.

■ REVIEW QUESTIONS

19. _____ and _____ _____ are two data processing functions performed by digital circuits.

20. Binary adders that have three inputs and two output lines, add binary numbers _____ column at a time.

21. The way in which the data are _____ before being added determines which of the four arithmetic functions are being performed.

22. A magnitude comparator indicates whether one number is _____ _____, _____ _____, or _____ to the other.

23. When binary numbers larger than 4 bits have to be compared, two 7485 ICs can be _____ to make an 8-bit comparator.

5.11 TROUBLESHOOTING

Combination circuits typically have an output of one gate driving the inputs of one or more gates, as shown in Fig. 5-53. This circuit shows several types of faults and how the pulser and probe are used to find them. G_1 is good, so the pulse string from the pulser is passed on to the top input of each gate to which it is feeding its signal.

The fault at NAND gate G_2 is an open at the top input. The signal that G_2 receives at this input may be in the invalid region or it may be picking up noise. How the gate responds to the problem is unpredictable. Therefore, it is possible for the probe to react in one of the three ways shown. Other types of faults are shown at gates G_3 through G_7.

Many combination circuits are constructed in IC form. For those, the best way to troubleshoot is to use their truth tables as a reference. Use that information when observing what the output signals should be for each combination of inputs.

Testing an Encoder

Figure 5-54(a) shows the block diagram of a 74147 IC encoder that encodes a keypad consisting of keys 1–9 into an equivalent BCD output. Both the input and output pins are active low. Therefore, in its inactive resting state, all inputs and outputs are high.

To test the circuit, a logic probe can be used. To find out if input 6 is properly encoded, connect the pulser to pin 6 and inject a pulse. The pulser senses a high and causes the node at pin 6 to go to a temporary low. Each time a pulse is injected, output pins A and D should remain high and B and C should go low, as shown in the truth table of Fig. 5-54(b). All of the other inputs can be tested in the same way.

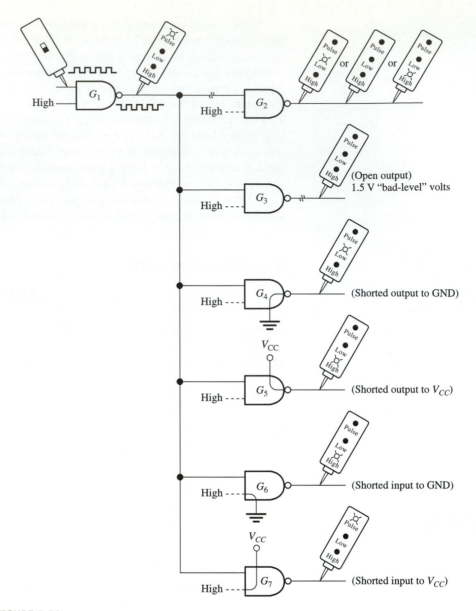

FIGURE 5.53
How logic gates react to various types of malfunctions.

Testing a Demultiplexer

Figure 5-55(a) shows the block diagram of a DM8223 1-line-to-8-line demultiplexer. The left portion of the truth table of Fig. 5-55(b) provides the input and output states of the device when it operates properly. The other sections show how the demultiplexer operates under certain faulty conditions.

Faulty Condition 1.   Data-select line *A* is shorted to ground. Therefore, when a 1 is applied to data-select line *A,* it is recognized as a 0.

Faulty Condition 2.   Data-select line *B* is shorted to V_{CC}. Therefore, when a 0 is applied to data-select line *B,* it is recognized as a 1.

Troubleshooting a BCD-to-Seven-Segment Decoder and Display

Suppose that segment *f* does not light (see Fig. 5-56). This symptom is first observed visually.

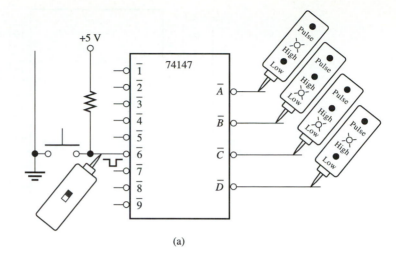

(a)

74147												
Inputs									Outputs			
$\bar{1}$	$\bar{2}$	$\bar{3}$	$\bar{4}$	$\bar{5}$	$\bar{6}$	$\bar{7}$	$\bar{8}$	$\bar{9}$	$\bar{D}$	$\bar{C}$	$\bar{B}$	$\bar{A}$
H	H	H	H	H	H	H	H	H	H	H	H	H
H	H	H	H	H	H	H	H	L	L	H	H	L
H	H	H	H	H	H	H	L	H	L	H	H	H
H	H	H	H	H	H	L	H	H	H	L	L	L
H	H	H	H	H	L	H	H	H	H	L	L	H
H	H	H	H	L	H	H	H	H	H	L	H	L
H	H	H	L	H	H	H	H	H	H	L	H	H
H	H	L	H	H	H	H	H	H	H	H	L	L
H	L	H	H	H	H	H	H	H	H	H	L	H
L	H	H	H	H	H	H	H	H	H	H	H	L

(b)

FIGURE 5.54
Testing a 74147 IC encoder: (a) Block diagram. (b) Function table.

Step 1. To verify that segment f does not light, connect a temporary jumper wire from ground to the lamp-test ($\overline{LT}$) pin. When activated, the $\overline{LT}$ pin is supposed to cause all segments to light. **Result:** Segment f still does not light.

Step 2. With the $\overline{LT}$ pin still activated, use a probe to check the logic levels at the output of the 7447 chip. They should all read low. **Result:** Output pins $\bar{a} - \bar{g}$ are all low.

Step 3. With the $\overline{LT}$ pin still activated, use a probe to check the logic levels at the display side of the current-limiting resistors. **Result:** All pins are high except pin $\bar{f}$, which is low.

The results indicate that both sides of the current-limiting resistor located between the $\bar{f}$ and f terminals is low. Therefore, there is an open circuit in segment f of the seven-segment display.

■ SUMMARY

■ An encoder is a device that converts data into a form that can be interpreted by a digital circuit. The most common type of encoder is one that converts a nonbinary code into binary.

■ A decoder converts a binary code applied to its inputs into an active state at one or more of its outputs. The most common types of decoders are those that convert binary codes into decimal and BCD counterparts.

■ Three common types of decimal display sources are incandescent filaments, LEDs, and LCDs.

■ A multiplexer is a digital device that has several inputs and one output, and it can select which input to connect to the output.

■ A demultiplexer is a digital device that has one input and several outputs, and it can select which output to connect to the input.

■ A parity circuit is used to determine if an error in the transmission of data has occurred.

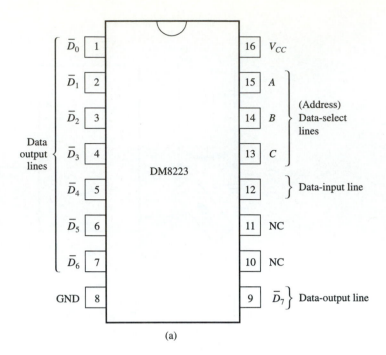

(a)

Data-input line	Data-select lines			Data-output lines																						
				Normal condition								Fault condition 1								Fault condition 2							
	C	B	A	$\bar{D}_0$	$\bar{D}_1$	$\bar{D}_2$	$\bar{D}_3$	$\bar{D}_4$	$\bar{D}_5$	$\bar{D}_6$	$\bar{D}_7$	$\bar{D}_0$	$\bar{D}_1$	$\bar{D}_2$	$\bar{D}_3$	$\bar{D}_4$	$\bar{D}_5$	$\bar{D}_6$	$\bar{D}_7$	$\bar{D}_0$	$\bar{D}_1$	$\bar{D}_2$	$\bar{D}_3$	$\bar{D}_4$	$\bar{D}_5$	$\bar{D}_6$	$\bar{D}_7$
1	0	0	0	0	1	1	1	1	1	1	1	0	1	1	1	1	1	1	1	1	1	0	1	1	1	1	1
1	0	0	1	1	0	1	1	1	1	1	1	0	1	1	1	1	1	1	1	1	1	1	0	1	1	1	1
1	0	1	0	1	1	0	1	1	1	1	1	1	1	0	1	1	1	1	1	1	1	0	1	1	1	1	1
1	0	1	1	1	1	1	0	1	1	1	1	1	1	0	1	1	1	1	1	1	1	1	0	1	1	1	1
1	1	0	0	1	1	1	1	0	1	1	1	1	1	1	1	0	1	1	1	1	1	1	1	1	1	0	1
1	1	0	1	1	1	1	1	1	0	1	1	1	1	1	1	0	1	1	1	1	1	1	1	1	1	1	0
1	1	1	0	1	1	1	1	1	1	0	1	1	1	1	1	1	1	0	1	1	1	1	1	1	1	0	1
1	1	1	1	1	1	1	1	1	1	1	0	1	1	1	1	1	1	0	1	1	1	1	1	1	1	1	0

(b)

FIGURE 5.55
Testing a DM8223 demultiplexer. (a) Pin diagram. (b) Function table.

■ A binary adder adds three 1-bit binary numbers and generates a sum and a carry output.

■ A magnitude comparator compares two binary numbers and determines if one is less than, greater than, or equal to the other.

■ PROBLEMS

1. The binary adder has _____ input ports and _____ output lines. (5-9)

2. Parity is a method of detecting _____. (5-8)

3. The process of converting a nonbinary code to a binary code is usually called _____; likewise, converting a binary to nonbinary code is called _____. (Encoders, Decoders)

4. List the functional logic circuits that are available in standard IC packages. (Encoders, Decoders)

5. A binary adder adds binary numbers _____ column at a time. (5-9)

6. Temporary storage devices called _____ transfer binary bits into and out of binary adders _____ column at a time. (5-9)

7. A (multiplexer or demultiplexer) has one input and several outputs. (5-8)

8. A (multiplexer or demultiplexer) is also called a data selector, and a (multiplexer or demultiplexer) is also called a data distributor. (5-6, 5-7)

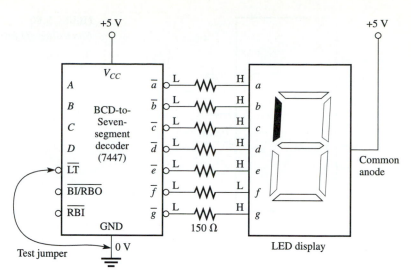

FIGURE 5.56
Troubleshooting a faulty decoder/LED display circuit.

9. One type of digital data word that consists of 4 bits is called a _____. (5-8)

10. The _____ input overrides all other inputs of a priority encoder if they are activated simultaneously. (5-3)

11. What three decisions does a magnitude comparator perform when comparing two different binary numbers? (5-10)

12. The sum and carry produced by a binary adder when binary bits 0 + 1 + carry 1 are applied to the inputs is _____. (5-9)

13. A demultiplexer can perform the conversion of (serial or parallel) data to (serial or parallel) data transfer. (5-7)

14. With five selector lines used by a multiplexer, how many separate input lines can be routed to one output line? (5-6)

15. Which segments of the seven-segment display illuminate when the following decimal numbers appear? (5-5)
    (a) 2
    (b) 4
    (c) 8

16. A common-anode lead of the MAN-72 is connected to _____. (5-5)

17. Encoders are usually found at the _____ section of digital equipment, whereas decoders are usually found at the _____ section. (Encoders, Decoders)

18. Insert the correct logic states (0 or 1) in the binary adder circuit of Fig. 5-57 for the problem 1 + 0 + carry 1. (5-9)

19. Another name for the pattern of bits applied to the data-select input lines of a multiplexer is the _____. (5–7)

20. Which segments of a LED display will light if the following binary bits are applied to the input pins of a 7447 IC: $A = 1$, $B = 0$, $C = 0$, and $D = 1$? (5-5)

21. An odd-parity circuit produces a parity bit to make the number of 1s in the nibble and parity line an (odd or even) number. (5-8)

22. An unwanted signal called _____ may come from the magnetic field of a nearby motor. (5-8)

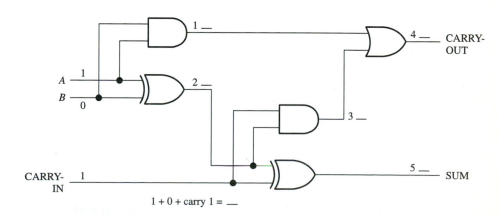

FIGURE 5.57
Logic gates for Problem 18.

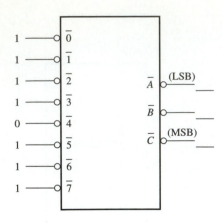

FIGURE 5.58
Block diagram for Problem 30.

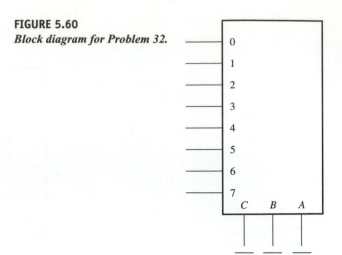

FIGURE 5.60
Block diagram for Problem 32.

23. An input lead with an overbar over the identification number or letter is activated by a (low or high) input. (5-2)

24. Describe one practical application for each of the following standard combination circuits:
   (a) Decimal-to-binary encoder (5-3)
   (b) Multiplexer (5-6)
   (c) Binary adder (5-9)
   (d) Magnitude comparator (5-10)

25. The Xs listed in the input section of truth tables for combination circuits represent _____ conditions. Define what this means. (5-2)

26. When troubleshooting a combination circuit, compare the input/output readings to its _____ _____ reference. (5-11)

27. From the logic diagram of Fig. 5-16(B), note that the output gates have circles at each output lead. What information do they provide the troubleshooting technician? (5-4)

28. The $\overline{LT}$ line of the 7447 IC performs what function? (5-5)

29. Describe which two conditions cause all the outputs of a 7447 IC to go low. (5-5)

30. Place 0s and 1s at the output of the 74148 IC when the bit patterns of Fig. 5-58 are applied at the input leads. (5-2)

31. Place 0s and 1s at the output of the 74147 IC when the bit patterns for Fig. 5-59 are applied at the input leads. (5-3)

32. Place 1s and 0s at the data-select lines of the multiplexer of Fig. 5-60 to enable input line 4 to be connected to the output. (5-6)

33. Which lamp turns on when the binary bits shown in Fig. 5-61 are applied to the inputs of the magnitude comparator? (5-10)

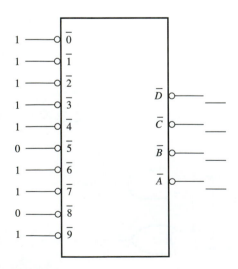

FIGURE 5.59
Block diagram for Problem 31.

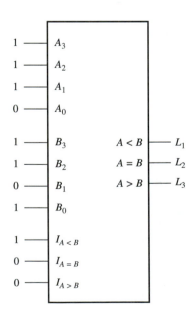

FIGURE 5.61
Block diagram for Problem 33.

FIGURE 5.62
Timing diagram for Problem 34.

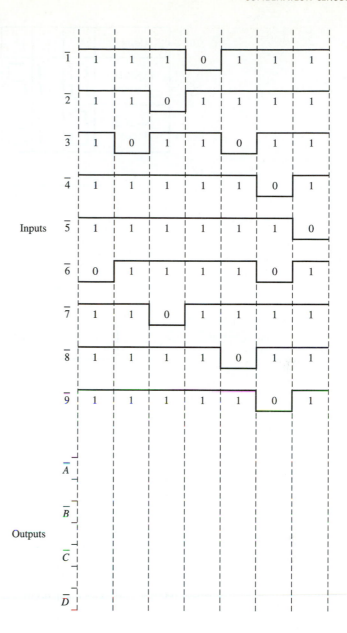

34. For the 74147 priority encoder IC, draw the output waveforms for the inputs shown in the timing diagram of Fig. 5-62 (5-3)

TROUBLESHOOTING

35. After wiring a BCD counter to a BCD decoder/driver and LED display, a technician observes the patterns of Fig. 5-63 on the display. What is the most likely cause of the fault? (5-11)
   (a) Connections *b* and *c* from the decoder/driver to the display are crossed.
   (b) Input connections *B* and *C* to the decoder/driver are crossed.
   (c) Input B of the decoder/driver is open.
   (d) All of the above.

36. The logic states at each pin of a defective 74150 multiplexer are found in Table 5-12. Why are the readings incorrect? (5-6,11)

TABLE 5.12
Logic states of a defective 74150 multiplexer for Problem 36

PIN	LOGIC STATE	PIN	LOGIC STATE
1	0	13	1
2	1	14	0
3	0	15	1
4	0	16	1
5	1	17	0
6	0	18	1
7	1	19	0
8	1	20	1
9	1	21	1
10	0	22	0
11	0	23	1
12	0	24	1

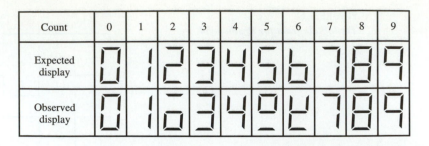

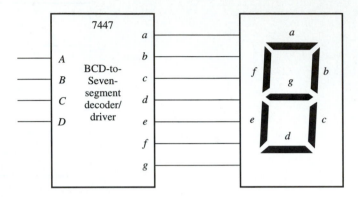

FIGURE 5.63
Display and block diagram for Problem 35.

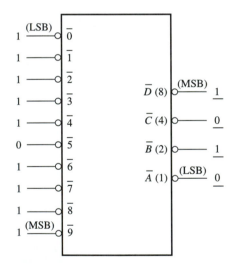

FIGURE 5.64
Answer to Reviews Question 6.

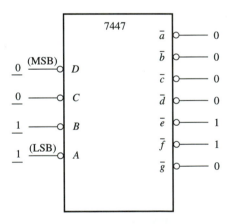

FIGURE 5.65
Answer to Review Question 9.

ANSWERS TO REVIEW QUESTIONS

1. Encoders   **2.** Don't-care   **3.** Bubble, overbar
4. enable   **5.** priority   **6.** See Fig. 5-64   **7.** output
8. Enable or strobe   **9.** See Fig. 5-65   **10.** Segments *a*, *b*, *c*, *d*, and *g*.   **11.** +5V   **12.** Multiplexer

13. Demultiplexer   **14.** Data   **15.** 2, 16 (0–15)
16. Wiring connections   **17.** Address   **18.** Parity
19. Arithmetic, decision making   **20.** one   **21.** manipulated   **22.** less than, greater than, equal   **23.** cascaded

CHAPTER 6

INTEGRATED CIRCUITS

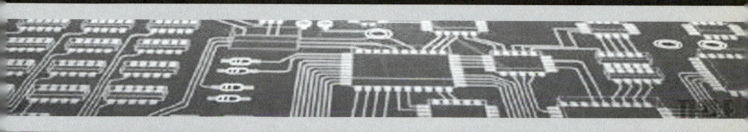

When you complete this chapter, you will be able to:

1. Distinguish between small, medium, large, very large, and ultra large-scale integration.
2. List the advantages of digital IC technology.
3. Describe the fabrication techniques used to manufacture bipolar ICs.
4. List the different IC families and describe the distinguishing characteristics of each.
5. List and describe the logic IC characteristics, such as logic level voltages and currents, propagation delay, power dissipation, noise immunity, operating temperatures, and fan-out and fan-in.
6. Determine the specifications, limiting conditions, and operating characteristics listed on a data sheet for an IC for design and troubleshooting purposes.
7. Select an IC from a family that provides the operating characteristics needed for a specific application.
8. Interface ICs from different families.

6.1 EVOLUTION OF ICs

Integrated circuits:

A miniature electronic circuit that contains transistors, diodes, resistors, capacitors, and interconnecting conductors.

Chip:

Another name used for an integrated circuit.

In Chapter 3, basic gates were introduced. Their symbols and circuit operations were described in detail. The first electrical logic gates were made from relays. Then they were made from electron tubes, followed by transistors. Today, most logic gate functions are performed by **integrated circuits (ICs).** Also known as a **chip,** the IC is a miniaturized circuit that contains the following:

- transistors
- diodes
- resistors
- capacitors
- interconnecting conductors

It is so small that most of the processing functions of a microcomputer can be performed by a single chip that takes up about the same amount of space as a few grains of salt. See Fig. 6-1. The first working IC was conceived and constructed in 1958 by Texas Instruments employee Jack Kilby. Robert Noyce is credited for further developing the IC for specialized use, especially in industrial applications. The contributions of these men have resulted in the emergence of a new technology that provides many new capabilities for electronic circuitry.

FIGURE 6.1
A microcomputer chip with its etched circuits compared to grains of salt, indicating the extent of miniaturization.

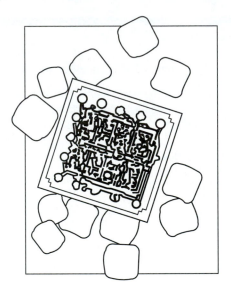

The widespread use of the integrated circuit began in the early 1960s. Since then, IC technology advancements have enabled this device to evolve through several significant stages. They are as follows:

Small-scale integration (SSI):

ICs that contain 20 to 50 components on a single silicon chip.

Early 1960s **Small-scale integration (SSI):** The first ICs developed in the early 1960s consisted of 20 to 50 components on each chip that enabled as many as 12 logic gate functions to be performed. Basic logic gates and flip-flop functions are performed by these chips.

Medium-scale integration (MSI):

ICs that contain 500 components on a single chip.

Mid-1960s **Medium-scale integration (MSI):** IC technology advancements enabled 500 components to be placed on a single chip. As a result, logic functions such as decoding and multiplexing, which require between 12 and 100 gates, were made possible.

Large-scale integration (LSI):

ICs that contain 20,000 components on a single silicon chip.

Late 1960s to Mid-1970s **Large-scale integration (LSI):** IC fabrication techniques advanced even further in the late 1960s when single chips with 20,000 components were produced that allowed between 100 and 10,000 logic functions. With that number of gates able to perform the logic of an entire digital system, the microprocessor was developed. It also provided enough storage space to enable IC memory devices to be developed.

Very large-scale integration (VLSI):

ICs that contain between 20,000 and 100,000 components on a single silicon chip.

Ultra large-scale integration (ULSI):

ICs that contain over 100,000 components on a single silicon chip.

Mid-1970s   **Very large-scale integration (VLSI):** The development of electron-beam lithography and other related IC manufacturing techniques resulted in allowing 20,000 to 100,000 components to be placed on a single chip. As a result, a new generation of micro-computer, consisting of both the central processing unit and memory, emerged.

Mid-1980s to the Present   **Ultra large-scale integration (ULSI):** As IC manufacturing technology continues to evolve, the number of components placed on a chip increases. The result is that complex devices such as the microprocessor become more powerful and are able to process larger amounts of information. Integrated circuits that contain over 100,000 components on a single chip are classified as ULSI.

ICs from all these stages of development are in use today. Because of their small size, low cost, low power consumption, and reliability, the IC has made a significant impact on digital circuit technology.

■ REVIEW QUESTIONS

1. Which of the following are advantages of ICs over discrete circuits?
   (a) Small size.
   (b) Low cost.
   (c) Low power consumption.
   (d) High reliability.
   (e) All of the above.
2. _____-scale-integration refers to fewer than 12 gates on the same chip. _____-scale-integration has 12 to 100 gates per chip, and _____-scale-integration means between 100 and 1000 gates per chip.
3. The placement of a microcomputer central processing unit on a single chip is the result of _____ _____-scale integration.
4. Which of the following is not formed on an IC chip?
   (a) Resistors.
   (b) Diodes.
   (c) Capacitors.
   (d) Transistors.
   (e) Inductors.
   (f) Interconnecting conductors.

6.2 MANUFACTURING AN INTEGRATED CIRCUIT

All types of ICs are produced by similar methods using the same basic materials and processes. These basic materials are the same P-type and N-type regions used in transistors. The operation of an IC is dependent on the interaction between P-type and N-type regions of silicon.

IC Logic Circuit Families

IC manufacturing techniques were initially developed in the 1960s. Early types of digital ICs were based on discrete-component circuit designs, but their operating characteristics were very limited. They were replaced by more sophisticated types as technology rapidly advanced.

During that time, many IC manufacturing companies were organized. Because the competition was very intense, a manufacturer would try to gain on the competition by developing new ICs that had characteristics ideal for certain applications. The result was that many different competitive logic circuit types appeared on the market.

Today, many of the earlier types of ICs are not produced because of obsolescence or because the companies that produced them no longer exist. Yet, there are presently dozens of different categories (called families) of logic ICs available. Some of these devices will soon become obsolete, and others will be improved because of new technological developments, which frequently occur. Because IC technology changes so rapidly, it is not practical to list all of the families and their characteristics. Instead, some of the more popular IC families will be examined.

A method of classifying digital IC families is to divide them into two different categories: **bipolar** and **metal-oxide semiconductors (MOS).** The difference between the two is based on the type of transistor formed on the substrate. The internal components of ICs with bipolar transistors are based on a design similar to the common NPN or PNP transistor, which consists of two PN junctions in close proximity, to perform their logic operation. MOS ICs use a different type of transistor, called a field-effect transistor or FET (also known as MOSFET), which consists of a single PN junction, to perform their logic operations. Figure 6-2 shows the basic hierarchy of digital IC families.

Bipolar:

The type of integrated circuit that consists of miniature internal NPN or PNP transistors.

Metal-oxide semiconductor (MOS):

The type of integrated circuit that consists of miniature internal FETs.

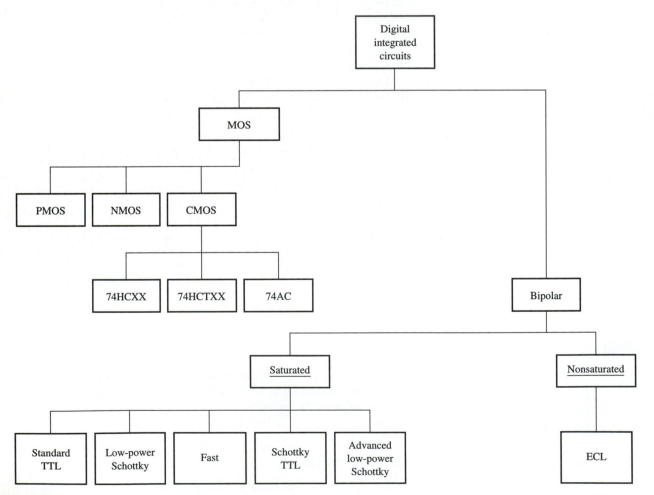

FIGURE 6.2
Basic hierarchy of digital IC families.

6.3   BIPOLAR FAMILIES

There are two types of *bipolar* families: *saturated* and *nonsaturated*.

Saturated

Transitor–transitor logic (TTL):

A popular IC that uses transistors at both the inputs and the outputs.

Of the saturated type, the **transistor–transistor logic** (TTL or T²L, "T-squared L") 7400 family is the most popular. It uses transistors at both the inputs and outputs (see Fig. 6-3). Because of its popularity, almost every IC manufacturer produces this family. Besides being used in SSI, it is also used in MSI and LSI.

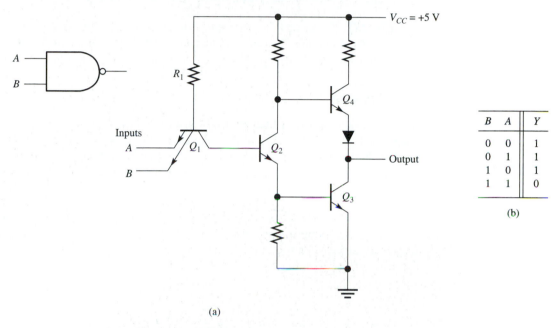

(a)

B	A	Y
0	0	1
0	1	1
1	0	1
1	1	0

(b)

FIGURE 6.3
(a) IC (TTL) internal circuitry of a NAND gate. (b) NAND-gate truth table.

The TTL family is made up of several subfamilies. The primary difference between them is the speed at which they operate and the amount of power they consume. Each one has a specific numerical or alphanumeric listing for identification, which starts with the number 74 to indicate they are a member of the 7400 series.

Figure 6-3(a) shows the internal circuitry of a two-input NAND gate. All of the transistors are NPN configurations that operate in either **saturation** or **cutoff.** When in saturation, the transistors act like a closed switch. When in cutoff, they act like an open switch.

Saturation:

Occurs when a transistor is in a state similar to a closed switch because almost all the current flows through it.

Cutoff:

Occurs when a transistor is in a state similar to an open switch because current does not flow through it.

Standard TTL (74XX)   For many years, the standard TTL type has been the most significant subfamily, but it has become obsolete from a performance standpoint because of its relatively slow switching times and high current consumption. It is currently available from suppliers but will eventually be discontinued. Standard TTL devices are the type of IC referred to throughout this book and are used for most of the experiments in this text.

Schottky TTL (74SXX)   The transistor configuration in Schottky TTL ICs is the same as in standard TTL circuits except that a special type of diode, known as a Schottky barrier diode, is diffused between the base and collector of each transistor. This diode reduces the saturation voltage when it changes to the on state. Therefore, it enables the transistor to turn on or off faster when it is switched from the opposite state.

Low-Power Schottky (74LSXX)   A version of the Schottky diode is the low-power Schottky. It uses larger-size resistors, which reduces the amount of current flow. Therefore, less power is dissipated by this subfamily device compared to the standard Schottky. However, the larger resistors increase the *R-C* time constants of the circuit components and therefore causes the device to operate slower.

Advanced Low-Power Schottky (74ALSXX)   Further improvements in IC design have resulted in the development of the advanced low-power Schottky. This device switches twice as fast and consumes half the power of the low-power Schottky.

Fast (74FXX)   By using a new process of integration called *oxide isolation,* a new subfamily, called *FAST TTL,* has been developed. Inside the IC, the components are smaller. The result is that the switching time is faster.

Nonsaturated

Instead of using transistors that operate in saturation or cutoff to produce 1 and 0 states, it is possible to design them so that they operate in the linear region of the transistor curve. As a result, this eliminates the recovery time of a saturated transistor, which allows faster switching between the two logic levels. There are two primary types of nonsaturated bipolar families.

Emitter-Coupled Logic (ECL)   ECL is the fastest operating TTL subfamily type. However, its power consumption is much higher. It is usually used to perform complex functions or for special-purpose applications.

6.4   MOS FAMILY IC

Field-effect transistor (FET):

A type of transistor used in the internal circuitry of CMOS ICs.

When the need for greater density occurred during the IC evolution period, it became difficult to miniaturize circuitry using the bipolar design. A simpler process was needed to meet yield and manufacturing cost objectives. That is when the first of a long series of metal-oxide semiconductor (MOS) processes was developed that involved fewer processing steps than are needed to fabricate bipolar ICs.

Unlike the bipolar IC transistor families, the MOS IC families use a **field-effect transistor (FET)** design for their internal circuitry. The attractive features of these devices are that they consume very little power and can be made very small. However, their slower switching speeds and susceptibility to static electricity have been their primary drawbacks.

The following describes the more common types of MOS families.

P-Channel MOS (PMOS)

The first MOS device used in digital circuit applications was made from the PMOS family. P-type MOSFETs utilize heavily doped P-channel field-effect transistors and provide excellent densities. This family is still used for memory chips where speed is not vital, such as for calculators. It is the slowest of the MOS families and requires two power supplies.

N-Channel MOS (NMOS)

The NMOS type of IC is made of N-channel field-effect transistors. Because current flows through N-doped semiconductor material faster than P-doped, the NMOS device is capable of twice the speed. It also provides larger currents for a given geometry than can be obtained with a PMOS IC. Most present-day LSI microprocessors and microcomputers are made from this family.

Complementary MOS (CMOS)

The CMOS family combines both P- and N-channel devices, so, therefore, it has speed and packing-density characteristics somewhere between those of PMOS and NMOS. CMOS ICs are ideal for battery-powered digital circuitry, such as in digital watches and aerospace applications. Because they have good immunity to magnetic fields, they are commonly used in factory equipment and automobile circuits, where large voltage transients are common. The big disadvantage of the MOS is its relatively slow operating speed. Of all the MOS families, CMOS ICs are by far the most frequently used.

CMOS devices made from older technology use metal gate construction and are numbered as the 4000 series. This series, for the most part, has been made obsolete by the newer CMOS devices that use silicon gate technology, because of their higher operating speed and lower power consumption. Numbered as the 74 series, these newer-generation CMOS devices use the same numbering system as TTL devices to identify the chip function, and the same pin numbers for input, outputs, supply voltages, and ground. The more common new-generation CMOS IC subfamilies are

74HCXX This subfamily was designed in the early 1980s to operate at faster speeds than the 74CXX subfamily it replaced. Its electrical characteristics are comparable to the TTL low-power Schottky IC. It is designed to connect its output terminals to the input leads of a low-power Schottky IC and is capable of driving 10 74LS loads.

74HCTXX The electrical characteristics of this subfamily are very similar to the 74HC subfamily. It is designed for applications where a TTL device must drive a CMOS device.

74AC This subfamily is designed to interface with the TTL 74ALS ICs.

Compared to the bipolar families, MOS ICs require extra care in handling. A thin insulating material located on the substrate can easily be punctured by an excessive electrostatic charge, such as that built up after walking across a carpet. For this reason, certain handling precautions should be taken. They are listed in Table 6-1.

TABLE 6.1
Handling precautions for MOS integrated circuits

1. MOS ICs should be stored or transported in conductive carriers, such as plastic trays.
2. When working with MOS ICs, place them on a grounded bench surface. Any tools or soldering irons used on MOS ICs should also be grounded.
3. Technicians or operators should also be grounded, usually with a wrist strap in series with a 1-megaohm resistor to ground.
4. Wear only antistatic clothing, such as cotton, when working with MOS ICs. Any synthetic material, such as nylon, should be avoided.
5. Always install a MOS device immediately into its circuit after it is removed from its protective carrier.
6. Try to avoid touching the pins.
7. Shorting straps should be placed across the edge connector of printed circuit boards containing MOS ICs when they are carried or transported.

Each family of integrated circuits is designed for a different application. Each has its own special capabilities and limitations. Designers usually select a given logic family for a certain application on the basis of its characteristics. For example, a logic family whose circuits switch from one logic state to another very rapidly would be used in a device such as a computer, which relies on speed. Another family with low power-consumption characteristics would be selected for a digital watch that operates on batteries for an extended period of time.

When making comparisons among logic families, their characteristics can be obtained from data sheets supplied by the IC manufacturers.

■ REVIEW QUESTIONS

5. Digital IC families are divided into two major categories: _____ and _____ .

6. There are two different types of bipolar families: _____ and _____ .

7. Generally, when fast speed is required, it is traded off with _____ power consumption.

8. A saturated bipolar transistor is driven into saturation and cutoff to develop _____ and _____ state output levels.

9. A Schottky IC has a _____ clamped across its base and collector, which allows it to consume less power and operate faster than standard TTL ICs.

10. Nonsaturated bipolar ICs operate in the _____ region of the transistor curve instead of reaching saturation and cutoff.

11. Which of the following is the fastest logic family used for high-speed applications?
    (a) Fast
    (b) ECL
    (c) Low-power Schottky
    (d) Schottky

12. Which of the following dominates the LSI field?
    (a) TTL
    (b) CMOS
    (c) NMOS
    (d) PMOS

13. Which of the following is the most popular MOS family?
    (a) CMOS
    (b) PMOS
    (c) NMOS

USING IC DATA SHEETS

Data sheets:

Literature written by the IC manufacturer that provides information about the minimum and maximum operation conditions of each type of IC.

To effectively work with integrated circuits, it is important to know how to use **data sheets.** These provide the necessary information about the minimum and maximum operating conditions that must be observed when working with each particular IC family. In all cases, the minimum or maximum worst-case rating is provided because this value is guaranteed by the manufacturer.

The next few sections provide guidelines for using IC specification data sheets. The material is presented in such a way as to enable the reader to extract information about IC devices from most manufacturers' data sheets. Because standard TTL IC family devices are referred to throughout this text and are used in the laboratory experiments throughout this book, a data sheet for a standard TTL device is used as a sample, as shown in Fig. 6-4.

At the top of the first data sheet page is the name of the device and its assigned part number.

QUAD 2-INPUT NAND GATE

DM7400

GENERAL DESCRIPTION

This device contains four independent gates each of which performs the logic NAND function.

CONNECTION DIAGRAM

Dual in-line package

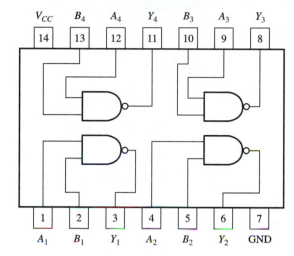

DM5400 (*J*) DM7400 (*N*)

FUNCTION TABLE

Inputs		Output
A	B	Y
L	L	H
L	H	H
H	L	H
H	H	L

H = High logic level
L = Low logic level

FIGURE 6.4
Page 1 of sample specification data sheet.

6.5 NAME

"Quad 2-Input NAND Gate" describes the device:

- *Quad* means there are four gates in the IC package.
- *2-Input* means that each gate has two inputs.
- *NAND* describes the type of logic device in this package.

6.6 PART NUMBER

Each manufacturer prints a label on the case of each IC package. This label represents useful information about the IC. The full designation for one Texas Instruments TTL IC is shown in Fig. 6-5: This labeling system has been universally adopted by all IC manufacturers.

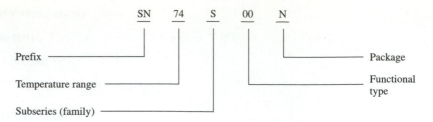

FIGURE 6.5
Part number.

Prefix

The prefix identifies the manufacturer of the IC. It consists of one or more letters, which are the first characters listed in the part number.

SN:	Texas Instruments
DM:	National
MC:	Motorola
P:	Philips
CD:	Harris
AM:	Advanced Micro Devices

Temperature Range

The 74 represents an IC that is manufactured for commercial–industrial use because its operating temperature range is 70°C. A 54 signifies that the IC is built for military use because its operating temperature range is 180°C.

Subseries (Family)

The subseries of the IC is designated by letters, which identify the subseries or integrated circuit family. For example:

F:	Fast TTL IC
S:	Schottky-clamped TTL IC
LS:	Low-power Schottky TTL IC
ALS:	Advanced low-power Schottky IC
No Letter:	Standard TTL IC

Functional Type

The functional type is identified by either two or three digits over a range from 00 to over 300 (new numbers are continually being added). They identify the type of circuit function. For example:

00:	Indicates a NAND gate
02:	Indicates a NOR gate
151:	Indicates a multiplexer

Package

The final designation on the label specifies the type of package. Integrated circuit packages are classified according to the way they are mounted on printed circuit boards (PCB). There are two mounting techniques, *through-hole,* and *surface mount technology* (SMT). The most popular through-hole type package, shown in Fig. 6-6(a) is known as the dual-in-line (DIP) package. Its leads are inserted through holes on the PC board, which are soldered on the opposite side. The DIP package is made with a ceramic or plastic case. Ceramic packages are often used for the 5400 series ICs because ceramic dissipates heat better than plastic. The pins of the surface mount package, shown in Fig. 6-6(b), are soldered directly to conductors on one side of the board. This style is referred to as the SOIC (small outline integrated circuit). Three other types of SMT technology packages are the PLCC (plastic leaded chip carrier), the LCCC (leadless ceramic chip carrier), and the flat pack, also shown in Fig. 6-6. Letters commonly used to indicate the type of package are:

P: Plastic dual-in-line package

F: Ceramic flat pack

L: Leadless ceramic chip carrier

S: Small outline IC

J: Ceramic dual-in-line package

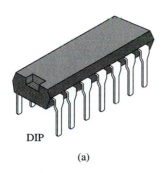

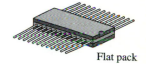

| DIP | SOIC | PLCC | LCCC | Flat pack |

(a)   (b)

FIGURE 6.6
IC packaging.

EXAMPLE 6.1

Identify all of the information that the following part number provides: DM5402P

Solution

DM: National Semiconductor (the manufacturer)

54: Temperature range for military use

__: No subseries letter signifies standard TTL family

02: Indicates a NOR-gate function

P: Plastic dual-in-line package

This package labeling system using the 7400 series is most commonly used by most manufacturers for the TTL IC families. However, some IC manufacturers use a different system.

6.7   GENERAL DESCRIPTION

Also located on the front page of the data sheet (Fig. 6-4) is a functional (general) description of the logic device.

6.8   CONNECTION DIAGRAM

A pin diagram is provided to show how the logic devices inside the IC package are connected to the terminal pins.

6.9   FUNCTION TABLE

A function table describes the operation of the device in a graphical format by providing information about which output logic states are generated by the combination of logic states applied to the inputs.

■ REVIEW QUESTIONS

14. Data sheets provide information about the _____ and _____ operating conditions of each IC family.

15. A dual four-input NAND gate describes:
    (a) Dual:
    (b) Four-input:
    (c) NAND:

16. An IC prefix consists of one or two letters that indicate what type of information?

17. The IC part number that starts with a 74 signifies that it is used for _____ applications, whereas the 54 indicates that it is used for _____ applications.

18. Define each of the following IC subseries identifiers:
    (a) F:
    (b) ALS:
    (c) S:
    (d) LS:
    (e) No letter:

19. Dual-in-line (DIP) packages are constructed from two materials, _____, which is identified on the part number by the letter J, and _____, which is identified by the letter P.

20. What information do the following sections of the data sheet provide?
    (a) General description
    (b) Connection diagram
    (c) Function table

6.10   ABSOLUTE MAXIMUM RATINGS TABLE

The Absolute Maximum Ratings Table (Fig. 6-7) provides information about those values beyond which the safety of the device cannot be guaranteed. If the device is operated at or beyond these limits for a certain length of time, it is likely that it will be permanently destroyed. Typically, these ratings include the following:

Supply Voltage   The power supply voltage connected to pin 14 must not exceed 7 volts.

V_{IN}, **Input Voltage**   A logic 1 signal at 5.5 volts or more should not be applied to a gate input.

ABSOLUTE MAXIMUM RATINGS TABLE

Supply voltage	7 V
Input voltage	5.5 V
Storage temperature range	−10°C to +75°C

RECOMMENDED OPERATING CONDITIONS TABLE

Symbol	Parameter	DM7400 MIN	DM7400 NOM	DM7400 MAX	Units
V_{CC}	Supply voltage	4.75	5	5.25	V
V_{IH}	High-level input voltage	2			V
V_{IL}	Low-level input voltage			0.8	V
I_{OH}	High-level output current			−0.4	mA
I_{OL}	Low-level output current			16	mA
T_A	Free-air operating temperature	0		70	°C

DC ELECTRICAL CHARACTERISTICS TABLE

Electrical characteristics over recommended operating free-air temperature

Symbol	Parameter	Conditions		MIN	TYP	MAX	Units
V_{OH}	High-level output voltage	V_{CC} = MIN, I_{OH} = MAX V_{IL} = MAX		2.4	3.4		V
V_{OL}	Low-level output voltage	V_{CC} = MIN, I_{OL} = MAX V_{IH} = MIN			0.2	0.4	V
I_{IH}	High-level input current	V_{CC} = MAX, V_I = 2.4 V				40	μA
I_{IL}	Low-level input current	V_{CC} = MAX, V_I = 0.4 V				−1.6	mA
I_{OS}	Short-circuit output curent	V_{CC} = MAX	DM54	−20		−55	mA
			DM74	−18		−55	
I_{CCH}	Supply current with outputs high	V_{CC} = MAX			4	8	mA
I_{CCL}	Supply current with outputs low	V_{CC} = MAX			12	22	mA
	Noise immunity					0.4	V

FIGURE 6.7
Page 2 of sample specification data sheet.

Storage Temperature Range   During storage, the IC should never be exposed to a temperature less than −10°C or higher than +75°C.

6.11   RECOMMENDED OPERATING CONDITIONS

This table, located in Fig. 6-7, lists the recommended operating conditions for the 7400 NAND gate IC made by the manufacturer. These include the following:

V_CC, Supply Voltage   The minimum to maximum power supply voltage limits used for the IC are listed along with the recommended supply level of 5 volts.

V_IH, High-Level Input Voltage   This is the minimum input logic level voltage (2.0 V) recognized by the IC as a valid logic 1 state signal. See Fig. 6-8.

V_IL, Low-Level Input Voltage   This is the maximum input logic level voltage (0.8 V) recognized by the IC as a valid logic 0 state signal. Any voltage above this is not accepted as a low by the logic circuit. See Fig. 6-8.

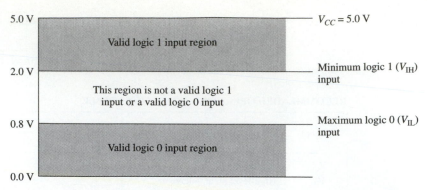

FIGURE 6.8
Valid and invalid input voltage levels.

Invalid region:

The voltage range between the minimum logic voltage level for a high state and the maximum logic voltage level for a low state.

Invalid Region: The voltage levels between V_{IH} and V_{IL} are called the **invalid region.** Any voltage applied to the gate inputs within this range could cause an unpredictable operation of the gate. The gate output may switch to the opposite state as soon as an input signal voltage enters the region or it might jump back and forth uncontrollably.

EXAMPLE 6.2

An IC has the following recommended operating condition values:

$$V_{IH} = 1.8 \text{ V}$$
$$V_{IL} = 0.6 \text{ V}$$

What are the valid voltage ranges for the 0 and 1 states applied to the input? What is the invalid voltage range?

Solution

0 state = 0 to 0.6 V

1 state = 1.8 to 5 V

Invalid = 0.7 to 1.8 V

Unused Input Connections: Figure 6-9(a) shows a NAND gate used as an inverter. This can be accomplished by applying a constant high at terminal A and using terminal B as the inverter input. One method of placing a high at the terminal A input is by leaving it disconnected, or floating. The problem with using a disconnected lead as a logic 1 input is that it can act as an antenna and pick up stray noise. Therefore, the recommended practice is to connect the lead to V_{CC}. *If a permanent 0 state is desired, a lead should be connected to ground.* See Figs. 6-9(b) and (c).

I_{OH}, High-Level Output Current   This is the amount ($-400 \mu A$) of current (conventional current) that flows from a 1 state output under specified load conditions. This is also known

Current sourcing:

Occurs whenever an output lead of a digital device has conventional current flowing out of it.

as **current sourcing.** (*Note:* When a minus sign appears on a current value, it indicates that the direction of conventional current is out of the device.)

   Current Sourcing   Whenever a gate output lead has current flowing out of it, the gate lead is said to *source current*. Figure 6-10 illustrates the current-sourcing action.

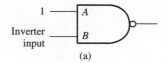

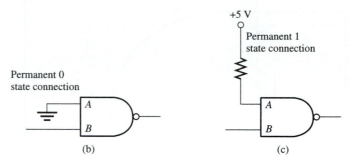

(a)

(b)   (c)

FIGURE 6.9
Unused input connections.

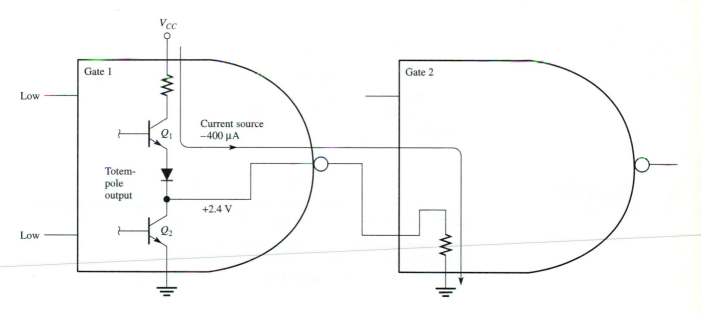

FIGURE 6.10
Current sourcing: the driving gate supplies current to the load gate in the high state.

When the output of gate 1 is in the high state, it supplies a current to gate 2. Therefore, the output of gate 1 acts as a current source for the input of gate 2.

I_{OL}, **Low-Level Output Current**   This is the amount (16 mA) of current (conventional current) that flows into a 0 state output under specified load conditions. This is also known as *current sinking.*

Current sinking:

Occurs whenever an output lead of a digital device has conventional current flowing into it.

   Current Sinking   Whenever a gate output lead has current flowing into it, the gate is said to *sink current*. Figure 6-11 illustrates the current-sinking action. When the output of

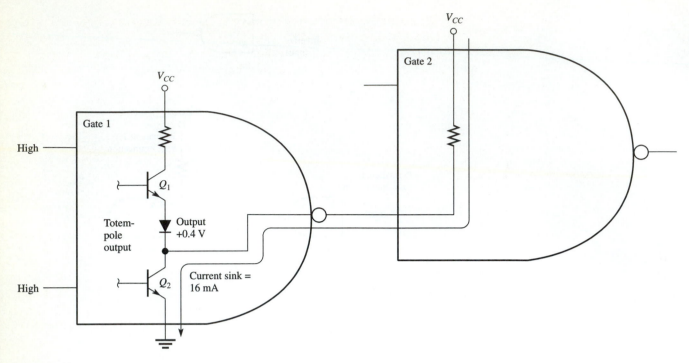

FIGURE 6.11
Current sinking: the driving gate receives current from the load gate in the low state.

gate 1 goes low, Q_2 turns on and provides a straight path for current to flow from the gate 1 output lead to ground. At the same time, the input circuitry of gate 2 acts like a resistance tied to $+V_{CC}$. Therefore, current flows from the input of gate 2 back through the output resistance of gate 1 to ground. In other words, the low state of the circuit driving an input of gate 2 must be able to sink a current coming from that input.

EXAMPLE 6.3

For the devices in Fig. 6-12, which is sinking and which is sourcing current? Which device causes the LED to illuminate the brightest and why?

Solution Device A is sourcing and device B is sinking. Device B causes the LED to illuminate the brightest because the low-level output current (I_{OL}) is 16 mA. With device A, the high-level output current (I_{OH}) is only -0.4 mA.

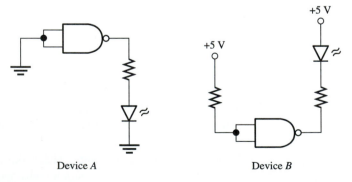

Device A Device B

FIGURE 6.12
Devices for Example 6.3.

T_A, Free-Air Operating Temperature This is the ambient temperature range of 0°C to 70°C that the IC should operate within.

The recommended operating conditions are values primarily controlled by the circuit designers. For example, the temperature range to which the IC is exposed is dictated by the designer. The designer also has control of what voltages are applied to the input leads. For example, if a transistor circuit is connected to the input of a gate, transistor circuit values should be selected that will generate an output voltage no higher than 0.8 volts for a (V_{IL}) low state or less than 2.0 volts for a (V_{IH}) high state. The designer also has control over how much current is sourced or sunk by selecting the resistance value of the load.

■ REVIEW QUESTIONS

Refer to the Absolute Maximum Ratings Table and the Recommended Operating Conditions Table of Fig. 6-7 to answer some of the questions.

21. The _____ _____ _____ Table provides information on the values beyond which the safety of the device cannot be guaranteed.

22. If the supply voltage exceeds _____ volts, the IC will be permanently damaged.

23. The input voltage of _____ volts or greater should never be applied to a gate input.

24. A 7400 IC should never be stored at temperatures below _____ or above _____ .

25. Under normal operating conditions, V_{CC} should never be below _____ volts or above _____ volts.

26. The minimum logic 1 state level voltage applied to a gate input is _____ volts, and the maximum logic 0 state level voltage applied to an input is _____ volts.

27. Any voltage applied to gate inputs within the _____ _____ could cause unpredictable operation of the gate.

28. An unused input should either be connected to _____ or _____ .

29. Whenever a gate lead has conventional current flowing out of it, the gate is said to _____ current. Whenever a gate lead has conventional current flowing into it, the gate is said to _____ current.

30. During operation, the IC should never be exposed to temperatures below _____ and above _____ .

6.12 DC ELECTRICAL CHARACTERISTICS TABLE

This table, located at the bottom portion of Fig. 6-7, lists the DC electrical characteristics of the 7400 NAND gate. These values are not normally controlled by the logic circuit designer. Instead, they are operation parameters guaranteed by the manufacturer if the values listed in the Recommended Operating Conditions Table are observed. These DC characteristics are as follows:

V_{OH}, High-Level Output Voltage This is the minimum voltage level (2.4 V) at a logic circuit output producing a logic 1 state. See Fig. 6-13. This specification is given only for active pull-up devices, which are also known as **totem-pole ICs.**

Totem Pole ICs identified as totem pole have a pair of internal transistors connected with their collector-to-emitter circuits in series, so one is on top of the other. The output terminal of the IC is connected to the junction between the two transistors. The totem-pole output design allows fast switching. The disadvantage of ICs with totem-pole output

Totem pole:

A pair of miniature transistors in an IC with their collector-to-emitter circuits in series to provide fast switching capabilities.

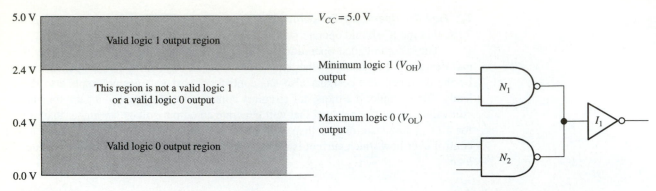

FIGURE 6.13
Valid and invalid output voltage levels.

FIGURE 6.14
Multiple outputs connected together.

configurations is that two or more of their outputs cannot be connected together as shown in Fig. 6-14.

It is sometimes desirable to connect the outputs of several logic gates in order to do the following:

1. Increase the number of gates it can drive (fan-out).
2. Perform some desired logic function.
3. Connect several devices to a common line (called a bus).

These are accomplished by using an *open-collector* IC.

Open Collector Integrated circuits with open collectors are specifically designed so that their outputs can be connected together. Figure 6-15 shows an open-collector inverter IC. It has the same function table as a regular inverter. However, the output is open when a logic 0 is applied to the input and the circuit is not completed unless an external pullup resistor of about 2.2 kiloohms is connected between the output and the supply voltage. Without the resistor, the IC will only produce a logic 0 state when a high is at the input, and cannot produce a logic 1 state when a 0 is at the input. Open-collector ICs can have more devices connected to their output than their totem-pole counterparts. They also allow multiple open-collector outputs to be connected together, as shown in Fig. 6-15. This process is called **wire anding.** These open-collector ICs can also be connected to a +12-volt V_{CC} through the pullup resistor, so that a +12-volt input can be the input of a CMOS IC.

Wire anding:

Occurs when multiple open collector outputs are connected together.

FIGURE 6.15
The 7405 hex inverter (open-collector output).

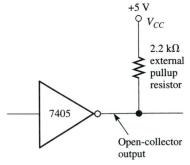

The disadvantages of an open-collector IC are as follows:

1. A discrete resistor is required.
2. The operating speeds are slower.
3. It is less immune to noise.
4. It is less able to drive large capacitive loads.

EXPERIMENT

Open-Collector Logic Devices

Objective

■ To observe the operation of an open-collector logic device.

Materials

1—+5-V DC Power Supply

1—7405 TTL IC

2—SPDT Switches

1—SPST Switch

2—LEDs

1—2.2-kΩ Resistor

1—68-Ω Resistor

1—220-Ω Resistor

Background Information

An inverter such as the one in Fig. 6-16 is designed to produce a logic state at its output that is opposite to that applied to its input. However, the 7405 IC inverter is classified as an open-collector device. A pullup resistor must be connected between its output and a +5-V supply. Without it, the inverter output will not be able to supply a logic high state when a low is applied to its input. Instead, it produces no logic signal, and functions as if there is an open output lead.

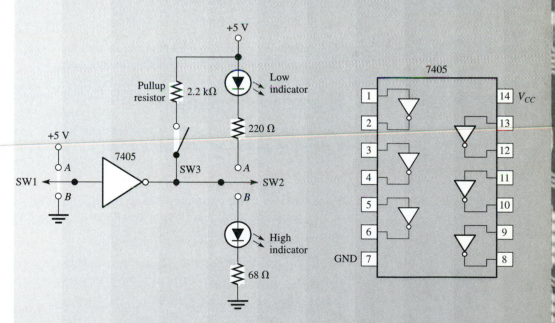

FIGURE 6.16

Both inverters and logic gates are available with open collectors. They are used primarily to allow greater flexibility in design, such as when a number of outputs are tied together. Regular logic devices operate unpredictably when their outputs are connected to a common line.

Step 1. Assemble the circuit shown in Fig. 6-16.

Procedure Information

The purpose of the LEDs is to indicate which type of logic state is produced at the output lead of the inverter. For example, by placing SW2 at position *A*, the LED labeled "Low Indicator" will illuminate if a 0 state is present at the inverter output. Placing SW2 at position *B* will cause the LED labeled "High Indicator" to illuminate if a 1 state is produced by the inverter. If an LED does not turn on when SW2 is placed in one of the *A* or *B* positions, the inverter functions as if there is an open output lead.

Step 2. To observe how an open-collector device functions without a pull-up resistor, open switch 3 (SW3).
Step 3. Place SW1 in the *A* position. A logic _____ (low, high) is at the inverter input. Observe the low and high indicators as you place SW2 in positions *A* and *B*. A _____ (low, high) is at the inverter output.
Step 4. Place SW1 in the *B* position. A logic _____ (low, high) is at the inverter input. Observe the low and high indicators as you place SW2 in positions *A* and *B*. A/n _____ (low, high, open) is at the inverter output.
Step 5. To observe how an open-collector device functions with a pull-up resistor connected, close SW3.
Step 6. Place SW1 and SW2 in the *A* position. A logic _____ (low, high) is at the inverter input, and a _____ (low, high) is at the inverter output.
Step 7. Place SW1 and SW2 in the *B* position. A logic _____ (low, high) is at the inverter input, and a _____ (low, high) is at the inverter output.

The inverter operates normally when a pull-up resistor is connected. Without a pull-up resistor, the inverter's output is an open when a logic low is applied to the input, and a logic low when a high is at the input.

Procedure Question 1

Experiment Steps 2–4 show that the inverter cannot produce a logic _____ (low, high) at its output when the pullup resistor is not connected.

■ EXPERIMENT QUESTION

1. An open-collector inverter without a pullup resistor produces a(n) _____ (low, high, open) at its output when a logic low is applied to its input.

V_{OL}, Low-Level Output Voltage   This is the maximum voltage level (0.4 V) at a logic circuit output producing a logic 0 state. See Fig. 6-13. (*Note:* Any voltages between V_{OL} and V_{OH} are considered invalid.)

I_{IH}, High-Level Input Current   This is the maximum input current (conventional current) that flows into the IC (40μA) when the input logic level voltage equals a 1 state of 2.4 volts.

I_{IL}, Low-Level Input Current   This is the current (−1.6 mA) that flows out of the input (conventional current) when the input logic level voltage equals a 0 state of 0.4 volts.

I_{OS}, Short-Circuit Output Current   With the IC output at a 1 state, the device that it is connected to may short to ground. This specification provides information about how much output current (conventional current) (−20 to −55 mA) can be allowed to flow out of the gate output when such a short-circuit condition exists.

I_{CCH}, Supply Current with Output High   This is the amount of current the dc power supply (V_{CC}) must furnish when one output is at a high state (see Fig. 6-17).

FIGURE 6.17
Supply current at logic 1.

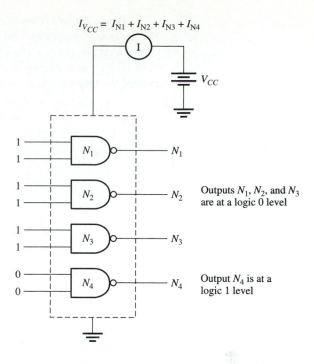

$$I_{V_{CC}} = I_{N1} + I_{N2} + I_{N3} + I_{N4}$$

Outputs N_1, N_2, and N_3 are at a logic 0 level

Output N_4 is at a logic 1 level

EXAMPLE 6.4

If I_{CCH} is 8 mA, what will be the supply current if three gate outputs are high?

Solution   When three outputs are high, the power supply must provide three times as much current as the value given on the data sheet for I_{CCH}. Therefore,

$$3 \times 8 \text{ mA} = 24 \text{ mA}$$

I_{CCL}, **Supply Current with Output Low**   This is the amount of current the DC power supply (V_{CC}) must furnish when one output is at a low state (see Fig. 6-18). When two outputs are low, the power supply must provide twice as much current as the value given by the data sheet.

FIGURE 6.18
Supply current at logic 0.

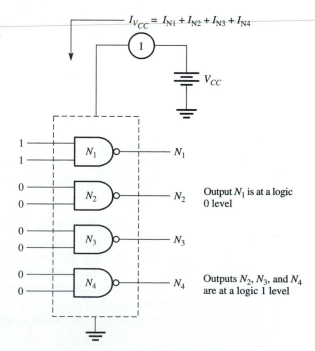

$$I_{V_{CC}} = I_{N1} + I_{N2} + I_{N3} + I_{N4}$$

Output N_1 is at a logic 0 level

Outputs N_2, N_3, and N_4 are at a logic 1 level

Power Supply Noise: In response to any change in the output voltage level of a TTL IC, the power supply current (I_{CC}) can fluctuate widely. Figure 6-19(a) illustrates the I_{CC} surges that are created whenever the output changes. These fluctuations are the result of both IC transistors being partially on or off for a brief period of time. If these glitches exceed the tolerance level of the power supply, the IC may not function properly.

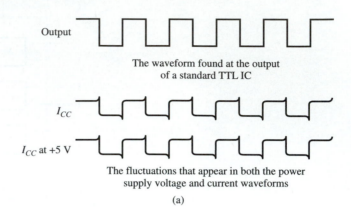

Output

The waveform found at the output
of a standard TTL IC

I_{CC}

I_{CC} at +5 V

The fluctuations that appear in both the power
supply voltage and current waveforms

(a)

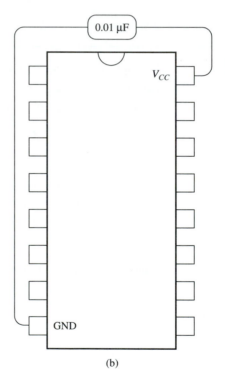

0.01 µF

V_{CC}

GND

(b)

FIGURE 6.19
(a) The waveforms and fluctuations for a square wave. (b) A capacitor connected between V_{CC} and ground to eliminate glitches.

To eliminate these glitches, a bypass capacitor is connected between the +5-V and ground of the IC. A rule of thumb is to place one 0.1-µF capacitor across these lines for every three ICs. See Fig. 6-19(b).

Noise Immunity: Stray electromagnetic fields can induce voltages on the connecting circuit paths between logic gates. These unwanted voltage signals are called *noise*. The primary source of noise that affects digital circuits originates externally and is transferred into the equipment by power supply lines. Other sources of noise come from static discharge that creates electromagnetic radiation. Each IC logic device has some noise immunity, which is its ability to tolerate noise voltages on its inputs. These noise signals can cause problems for gate devices.

Figure 6-20 shows the difference between the output logic voltage levels and the input logic voltage levels of a 7400 TTL IC. The minimum logic 1 output voltage is 2.4 volts (V_{OH}) and the minimum input logic voltage level is 2.0 volts (V_{IH}). A voltage range of 0.4 volts, called the *noise margin,* exists between the output and input levels. This margin allows for noise up to 0.4 volts to be induced into the output-to-input circuit path. For example, suppose a negative-going noise signal of -0.5 volts cancels a positive 2.4-volt logic 1 output to 1.9 volts. It would apply a disallowed voltage to the input of the next gate. Likewise, a similar problem can exist when the output of a gate sends a low to an input of the next gate. The maximum logic 0 output voltage is 0.4 volts (V_{OL}) and the maximum logic 0 input voltage is 0.8 volts (V_{IL}), resulting in a noise margin of 0.4 volts. If a positive-going noise signal of 0.5 volts is added to a positive 0.4-volt logic 0 output (V_{OL}), it would cause a 0.9-volt disallowed voltage to be applied to the input of the next gate.

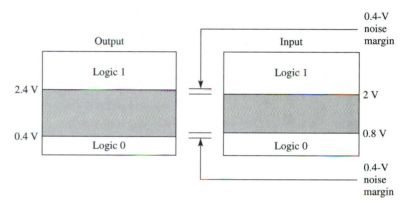

FIGURE 6.20
Noise margin between a TTL NAND-gate input and output (known as noise immunity).

The noise margin can be improved by circuit engineers if they design circuits so that a logic 1 output (V_{OH}) is kept well above V_{IH} of the next gate and a logic 0 output (V_{OL}) is kept well below V_{IL} of the next gate.

EXAMPLE 6.5

What is the noise margin of the two circuits of Fig. 6-21?

FIGURE 6.21
Circuits for Example 6.5.

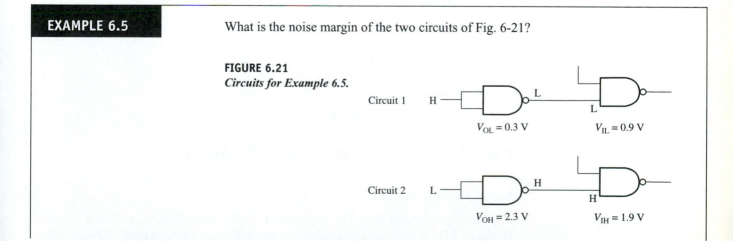

Solution   For circuit 1,

$$V_{IL} = 0.9 \text{ V}$$
$$V_{OL} = \underline{-0.3 \text{ V}}$$
$$0.6 \text{ V or } 600 \text{ mV}$$

For circuit 2,

$$V_{OH} = 2.3 \text{ V}$$
$$V_{IH} = \underline{-1.9 \text{ V}}$$
$$0.4 \text{ V or } 400 \text{ mV}$$

EXAMPLE 6.6

What is the largest negative-voltage noise spike that can be induced into the line connecting the output of one gate with a V_{OH} rating of 2.3 volts to the input of the next gate with a V_{IH} rating of 2.1 volts without causing the circuit to operate unpredictably?

Solution   The noise margin is determined by subtracting V_{IH} from V_{OH}.

$$V_{OH} = 2.3 \text{ V}$$
$$V_{IH} = \underline{-2.1 \text{ V}}$$
$$0.2 \text{ V negative spike}$$

EXPERIMENT

Input and Output Logic Voltage Levels

Objectives

- To obtain information about the operational characteristics of ICs from a data manual.
- To measure the input and output logic voltage levels of an IC.
- To determine the noise margin.

Materials

    1—Logic Data Manual for TTL ICs

    1—7404 IC

    1—LED

    1—10-Ohm Resistor

    1—68-Ohm Resistor

    1—300-Ohm Potentiometer

    1—+5-V DC Power Supply

    1—Multimeter

    1—Dual-Trace Oscilloscope

PART A—INPUT/OUTPUT VOLTAGE LEVELS

Background Information

The ideal voltage value for a logic high state is 5 volts, and 0 volts is ideal for a low state. However, TTL logic devices are designed to operate within a certain range of these voltage values. These voltage ranges are different for input logic states compared to output logic states.

Input Logic States

Step 1. Examine the data sheet for a TTL 7404 hex inverter IC. The minimum voltage at which one of the inverters accepts a logic high state at its input is identified by V_{IH}. This abbreviation indicates "voltage input when high." The minimum logic high input voltage listed is +2 V.

The maximum voltage at which the 7404 IC inverter accepts a logic low state at its input is identified by V_{IL}. This abbreviation indicates "voltage input when low." What is the maximum logic low input voltage listed in the manual?

_____ V

Output Logic States

Step 2. The minimum voltage at which a 7404 IC produces a logic high state at its output is identified by V_{OH}. This abbreviation indicates "voltage output when high." The minimum logic high output voltage listed in the data sheet is

_____ V

Procedure Question 1

The maximum voltage at which a 7404 IC produces a logic low state at its output is identified by the letters _____, which indicates _____ _____ _____ _____.

The maximum logic low output voltage listed in the data sheet is _____.

These voltage levels have been established by the manufacturer as guaranteed minimum and maximum values. However, you may find that the ICs operate at voltages outside these parameters.

Procedure

Step 3. Assemble the circuit shown in Fig. 6-22.

Figure 6.22

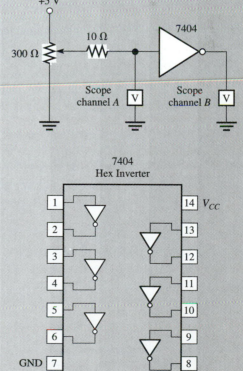

Step 4. Turn the power on.

Determining V_{OH} and V_{OL}

Step 5. Adjust the potentiometer until the horizontal line on the oscilloscope at the IC output, which is a DC voltage, is at its maximum reading. This is V_{OH}.

$$V_{OH} = \underline{\hspace{2cm}}$$

Step 6. Adjust the potentiometer until the horizontal line on the oscilloscope at the IC output is at its minimum value. This is V_{OL}.

$$V_{OL} = \underline{\hspace{2cm}}$$

Determining V_{IL}

Step 7. Assemble the circuit shown in Figure 6-23.

Figure 6.23

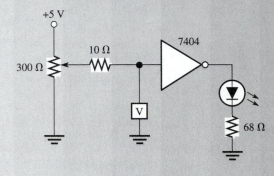

Step 8. Turn the power on.

Step 9. Test the circuit by adjusting the potentiometer so that the voltmeter reads +5 volts. The LED should be off. Now adjust the potentiometer until the voltmeter reading is at its lowest value. The LED should be on.

Step 10. Start with the potentiometer set so that 0 volts is read by the voltmeter. The LED should be on. Slowly turn the potentiometer to increase the voltage at the inverter input. Carefully observe the LED. When it begins to dim, record the voltage reading for V_{IL}.

$$V_{IL} = \underline{\hspace{2cm}}$$

When the LED begins to dim, the inverter is starting to switch to the opposite state. This condition develops because the voltage applied by the potentiometer is rising above the maximum voltage level of an acceptable low at the inverter.

Determining V_{IH}

Step 11. Assemble the circuit shown in Fig. 6-24.

FIGURE 6.24

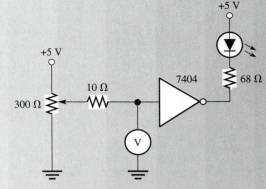

Step 12. Turn the power on.

Step 13. Test the circuit by adjusting the potentiometer so that the voltmeter reads +5 volts. The LED should be on. Now adjust the potentiometer until the voltmeter reading is at its lowest value. The LED should be off.

Step 14. Start with the potentiometer set so that the input is at +5 volts. Turn the potentiometer to reduce the voltage. Carefully observe the LED. When it first begins to dim, record the voltage reading for V_{IH}.

$$V_{IH} = \underline{\hspace{2cm}}$$

When the LED begins to dim, the inverter is starting to switch to the opposite state. This condition exists because the voltage applied by the potentiometer is dropping below the minimum voltage level of an acceptable high at the inverter input.

Step 15. Using a data manual for logic TTL ICs, find the following values for the 7404 IC inverter. Compare them to the measured values obtained by your findings.

Data Book Values	Measured Values
V_{OH} ——————	V_{OH} ——————
V_{OL} ——————	V_{OL} ——————
V_{IL} ——————	V_{IL} ——————
V_{IH} ——————	V_{IH} ——————

Note: Consult your instructor if your readings are not within the voltage ranges specified by the manufacturer.

PART B—NOISE MARGINS

Background Information

Logic devices are designed to operate so that the V_{OH} value is higher than the V_{IH} value. For example, the rated V_{OH} value of 2.4 volts is higher than the rated V_{IH} value of 2.0 volts. This margin allows stray voltages of up to −0.4 volts to be induced into the output-to-input circuit path between the gates. Likewise, the V_{OL} rating of 0.4 volts is lower than the V_{IL} rating of 0.8 volts. This margin allows stray voltages of up to +0.4 volts to be induced into the output-to-input circuit paths between the gates. These unwanted signals are called *noise.* Noise is caused by electric devices such as motors, which produce strong magnetic fields around them. The difference between the output and input high and the output and input low is called the *noise margin.*

Step 16. Determine the following noise margins from the measurements obtained in Steps 5, 6, 10, and 14.

Logic High Noise Margin

$$V_H = V_{OH} - V_{IH} = \underline{\hspace{2cm}} - \underline{\hspace{2cm}} = \underline{\hspace{2cm}}$$

Logic Low Noise Margin

$$V_L = V_{IL} - V_{OL} = \underline{\hspace{2cm}} - \underline{\hspace{2cm}} = \underline{\hspace{2cm}}$$

■ EXPERIMENT QUESTIONS

1. V_{OH} should be _____ (>, <, =) V_{IH}. V_{OL} should be _____ (>, <, =) V_{IL}.

2. Stray voltages induced into output-to-input circuit paths between gates are called _____.

3. If V_{OH} equals 2.2 volts and the V_{IH} rating is 1.9 volts, the noise margin is _____ volts.

■ REVIEW QUESTIONS

Refer to Fig. 6-7 to answer some of these questions.

31. The minimum logic 1 state level voltage produced at the gate output is _____ volts, and the maximum logic 0 state level voltage generated at the output is _____ volts.

32. Compared to an open-collector configuration, the totem-pole design allows (slower or faster) switching speed.

33. The open-collector configuration utilizes an _____ resistor so that several gate _____ can be connected to one gate _____.

34. The maximum input current that flows into the gate input when the input logic level input equals 2.4 volts is _____.

35. The allowable current that flows out of the gate output lead when a short-circuit condition exists is from _____ to _____.

36. The maximum amount of current the DC power supply (V_{CC}) must furnish when one output is at a high state is _____.

37. Power supply noise problems can be eliminated by placing a capacitor across _____ and _____ pins of an IC.

38. The amount of stray voltage that may be induced into the line connecting an output of one gate to an input of another gate is called _____ _____.

6.13 SWITCHING CHARACTERISTICS TABLE

The Switching Characteristics Table (Fig. 6-25) provides the switching, or AC, characteristics of the standard TTL 7400 IC with V_{CC} at +5 volts and the T_A at 25°C. The values given are referred to as *propagation delay* times.

SWITCHING CHARACTERISTICS TABLE
Switching characteristics at $V_{CC} = 5$ V and $T_A = 25°C$

Parameter	$C_L = 15$ pF $R_L = 400\ \Omega$			Units
	MIN	TYP	MAX	
t_{PLH} propagation delay time low-to-high level output		12	22	ns
t_{PHL} propagation delay time high-to-low level output		7	15	ns

INPUT AND OUTPUT LOADING TABLE

Parameter	MIN	MAX
Fan-out		10
Fan-in		1

FIGURE 6.25
Page 3 of sample specification data sheet.

Propagation delay:

The delay time that takes place between when an input transition of an IC takes place until the output changes states.

Propagation Delay Integrated circuits do not react instantaneously to a signal applied to an input. Instead, some delay exists between when a logic transition takes place at an input until a change is made at the output. This reaction time, called *propagation delay,* determines the maximum speed of operation of an IC, which is an important characteristic to consider when designing digital circuits. Using an IC that cannot respond fast enough to a high-frequency square wave applied to its input would likely create problems.

Figure 6-26 graphically illustrates how a logic signal experiences a delay in going through a logic circuit. The propagation delay is measured from the point when the input logic voltage reaches 50% of its change to the point when the output logic voltage has attained 50% of its total change. There are two propagation delay times to consider and they are listed in the Switching Characteristics Table:

T_{PLH}, **Propagation-Delay-Time Low-to-High-Level Output**   This is the delay time that takes place from the point when an input transition takes place until the output goes from a logic 0 to a logic 1 state (low to high).

T_{PHL}, **Propagation-Delay-Time High-to-Low-Level Output**   This is the delay time that takes place from the point when an input transition takes place until the output goes from a logic 1 to a logic 0 state (high to low).

FIGURE 6.26
Propagation delay times.

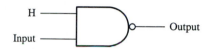

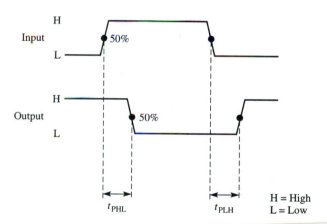

Generally, T_{PHL} and T_{PLH} are not the same value, and both vary depending on load conditions. Typically, propagation delay is measured in nanoseconds. A circuit with a short delay, such as 20 ns, operates faster than one with a 40-ns delay. Another term commonly used for propagation delay is *switch time*.

When several logic devices are cascaded, it often becomes necessary to determine the circuit's total propagation delay time, T_{PD}. This value is found by adding the worst-case delay for each logic device.

EXAMPLE 6.7

For the circuit of Fig. 6-27, the following propagation delay times are given:

$$\text{NOR gate: } T_{PHL} = 30 \text{ ns}$$
$$T_{PLH} = 25 \text{ ns}$$
$$\text{AND gate: } T_{PHL} = 17 \text{ ns}$$
$$T_{PLH} = 14 \text{ ns}$$

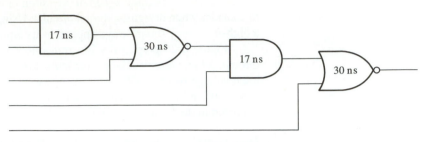

FIGURE 6.27
Circuit for Example 6.7.

Calculate the T_{PD}.

Solution This value is found by adding the longest, or worst-case, delay for each gate.

NOR-gate worst-case delay: 30 ns $\times$ 2 NOR gates = 60 ns
AND-gate worst-case delay: 17 ns $\times$ 2 AND gates = 34 ns

T_{PD} = 60 ns + 34 ns = 94 ns

EXPERIMENT

Propagation Delay

Objective

- To obtain information about the operational characteristics of ICs from a data manual.
- To determine propagation delay of an IC.

Materials

1—+5-V DC Power Supply

1—Square Wave Generator

1—Dual-Trace Oscilloscope

1—7404 IC

Background Information

Integrated circuits do not react instantaneously to a signal applied to an input. Instead, some delay exists from when a logic transition takes place at an input until a change is made at the output. This reaction time, called *propagation delay,* determines the maximum speed at which an IC can operate. There are two different propagation delay times to consider:

T_{PLH} = Propagation delay time low-to-high *output.*
T_{PHL} = Propagation delay time high-to-low *output.*

Generally, T_{PLH} and T_{PHL} are not the same value; however, they are very close. Typically, propagation delay is measured in nanoseconds. The lower the number, the faster the operation.

Procedure Question 1

The T_{PLH} and the T_{PHL} values pertain to the _____ (input, output) of the logic device.

Step 1. Connect the six inverters in series as shown in Fig. 6-28(a).

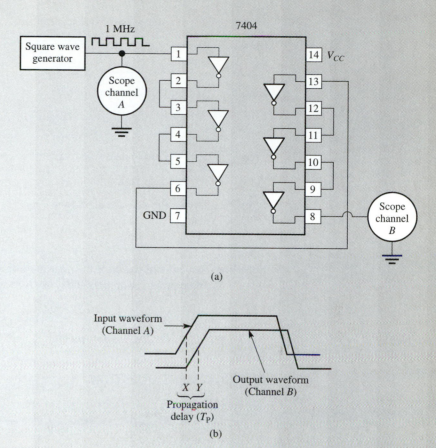

(a)

(b)

FIGURE 6.28

Step 2. Turn the power on.
Step 3. Apply a 1-MHz square wave signal to the input of the first inverter.
Step 4. Connect channel A of the oscilloscope to the input of the first inverter and channel B to the output of the sixth inverter. Make the necessary scope adjustments to obtain a waveform similar to the one shown in Fig. 6-28(b).

Procedure Information

Because the propagation delay of the inverter is only a few nanoseconds, it is difficult to detect such a short time interval on the oscilloscope. By cascading six inverters in series, the cumulative propagation delays of all six can be seen on the scope. Three of them are T_{PHL}, and the other three are T_{PLH}. By dividing the cumulative time interval by six, the propagation delay of one inverter can be found. This value is not T_{PHL} or T_{PLH}, but the average of the two.

Step 5. The following sequence of steps must be observed when reading the propagation delay of the six inverters:

    (a) Count the horizontal divisions between the two measured points, X and Y, as shown in Fig. 6-28(b).
    (b) Multiply the number of divisions between the two measured points X and Y times the TIME/DIV control setting to determine the total propagation delay of the six inverters.

Step 6. Divide the time delay of the six inverters by 6 to find the time delay for one inverter.

Time delay of one inverter _____

Step 7. Using the data book, find the T_{PHL} and T_{PLH} value of the 7404 IC.

$T_{PHL} =$ _____ $T_{PLH} =$ _____

Step 8. Multiply each propagation delay by 3.

$3 \times (T_{PHL}) =$ _____
$3 \times (T_{PLH}) =$ _____

Step 9. Add the products of the two numbers multiplied in Step 8.

_____

Step 10. Divide the sum found in Step 9 by 6 to obtain the average of the T_{PHL} and T_{PLH} values:

_____

Step 11. Record and compare the measured propagation time delay from Step 6 and the rated propagation time delay from Step 10.

Measured T_P Data Manual T_P

_____ nanoseconds   _____ nanoseconds

■ EXPERIMENT QUESTION

1. The time it takes for a gate output to change states after a logic transition at its input occurs is called _____, and it is measured in _____ seconds.

6.14   INPUT AND OUTPUT LOADING AND FAN-OUT TABLE

Fan-Out

Fan-out:

The number of digital inputs that one output of a digital device can reliably drive.

In many situations, a logic circuit is required to drive several logic inputs. The maximum number of logic inputs that one output can drive reliably is called **fan-out.** The fan-out of the 7400 TTL NAND gate is 10 because the I_{OL} or I_{OH} rating is exceeded if more than 10 inputs are connected to the output. Figure 6-29 illustrates this rule by showing 10 gate inputs connected to one gate output.

The I_{OL} sink rating is 16 mA for gate 1, and the I_{IL} rating for individual gates 2 through 11 is -1.6 mA. Therefore, the output of one gate can sink the input currents of 10 gates. The fan-out rating can also be calculated by dividing the output current by the input current.

EXAMPLE 6.8

For logic 0, output current $I_{OL} = 16$ mA and input current $I_{IL} = -1.6$ mA. Find the fan-out (0 state).

Solution

$$\frac{\text{Output current } (I_{OL})}{\text{Input current } (I_{IL})} = \frac{16 \text{ mA}}{-1.6 \text{ mA}} = 10$$

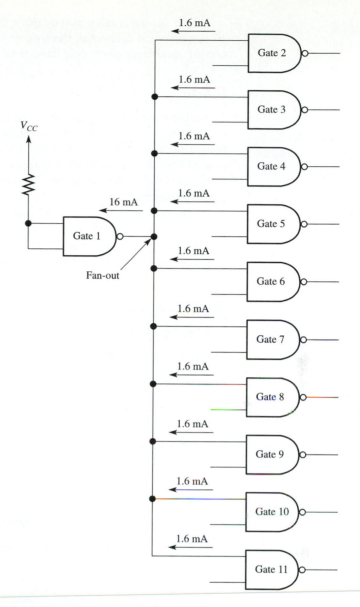

FIGURE 6.29
Fan-out of 10, showing origins of all currents. Fan-out = 16 mA/−1.6 mA = 10 devices.

A 7400 TTL IC output becomes a current source when it is at a high (I_{OH}) and outputs −400 μA. When a 7400 input is driven high (I_{IH}), its current is 40μA. Therefore, a high output of one gate can source the input current of 10 gates.

EXAMPLE 6.9	For logic 1, output current I_{OH} = 400 μA and input current I_{IH} = 40 μA. Find the fan-out (1 state).

Solution

$$\frac{\text{Output current } (I_{OH})}{\text{Input current } (I_{IH})} = \frac{-400 \text{ μA}}{40 \text{ μA}} = 10$$

The fan-out is calculated by using the maximum output current rating. If more than 10 outputs were connected, the sink or source current at the output lead would be exceeded, resulting in permanent damage to a gate output.

Fan-In

Fan-in:

The number of devices that can be connected to the input on a digital device.

This is the number of devices that can be connected to the gate input. Generally, **fan-in** is 1.

To work effectively, users of digital ICs need to know and understand how to use these data sheets. Engineers use this information when designing circuits. Technicians who troubleshoot these devices must also rely on this information. Without specifications with which to compare measurements, a troubleshooter has no way to tell if a device is performing properly.

The preceding sections on IC specification sheets provide an abbreviated example of the information commonly found in data manuals. These do not provide, however, all the possible information about an IC device. Additional information can be obtained by contacting IC manufacturers, who are often willing to give advice by telephone.

■ REVIEW QUESTIONS

Refer to Fig. 6-25 to answer some of these questions.

39. The delay time between when an input transition takes place until the output goes from a logic 1 to a logic 0 state is called T _____, _____ _____ time.
40. A circuit that operates at a switching time of 30 ns is (faster or slower) than one that operates at 40 ns.
41. T_{PHL} is normally different from T_{PLH}. True or False?
42. The number of gate inputs that can be connected to one gate output is called _____ and the maximum number for a 7400 TTL IC device is _____.
43. The fan-in of the 7400 IC is _____.

EXPERIMENT

Input and Output Current Ratings

Objectives

- To obtain information about the operational characteristics of ICs from a data manual.
- To measure the sinking and sourcing values of an IC.
- To determine fan-out.

Materials

       1—+5-V Power Supply

       1—Ammeter

       1—120-Ohm resistor

       4—LEDs

       2—7400 NAND Gate ICs

Background Information

The TTL data manuals specify that a logic device such as the 7400 NAND gate is capable of providing -400 μA of output current (labeled I_{OH}) when it produces a logic high state at its output. This rating is based on conventional current, which flows from a positive to a more negative potential. Because current flows out of the output, this is called *current sourcing*. The negative sign indicates that current is flowing out of the terminal. The input that the output is driving also has a manufacturer's rating (labeled I_{IH}), which is 40 μA when a logic high is applied to it.

When the NAND gate output is low, conventional current flows into it. This is called *current sinking*. The manufacturer's rating for current flowing into it is 16 mA, and it is labeled I_{OL}. The input of the device that the NAND gate output is driving supplies the current. It has a current rating that is labeled I_{IL}, and it is −1.6. mA when a logic low is present.

Determining Current I_{IL}

Step 1. Assemble the circuit shown in Fig. 6-30. The LED at the output of any gate connected to N_1 will be off. The LED at the output of any gate not connected to N_1 will be on.

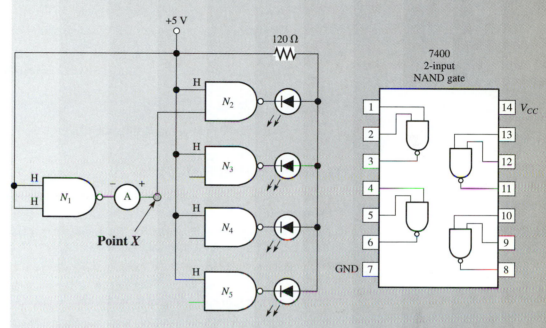

FIGURE 6.30

Step 2. Turn the power on.
Step 3. Record the I_{IL} current supplied to N_1 from N_2 by taking the meter reading on the ammeter.

$$N_2\ I_{IL} = \text{_____}$$

Procedure Question 1

The output device is current _____ (sourcing, sinking).

Step 4. Connect a jumper wire from point X to the bottom lead of N_3 and record the I_{IL} current supplied to N_1 from N_2 and N_3.

$$N_2 + N_3\ I_{IL} = \text{_____}$$

Step 5. Repeat Step 4 by connecting N_4 to point X.

$$N_2 + N_3 + N_4\ I_{IL} = \text{_____}$$

Step 6. Repeat Step 4 by connecting N_5 to point X.

$$N_2 + N_3 + N_4 + N_5\ I_{IL} = \text{_____}$$

Step 7. Using the data manual, find the rating of the 7400 IC for I_{OL}.

Data manual rating for $I_{OL} =$ _____

Procedure Question 2

Based on the I_{IL} current readings obtained in Steps 6 and 7, how many NAND gate inputs can be connected to N_1 before its I_{OL} rating is exceeded?

This value is referred to as *fan-out,* which is defined as the number of gate inputs that the output of a gate can drive.

Step 8. Using the data manual, obtain the values in the following formula and calculate the fan-out rating for the 7400 IC when its output is low.

$$\text{Fan-out} = \frac{I_{OL}}{I_{IL}}$$

Data Manual fan-out rating for a logic low state = _____

Step 9. Using the data manual, obtain the values in the following formula and calculate the fan-out rating for the 7400 IC when its output is high.

$$\text{Fan-out} = \frac{I_{OH}}{I_{IH}}$$

Data Manual fan-out rating for a logic high state = _____

■ EXPERIMENT QUESTIONS

1. Conventional current that flows out of the output is called current _____ (sourcing, sinking).
2. An I_{IL} current value with a negative sign indicates that conventional current is flowing _____ (into, out of) the gate input terminal.
3. The number of gate inputs that the output of a gate can drive is called _____.
4. Suppose the I_{OL} rating of gate A is 12 mA. If the inputs of each gate connected to its output use 2 mA of current, what is the fan-out of gate A?

6.15   SELECTING ICs

There is no single IC family that can achieve maximum performance in every one of the categories listed in the sample data sheet. Each family is usually strong in some areas and weaker in others. Often, to strengthen a performance characteristic of one IC, another has to suffer. For example, the reduced power consumption of a TTL subfamily IC usually results in reduced speed. Thus, each IC family has its own predominant characteristics.

TABLE 6.2
IC family comparison chart

Characteristic Parameters	Std TTL 74XX	Schottky TTL 74SXX	Low-Pwr Schottky TTL 74LSXX	Adv Low-Pwr Schottky TTL 74ALSXX	Fast TTL 74FXX	ECL	CMOS 4000	CMOS 74HCT	CMOS 74AC	PMOS	NMOS
Supply voltage	5	5	5	5	5	−5.2	+3 to 18	2 to 5	2 to 5	±12	+5
Propagation delay (ns)	9	3	9.5	4	2.75	1	25	10	4.7	300	50
Power dissipation (mW per gate)	10	19	2	1.25	5.5	60	0.01 to 1	1.75	2.1	1.7	1
Fan-out	10	10	20	20	33	10	≥50	10	60	20	>10

For a certain application, certain characteristics are desired, and a family is chosen that can meet those requirements. Table 6-2 provides a summary of the more important IC characteristics so that comparisons can be made between various IC families.

6.16   INTERFACING ICs

Interfacing:

Electronically making different circuits compatible with one another in terms of speed, power dissipation, and logic level voltages.

All of the devices in a logic family are compatible with each other. That is, they allow direct connection without the need for any special interfacing device. **Interfacing** in electronics means making different circuits compatible with one another in terms of speed, power dissipation, logic level voltages, etc. Different IC families often are not compatible with each other and, therefore, one cannot be directly interfaced to another. Yet, in practically every type of digital equipment, interfacing is required. For example, there are some situations when a particular type of digital equipment may require one section to operate differently than another. A memory section might require CMOS ICs because of their density and large storage capabilities. Another section that processes data may require high-speed TTL circuitry to perform fast calculations. Therefore, it is necessary to connect the ICs in one section to the ICs in the other section through some type of interface scheme.

There are many possible ways to interface circuits. For an idea of how this is accomplished, examples of how to interface the two most popular families, CMOS to TTL and TTL to CMOS, are provided. To find out how to interface other logic families, application notes or application handbooks that provide interface recommendations can be obtained from most manufacturers. Information about how to interface the different types of logic families to the outside world is also available in these handbooks.

Interfacing TTL to CMOS

Each of these IC families operates at different logic level voltages. Therefore, at the lead that joins them, additional electronic components need to be connected to make them compatible.

When interfacing a TTL to a CMOS IC using the same +5-V power supply, a pull-up resistor is required, as shown in Fig. 6-31. The pull-up resistor is used to ensure that a logic 1 is of sufficient voltage amplitude to operate the CMOS IC properly. This is because the logic level voltage is greater for a CMOS than for a TTL logic 1.

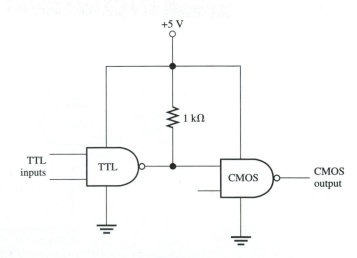

FIGURE 6.31
Interfacing TTL to CMOS using one power supply.

When two different power supplies are used, a transistor circuit may be used to isolate, or buffer, the TTL IC from the CMOS IC (see Fig. 6-32). The transistor is also needed to drive the CMOS IC input.

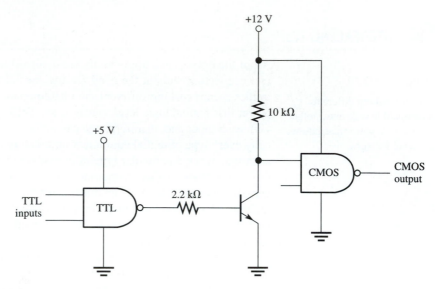

FIGURE 6.32
Interfacing TTL to CMOS using two power supplies.

Interfacing CMOS to TTL

CMOS ICs can directly drive low-power TTL ICs. However, when interfacing CMOS with the other TTL ICs, it is probably easier to use a special buffer/converter IC, such as the CD4049 or CD4050.

With the same +5-V power supply, the CMOS buffer/converter and TTL ICs are connected to the same voltage, as shown in Fig. 6-33. When using different power supply voltages, the buffer/converter IC is connected to the voltage used with the TTL IC, as shown in Fig. 6-34.

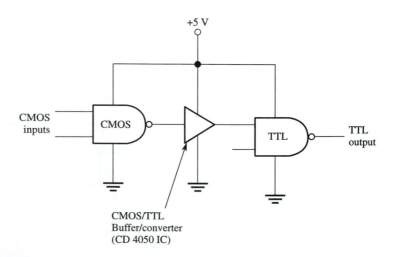

FIGURE 6.33
Interfacing CMOS to TTL using one power supply.

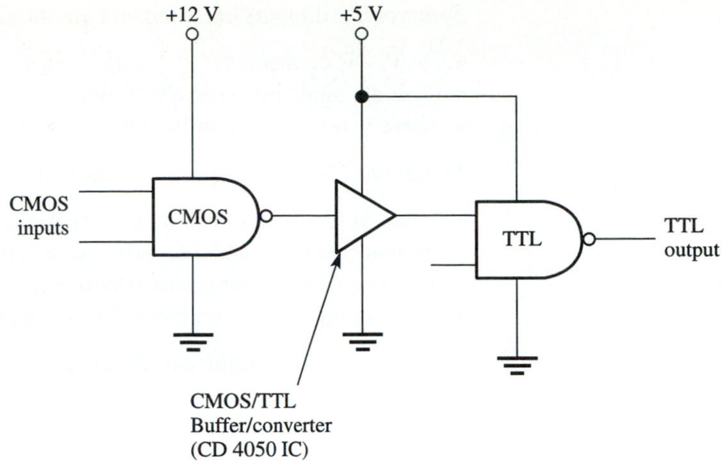

FIGURE 6.34
Interfacing CMOS to TTL using two power supplies.

■ REVIEW QUESTIONS

44. TTL devices are best suited for which of the following?
(a) Low power consumption
(b) Fast operating speeds
(c) Good noise immunity
(d) Fan-out
(e) Circuit density

45. CMOS circuitry is especially weak in which of the following areas?
(a) Dense complex circuitry
(b) Very short propagation delays
(c) Low power consumption
(d) Fan-out
(e) Good noise immunity

46. Which of the following families provides the shortest propagation delays?
(a) Low-power Schottky
(b) Standard TTL
(c) ECL

47. Which of the following is the fastest operating MOS family?
(a) NMOS
(b) PMOS
(c) CMOS

48. What is the method used to connect TTL ICs and CMOS ICs together called?
(a) Buffering
(b) Current sourcing
(c) Current sinking
(d) Interfacing
(e) Coupling

6.17 TROUBLESHOOTING

Some of the defects that are commonly associated with integrated circuits are as follows:

Symptom: Chip very hot. *Problem:*

■ IC is inserted in the wrong direction.
■ IC is internally shorted.

Symptom: All inputs are good and all outputs show nothing. *Problem:*

- No V_{CC} is connected.
- IC is damaged by static electricity.
- There is no external pullup resistor for the open-collector IC.

Symptom: Circuit operation intermittently changes from good to bad. *Problem:*

- There is a slight crack in the internal IC interconnecting circuit path. Spray circuit coolant for one to two seconds and the circuit should change its symptom before returning to its original symptom after it warms.
- Floating inputs pick up noise. Tie leads to V_{CC} or ground.

Various symptoms can occur if one of the leads is bent as it is inserted into the IC socket.

■ SUMMARY

- Present-day logic families consist of ICs that are internally made up of miniature circuits containing transistors, diodes, resistors, capacitors, and interconnecting conductors.
- Integrated circuits offer the advantages of small size, low cost, low power consumption, and high reliability.
- Integrated circuits are manufactured by a process called *photolithographic fabrication.*
- Integrated circuit families are divided into two different categories: bipolar and MOS (FETs), determined by the type of transistor used in the internal circuitry. Each category is further divided into logic IC families, each having different characteristics.
- Bipolar ICs are divided into two categories: saturated and nonsaturated. Saturated subfamilies include standard TTL, Schottky, low-power Schottky, advanced low-power Schottky, and fast. A nonsaturated subfamily is the ECL.
- MOS ICs include NMOS, PMOS, and CMOS.
- The most important specifications for any of the logic families are operating voltages, power dissipation, logic level voltages, fan-out, propagation delay, noise immunity, and operating temperature range.
- Each IC family has its own advantages and disadvantages over other families. The family chosen is determined by the characteristic needed.

BIPOLAR		MOSFET	
Advantages	*Disadvantages*	*Advantages*	*Disadvantages*
Fast	High power consumption	Low power consumption	Slow
Large variety		High density	Handling precautions needed

- All devices in the same logic family can be directly connected to each other. Special circuit configurations are needed when interconnecting ICs of different families.

■ PROBLEMS

1. What is the number of components fabricated on the following IC integration categories? Give an application of each. (6-1)
   (a) SSI
   (b) MSI
   (c) LSI
   (d) VLSI
   (e) ULSI
2. During the evolution of digital electronics, logic gates have been made from what devices? (6-1)
3. What are the most significant advantages offered by IC technologies over discrete circuit designs? (6-1)
4. Combination circuits, such as decoders and multiplexers, are constructed on a _____-scale integrated IC package. (6-1)
5. There are dozens of categories of logic ICs, called _____. (6-2)
6. The two basic technologies for digital ICs are _____ and _____. (6-2)
7. Explain what the difference is internally between the two major categories of ICs called bipolar and MOS. (6-2)
8. Which IC subfamily would a digital watch most likely be made from? Why? (6-4)
9. The operating speed of saturated bipolar devices is (faster or slower) than that of nonsaturated bipolar devices. (6-3)
10. What are the bipolar logic family ICs that fall under the following categories: saturated and unsaturated? (6-3)
11. The packing density of MOS ICs is better than that of bipolar ICs because more _____ can be fabricated on the same chip area. (6-4)
12. CMOS devices combine _____ and _____ subfamilies to form their internal circuitry. (6-4)
13. Compare bipolar and MOS devices with respect to the following characteristics (6-15):
    (a) Speed
    (b) Power consumption
    (c) Noise immunity
    (d) Cost per gate
    (e) Fan-out

14. Decreasing the propagation delay of a logic circuit usually results in an increase in which of the following (6-3):
    (a) Fan-out
    (b) Noise immunity
    (c) Power dissipation
    (d) Package size

15. Fill in the blanks to identify what the following part number represents. (6-6)

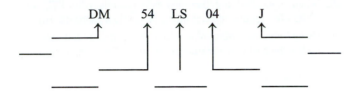

DM   54   LS   04   J

16. The difference between a 7404 and a 5404 TTL IC is their _____ _____. The 7404 IC case is usually made of _____ and the 5404 IC package is usually made of _____. (6-6)

17. What type of information does a part number prefix on the IC case stand for? (6-6)

18. To which family does each of the following belong? (6-6)
    (a) 7400
    (b) 74F00
    (c) 74S00
    (d) 74ALS00
    (e) 74LS00

19. Describe what information is provided by the Absolute Maximum Ratings Table. (6-10)

20. If the power supply voltage is 5.35 volts during the operation of a 7400 IC, will it be destroyed? Explain. (6-10)

21. Any input logic voltage signals between 0.8 and 2.0 volts fall within the _____, and causes _____ operation of the 7400 gate. (6-11)

22. Unused input connections should be tied either to _____ or _____. (6-11)

23. The NAND gates in the circuit of Fig. 6-35 have the following ratings: I_{OL} = 12 mA, I_{IH} = 30 μA, I_{OH} = −3 mA, and I_{IL} = −2 mA. What is the maximum number of NAND gates (N_2–N_8) that can be connected by switches to N_1? This number is also referred to as _____. Is N_1 sourcing or sinking current when only one switch is closed? (6-11, 6-14)

24. What does the minus sign for the I_{OH} value of Problem 23 represent? (6-11)

25. Whenever a gate output lead has conventional current flowing (into or out of) it, the gate lead is said to source current. (6-11)

26. Is the output of the gate of Fig. 6-36 sourcing or sinking? (6-11)

27. What circuit configuration is required when using an IC with an open collector? (6-12)

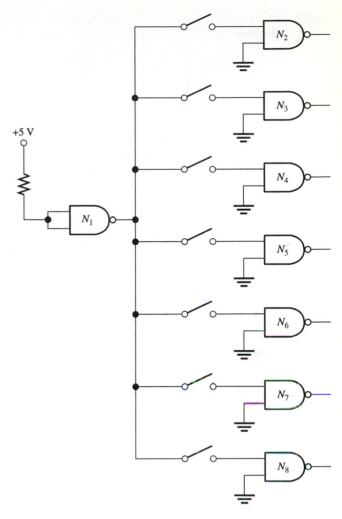

+5 V

FIGURE 6.35
Circuit for Problem 23.

Refer to the following data sheet characteristics to answer Problems 28 to 30. Two different logic family ICs have the following characteristics:

	IC FAMILY A	IC FAMILY B
V_{IH} (V)	1.5	1.9
V_{IL} (V)	0.9	0.7
V_{OH} (V)	2.2	2.3
V_{OL} (V)	0.2	0.3
T_{PLH} (ns)	9	13
T_{PHL} (ns)	7	10

28. Which family has the best noise immunity? (6-12)

29. Which device can operate at higher frequencies? (6-13)

30. Which of the following improves noise immunity? (6-12)
    (a) Logic 1 output voltage is decreased
    (b) Logic 0 output voltage is increased.

FIGURE 6.36
Circuit for Problem 26.

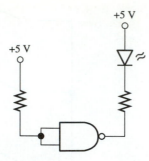

(c) Logic 1 output voltage is decreased and logic 0 output voltage is increased.

(d) Logic 1 output voltage is increased and logic 0 output voltage is decreased.

31. Describe how to eliminate any possibility of power supply noise. (6-12)

32. Another term commonly used for propagation delay is _____ _____. (6-13)

33. What are the acceptable input and output logic level voltage ranges for a 7400 TTL IC? (6-11)

INPUT	OUTPUT
Low:	Low:
High:	High:

34. Under normal operating conditions, a NAND gate package has an I_{CCL} rating of 12 mA. When two gate outputs are low, the power source must supply how much current? (6-12)

35. From the Absolute Maximum Ratings Table portion of Fig. 6-7, what is the "input voltage" rating of the 7400 IC? What does it mean? (6-10)

36. The inverters of the circuit in Fig. 6-37 have a T_{PLH} rating of 12 ns and T_{PHL} of 7 ns. When will the circuit operate faster: when the switch changes from position 1 to 2 or position 2 to 1? (6-13)

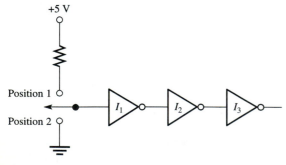

FIGURE 6.37
Circuit for Problem 36.

37. _____ operating conditions are values primarily controlled by the circuit designer. (6-11)

38. DC electrical characteristics are operation parameters guaranteed by the manufacturer if the values listed in the _____ _____ _____ Table are observed. (6-12)

39. (a) What is the advantage of a totem-pole IC over an open-collector IC? (b) What are two advantages of an open-collector IC over a totem-pole IC? (6-12)

40. When connected together, list two reasons why TTL and CMOS ICs require special interface configurations. (6-16)

41. Using Table 6-2, determine which integrated circuit subfamily has the best performance rating in the following categories (6-15):
    (a) Fan-out
    (b) Fast operation
    (c) Lowest power consumption

42. A technician breadboards a logic circuit for testing the operation of its design. After troubleshooting it because it does not work, the technician finds that an inverter from a 7405 TTL IC chip does not generate a logic high. After the chip is replaced with another 7405 IC, the same problem occurs. Which of the following is the most likely cause? (6-17)
    (a) The input is shorted to a high.
    (b) The output is shorted to a low.
    (c) A pullup resistor is needed for the open-collector output.
    (d) All of the above.

43. A technician assembles a logic circuit for testing the operation of its design. When it is turned on, some of the logic devices change state erratically. After the wiring connections are checked and the ICs are replaced, it still does not work properly. Finally, a waveform on a scope shows noise-distortion spikes on the V_{CC} reading. Which of the following should be done to correct the fault? (6-17)
    (a) Place a capacitor between V_{CC} and ground.
    (b) Add a regulator to the power supply circuit.
    (c) Replace the power supply.
    (d) All of the above.

■ ANSWERS TO REVIEW QUESTIONS

1. (e)   2. Small, medium, large   3. very large   4. (e)
5. bipolar, MOS   6. saturated, nonsaturated   7. high
8. low (0), high (1)   9. diode   10. linear   11. (b)
12. (d)   13. (a)   14. minimum, maximum
15. (a) Dual: Two gates in the IC package   (b) Four-input: Each gate has four inputs   (c) NAND: Type of logic device in the package   16. The manufacturer's identification.
17. commercial–industrial, military   18. (a) F: fast
(b) ALS: advanced low-power Schottky   (c) S: Schottky clamped TTL   (d) LS: low-power Schottky TTL
(e) No letter: standard TTL   19. ceramic, plastic
20. (a) General description: functional description of the logic device.   (b) Connection diagram: diagram that shows the internal circuitry of an IC and the external input and output connections.   (c) Function table: truth table of the IC that graphically illustrates its operation.   21. Absolute Maximum Ratings   22. 7   23. 5.5   24. −10°C, +75°C

25. 4.75, 5.25　　**26.** 2.0, 0.8　　**27.** invalid region
28. V_{CC}, ground　　**29.** source, sink　　**30.** 0°C, 70°C
31. 2.4, 0.4　　**32.** faster　　**33.** external, outputs, input
34. 40 μA　　**35.** −20 to −55 mA (5400 series) or −18 to −55 mA (7400 series)　　**36.** 8 mA　　**37.** V_{CC}, ground

38. noise margin (or noise immunity)　　**39.** T_{PHL}, propagation delay　　**40.** faster　　**41.** True　　**42.** fan-out, 10
43. 1　　**44.** (b)　　**45.** (b)　　**46.** (d)　　**47.** (a)
48. (d)

CHAPTER 7

BASIC STORAGE ELEMENTS: LATCHES AND FLIP-FLOPS

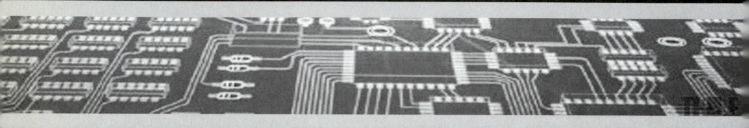

When you complete this chapter, you will be able to:

1. Describe the difference between a gate and a latch or a flip-flop.
2. Describe what is meant by the set and reset states of a latch or a flip-flop and explain how the state can be determined.
3. Explain the internal operation of latches and flip-flops constructed from logic gates.
4. Describe the difference between an asynchronous and a synchronous latch.
5. Identify the latches and flip-flops from the logic diagram symbols.
6. Explain how the clear and preset conditions of a flip-flop are obtained.
7. Develop a truth table consisting of 0 and 1 input and output states and modes of operation for latches and flip-flops.
8. Draw or recognize the output waveform on a timing diagram of the latches and flip-flops using truth tables.
9. Describe the difference between a latch and a flip-flop.
10. Identify edge-triggered flip-flops and level-triggered latches.
11. Draw or recognize a J–K flip-flop configured to operate like a gated R–S latch, D flip-flop, and T flip-flop.
12. Give a practical application of the latches and flip-flops.
13. Use pin diagrams to properly operate flip-flops in integrated circuit form.
14. Troubleshoot flip-flops in integrated circuit form.

7.1 INTRODUCTION

The previous chapters provided information about logic devices such as gates and combination circuits. These devices are capable of making decisions ranging from extremely simple to highly complex. One characteristic that these logic devices share is that their outputs at any instant are determined solely by the inputs applied at that instant.

Certain types of digital circuit operations require the capability of producing outputs resulting from input signals that were applied at a previous time. For example, to perform a simple addition operation, it is necessary for a computer to have a means of storing numbers before they are added. The computer also requires a place to store the answer after the addition has been completed. In other words, there is a need for circuits that have some memory capabilities.

The basic device in a digital circuit capable of the memory function is a **storage element.** It comes in two forms: a **latch** and a **flip-flop.** It is capable of storing one bit of binary information, a high or a low at an output. Similar to the logic gate, which is the building block of combination circuits, the storage element is the building block of devices called **sequential circuits,** which are covered in Chapters 8 and 9. Sequential circuits use as many storage elements as necessary to store a desired number of bits. These sequential circuits include counters, storage registers, and shift registers.

Storage element:

A basic digital circuit capable of the memory function.

There are many types of latches and flip-flops available. They are all capable of storing data indefinitely. However, the difference between them is that some require far more input information than others before they recognize the command to store data. Which type of storage element to use is usually determined by the complexity of the sequential circuit. Because most sequential circuit operations are performed by only a few different kinds of storage elements, these are described in detail. Flip-flops are normally available in integrated circuit form. They can also be constructed from discrete logic gate configurations. To best illustrate their operation, the logic gate types are described first. Schematic symbols and truth tables for each type are examined. Pin diagrams of latches and flip-flops available in IC form are also provided.

7.2 BASIC STORAGE-ELEMENT CONCEPTS

Latch:

A basic type of storage element capable of storing one bit of information when its clock input is at the required logic level.

The basic logic symbol for a storage element is shown in Fig. 7-1. Unlike logic gates that have one output, latches and flip-flops have two outputs labeled Q and $\overline{Q}$. Unless otherwise noted, the Q output is the primary, or reference, output. The $\overline{Q}$ output provides the complement, or inversion, of the Q output.

Flip-flop:

A basic type of storage element capable of storing one bit of information during the transition of the enable signal applied to the clock input.

FIGURE 7.1
Basic storage element symbol.

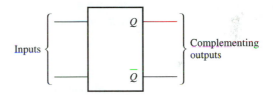

Sequential circuit:

A circuit made up of several flip-flops that are classified into two categories, counters and registers.

The storage element is said to be in the set (also known as preset) state when:

Output Q = logic 1 (high)

Output $\overline{Q}$ = logic 0 (low)

or in the reset (also known as clear) state when:

Output Q = logic 0 (low)

Output $\overline{Q}$ = logic 1 (high)

Also, a storage element has one or more inputs, and which binary data bit is stored is dependent on the type of logic signal applied.

■ REVIEW QUESTIONS

1. Logic gates are the basic building blocks of _____ circuits; latches and flip-flops are the basic building blocks of _____ circuits.

2. A functional characteristic that a latch or a flip-flop has that a logic gate does not have is its _____ capability.

3. One storage element can store _____ bit(s) of binary data.

4. A latch or a flip-flop has two outputs that normally _____ each other.

5. Unless specified otherwise, the _____ output is the primary output of a latch or a flip-flop.

6. A storage element is set when Q = _____ and reset when $\overline{Q}$ = _____.

7. Another common term used instead of set is _____, and another term for reset is _____.

7.3 NOR R–S LATCH

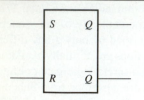

FIGURE 7.2
Schematic symbol of a NOR R–S latch.

The most basic type of storage element is an *R–S latch*. Its schematic symbol is shown in Fig. 7-2, which shows two inputs labeled *S* and *R* on the left and two complementary outputs on the right labeled *Q* and $\overline{Q}$. Its name is derived from its two basic operating states: reset (*R*) and set (*S*). When it is in the set state, it stores a 1 at its primary output *Q*, and when in the reset state, a 0 is stored at *Q*. Figure 7-3 illustrates how two NOR gates are connected to operate as an *R–S* type latch.

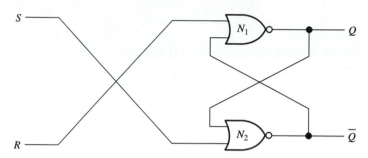

FIGURE 7.3
NOR gate R–S latch.

Because there are only two inputs where external signals from outside the latch are applied, only four input conditions are possible. The latch becomes set when a 1 is applied to the S input and a 0 at the R input. This is called the set mode of operation. It becomes reset when a 0 is applied to the S input and a 1 at the R input. This is called the reset mode of operation. If 0s are applied to both the S and R inputs simultaneously, the latch is in the hold mode of operation and retains the binary state previously stored into it. If 1s are applied to both the S and R inputs simultaneously, the latch is in the prohibit mode. Both of the outputs are at a "0" state, so the latch is neither set nor reset. As this condition may produce unpredictable results, it should be avoided.[†]

Like logic gates, latches also have truth tables to describe their operation, as shown in Table 7-1.

TABLE 7.1
NOR-Latch truth table

S	R	MODE OF OPERATION	Q	$\overline{Q}$
0	0	Hold	X*	X
0	1	Reset	0	1
1	0	Set	1	0
1	1	Prohibit	0	0

*The X means the output conditions from the previous state remain the same.

[†]*Note:* Those readers who trace the signals throughout the gated latch should first assume a set or reset state. Normally, latches will go to one or the other state after power is applied because they aren't perfectly balanced. This is due to the varying tolerances of the internal components.

EXAMPLE 7.1

The S and R waveforms provided in Fig. 7-4(a) are applied to the inputs of the NOR R–S latch shown in Fig. 7-4(b). Write the mode of operation for each time period and draw the Q output waveform that results.

Solution See the bottom of Fig. 7-4(a).

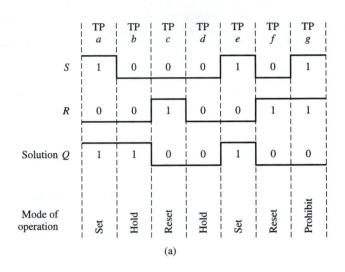

(a)

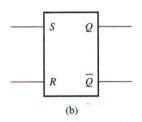

(b)

FIGURE 7.4
(a) S and R waveforms and (b) NOR R–S latch for Example 7.1.

Time Period

a S input = 1 and R input = 0. Mode of operation = set. Therefore, the Q output = 1.

b S input = 0 and R input = 0. Mode of operation = hold. Therefore, the Q output = 1.

c S input = 0 and R input = 1. Mode of operation = reset. Therefore, the Q output = 0.

d S input = 0 and R input = 0. Mode of operation = hold. Therefore, the Q output = 0.

e S input = 1 and R input = 0. Mode of operation = set. Therefore, the Q output = 1.

f S input = 0 and R input = 1. Mode of operation = reset. Therefore, the Q output = 0.

g S input = 1 and R input = 1. Mode of operation = prohibit. Therefore, the Q output = 0.

| 7.4 | NAND *R–S* LATCH |

Another way of constructing an *R–S* latch is by using two NAND gates, as shown in Fig. 7-5. Because the operation of a NAND latch differs from that of a NOR latch, the operation of the *R–S* latch made from these two gates is also different.

FIGURE 7.5
NAND-gate **R–S** *latch.*

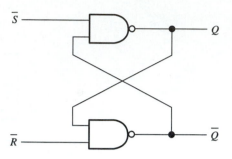

From the NAND *R–S* latch truth table (Table 7-2), note that to set the *Q* output high, a low is applied to the $\overline{S}$ input and a high to the $\overline{R}$ input. To reset the latch ($Q = 0$), a high is applied to the $\overline{S}$ input and a low to the $\overline{R}$ input. When a high is applied to both inputs, it is in the *hold* mode, and when both inputs are low, it is in the *prohibit* mode. In comparing the truth table of the NOR and NAND *R–S* latches, note that all input and output conditions are opposite.

TABLE 7.2
NAND-Latch truth table

$\overline{S}$	$\overline{R}$	MODE OF OPERATION	Q	$\overline{Q}$
0	0	Prohibit	1	1
0	1	Set	1	0
1	0	Reset	0	1
1	1	Hold	X	X

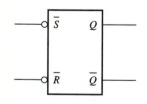

FIGURE 7.6
Schematic symbol of a NAND R–S latch.

The schematic symbol for the *R–S* latch is shown in Fig. 7-6. Note that the input* leads have small circles on them, called *bubbles*. Recall that a logic gate with a bubble represents an active-low symbol. This means that the output is activated by applying a low signal instead of a high. The same rule applies to a latch.

---

*The bars at the inputs mean that the signals applied to them during their resting state are normally high. To activate the device, the signal must be made low.

| **EXAMPLE 7.2** | The $\overline{S}$ and $\overline{R}$ waveforms in Fig. 7-7(a) are applied to the inputs of the NAND *R–S* latch shown in Fig. 7-7(b). Write the mode of operation for each time period and draw the *Q* output waveform that results.

Solution See the bottom of Fig. 7-7(a). |

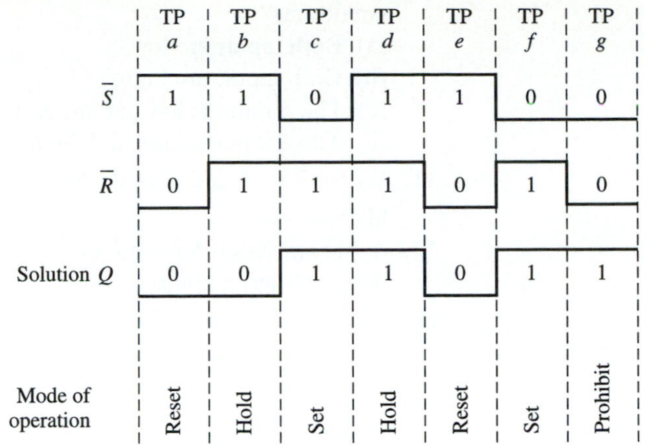

	TP a	TP b	TP c	TP d	TP e	TP f	TP g
$\overline{S}$	1	1	0	1	1	0	0
$\overline{R}$	0	1	1	1	0	1	0
Solution Q	0	0	1	1	0	1	1
Mode of operation	Reset	Hold	Set	Hold	Reset	Set	Prohibit

Time period

a $\overline{S}$ input = 1 and $\overline{R}$ input = 0. Mode of operation = reset. Therefore, the Q output = 0.

b $\overline{S}$ input = 1 and $\overline{R}$ input = 1. Mode of operation = hold. Therefore, the Q output = 0.

c $\overline{S}$ input = 0 and $\overline{R}$ input = 1. Mode of operation = set. Therefore, the Q output = 1.

d $\overline{S}$ input = 1 and $\overline{R}$ input = 1. Mode of operation = hold. Therefore, the Q output = 1.

e $\overline{S}$ input = 1 and $\overline{R}$ input = 0. Mode of operation = reset. Therefore, the Q output = 0.

f $\overline{S}$ input = 0 and $\overline{R}$ input = 1. Mode of operation = set. Therefore, the Q output = 1.

g $\overline{S}$ input = 0 and $\overline{R}$ input = 0. Mode of operation = prohibit. Therefore, the Q output = 1.

(a)

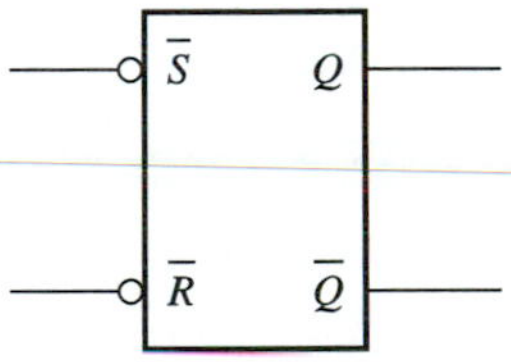

(b)

FIGURE 7.7
(a) S and R waveforms and (b) NAND R–S latch for Example 7.2.

The types of latches and flip-flops that are examined in this chapter all have one thing in common: They all begin with an R–S latch. Extra gates and connections are made to an R–S latch to form the other types.

■ REVIEW QUESTIONS

8. The R–S NOR latch is (set or reset) when $S = 0$ and $R = 1$.

9. The NOR latch has active-_____ (low, high) inputs.

10. The prohibit state of a NOR-gate latch occurs in which of the following input conditions?
    (a) Both inputs are low.
    (b) Both inputs are high.
    (c) The S input is low and the R input is high.
    (d) The S input is high and the R input is low.

11. When $\overline{R} = 1$ and $\overline{S} = 0$ are applied to the R–S NAND latch, $Q =$ _____ and $\overline{Q} =$ _____.

12. A NAND latch has active-_____ (low or high) inputs and active-_____ (low or high) outputs.

13. The prohibit state of the NAND latch occurs when both inputs are simultaneously _____ (low, high).

7.5   GATED R–S LATCH

The operation of the R–S latch is the basis for almost every other type of latch or flip-flop. One characteristic that an R–S latch has is that its outputs immediately change states when incoming signals are applied to its inputs. However, in some situations, this property is undesirable. For example, while processing data, many types of digital circuits generate a string of 1s and 0s at their outputs. If these output lines are connected to semiconductor memories, for example, it would be usually desirable to ignore the intermediate data before storing the final result. Because R–S latches respond to all of the signals applied to them, they are seldom used for memory storage. For the same reason, they are seldom used in many other types of digital circuit applications.

To control *when* the input affects the output, a modification of the R–S circuit is made. This change involves the addition of two NAND gates, called *enable gates*. As shown in Fig. 7-8, their outputs are connected to the original NAND R–S latches. Two inputs of the enable gates become the new R and S inputs, which are both active-high. A common connection, called an *enable input*, is connected to the other inputs of each gate.

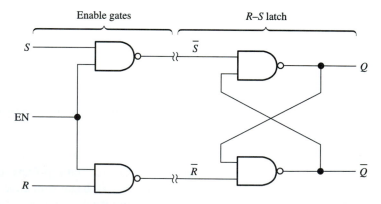

FIGURE 7.8
Adding enable gates to a NAND R–S latch.

When the enable input is held low, it causes both enable gates to produce a high at their outputs. This causes the R–S latch to be in the hold mode. Therefore, the circuit is disabled. When the enable input is high, the stream of incoming data is allowed to pass through and set or reset the *R–S* latch to the desired state. This type of circuit is called a *gated RS latch*. Its operation is summarized by the truth table shown in Fig. 7-9(b). The schematic symbol for this device is shown in Fig. 7-9(a).

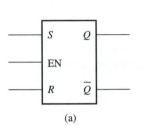

Inputs			Mode of operation	Outputs	
EN	S	R		Q	$\bar{Q}$
⊓	0	0	Hold	No change	
⊓	0	1	Reset	0	1
⊓	1	0	Set	1	0
⊓	1	1	Prohibit	1	1

(a)   (b)

FIGURE 7.9
*Gated **R–S** latch: (a) Schematic symbol. (b) Truth table.*

EXAMPLE 7.3

The *S, R,* and EN waveforms in Fig. 7-10(a) are applied to the inputs of the gated *R–S* latch shown in Fig. 7-10(b). Write the mode of operation for each time period and draw the *Q* output waveform that results.
From Fig. 7-10(a):

① and ③: A high applied to the EN input enables the circuit and allows it to respond to the *R* and *S* inputs.

② and ④: A low applied to the EN input disables the circuit, which causes the latch to be frozen.

Solution   See the bottom of Fig. 7-10(a).

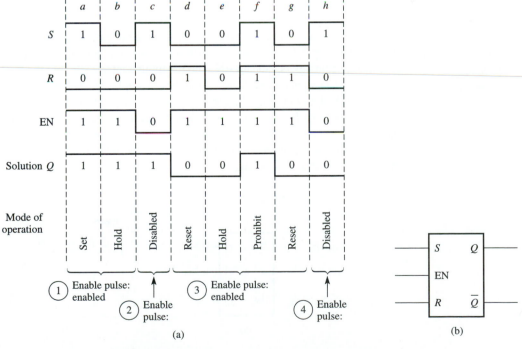

(a)   (b)

FIGURE 7.10
(a) S, R, and EN waveforms and (b) gated R–S latch for Example 7.3.

Asynchronous:

Latches that respond to signals as soon as they are applied to the inputs.

The basic *R–S* latch is termed **asynchronous** because it responds to inputs as soon as they occur. Any gated latch that is enabled by a clock signal is considered **synchronous** because its operation is synchronized with other latches by a clock signal applied to all of them simultaneously.

7.6 *D* LATCH

Synchronous:

Latches that respond to data signals applied to their inputs only when they are enabled by a clock signal.

Earlier in this chapter, it was explained that if highs are simultaneously applied to both the set and reset input lines of a NOR *R–S* latch, potentially unpredictable conditions would exist. Likewise, lows should not be simultaneously applied to an *R–S* NAND latch for the same reason. Therefore, it is recommended that putting the *R–S*-type latch in the prohibit mode should be avoided.

One way to prevent the prohibit mode from occurring is by connecting the *R* and *S* inputs with an inverter, as shown in Fig. 7-11. This ensures that the set and reset inputs are always opposite. Data are applied only to the *S* input, which is renamed the *D* (data) input.

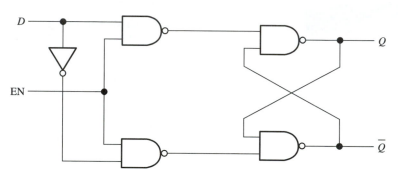

FIGURE 7.11
Logic gate **D** *latch.*

D latch:

A storage device that has an inverter connected between its *R* and *S* inputs to ensure that data applied to them is always opposite.

This device is called a ***D* latch.** Its schematic symbol is shown in Fig. 7-12(a) and its truth table in Fig. 7-12(b). It has the customary two complementary outputs, labeled *Q* and *Q̄*, and an enable inputs, labeled EN. The EN input controls whether the latch recognizes or ignores the data signal at the *D* input line. For example, the data applied to the *D* input are transferred to the *Q* output when the EN input is high. If the data at the *D* input change when the EN input is at a 1, identical changes are felt at the output as well. When the EN input goes low (binary 0), any changes on the *D* input are ignored. Instead, the binary information that was present at *D* when the transition occurred is retained at the *Q* output until the EN input goes high again.

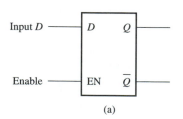

Inputs		Outputs	
D	Enable	*Q*	*Q̄*
0	0	No change	
0	1	0	1
1	0	No change	
1	1	1	0

(a) (b)

FIGURE 7.12
D latch: (a) Schematic symbol. (b) Truth table.

Data latch:

A storage device capable of holding data one bit in length; several connected together form a storage register.

Storage register:

A circuit made of several D flip-flops connected together that temporarily stores binary numbers and data.

This latch is also called a **data latch** and receives its name from its ability to hold *data* in its internal storage. Several D latches connected together are well suited for temporarily storing binary numbers and data in circuits called **storage registers.**

EXAMPLE 7.4

The D and EN waveforms in Fig. 7-13(a) are applied to the inputs of the data latch shown in Fig. 7-13(b). Draw the Q output waveform that results.

Solution See the bottom of Fig. 7-13(a).[*] The only time data are transferred from the D input to the Q output is when the enable pulse is in its high state.

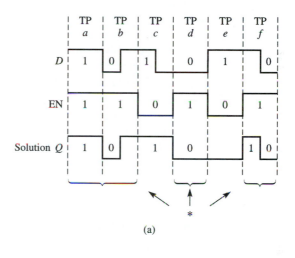

(a)

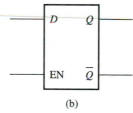

(b)

FIGURE 7.13
(a) D and EN waveforms and (b) the data latch for Example 7.4.

■ **REVIEW QUESTIONS**

14. The gated R–S latch is enabled when the EN input is (low or high) and disabled when the EN input is (low or high).

15. The gated R–S latch is reset when $S = $ (0 or 1), $R = $ (0 or 1), and the EN input is activated.

16. The R–S latch operates (asynchronously or synchronously) and the gated R–S latch operates (asynchronously or synchronously).

17. Which of the following types of latches does not have a prohibit state?
    (a) NOR latch
    (b) NAND latch
    (c) Gated *R–S* latch
    (d) *D*

18. If the *D* input of a *D* latch is low and the EN input is high, the latch is
    (a) set
    (b) reset
    (c) in the toggle state
    (d) in a hold state

7.7 *D* FLIP-FLOP

Sections 7-3 through 7-6 described the operation of the storage element called a latch. Another storage element is the **flip-flop.** It is similar to the latch in that it also has two stable states, two complementary outputs, and similar inputs. The difference is how they are enabled. A latch is enabled when the enable signal is at the active level. A flip-flop is enabled only during the transition of its enable signal, from low to high or high to low. The enable signal applied to a flip-flop is commonly referred to as a *clock* (CLK). The label CLK is used at the enable input instead of EN.

Figure 7-14(a) shows a modified *D* latch with a resistor-capacitor network (called a *differentiator*) placed between the CLK input and the enable gates. Its symbol is shown in Fig. 7-14(b). When a 1 state signal is applied to the CLK input, the *RC* network allows only a momentary positive pulse to exist at the enabler, as shown in Fig. 7-14(c). There-fore, instead of transferring data from input *D* to the *Q* output while the enable input signal is at a high level, it provides the transfer when the positive edge of the clock signal occurs. This device is called a **positive-edge-triggered flip-flop.** Because it is more selective, it is often preferred over the level-type latch. It is commercially available in IC form as a TTL 7474 IC. Its pin diagram and truth table are shown in Figs. 7-14(d) and (e), respectively. The functions of the asynchronous PS and CLR inputs are described in Section 7-11. Because the data at input *D* must wait until the clock input goes high, *D* latches and flip-flops are also called *delay* latches and flip-flops.

Positive-edge-triggered flip-flop:

A flip-flop that allows data applied to its input to be transferred to its output when the low-to-high transition of a signal occurs.

7.8 CLOCK-PULSE TRIGGERING

The clock signals that activated the gated *R–S* latch and the data latch described earlier in this chapter did so while the clock was high. By connecting a resistor–capacitor network to the clock input of the delay flip-flop, the circuit was activated only during the low-to-high transition of the clock pulse. It is possible to modify latches and flip-flops even further so that they are activated while the clock signal is low or during the high-to-low transition of the clock pulse. The result of latches and flip-flops responding to a clock input is called **clock-pulse triggering,** of which there are four types.

Clock-pulse triggering:

Occurs when latches or flip-flops respond to a clock signal.

Edge Triggering

A typical clock-pulse train is shown in Fig. 7-15. The horizontal distance on the waveform represents time and the vertical distance is voltage (+5 V for TTL ICs). Starting at the left, the signal is low (or 0 state at ground potential). As clock pulse *A* arrives, the pulse changes to a 1 state on the positive leading edge (also called low to high or L to H).

On the right side of pulse *A,* the waveform changes from +5 V (also called high to low) on the negative edge. Flip-flops that transfer data from their input to the output on the low-to-high edge are called *positive-edge-triggered* flip-flops. Figure 7-16 shows how the *Q* output of such a flip-flop changes to a high on the positive edge of the first clock pulse and back to a low on the positive edge of the second clock pulse (pulse *B*).

Edge triggering:

Occurs when data applied to the input leads of a flip-flop is transferred to its output during the low-to-high or high-to-low transition of a clock pulse.

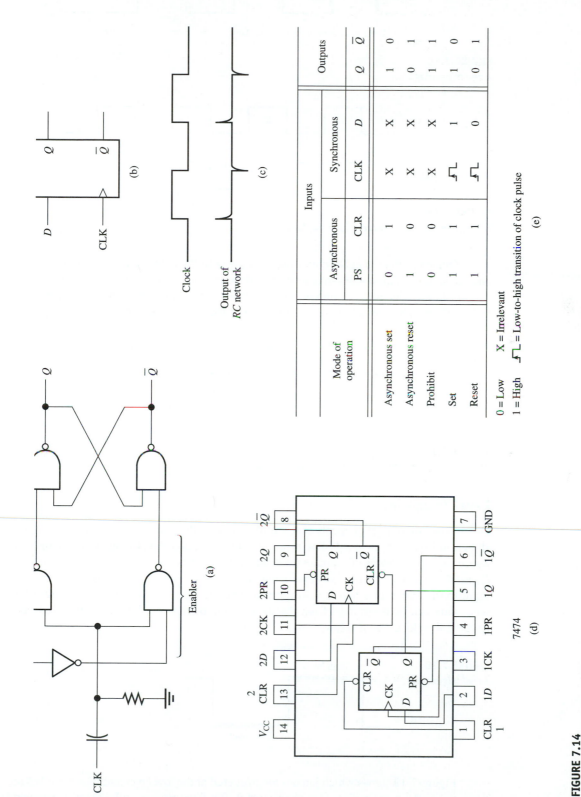

(a)

(b)

Clock

Output of RC network

(c)

	Inputs				Outputs	
Mode of operation	Asynchronous		Synchronous			
	PS	CLR	CLK	D	Q	$\bar{Q}$
Asynchronous set	0	1	X	X	1	0
Asynchronous reset	1	0	X	X	0	1
Prohibit	0	0	X	X	1	1
Set	1	1	⌐	1	1	0
Reset	1	1	⌐	0	0	1

0 = Low   X = Irrelevant
1 = High ⌐L = Low-to-high transition of clock pulse

(e)

7474

(d)

FIGURE 7.14

Positive-edge-triggered D flip-flop: (a) Logic gate diagram. (b) Schematic symbol. (c) Timing diagram: response of a resistor–capacitor network to a clock input. (d) Pin diagram of a 7474 TTL IC. (e) Truth table of a 7474 TTL IC.

229

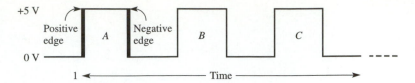

FIGURE 7.15
Edge triggering.

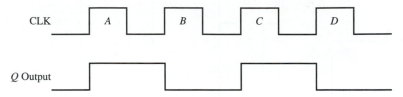

FIGURE 7.16
Waveform of positive-edge-triggered flip-flops.

Negative-edge-triggered flip-flop:

A flip-flop that allows data applied to its input to be transferred to its output when the high-to-low transition of a signal occurs.

Other flip-flops transfer data from their input to the output on the high-to-low edge and are called **negative-edge-triggered flip-flops.** Figure 7-17 shows how the Q output of such a flip-flop changes to a high on the high-to-low edge of the first clock pulse (pulse A). It then changes back to a low on the next clock-pulse negative-going edge (pulse B).

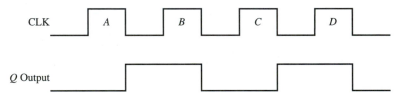

FIGURE 7.17
Waveform of negative-edge-triggered flip-flops.

Level Triggering

Level-triggered latch:

A latch that allows data applied to its input to be transferred to its output when the signal applied to the clock input is at a high or a low.

Other types of storage elements that are triggered while a square wave, such as the one shown in Fig. 7-18, is at a 1 state or at a 0 state are called **level-triggered latches.** The figure shows two levels of a square wave when these latches are triggered. There are *high-level-triggered* latches and *low-level-triggered* latches. Data can be transferred from the input to the output of a high-level-triggered latch anytime the clock signal is high. Likewise, data can be transferred from the input to the output of a low-level-triggered latch anytime the clock signal is low.

FIGURE 7.18
Level triggering.

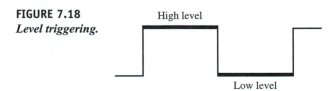

Figure 7-19 shows which latches are triggered at the low level and which latches are triggered at the high level. It also shows which flip-flops trigger at the positive edge and which ones trigger at the negative edge.

The reason the low-level latches and negative-edge-triggered flip-flops are triggered during a low or high-to-low transition of a clock pulse is because an internal inverter is

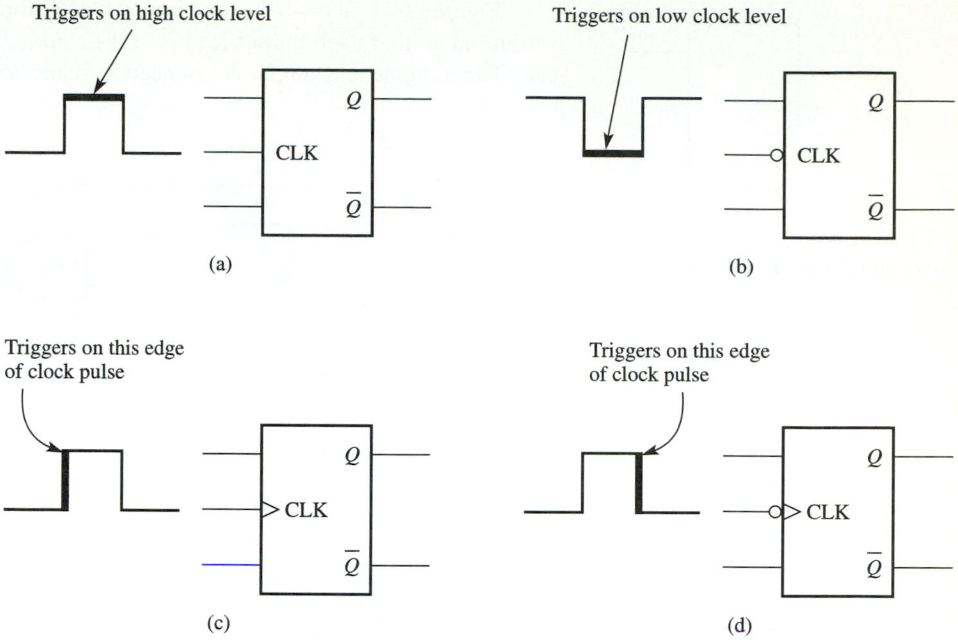

FIGURE 7.19
Four types of clock triggering: (a) High-level-triggered latch. A straight lead into the clock input indicates that the latch is triggered during the high level. (b) Low-level-triggered latch. A lead with a low state indicator at the clock input signifies that the latch is triggered during the low level. (c) Positive-edge-triggered flip-flop. A straight lead into the clock input with a triangle indicates that the flip-flop is triggered during the positive edge. (d) Negative-edge-triggered flip-flop. A lead with the low state indicator and triangle at the clock input signifies that the flip-flop is triggered during the negative edge.

used at their clock inputs. Otherwise, their circuitry is identical to their respective high-level and positive-edge-triggered counterparts.

Information regarding which storage elements are triggered at the edges and which are triggered at the levels is best obtained from data manuals.

■ **REVIEW QUESTIONS**

19. A flip-flop differs from a latch in the way it is _____.
20. What are the two different names given to the D latch or flip-flop?
21. Which of the following are D flip-flops primarily used for?
    (a) Storage registers
    (b) Demultiplexers
    (c) Counters
    (d) Encoders
22. What are four ways in which latches and flip-flops are triggered by clock pulses?

7.9 *T* FLIP-FLOP

T flip-flop:

Also known as a toggle flip-flop, it changes states on every triggering portion of a clock signal.

There are some applications that require a flip-flop to change its output state every time a clock pulse arrives at its input. A flip-flop that has this property is called a *T* (or *toggle*) **flip-flop.** The schematic symbol of this flip-flop is shown in Fig. 7-20. It shows one input, labeled *T*, where the clock pulse is applied and the usual two complementary outputs Q and $\overline{Q}$.

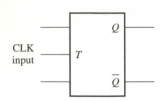

FIGURE 7.20
Schematic symbol of a T flip-flop.

Figure 7-21 shows the diagram of a T flip-flop made from logic gates. It is configured so that each output lead is cross-coupled back to an input of the other enable gate. The T input is a common connection made to the remaining input of each enable gate.

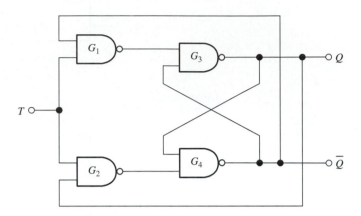

FIGURE 7.21
Logic gate diagram of a T flip-flop.

Assuming that the flip-flop is reset, the $\overline{Q}$ output is high and feeds back a 1 to the top input of G_1. At the same time, a 0 at Q is applied to the bottom lead of G_2. These feedback signals constitute a set command that the flip-flop toggles to after a positive level clock pulse arrives. This change causes the $\overline{Q}$ output to feed back a 0 to the G_1 input and the Q output to provide a 1 at G_2. As a result, the flip-flop toggles back to the reset state after the next clock pulse arrives.

Because it takes two clock pulses to cause the flip-flops to make a complete transition from ($Q =$) low to ($Q =$) high and back to low, these flip-flops are sometimes used as frequency dividers. Several cascaded T flip-flops are often used in counting circuits.

7.10   J–K FLIP-FLOP

J–K flip-flop:

A universal type of flip-flop that can function as most other types of flip-flops.

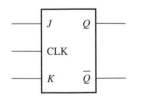

FIGURE 7.22
Schematic symbol of a J–K flip-flop.

The **J–K flip-flop** is the most versatile flip-flop available because it can function as a gated R–S latch, D flip-flop, or T flip-flop. Therefore, it is also referred to as the *universal* flip-flop. Figure 7-22 shows the schematic symbol of the J–K flip-flop. It has two data inputs, labeled J and K, a clock input, and the customary complementing Q and $\overline{Q}$ outputs. Even though the J input may be considered equivalent to the S input and the K input may be considered equivalent to the R input, the J and K designations are used so that the operation of the J–K flip-flop is not confused with the R–S latch. The operation of the J–K flip-flop differs slightly because it was designed to overcome the ambiguity (prohibit) mode of the R–S latch, and it can function when both data inputs are active, which the R–S device cannot do.

One possible way of constructing a J–K flip-flop using logic gates is shown in Fig. 7-23(a). It consists of a NOR R–S latch and a pair of three-input AND gates to perform the clock-enable function. One input of each AND gate is connected to an input of the other AND gate to form the clock input. Another input of each AND gate receives a feedback signal from the NOR gate it is driving. The third input of each AND gate, J or K, is the data input. If the J input remains low, the Q output cannot go high. Likewise, if the K input remains low, the $\overline{Q}$ output cannot go high. The truth table of Fig. 7-23(b) illustrates the four modes of operation.

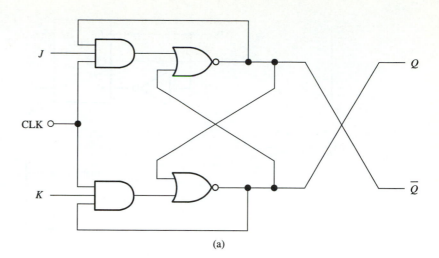

(a)

Inputs			Outputs			Mode of operation
CLK	J	K	Q	$\overline{Q}$	Effect on output Q	
⊓	0	0	No change		No change — disable	Hold
⊓	0	1	0	1	Reset or cleared to 0	Reset
⊓	1	0	1	0	Set to 1	Set
⊓	1	1	Toggle		Changes to opposite state	Toggle

(b)

FIGURE 7.23
J–K flip-flop: (a) Logic-gate diagram. (b) Truth table.

Four Modes of Operation

1. *J* input low, *K* input low: When the clock goes high, nothing happens. This is called the *hold* mode.
2. *J* input high, *K* input low: When the clock goes high, the 1 at *J* is passed to *Q*, and the 0 at *K* is passed to the $\overline{Q}$ output. This is called the *set* mode.
3. *J* input low, *K* input high: When the clock goes high, the 0 at *J* is passed to *Q*, and the 1 at *K* is passed to the $\overline{Q}$ output. This is called the *reset* mode.
4. *J* input high, *K* input high: When the clock input goes high, the outputs change to the opposite state. This is called the *toggle* mode.

Figure 7-24 shows how the *J–K* flip-flop can be configured to operate as *D* (data and delay) and *T* (toggle) flip-flops.

7.11 PRESET AND CLEAR INPUTS

When digital equipment is turned on by applying power, the logic output states of each flip-flop inside it cannot be predicted. Some *Q* outputs may be high, while others are low. A group of flip-flops that set or reset in a random pattern is seldom useful.

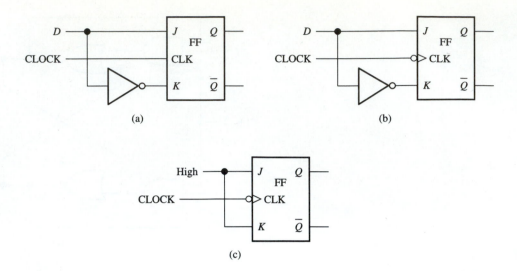

(a)   (b)

(c)

FIGURE 7.24
J–K flip-flop configured to operate as other types of flip-flops: (a) As a data flip-flop. (b) As a delay flip-flop. (c) As a T flip-flop.

In addition to the data and clock inputs, there are some types of flip-flops that also have preset and clear inputs. A logic gate *J–K* flip-flop configuration with these inputs is shown in Fig. 7-25. These inputs are used to initialize a group of flip-flops to ensure a desired starting point either by *clearing* all of them or by *presetting* desired data into them. The preset and clear inputs operate independently of the data and clock inputs and, if necessary, override them. Because they do not rely on the clock input, they are considered as asynchronous inputs.

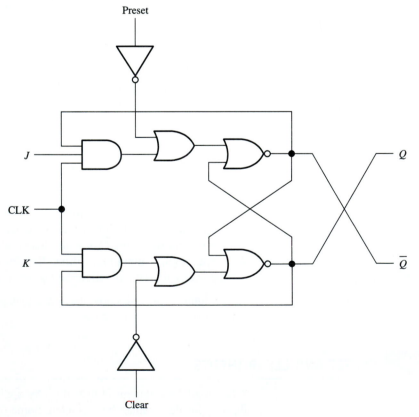

FIGURE 7.25
Logic gate diagram of a J–K flip-flop with preset and clear inputs.

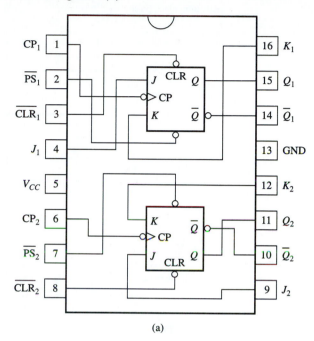

In Fig. 7-26, a flip-flop symbol shows the preset and clear inputs, which are most often active-low. Many applications do not require that these inputs be used. Therefore, they are often left floating (disconnected). To preset a flip-flop to a 1 state (at the Q output), a momentary 0 is applied to the PR input. To clear a flip-flop to a 0 state (at the Q output), a momentary 0 is connected to the CLR input.

A pin diagram of a negative-edge-triggered J–K flip-flop in IC form with PR and CLR inputs is shown in Fig. 7-27(a). It is available as a TTL IC with a number of 7476. Its truth table is shown in Fig. 7-27(b).

FIGURE 7.26
Schematic symbol of a **J–K**
flip-flop.

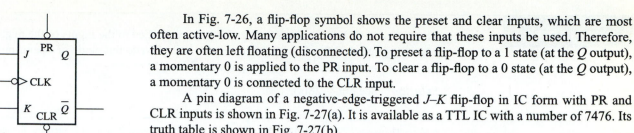

(a)

Inputs					Outputs		Mode of operation
Asynchronous		Synchronous					
PS	CLR	CLK	J	K	Q	$\overline{Q}$	
0	1	X	X	X	1	0	Asynchronous set
1	0	X	X	X	0	1	Asynchronous reset
0	0	X	X	X	1	1	Prohibit
1	1	⎍↓	0	0	No change		Hold
1	1	⎍↓	0	1	0	1	Reset
1	1	⎍↓	1	0	1	0	Set
1	1	⎍↓	1	1	Opposite state		Toggle

0 = Low   X = Irrelevant
1 = High   ⎍↓ = Negative-edge clock pulse

(b)

FIGURE 7.27
The 7476 J–K flip-flop IC: (a) Pin diagram. (b) Truth table.

EXAMPLE 7.5

The *J, K,* and CLK waveforms in Fig. 7-28(a) are applied to the inputs of the negative-edge-triggered *J–K* flip-flop shown in Fig. 7-28(b). Write the mode of operation* for each time period and draw the *Q* output waveform that results.

Solution   See Fig. 7-28(a).

* *Note:* The mode of operation of a flip-flop refers to what condition it is in as it prepares for the next activating clock pulse.

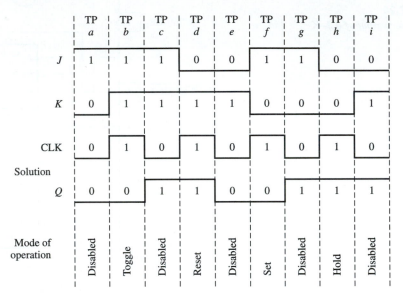

Time period

a The *Q* output becomes 0 prior to the time period. It remains in this state because a 0 at the clock has no effect on the flip-flop.

b The *J* and *K* inputs are high. The flip-flop is in the toggle mode as it prepares for the negative-edge of the clock pulse.

c The flip-flop has set and remains in this state because a 0 is present at the clock input.

d *J* = 0 and *K* = 1. The flip-flop is in the reset mode as it prepares for the negative edge of the clock pulse.

e The flip-flop has reset and remains in this state because a 0 is present at the clock input.

f *J* = 1 and *K* = 0. The flip-flop is in the set mode as it prepares for the negative edge of the clock pulse.

g The flip-flop has set and remains in this state because a 0 is present at the clock input.

h *J* = 0 and *K* = 0. The flip-flop is in the hold mode as it prepares for the negative edge of the clock pulse.

i The flip-flop remains set because it was in the hold mode before the negative edge transition of the clock pulse.

(a)

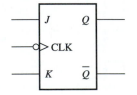

Note: The flip-flop goes into a mode of operation to prepare for the activating triggering portion of the clock pulse. When the negative edge occurs, the flip-flop then responds.

(b)

FIGURE 7.28
(a) J, K, and CLK waveforms, and (b) a negative-edge-triggered J–K flip-flop for Example 7.5.

EXPERIMENT

J–K *Flip-Flop*

Objectives

- To understand the four modes of operation of a *J–K* flip-flop.
- To understand how the *J–K* flip-flop operates as a *D* flip-flop, a *T* flip-flop, a storage device, and a frequency divider.

Materials

4—Logic Switches

2—150-Ohm Resistors

2—LEDs

1—7476 IC

1—Logic Clock Pulse Push Button (bounceless)

1—Clock Pulse Generator or Signal Generator

1—+5-V DC Power Supply

1—Oscilloscope

Introduction

A flip-flop is a digital component that is classified as a memory device. Like a logic gate, which is the most basic element of a combination circuit, the flip-flop is the most basic element of sequential circuits, which make up counters and registers.

The *J–K* flip-flop is the most versatile type of flip-flop. When configured in various ways, it is capable of operating like most other types of flip-flops. Therefore, it is also referred to as the *universal* flip-flop. The *J–K* flip-flop consists of five input leads and two output leads. The output leads are labeled Q and $\overline{Q}$ and complement each other. Unless otherwise stated, the Q terminal is the primary output. The *J–K* flip-flop is available in IC form. The TTL 7476 IC shown in Fig. 7-27 contains two individual *J–K* flip-flops in a single package.

Procedure

Step 1. Assemble the circuit shown in Fig. 7-29.

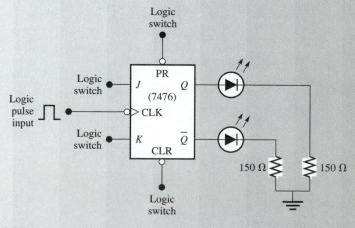

FIGURE 7.29

Background Information

Preset and Clear Inputs

The *J–K* flip-flop is capable of storing one bit of binary data. One way to insert data into the flip-flop is by parallel loading using the PR (Preset) or CLR (Clear) inputs.

When a 1 is loaded into the flip-flop, a logic high bit is stored at the primary terminal, called the Q output. The flip-flop is set when it stores a 1. At the same time, the opposite 0 state will be at the $\overline{Q}$ output. The flip-flop becomes set by applying a momentary active-low state to the PR input.

To load a 0 into the flip-flop, a momentary low is applied to the CLR input. The flip-flop is then in the clear or reset state when a 0 is at the Q output and a 1 is at the $\overline{Q}$ output.

An activated PR or CLR input overrides all synchronous inputs, such as the CLK, *J*, and *K* leads. The PR and CLK inputs should not both be activated at the same time.

Step 2. Place the four logic switches in the +5-V position.

Procedure Question 1

When a momentary low is applied to the PR input, is the flip-flop in its set or clear condition?

Procedure Question 2

When a momentary low is applied to the CLR input, is a 1 or a 0 produced at the Q output?

Step 3. Turn on the power and verify the answers for Procedure Questions 1 and 2.

Background Information

Synchronous Inputs

The three remaining input leads, labeled *J, K,* and CLK, are called *synchronous inputs.* The logic states applied to them cause the flip-flop to operate four different ways. These are called *modes of operation,* and they are classified as *toggle, hold, set,* and *reset.* Figure 7-30 shows the function table that summarizes the flip-flop's operation. Notice that there are four possible combinations of logic states applied to the *J* and *K* inputs. The logic states applied before the clock pulse cause the flip-flop to respond in one of four different ways after the clock pulse arrives.

Inputs			Outputs			Mode of operation
CLK	*J*	*K*	*Q*	$\overline{Q}$	Effect on output *Q*	
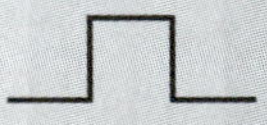 ⎍	0	0	No change		No change — disable	Hold
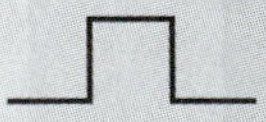 ⎍	0	1	0	1	Reset or cleared to 0	Reset
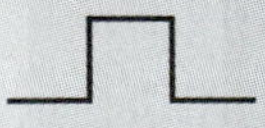 ⎍	1	0	1	0	Set to 1	Set
 ⎍	1	1	Toggle		Changes to opposite state	Toggle

FIGURE 7.30

Toggle Mode

When both the J and K inputs have highs applied to them at the same time before the clock pulse, the J–K flip-flop is in the *toggle* mode of operation. As clock pulses are applied to the CLK input, the flip-flop changes states on every negative edge of each square wave. This characteristic enables the J–K flip-flop to operate like a toggle (T) flip-flop. The T flip-flop is capable of dividing in half the square wave frequency applied to its input.

Step 4. Place the PR, CLR, J, and K input switches in the high states.

Procedure Question 3

What will happen to the Q and $\overline{Q}$ output leads of the J–K flip-flop every time the push button at the CLK input is pressed?

Step 5. Verify the answer to Procedure Question 3.
Step 6. Replace the logic pulse push button with a clock generator and apply 1 KHz to the CLK input.

Procedure Question 4

What should be the frequency at the Q output of the flip-flop?

Step 7. With the frequency counter or oscilloscope, verify the answer to Procedure Question 4.

Hold Mode

When both the J and K inputs have lows applied to them at the same time before the clock pulse, the flip-flop is in the *hold* mode of operation. If a clock pulse is applied, the output states will remain unchanged.

Step 8. Using the PR input, set the flip-flop with a momentary low. Place the logic switches connected to the J and K inputs to a low.

Procedure Question 5

What logic state will be present at the Q output after a clock pulse is applied?

Step 9. Verify your answer to Procedure Question 5.
Step 10. Using the CLR input, reset the flip-flop with a momentary low. Logic lows should still be applied to the J and K inputs.

Procedure Question 6

What logic state will be present at the Q output after a clock pulse?

Step 11. Verify the answer to Procedure Question 6.

D Flip-Flop Operation

To operate as a D flip-flop, the J and K inputs must be at opposite states. To transfer a logic 1 bit into the flip-flop, a high is applied to the J input before a clock pulse. The flip-flop is then in the *set* mode. After the clock pulse, the 1 is transferred to the Q output. A logic 0 bit is transferred into the flip-flop the same way. Before the clock pulse, a low is applied to the J input. The flip-flop is in the reset mode. After the clock pulse, a 0 is at the Q output.

Procedure Question 7

To transfer a logic 1 bit into the flip-flop, what logic state is applied to the following inputs before the clock pulse?

$J = \underline{\hspace{2cm}}$ $K = \underline{\hspace{2cm}}$

Step 12. Verify your answer to Procedure Question 7 by placing a high at the J input and a low at the K input. Then apply a clock pulse.

Step 13. Transfer a logic 0 bit into the flip-flop by placing a low at the J input and a high at the K input. Then apply a clock pulse.

Step 14. Fill in the blank spaces of Table 7-3 to summarize the operation of the J–K flip-flop.

TABLE 7.3

Inputs					Outputs				
		Synchronous			Before clock pulse		Mode of operation	After clock pulse	
Preset	Clear	CLK	J	K	Q	$\overline{Q}$		Q	$\overline{Q}$
1	1	⊓	0	0	1	0			
1	1	⊓	0	1	1	0			
1	1	⊓	1	0	0	1			
1	1	⊓	1	1	1	0			
0	1	⊓	0	0	1	0			
1	1	⊓	1	0	1	0			
1	1	⊓	1	1	1	0			
1	1	⊓	0	0	0	1			
1	1	⊓	0	1	0	1			
1	1	⊓	1	0	0	1			
1	0	⊓	0	0	0	1			
1	1	⊓	1	1	0	1			

Step 15. Verify your answer by placing the input switches of Fig. 7–28 in the positions listed on each line of the table. Observe the LEDs at the outputs to determine whether the flip-flop responds as you expected it to.

■ EXPERIMENT QUESTIONS

1. The primary output of a J–K flip-flop is the \underline{\hspace{1.5cm}} ($Q, \overline{Q}$) lead.
2. The flip-flop is set when the \underline{\hspace{1.5cm}} ($Q, \overline{Q}$) output is a 0.
3. The flip-flop is cleared when a \underline{\hspace{1.5cm}} (low, high) is applied to the PR lead and a (low, high) is at the CLR lead.
4. The \underline{\hspace{1.5cm}} and \underline{\hspace{1.5cm}} override all synchronous inputs.
5. To place the flip-flop in the toggle mode, list the logic states that the following input leads must be in:

PR =

CLR =

$J =$

$K =$

6. If 100 Hz is applied to a T flip-flop, what is the frequency at the Q output?

7. Suppose the flip-flop is reset and is in the hold mode. What logic state will be at the Q output after the next clock pulse?

8. When the J–K flip-flop is in the reset mode, a _____ (0,1) is at the Q output after the clock pulse is applied.

9. If the following signals are applied to the inputs of the J–K flip-flop before a clock pulse, which logic state will be at the Q output after the clock pulse? Why?

PR: 0

CLR: 1

J: 0

K: 1

7.12   MASTER–SLAVE J–K FLIP-FLOP

Glitch:

An unwanted logic pulse that often results when two or more logic components operate at different speeds rather than simultaneously.

Although there are several versions of the J–K flip-flop, a very common type is shown in Fig. 7-31. It is called a *master–slave* flip-flop, and a positive-level-triggered type made from two gated R–S latches is illustrated in this figure. On the left, it shows the master section of the circuit made up of a NOR-type R–S latch in schematic form and two enabler gates. An identical circuit on the right is the slave section of the device. An inverter is placed between the clock inputs of each section. The primary purpose of a master–slave flip-flop is to prevent race condition problems. Race conditions generate an unwanted logic pulse called a **glitch**.

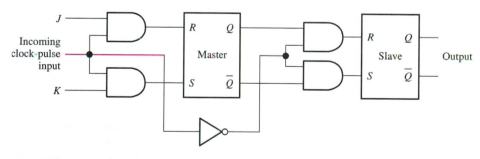

FIGURE 7.31
Master–slave flip-flop diagram.

Glitches

Figure 7-32 shows a NAND gate with its truth table and waveform diagram to demonstrate what a glitch is.

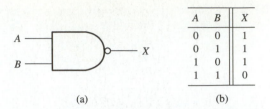

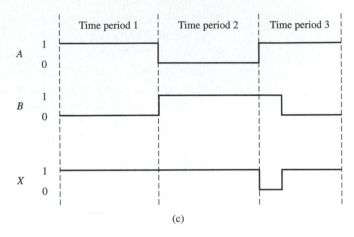

(c)

FIGURE 7.32
NAND gate: (a) Schematic symbol. (b) Truth table. (c) Timing diagram.

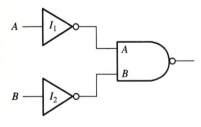

FIGURE 7.33

Two inverters and a NAND gate show how a glitch is formed.

Time Period 1   During time period 1, input A initially has a 1 state applied to it, while at the same time, input B has a 0 state. The output generated is a 1 state.

Time Period 2   At the beginning of time period 2, both inputs A and B change states ($A = 0$ and $B = 1$) at exactly the same time. The truth table verifies that the output of the NAND gate remains at a 1 state.

Time Period 3   At the beginning of time period 3, both inputs change again. However, suppose that input B changes state slightly slower than input A. For a brief moment, both inputs are a logical 1, and the output of the NAND gate changes from 1 to 0. The B input soon becomes a 0 state and, therefore, returns the output of the gate to a high. However, momentarily, the gate output was a 0 state, or an unwanted logic pulse called a *glitch*.

Glitches are often a result of two or more logic components that operate at different speeds because their propagation delays are not the same. For example, Fig. 7-33 shows an inverter connected to each input of the NAND gate that was described in Fig. 7-32. Suppose that two logic states were applied to the input of each inverter precisely at the same time. If the propagation delay of I_1 is shorter than the propagation delay of I_2, the glitch shown in the timing diagram of Fig. 7-32(c) would occur.

Despite being only a few nanoseconds in duration, many digital circuits (especially flip-flops) recognize glitches as valid binary information and respond by changing states when they are not supposed to.

Glitches can be 1 or 0 logic states and are a common problem found in digital circuits. Flip-flops can be protected from glitches by using a second latch, such as the one used in the master–slave configuration illustrated in Fig. 7-31.

Circuit Operation

(Refer to Fig. 7-31.)

1. Binary data is placed at the J and K inputs.
2. The leading edge of a clock pulse arrives, which passes the J and K inputs to the R and S inputs of the master latch.
3. During the time the incoming clock signal is a 1 state, the master latch has plenty of time to set up and allow for glitches to occur. For example, suppose that because of the propagation delays of the circuitry connected to the J or K inputs, an incorrect signal was applied before the leading edge of the clock pulse arrived. The desired master-latch condition can be corrected because the 1 state time duration of the clock pulse provides enough time to allow for glitches to pass.
4. Also, during the 1 state of the incoming clock pulse, the slave latch is disabled as it waits for the Q and $\overline{Q}$ outputs of the master latch, because the inverter holds its clock input low.
5. The trailing edge of the incoming clock pulse to the master latch arrives as it goes to a 0 state. This clock signal is complemented to a 1 state through the inverter, which enables the slave latch as the data are transferred from the master to the slave.

Even though some types of master–slave flip-flops are enabled during an active-level clock signal, they are still technically called flip-flops instead of latches.

■ REVIEW QUESTIONS

23. A T-type flip-flop is also called a _____ flip-flop and is capable of dividing the frequency of the clock signal applied to it by _____.
24. The T flip-flops are often used in
    (a) Storage registers
    (b) Multiplexers
    (c) Counters
    (d) Encoders
25. Activating the PR input of the circuit in Fig. 7-26 with a (low or high) effectively (sets or clears) the flip-flop so that a logical (0 or 1) appears at the Q output.
26. The _____-type flip-flop is the most versatile of all the flip-flops.
27. In a J–K flip-flop, if $J = 0$ and $K = 1$, what will be output Q after the next clock pulse: (0 or 1)?
28. Which of the following are ways in which the J–K flip-flop in Fig. 7-26 can be cleared?
    (a) Ground the clock input
    (b) Ground the clear input
    (c) Ground the preset input
    (d) Set J to 0 and K to 1, and apply the clock pulse
    (e) Set J to 1 and K to 0, and apply a clock pulse
29. Preset and clear inputs override all other inputs. True or false?
30. Fill in the truth table of Fig. 7-34 for the J–K flip-flop.

Inputs			Outputs			Mode of operation
CLK	J	K	Q	$\overline{Q}$	Effect on output Q	
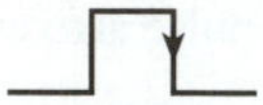 ⊓↓	0	0				
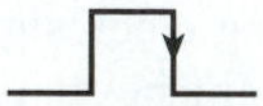 ⊓↓	0	1				
 ⊓↓	1	0				
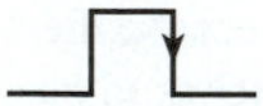 ⊓↓	1	1				

FIGURE 7.34
Truth table for Review Question 30.

7.13 TROUBLESHOOTING FLIP-FLOPS

An IC flip-flop tends to fail catastrophically; it either works or it does not. Faults develop internally or externally to the IC package, as described next.

External

- An open lead
- A short to V_{CC} or ground or between interconnections

Internal

- An open interconnect
- Shorted interconnects
- A short from V_{CC} to interconnect
- A short from ground to interconnect
- An internal defect that produces a "stuck-at" condition

The operation of the flip-flop can be checked in several ways. One procedure is to connect a logic pulser at the preset or clear inputs and apply a pulse to them. The logic probe can then be connected to the Q output to detect if the flip-flop sets or resets, as shown in Fig. 7-35.

Another procedure is to connect a 1 to the J and K inputs so that the flip-flop is in the toggle mode. Then apply clock pulses with a pulser to the clock input. The logic probe connected to the Q output should detect the flip-flop changing states every time a clock pulse is applied by the pulser, as shown in Fig. 7-36.

It is possible to test an in-circuit J–K flip-flop to determine if it is operating properly in the set or reset mode. To determine if it can set, place a jumper from ground to the K terminal. Then place a jumper between the J and clock inputs. When it is pulsed, the flip-flop sets. See Fig. 7-37(a). The reset operation can be checked by using the same procedure when one jumper wire is placed between ground and the J input and another jumper between the K and clock inputs. See Fig. 7-37(b).

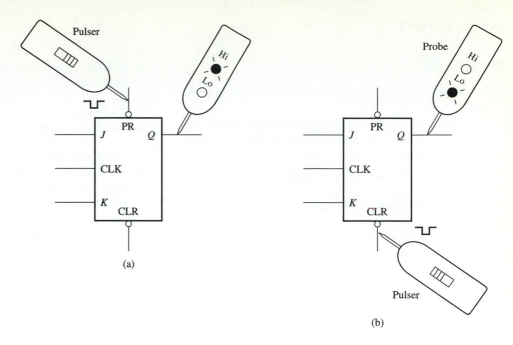

(a)

(b)

FIGURE 7.35
Testing the operation of a flip-flop: (a) Presetting the flip-flop. (b) Clearing the flip-flop.

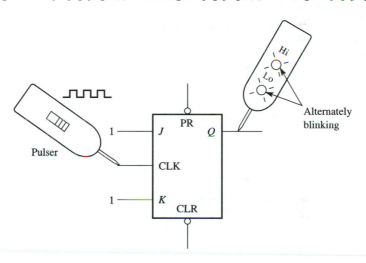

FIGURE 7.36
Determining if a flip-flop toggles.

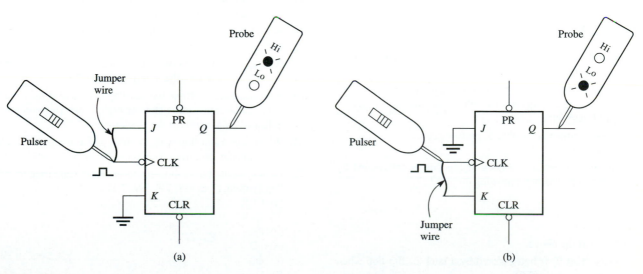

(a)

(b)

FIGURE 7.37
Testing the operation of a J–K flip-flop: (a) Testing the set mode of operation. (b) Testing the reset mode of operation.

■ SUMMARY

- Both latches and flip-flops are capable of storing 1 bit of binary data.

- Both latches and flip-flops have at least one input and two outputs, labeled Q and $\overline{Q}$, that complement each other. Unless specified otherwise, Q is the primary output.

- When a latch or flip-flop is set, Q is high, and when reset, Q is low.

- A flip-flop can be set or reset by using preset and clear inputs.

- The most elementary latch is the R–S type; its design is the basis on which most other latches and flip-flops are formed.

- Latches and flip-flops can be constructed from logic gates and are also available in IC form.

- Synchronous latches and flip-flops are activated by clock pulses four different ways: by the positive edge, positive level, negative edge, and negative level of the clock pulse.

- The operation of an asynchronous latch is determined by the signals applied to its data inputs. The operation of a synchronous latch or flip-flop is determined by the signals applied to the data inputs and the enable or clock input.

- The difference between latches and flip-flops is how they are enabled. A latch is enabled when the enable signal is at an active level. A flip-flop is activated during the L-to-H or H-to-L transition of its enable signal.

- A D flip-flop has two inputs, the D and CLK (clock). When the clock signal goes high, the data at the D input is transferred to the Q output. This type of flip-flop is used to temporarily store binary data.

- A T flip-flop has only one input and it toggles every time a clock pulse is applied to the input. This type of flip-flop is used in counters.

- A J–K flip-flop is called the universal flip-flop because it can be configured with external connections to operate like most other latches and flip-flops.

- A master–slave flip-flop is used to prevent race problems called glitches.

■ PROBLEMS

1. What is the primary difference between combination and sequential devices? (7-1)

2. A storage element comes in two forms, called a _____ and _____, that are capable of storing _____ bit(s) of binary information. (7-1)

3. A storage element is capable of performing the _____ function. (7-1)

4. The binary bit stored in a latch or flip-flop is determined by measuring which of the following outputs? (7-2)
   (a) Q
   (b) $\overline{Q}$
   (c) Q or $\overline{Q}$
   (d) None of the above.

5. A binary (1 or 0) is being stored by a latch or flip-flop when the $\overline{Q}$ output is low. (7-2)

6. Which of the following states is a flip-flop in when the $\overline{Q}$ output is high? (7-2)
   (a) Set
   (b) Reset

7. The (R–S or gated R–S) latch is categorized as an asynchronous type and the (R–S or gated R–S) latch is categorized as a synchronous type. (7-5)

8. A(n) (asynchronous or synchronous) latch responds to input data signals as soon as they arrive, and a(n) (asynchronous or synchronous) latch responds to input data signals only when a clock pulse is applied to it. (7-5)

9. When is a latch enabled? (7-7)

10. When is a flip-flop enabled? (7-7)

11. Which of the following flip-flops has only one input? (7-9)
    (a) R–S
    (b) T
    (c) D
    (d) J–K

12. Which of the following flip-flops always changes its outputs to the opposite state at every clock input pulse? (7-9)
    (a) R–S
    (b) T
    (c) D
    (d) J–K

13. Which of the following flip-flops is often referred to as the universal flip-flop because the operation of most other types of latches and flip-flops can be performed by this device? (7-10)
    (a) R–S
    (b) D
    (c) T
    (d) J–K

14. The primary advantage of the master–slave J–K flip-flop configuration is that it eliminates the unwanted signal condition called a _____. (7-12)

15. What are three ways that data can be loaded into a flip-flop? (7-11)

16. _____ and _____ inputs override the data and clock inputs. (7-11)

17. The bubble at the preset and clear inputs indicates that they are activated by a (low or high). (7-11)

18. A momentary low applied to the preset input of a J–K flip-flop makes output $Q = $ (0 or 1). (7-11)

19. A data latch can be built from a gated R–S latch by placing an _____ across the R and S inputs. (7-6)

20. An _____-triggered flip-flop is indicated by a triangle at the clock input and a (low or high) level-triggered latch is indicated by a bubble at the clock input. (7-8)

21. When does a flip-flop with a bubble and a triangle at its clock input transfer data from its input to its output? (7-8)

22. Identify each of the following acronyms that are used on a flip-flop symbol. (7-6, 7-7, 7-8, 7-9, 7-11)
    (a) CLR
    (b) CLK
    (c) PR
    (d) FF
    (e) D
    (f) T

23. What are the four modes of operation for the *J–K* flip-flop? (7-10)

24. Give a practical application for each of the following flip-flops. (7-6, 7-9, 7-10)
(a) *D*
(b) *T*
(c) *J–K*

25. Write the mode of operation for each time period and draw the *Q* output waveform for the latch of Fig. 7-38. (7-4)

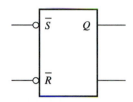

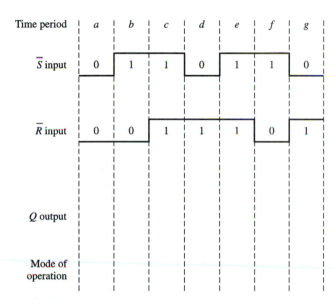

FIGURE 7.38
Latch and input waveforms for Problem 25.

26. In the table in Fig. 7-39, list the binary outputs for the 7474 *D* flip-flop IC. (7-7)

27. If a 16-kHz square wave is applied to the clock input of a *T* flip-flop, what would be the output frequency? (7-9)

28. List the binary value at the *Q* output of the 7476 flip-flop IC after each of the clock pulses in the timing diagram shown in Fig. 7-40.

29. Using the schematic symbol for the 7476 *J–K* flip-flop IC, draw this device to show how it can be configured to operate as the following: (7-10)
(a) *T* flip-flop
(b) *D* flip-flop

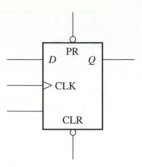

Time period	Inputs				Output
	D	CLK	PR	CLR	Q
a	1	1	1	1	
b	0	0	1	0	
c	1	1	0	1	
d	0	0	1	1	
e	0	1	1	1	
f	1	0	1	1	
g	1	1	1	1	
h	0	0	0	0	

FIGURE 7.39
D *flip-flop for Problem 26.*

Troubleshooting

30. By using a logic pulser and logic probe, as shown in Fig. 7-41, the following tests are made on a faulty *J–K* flip-flop: (7-13)

Test 1: Use the pulser to:

Set FFA ($Q_A = 1, \overline{Q_A} = 0$)
Clear FFB ($Q_B = 0, \overline{Q_B} = 1$)

After the first clock pulse to FFB,

Q_A = high
Q_B = high

After the second clock pulse to FFB,

Q_A = high
Q_B = high

Test 2: Use the pulser to:

Clear FFA ($Q_A = 0, \overline{Q_A} = 1$)
Clear FFB ($Q_B = 0, Q_B = 1$)

After the first clock pulse to FFB,

Q_A = low
Q_B = high

After the second clock pulse to FFB,

Q_A = low
Q_B = low

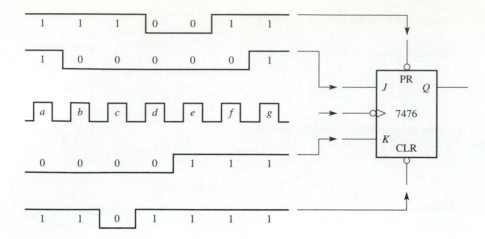

FIGURE 7.40
Timing diagram and flip-flop for Problem 28.

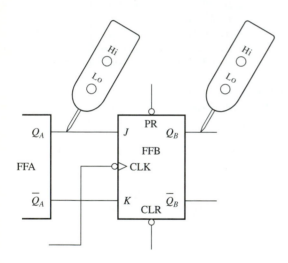

FIGURE 7.41
Faulty J–K flip-flop for Problems 30 to 32.

Tests 1 and 2 reveal that the most likely cause of the problem is which of the following?
(a) $\overline{Q_A}$ is shorted to ground.
(b) Open between Q_A and J input of FFB.
(c) FFB input J is shorted to GND.
(d) FFB input K is shorted to V_{CC}.

31. By using a logic pulser and logic probe, as shown in Fig. 7-41, the following test is made on a faulty J–K flip-flop: (7-13)

    Condition:

    FFA is set ($Q_A = 1, \overline{Q_A} = 0$)
    FFB is reset

    Symptom: After the first clock pulse, FFB remains reset.

    Which of the following conditions would cause FFB to remain unchanged?
    (a) The clock input is open.
    (b) The J input is shorted to ground.
    (c) The clear input is shorted to ground.
    (d) All of the above.

32. For the FFB in the circuit in Fig. 7-41, assume that the J and K input leads have no wires connected to them. Normally, the condition of no connections at the J and K input leads is

Inputs			Outputs			Mode of operation
CLK	J	K	Q	$\overline{Q}$	Effect on output Q	
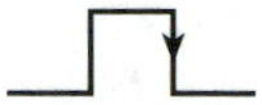 ⌐⌐⌐	0	0	No change		No change — disable	Hold
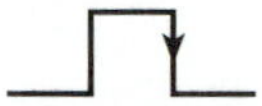 ⌐⌐⌐	0	1	0	1	Reset or cleared to 0	Reset
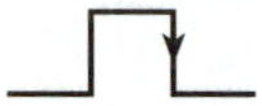 ⌐⌐⌐	1	0	1	0	Set to 1	Set
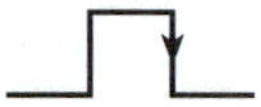 ⌐⌐⌐	1	1	Toggle		Changes to opposite state	Toggle

FIGURE 7.42
Answer to Review Question 30.

recognized as a high state. Observe the following test procedure and determine the most likely cause of failure. Pulses are applied by a logic pulser, and the Q output condition is observed by a logic probe. (7-13)

STEP	ACTION	RESULT
1	Apply a pulse to the PR input	Q = high: correct response
2	Apply a pulse to the CLR input	Q = low: correct response
3	Apply a pulse to the CLK input	Q = low: incorrect response
4	Apply a pulse to the PR input	Q = high: correct response
5	Apply a pulse to the CLK input	Q = low: correct response
6	Apply a pulse to the CLK input	Q = low: incorrect response

■ ANSWERS TO REVIEW QUESTIONS

1. combination, sequential   2. memory   3. 1
4. complement   5. Q   6. 1, 1   7. preset, clear
8. reset   9. High   10. (b)   11. 1, 0
12. low, high   13. Low   14. high, low   15. 0, 1
16. asynchronously, synchronously   17. (d)   18. (b)
19. enabled   20. Data and delay   21. (a)
22. (1) high-level, (2) positive-edge, (3) low-level,
    (4) negative-edge   23. toggle, 2   24. (c)
25. low, sets, 1
26. *J–K*   27. 0   28. (b) and (d)   29. True
30. See Fig. 7-42

CHAPTER 8

COUNTERS

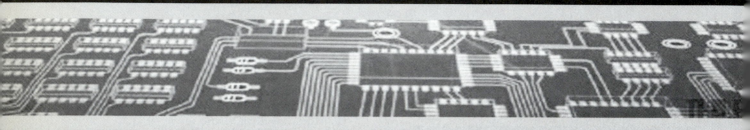

When you complete this chapter, you will be able to:

1. List the two classifications of counters and describe their characteristics.
2. Identify the types of counter configurations from a schematic diagram.
3. Determine the maximum decimal count for a specified modulo counter.
4. List practical applications of each type of counter.
5. List the correct outputs of a counter when given specific input signals.
6. Calibrate the output frequency of a frequency divider given an input frequency.
7. Determine the operation of counters that are in IC form.
8. Perform basic troubleshooting of counters.

INTRODUCTION TO SEQUENTIAL CIRCUITS

The different types of logic gate devices described in Chapter 3 are capable of making basic decisions by producing an output signal resulting from the signals applied to the inputs. In Chapter 5, these building block devices were connected together in various configurations called combination circuits. Combining gates makes it possible to perform operations that are more complex than those performed by single gates.

Another building block device called a flip-flop was introduced in Chapter 7. Alone, it is capable of two primary functions: storing one bit of binary data and dividing in half a square wave frequency applied to its input. A circuit constructed of several flip-flops is capable of performing operations more complex than those performed by a single flip-flop. When flip-flops are connected together, they are called *sequential circuits* which is the subject of this and the next chapter. They are called sequential circuits because they are either controlled by, or used for controlling, other circuitry in a specific sequence dictated by a clock or an enable/disable signal.

Sequential circuits are classified into two major categories, *counters* and *registers*. These circuits are vital in the operation of digital equipment.

8.1 INTRODUCTION TO COUNTERS

Counter:

A common digital circuit made up primarily of flip-flops that tallies the number of pulses arriving at its input.

A **counter** is a common digital circuit made up of flip-flops and, in some cases, gates that tally the number of pulses arriving at their inputs. This number can be determined at any time because the flip-flops of the counter change state in a manner that is representative of the exact number of pulses applied to the input. Some counters increment and other counters decrement each time a pulse is received.

These input pulses can occur randomly or at fixed intervals. Random pulses can originate from converted electrical signals that represent objects as they are counted. They can also represent random events that occur over a period of time, such as scoring points in a game. Pulses that occur at precise fixed intervals are commonly called **clock pulses.** These input pulses are often used by timing devices, such as digital clocks, stop watches, appliances, and traffic control systems such as lights at an intersection. Computers use counters to control the order of events and operations. When a program is entered into a computer, an internal counter checks that one step is completed before allowing the computer to advance to the next operation.

Counters may be divided into two general categories, *asynchronous* and *synchronous*.

Clock pulses:

Pulses that are applied to digital circuits, usually at precise fixed intervals.

Asynchronous counter:

A basic counter consisting of several cascading J–K flip-flops that receive their clock signal from the output of the preceding flip-flop.

Ripple counter:

A counter consisting of several cascading J–K flip-flops that receive their clock signal from the output of the preceding flip-flop.

Asynchronous Counter

The **asynchronous counter,** or **ripple counter,** is the simplest and most basic counter. It can be constructed with J–K flip-flops. The clock for each successive flip-flop is obtained from the output of the preceding flip-flop. The term *ripple* is used because clock pulses ripple from one flip-flop to the next in a sequential order, causing the count to propagate down the line. There are several types of asynchronous counters. The ones described in this chapter are used in the circuitry covered in subsequent chapters.

8.2 BINARY UP-COUNTER

Up-counter:

A counter consisting of several cascading flip-flops that increases its count by 1 on every clock pulse applied to its input.

A common type of asynchronous counter is the binary **up-counter,** a device capable of counting the natural binary code in a sequential fashion. A binary counter such as the one shown in Fig. 8-1(a) consists of four J–K flip-flops connected in cascade and counts in binary from 0_{10} (0000) to 15_{10} (1111). The individual Q outputs of each flip-flop represent one of the weighted-value positions in a binary number.

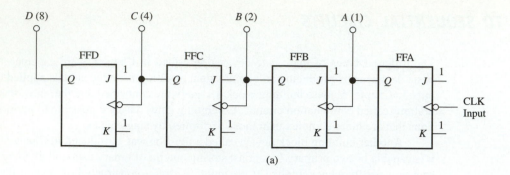

(a)

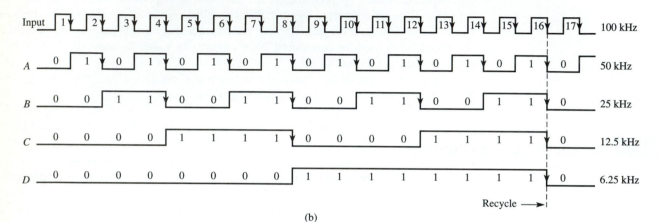

(b)

Pulse no.	D	C	B	A	
	0	0	0	0	
1	0	0	0	1	
2	0	0	1	0	
3	0	0	1	1	
4	0	1	0	0	
5	0	1	0	1	
6	0	1	1	0	
7	0	1	1	1	Recycle
8	1	0	0	0	
9	1	0	0	1	
10	1	0	1	0	
11	1	0	1	1	
12	1	1	0	0	
13	1	1	0	1	
14	1	1	1	0	
15	1	1	1	1	
16					

(c)

FIGURE 8.1
Four-bit binary up-counter: (a) Logic diagram. (b) Timing diagram. (c) Count sequence.

The clock input connected at FFA is the input of the counter, as each incoming clock pulse represents bits to be counted. The Q output of FFA serves as the 1s output of the counter and also is the input clock to FFB. The Q output of FFB acts as the 2s output of the counter and also is the input clock to FFC. The Q output of FFC serves as the 4s output of the counter and also is the input clock to FFD. Likewise, the Q output of FFD acts as the 8s output of the counter.

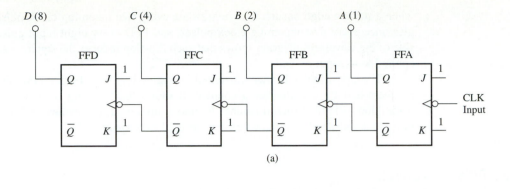

(a)

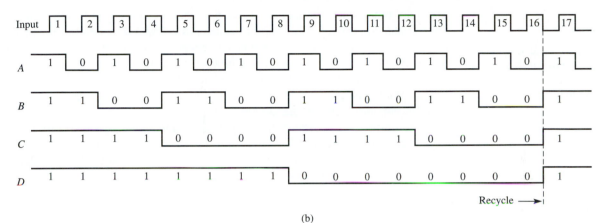

(b)

D	C	B	A	
1	1	1	1	
1	1	1	0	
1	1	0	1	
1	1	0	0	
1	0	1	1	
1	0	1	0	
1	0	0	1	
1	0	0	0	Recycle
0	1	1	1	
0	1	1	0	
0	1	0	1	
0	1	0	0	
0	0	1	1	
0	0	1	0	
0	0	0	1	
0	0	0	0	

(c)

FIGURE 8.2
Four-bit binary down-counter: (a) Logic diagram. (b) Timing diagram. (c) Count sequence.

All of the J and K inputs are at a permanent 1 state. An inverter bubble and triangle are located at each clock input. As a result, each flip-flop toggles when a trailing edge of a pulse arrives at its clock input.

The operation of this counter can be illustrated by the timing diagram of Fig. 8-1(b), which includes the input to the counter and the outputs of each flip-flop. The counter begins at a count of 0000. The timing diagram shows that output A toggles back and forth between 0 and 1 every time the trailing edge of each incoming pulse arrives at the counter input. Because the clock input of FFB is connected to output A and because it toggles only when a trailing edge occurs, it switches state every two incoming clock pulses, FFC toggles once every four incoming clock pulses, and FFD every eight input pulses. The right side of the waveform diagram shows that each flip-flop reduces the square wave frequency applied to its input by half.

The operation is further illustrated in Fig. 8-1(c) by a binary table that shows the numbers that are actually being counted. It shows the counting sequence of this circuit from 0000 to 1111. When it reaches its maximum count, it recycles back to 0000 on the next clock pulse to begin the count sequence all over again.

8.3 BINARY DOWN-COUNTER

Down-counter:

A counter consisting of several cascading flip-flops which decreases its count by one on every clock pulse applied to its input.

There are times where an application requires that a counter count from some predetermined value toward zero. This type of counter is called a **down-counter.** The logic diagram configuration of a 4-bit down-counter appears in Fig. 8-2(a). It looks very similar to the binary up-counter. The Q outputs of FFA-FFD remain as the 8-4-2-1 binary positional-value output connections of the counter. What makes it operate as a down-counter is that instead of the Q outputs of FFA, B, and C providing clock signals to the next more significant flip-flop, the $\overline{Q}$ outputs are used instead.

To illustrate its operation, a timing diagram and count sequence table are shown in Figs. 8-2(b) and (c), respectively. Each flip-flop is set and the counter begins at 1111. At the end of the fifteenth clock pulse, the counter decrements to 0000. After the sixteenth clock pulse, it recycles to 1111.

8.4 CONTROLLING COUNTER FUNCTIONS

It is often necessary to begin either the count-up or count-down sequence of a counter starting from a desired number. This number can be inserted into the counter by using either the reset or preset leads of each flip-flop, as shown in Fig. 8-3.

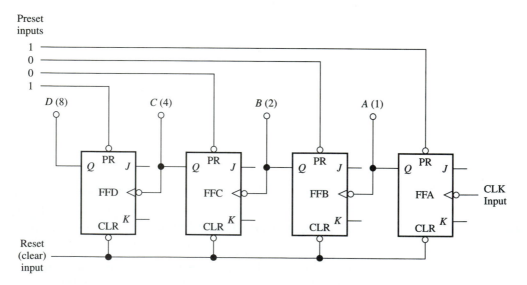

FIGURE 8.3

Presetting a 6_{10} into the counter.

The reset operation is also known as the *clear function.* For the counter to operate normally, the reset line must be open or held high. As soon as the reset line is brought low, however, the entire counter resets to 0000.

The preset operation is also referred to as the *set function*. The preset inputs are used to load a specific number other than 0000 into the counter. To do so, the counter is first cleared. Then a low must be applied to the preset input of any flip-flop that is to be set to 1 and a high to any flip-flop that is to remain at 0. For example, for the decimal number 6 to be loaded into the counter in Fig. 8-3, a 1001 is applied to the preset inputs. Before the counter sequence begins, these lines must then be brought high.

EXAMPLE 8.1

Prepare a timing diagram using the clear and preset inputs of FFA-FFD to show how a 4_{10} is loaded into the counter circuit in Fig. 8-3.

Solution See Fig. 8-4.

FIGURE 8.4
Timing diagram for Example 8.1.

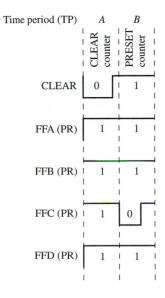

Time Period (TP)

A   A momentary low at the clear input makes the contents of the counter 0000. This action removes any number stored in the counter before a new value is entered.

B   A momentary low at the PR input of flip-flop *C* causes it to set while the other flip-flops remain reset. The contents of the counter become 0100_2.

■ REVIEW QUESTIONS

1. When several logic gates are connected together, they are called _____ circuits. When several flip-flops are connected together, they are called _____ circuits.

2. Sequential circuits are divided into two major categories: _____ and _____ .

3. The two major classifications of counters are _____ and _____ .

4. A(n) _____ counter uses the output of one flip-flop to clock the input of the next flip-flop.

5. Pulses applied to an input of a counter at precise intervals are commonly called _____ pulses.

6. In a binary asynchronous up-counter, the _____ output of one flip-flop is connected to the clock input of the next flip-flop. In a binary asynchronous down-counter, the _____ output of a flip-flop is connected to the clock input of the next flip-flop.

8.5 MODULO COUNTERS

Modulo:

The maximum number of counts a counter is capable of making.

Counters are sometimes identified by the maximum number of counts they are capable of making. The maximum count is designated by the word **modulo.** For example, a 3-bit pure binary counter has a natural modulus of 8_{10} (0–7), a 4-bit binary counter is capable of 16 counts (0–15), and a binary 5-stage counter 32_{10} (0–31).

Counters can be made to count numbers that are not pure binary. This is accomplished by using a counter with a natural modulus and modifying it so that it skips the proper number of steps.

8.6 MOD-10 DECADE UP-COUNTER

Decade counter:

An up-counter that can count ten different numbers, 0 through 9.

BCD counter:

Also known as a decade counter, it counts in BCD from 0000 to 1001.

The most popular modulo counter is the **decade counter.** It is also commonly known as a **BCD counter** because it is used to count the standard 8–4–2–1 BCD code ranging from 0000 to 1001. The fact that it counts 10 clock pulses before recycling makes it ideal for many applications that require decimal digit displays. There are several ways to make a counter recycle after 10 clock pulses. One method is shown in Fig. 8-5. This counter consists of four *J–K* flip-flops and one AND gate. FFA is the LSB flip-flop (also the counter input) and toggles on the trailing edge of each incoming clock pulse.

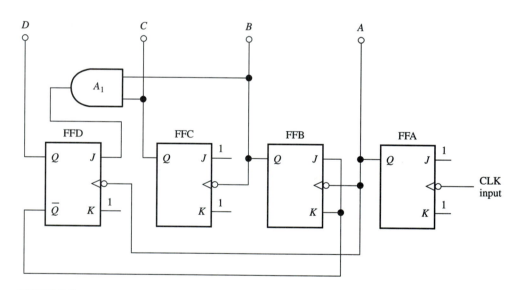

FIGURE 8.5
A mod-10 up-counter. (Before any clock pulse arrives, the decade counter is cleared to 0000.)

Before the counter begins to count, all flip-flops are reset. As a result, the $\overline{Q}$ output of FFD is a high, which is fed back to the *J* and *K* inputs of FFB, putting it in the *toggle* mode. Because the *J* and *K* inputs of FFA and FFC are at permanent highs, they are also in the *toggle* mode. The *Q* outputs of FFB and FFC are low, which produces a low at the output of A_1. With a 0 at the *J* input of FFD and a 1 at the *K* input, it is in the *reset* mode. Therefore, this circuit operates identically to the mod-16 binary ripple up-counter during the first seven incoming clock pulses.

When 7_{10} is in the counter (0111_2), FFB and FFC *Q* outputs are both high, which enables the AND gate. With a 1 state at the output of the AND gate, a high is applied to the *J* input of FFD, putting it into the *toggle* mode for when the trailing edge of clock pulse 8 arrives.

The counter is next incremented to 1000 (8_{10}). This causes the $\overline{Q}$ output of FFD to apply lows to the J and K inputs of FFB, putting it into the *no-change* mode. Also, the AND gate is disabled, which puts a 0 state at the J input of FFD, placing it in the *reset* mode.

As the trailing edge of the ninth clock pulse arrives, FFA toggles again so that the counter increments to 1001_2.

On the tenth clock pulse, FFA toggles so that its Q output goes to a 0 state, which generates a trailing-edge signal that is applied to FFB and FFD. FFB is not affected because it is in the *no-change* mode because the J and K inputs are low. It does, however, toggle FFD because it is in the *reset* mode. As a result, the counter is restored to 0000 on the trailing edge of the tenth clock pulse. This decade counter is used in subsequent chapters.

8.7 MOD-6 UP-COUNTER

Mod-6 up-counter:

An up-counter capable of counting six counts from 0 to 5 before recycling back to 0.

A **mod-6 up-counter** is capable of counting from 0 to 5 before recycling back to 0. This type of counter is used to count the most significant digit of minutes and seconds in a clock. A mod-6 counter can be formed by combining a mod-3 counter with a mod-2 counter, as shown in Fig. 8-6.

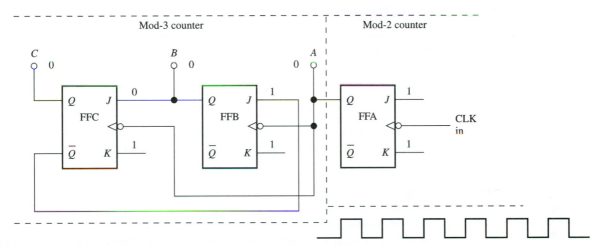

FIGURE 8.6
A mod-6 up-counter (before the first clock pulse arrives).

The single flip-flop mod-2 (divide-by-2) counter is placed at the input stage and its output provides the subordinate clock pulse to the mod-3 counter. The mod-3 stage follows its normal count pattern of 0–2, except that it remains in each count state for two counts instead of incrementing at every count as it does when used alone.

Count Sequence for the Mod-6 Up-Counter

Before the First Clock Pulse

The diagram in Fig. 8-6 shows all of the flip-flops reset and the logic states that are applied to their J and K inputs.

After the First Clock Pulse

The negative edge of the first clock pulse causes FFA to toggle. The high at its output creates a count of 001_2. The logic states applied to the J and K inputs of FFB and FFC remain the same.

After the Second Clock Pulse

The negative edge of the second clock pulse causes FFA to toggle. The high-to-low transition of its Q output causes FFB to also toggle. FFC does not change because it was in the reset mode prior to the clock pulse. The result is a count of 010_2 is present in the counter. The high at the output of FFB is now applied to the J input of FFC, which puts it into the toggle mode.

After the Third Clock Pulse

The negative edge of the third clock pulse causes FFA to toggle and increment the count to 011_2. The other flip-flops remain unchanged and are in the toggle mode for the next clock pulse.

After the Fourth Clock Pulse

The negative edge of the fourth clock pulse causes all flip-flops to toggle, resulting in a count of 100_2. A low applied to the J inputs of FFB and FFC cause them to go into the reset mode.

After the Fifth Clock Pulse

The negative edge of the fifth clock pulse causes FFA to toggle and increment the count to 101_2. Flip-flops B and C remain unaffected because they did not receive a negative-edge clock input from FFA.

After the Sixth Clock Pulse

The negative edge of the sixth clock pulse causes FFA to toggle to a 0. The high-to-low transition of its output causes FFC to reset, and the counter recycles to 000_2.

8.8 ■ MOD-10 DOWN-COUNTER

Mod-10 down-counter:

A counter that makes ten counts backwards from 9 to 0 before recycling back to 9.

A **mod-10 down-counter** is capable of counting backwards from 9 to 0 before recycling back to its maximum count of 9. This type of counter could be used to decrement the least significant minutes and seconds digits on a scoreboard.

The circuit configuration of a mod-10 down-counter consists of four J–K flip-flops, a NOR gate, and a NAND gate. Figure 8-7 shows the schematic diagram of a mod-10 down-counter.

Count Sequence for the Mod-10 Down-Counter

Before the First Clock Pulse Applied to FFA (always in the toggle mode)

All four flip-flops are in the toggle mode.

After the First Clock Pulse

The negative edge causes FFA to toggle. The low-to-high transition from its $\overline{Q}$ output applied to the clock inputs of FFB and FFC have no effect. The count decrements to $8(1000_2)$.

After the Second Clock Pulse

The negative edge causes FFA to toggle. The high-to-low transition from its $\overline{Q}$ output causes FFB and FFD to toggle. The high-to-low transition from the $\overline{Q}$ output of FFB causes FFC to toggle. The count decrements to $7(0111_2)$. The lows at the $\overline{Q}$ outputs of FFB

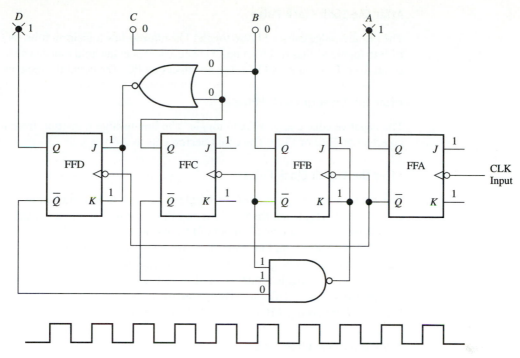

FIGURE 8.7
A mod-10 down-counter (before the first clock pulse arrives).

and FFC cause the NAND gate output to remain at a high. FFB remains in the toggle mode. The highs at the Q outputs of FFB and FFC cause the NOR gate output to apply a low at the J and K inputs of FFD. It is in the hold mode.

Until the number in the counter decrements to 1, at least one low state will be applied to an input of the NAND gate, causing it to produce a high at its output. This logic 1 is applied to the J and K inputs of FFB, causing it to be in the toggle mode until after the eighth clock pulse is applied to the counter.

Until the number in the counter decrements to 1, at least one high state will be applied to an input of the NOR gate, causing it to produce a low at its output. This logic 0 is applied to the J and K inputs of FFD, causing it to be in the hold mode until after the eighth clock pulse is applied to the counter.

After the Third Clock Pulse

The negative edge causes FFA to toggle. The low-to-high transition from its $\overline{Q}$ output has no effect on FFB or FFD. The count decrements to $6(0110_2)$.

After the Fourth Clock Pulse

The negative edge causes FFA to toggle. The high-to-low transition from its $\overline{Q}$ output causes FFB to toggle but has no effect on FFD because it is in the hold mode. The low-to-high transition from the $\overline{Q}$ output of FFB has no effect on FFC. The count decrements to $5(0101_2)$.

After the Fifth Clock Pulse

The negative edge causes FFA to toggle. The low-to-high-transition from its $\overline{Q}$ output has no effect on FFB or FFD. The count decrements to $4(0100_2)$.

After the Sixth Clock Pulse

The negative edge causes FFA to toggle. The high-to-low transition from its $\overline{Q}$ output causes FFB to toggle but has no effect on FFD because it is in the hold mode. The low-to-high transition from the $\overline{Q}$ output of FFB has no effect on FFC. The count decrements to $3(0011_2)$.

After the Seventh Clock Pulse

The negative edge causes FFA to toggle. The low-to-high transition from its $\overline{Q}$ output has no effect on FFB or FFD. The count decrements to $2(0010_2)$.

After the Eighth Clock Pulse

The negative edge causes FFA to toggle. The high-to-low transition from its $\overline{Q}$ output causes FFB to toggle but has no effect on FFD because it was in the toggle mode. The low-to-high transition from the $\overline{Q}$ output of FFB has no effect on FFC. The count decrements to $1(0001_2)$. With FFB and FFC reset, lows from their Q outputs are applied to the NOR gate. The output of the NOR gate is a high, which is applied to the J and K inputs of FFD, causing it to be in the toggle mode. With FFB, FFC and FFD reset, highs from their $\overline{Q}$ outputs are applied to the NAND gate. The output of the NAND gate is a low, which is applied to the J and K inputs of FFB, causing it to be in the hold mode.

After the Ninth Clock Pulse

The negative edge causes FFA to toggle. The low-to-high transition from its $\overline{Q}$ output has no effect on FFB or FFD. The count decrements to $0(0000_2)$.

After the Tenth Clock Pulse

The negative edge causes FFA to toggle. The high-to-low transition from its $\overline{Q}$ output causes FFD to toggle. FFB does not change because it is in the hold mode. The count recycles to $9(1001_2)$.

8.9 MOD-6 DOWN-COUNTER

Mod-6 down-counter:

A counter that makes six counts backwards from 5 to 0 before recycling back to 5.

A **mod-6 down-counter** is capable of counting backwards from 5 to 0 before recycling back to its maximum count of 5 (101_2). This type of counter could be used to decrement the most significant minutes and seconds digits on a scoreboard.

The circuit configuration of a mod-6 down-counter consists of three J–K flip-flops and an OR gate. Figure 8-8 shows the schematic diagram of a mod-6 down-counter.

Count Sequence for the Mod-6 Down-Counter

Before the First Clock Pulse

All three flip-flops are in the toggle mode.

After the First Clock Pulse

The negative edge causes FFA to toggle. The positive transition of its $\overline{Q}$ output applied to the clock inputs of FFB and FFC does not cause them to toggle. The count decrements to $4(100_2)$.

After the Second Clock Pulse

The negative edge causes FFA to toggle. The negative transition of its $\overline{Q}$ output applied to the clock inputs of FFB and FFC cause them both to toggle. The count decrements to $3(011_2)$. The Q output of FFB applies a 1 to the OR gate. The 1 from the OR gate output is

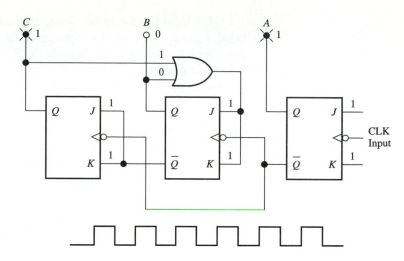

FIGURE 8.8
A mod-6 down-counter (before the first clock pulse arrives).

applied to the J and K inputs of FFB, causing it to remain in the toggle mode. The $\overline{Q}$ output of FFB applies a 0 to the J and K inputs of FFC, putting it in the hold mode.

After the Third Clock Pulse

The negative edge causes FFA to toggle. The positive transition of its $\overline{Q}$ output applied to the clock inputs of FFB and FFC does not cause them to change. The count decrements to $2(010_2)$.

After the Fourth Clock Pulse

The negative edge causes FFA to toggle. The negative transition of its $\overline{Q}$ output applied to FFB causes it to toggle. FFC remains unchanged because it was in the hold mode. The count decrements to $1(001_2)$. Two lows applied to the OR gate causes it to feed a low at the J and K inputs of FFB, which puts it in the hold mode. The high at the $\overline{Q}$ output of FFB is fed to the J and K inputs of FFC, putting it in the toggle mode.

After the Fifth Clock Pulse

The negative edge causes FFA to toggle. The low-to-high clock transition from its $\overline{Q}$ output does not cause FFC to toggle. The count decrements to 000_2. FFA and FFC are in the toggle mode, and FFB is in the hold mode.

After the Sixth Clock Pulse

The negative edge causes FFA to toggle. The high-to-low transition of its $\overline{Q}$ output causes FFC to toggle. FFB remains unchanged because it is in the hold mode. The count recycles to $5(101_2)$.

Asynchronous counter circuits are available in standard medium-scale IC packages. Two widely used counter ICs are the 7493 binary counter and the 7490 decade counter.

8.10 THE 7493 BINARY COUNTER

Binary counter:

An up-counter that is capable of counting in pure binary.

The 7493 integrated circuit operates as a binary up-counter. It contains four negative-edge-triggered J–K flip-flops, and internally it is made from a mod-2 and a mod-8 counter. Each counter can be used separately or connected externally to form a mod-16 counter.

The logic diagram of this device appears in Fig. 8-9(a). It shows two clock inputs labeled A and B. Clock input A is the clock for the mod-2 counter, and clock input B is the clock for the mod-8 counter. To make the IC operate as a mod-16 counter, the clock input is connected to input A and an external jumper is connected from the mod-2 output, Q_A, to the clock input of the mod-8 counter, input B.

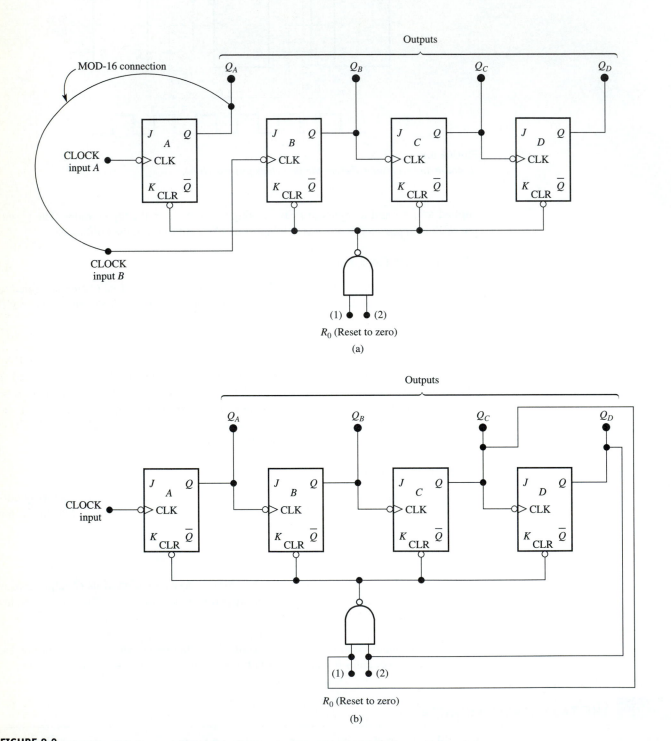

FIGURE 8.9
The 7493 binary up-counter IC: (a) Mod-16 binary up-counter. (b) Mod-12 up-counter. (c) Pin diagram. (d) Count sequence.

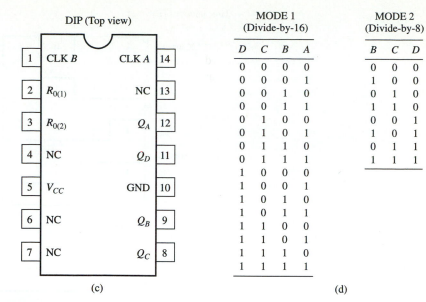

DIP (Top view)	
1 CLK B	CLK A 14
2 $R_{0(1)}$	NC 13
3 $R_{0(2)}$	Q_A 12
4 NC	Q_D 11
5 V_{CC}	GND 10
6 NC	Q_B 9
7 NC	Q_C 8

(c)

MODE 1 (Divide-by-16)					MODE 2 (Divide-by-8)		
D	C	B	A		B	C	D
0	0	0	0		0	0	0
0	0	0	1		1	0	0
0	0	1	0		0	1	0
0	0	1	1		1	1	0
0	1	0	0		0	0	1
0	1	0	1		1	0	1
0	1	1	0		0	1	1
0	1	1	1		1	1	1
1	0	0	0				
1	0	0	1				
1	0	1	0				
1	0	1	1				
1	1	0	0				
1	1	0	1				
1	1	1	0				
1	1	1	1				

(d)

FIGURE 8.9
(Continued)

Also included are two reset leads, labeled $R_{0(1)}$ and $R_{0(2)}$. Either or both R_0 leads must be connected low for normal counting. When both R_0 leads are at active-high states, they reset both counters at any point in the standard count order. This feature enables the 7493 mod-16 counter to operate at a different modulus count. For example, to operate as a mod-12 counter, Q_C and Q_D output leads are connected to $R_{0(1)}$ and $R_{0(2)}$, as shown in Fig. 8-9(b). As soon as the binary count of 1100_2 (12_{10}) is reached, $R_{0(1)}$ and $R_{0(2)}$ are both high, which causes the counter to reset to 0000_2 (00_{10}). Because the propagation delay for this device is only 40 nanoseconds, the count of 12 is present faster than can be detected by the human eye. Therefore, 12 is not considered a valid count.

EXAMPLE 8.2

To operate as a mod-6 counter, which output leads of the 7493 IC are connected to pins $R_{0(1)}$ AND $R_{0(2)}$?

Solution Outputs Q_B and Q_C.

Figure 8-9(c) shows the pin diagram and Fig.8-9(d) shows the function table of this IC.

EXAMPLE 8.3

The input waveforms in Fig. 8-10 are applied to the 7493 mod-16 counter. Draw the output waveforms that result. Initially, set $Q_A = 1$, $Q_B = 0$, $Q_C = 0$, and $Q_D = 1$.

Solution See Fig. 8-10.

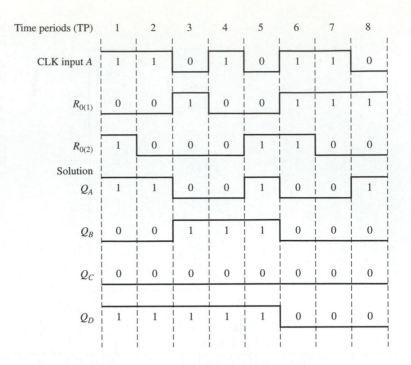

Time periods (TP)	1	2	3	4	5	6	7	8
CLK input A	1	1	0	1	0	1	1	0
$R_{0(1)}$	0	0	1	0	0	1	1	1
$R_{0(2)}$	1	0	0	0	1	1	0	0
Solution Q_A	1	1	0	0	1	0	0	1
Q_B	0	0	1	1	1	0	0	0
Q_C	0	0	0	0	0	0	0	0
Q_D	1	1	1	1	1	0	0	0

Time period (TP)

1 Both R_0 inputs are not high and the clock input is at a high state. The condition of the counter's outputs remains unchanged.

2 Both R_0 inputs are not high and the clock input is at a high state. The outputs remain unchanged.

3 Both R_0 inputs are not high and the clock input changes from a high to a low state. The count in the counter increments from 9 to 10.

4 Both R_0 inputs are not high and the clock input changes from a low to a high state. The outputs remain unchanged.

5 Both R_0 inputs are not high and the clock input changes from a high to a low state. The count in the counter increments from 10 to 11.

6 Both R_0 inputs become high and the count goes to zero.

7 Both R_0 inputs are high and the clock input is a high. The outputs remain unchanged.

8 Both R_0 inputs are not high and the clock input changes from a high to a low state. The counter increments to 1.

FIGURE 8.10
Input and output waveforms for Example 8.3.

EXPERIMENT

Mod-16 Binary Up-Counter

Objectives

■ To construct and operate a counter that counts from 0 to 15.

■ To connect a decoder to the counter to cause one of 16 LEDs to illuminate on each count.

■ To modify the circuit so that it counts in several different sequences.

Materials

1—7493 IC

1—74154 IC

16—LEDs

1—150-Ohm Resistor

1—Logic Pulse Push Button (bounceless)

1—Signal Generator

1—+5-V DC Power Supply

1—Oscilloscope or Frequency Counter

1—Voltmeter or Logic Probe

Introduction

The 7493 IC is primarily a 4-bit binary asynchronous counter that is capable of counting from 0 to 15. The counter is activated by a clock pulse signal. Each time a pulse is applied to its clock input, the counter increments one count.

The counter shown in Fig. 8-11 has four output leads, labeled A, B, C, and D. Each one represents a pure binary weighted value. A 1 at output $D = 8$, at output $C = 4$, at output $B = 2$, and at output $A = 1$. Sixteen different combinations of 1s and 0s at the outputs produce the decimal equivalent counts 0 to 15.

The purpose of this experiment is to demonstrate how a mod-16* binary up-counter repeatedly increments from 0 to 15. The four outputs of the counter are connected to a

---

*A mod- or modulus number refers to the number of different counts made by a counter. The modulus number is always one number higher than the highest count of a counter. For example, the highest count of the mod-16 counter is 15.

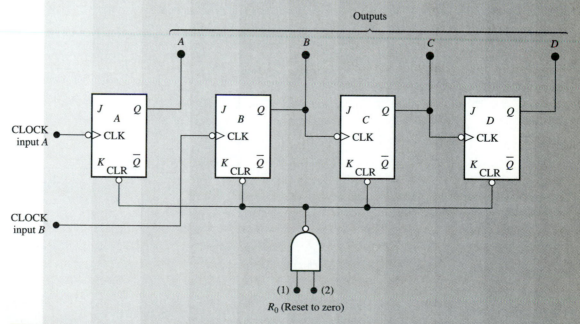

FIGURE 8.11

1-of-16 decoder. The decoder converts the binary input and drives 16 different LEDs to indicate each count. To illustrate the flexibility of the 7493 IC, the circuit will be modified to operate at several different modulus counts. It will further be altered to count in different sequences.

Procedure

Step 1. Construct the circuit shown in Fig. 8-12.

Step 2. Apply clock pulses and observe the LEDs as the counter increments through each count.

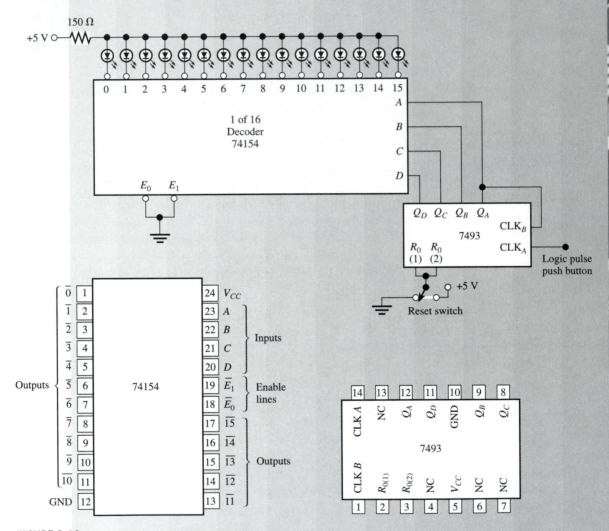

FIGURE 8.12

Step 3. For each LED count that follows, use a logic probe or voltmeter to measure the logic states present at each lead of the counter. These output states indicate the binary number in the counter.

LED Count	Counter Output Readings			
	D	C	B	A
4				
9				
13				

Procedure Question 1

Does each binary count match the LED that is lit? (Yes, No)

Background Information

$R_{0(1)}$ and $R_{0(2)}$ are reset pins used to clear the counter to zero when both pins, 2 and 3, are brought high. By using these pins, the counter can be modified to increment to a count other than 15 before recycling to zero. For example, suppose output A is connected to one reset pin and output C to the other reset pin. When the counter reaches a count of 5, outputs A and C go high. The two 1s at the reset pins clear the counter to zero. The count of 5 cannot be seen by the operator because it only lasts for several nanoseconds. Therefore, the counter repeatedly scans LEDs 0 to 4.

Step 4. Make the necessary connections to the 7493 IC so that it operates as a mod-5 counter (count 0–4).

Procedure Question 2

How can the circuit be modified to operate as the following types of counters?

Mod-6:

Mod-10:

Background Information

One function that a binary counter is capable of performing is dividing the frequency applied to it by a power of 2. One flip-flop divides by 2 (2^1), two flip-flops divide by 4 (2^2), three flip-flops divide by 8 (2^3), etc. Therefore if 100 Hz is applied to the first flip-flop, 50 Hz will be produced at its output. By connecting several flip-flops in a binary counter together, the division can be made much greater.

Procedure Question 3

If 240 Hz is applied to the input at pin 14 of the counter in Fig. 8-12, what is the frequency at the following outputs:

FFA:

FFB:

FFC:

FFD:

Step 5. Verify your answer by connecting a frequency counter or oscilloscope to the outputs of the 7493 IC configured as a mod-16 counter. Apply 240 kHz to the input at pin 14 of the counter.

Procedure Question 4

If 160 Hz is applied to the counter input, what is the output at FFD? _____ Hz. Therefore the counter becomes a divide-by-_____ counter.

ADVANCED SECTION

Background Information

The 7493 IC is made up of a mod-2 and a mod-8 counter. To operate as a mod-16 counter, the output of the mod-2 counter (Q_A) must be connected to clock input CLK B of the mod-8 counter.

Procedure Question 5

How can the 7493 IC be connected to the 74154 decoder to show its operation as a mod-2 counter?

Step 6.   Reassemble the circuit in Fig. 8-2. Remove the jumper wire between pin 1 and pin 12 to verify your answer to Procedure Question 5. (You may have to clear the counter by applying a momentary high to the R_0 inputs simultaneously.)

Procedure Question 6

How can the 7493 IC be connected to the 74154 decoder to show its operation as a mod-8 counter?

Step 7.   Make the appropriate connections to verify your answer. (You may have to clear the counter by applying a momentary high to the R_0 inputs simultaneously.)

■ EXPERIMENT QUESTIONS

1. What does mod-, or modulus, mean?
2. The 7493 IC is primarily made up of three counters which are mod-_____, mod-_____, and mod-_____.
3. The R_0 inputs are used to _____ the counter when both R_0 pins are _____ (low, high).
4. The 7493 IC can be modified to operate as a mod-12 counter by connecting which outputs to the R_0 pins?
5. What would be a practical application of the 7493 IC configured to operate as the following counters:

    Mod-6:

    Mod-10:

6. If 200 Hz is applied to the CLK B input of the 7493 IC, what is the frequency at output Q_D?

Advanced Questions

7. Without connecting any external logic devices to the 7493 IC, list all of the possible modulus count sequences it can make.
8. Using a two-input AND gate and a two-input NAND gate, how can the mod-16 counter be modified so that it counts from 0 to 6 and stops?

9. Suppose that the mod-16 7493 counter repeatedly counts from 0 to 1. What would *not* cause the problem?

    A. Output *B* is shorted to GND.

    B. Pin 1 is shorted to GND.

    C. An open exists between pins 1 and 12.

    D. FFB is defective.

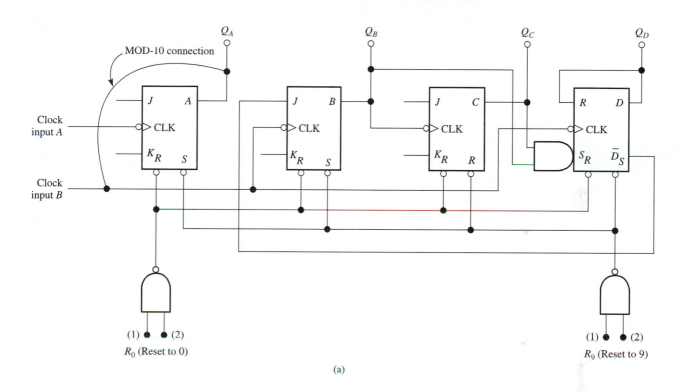

(a)

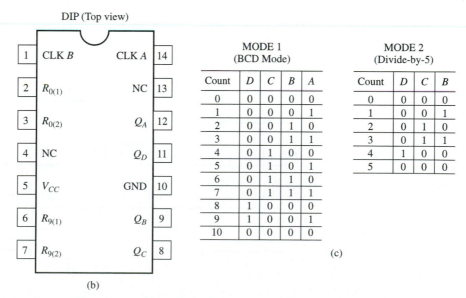

Count	D	C	B	A
0	0	0	0	0
1	0	0	0	1
2	0	0	1	0
3	0	0	1	1
4	0	1	0	0
5	0	1	0	1
6	0	1	1	0
7	0	1	1	1
8	1	0	0	0
9	1	0	0	1
10	0	0	0	0

MODE 1 (BCD Mode)

Count	D	C	B
0	0	0	0
1	0	0	1
2	0	1	0
3	0	1	1
4	1	0	0
5	0	0	0

MODE 2 (Divide-by-5)

(b)

(c)

FIGURE 8.13
The 7490 decade counter IC: (a) Logic diagram. (b) Pin diagram. (c) Count sequence.

THE 7490 DECADE COUNTER

The 7490 IC counts through a BCD count sequence. It contains four negative-edge-triggered flip-flops, and internally it is made from a mod-5 and a mod-2 counter. Each counter can be used separately or connected externally to form a decade counter.

The logic diagram for this device appears in Fig. 8-13(a) on the previous page. It shows two clock inputs labeled A and B. Clock input A is the clock for the mod-2 counter, and clock input B is the clock for the mod-5 counter. To make the IC operate as a mod-10 counter, the clock input is connected to input A and an external jumper is connected from the mod-2 output, Q_A, to the clock input of the mod-5 counter, input B.

Also included are two separate pairs of leads, labeled $R_{0(1)}$, $R_{0(2)}$ and $R_{9(1)}$, $R_{9(2)}$. Either or both R_0 pins or R_9 leads must be connected low for normal counting. When the active-high R_0 leads are both logic 1, the counter output resets to decimal 0, or DCBA = 0000. If the active-high R_9 leads are both logic 1, the counter output presets to decimal 9, or DCBA = 1001. Figure 8-13(b) shows the pin diagram and Fig. 8-13(c) shows the function table of this IC.

EXAMPLE 8.4

The input waveforms in Fig. 8-14 are applied to the 7490 decade counter. Draw the output waveforms that result.

Solution See Fig. 8-14.

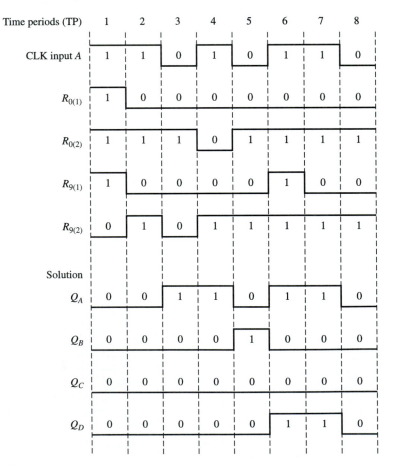

FIGURE 8.14
Input and output waveforms for Example 8.4.

Time period (TP)

1   Both R_0 inputs are high and the count in the counter becomes 0.

2   Neither both the R_0 nor both the R_9 input pins is high, and the clock's input is at a high state. The count produced at the output pins remains unchanged.

3   Neither both the R_0 nor both the R_9 input pins is high, and the clock's input changes from a high to a low state. The count at the output pins increments from 0 to 1.

4   Neither both the R_0 nor both the R_9 input pairs is high, and the clock's input changes from a low to a high state. The clock's output remains unchanged.

5   Neither both the R_0 nor both the R_9 input pins is high, and the clock's input changes from a high to a low state. The clock's output increments from a 1 to a 2.

6   Both R_9 inputs are high, and the count at the outputs become a 9.

7   Neither both the R_0 nor both the R_9 input pins is high, and the clock's input remains high. The counter's outputs remain unchanged at 9.

8   Neither the R_0 nor the R_9 input pairs is high, and the clock's input changes from a high to a low state. The counter recycles from a 9 to a 0.

FIGURE 8.14
(Continued)

■ REVIEW QUESTIONS

7. The entire counting sequence of a mod-4 counter counts _____ different numbers, which are _____ to _____.

8. A counter that counts from 111_2 to 000_2 is called a _____ _____-counter.

9. A mod-16 counter requires _____ flip-flops.

10. A binary 00000 is the next count of a mod-32 up-counter after the count of binary _____.

11. A decade counter increments from a count of 0000 to 1001 before it _____ back to 0000.

12. A mod-8 up-counter can be constructed from a 7493 binary counter by connecting the clock input to input _____ and taking the output from outputs _____.

13. A 7490 IC is made up of a mod-_____ and mod-_____ counter; to make it a decade counter, an external jumper wire is connected from the mod-2 output _____ to the mod-5 _____ _____ B.

Synchronous

Asynchronous ripple counters are the least expensive to use because they require the least number of components to operate. However, the primary disadvantage of these types of counters is that they can only operate at lower frequencies. The reason for this limitation is illustrated in Fig. 8-15, which shows an asynchronous mod-8 ripple counter and its timing diagram. From the timing diagram, note that there is a response delay of each flip-flop to the trailing edge of a pulse from the output of the previous flip-flop. This is a result of the propagation delay of each flip-flop. At lower frequencies, this situation is not a problem as long as the total propagation delay of the counter is shorter than the period between input pulses. However, Fig. 8-15(b) shows what happens when an incoming signal at a higher frequency is applied to the counter. In this situation, a problem arises as a result of the accumulation of propagation delays in the three flip-flops. Each flip-flop has a propagation delay of 30 ns. The square wave period is 80 ns, so the 30-ns delay by FFA

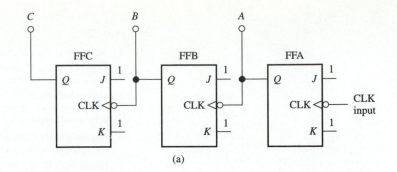

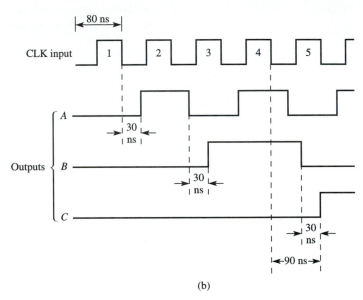

FIGURE 8.15
An asynchronous counter: (a) Logic diagram. (b) Response delays.

is no problem. The propagation delay of FFB in response to the output of FFA is also 30 ns. Added together, the propagation delay of FFA and FFB is 60 ns, which is still less than the 80-ns time period of the clock signal. Therefore, the circuit operates properly at output *B*. However, the diagram illustrates that the counter increments to the number 4 as the FFC output goes high 90 ns after the trailing edge of the fourth clock pulse. This exceeds the 80-ns time period of the clock signal, which causes the counter not to increment to 4 until after the fifth clock pulse has arrived. Serious problems would arise from such a situation and would be compounded with additional flip-flops in this type of ripple counter.

Synchronous counter:

A counter in which all of its flip-flops are triggered simultaneously by the same clock pulse.

This problem can be overcome with the use of **synchronous counters,** in which all the flip-flops are triggered simultaneously by the same clock pulse. Because the clock pulse arrives at all the flip-flops simultaneously, a method must be used to control when a particular flip-flop is to toggle or remain unaffected. This is accomplished by using the *J* and *K* inputs of *J–K* flip-flops.

Figure 8-16 shows the logic diagram and truth table for a mod-16 synchronous counter. The counter includes four *J–K* flip-flops and two AND gates. The key to the operation of this circuit is that the *J* and *K* inputs of each flip-flop are controlled separately before the arrival of the oncoming synchronous clock pulse.

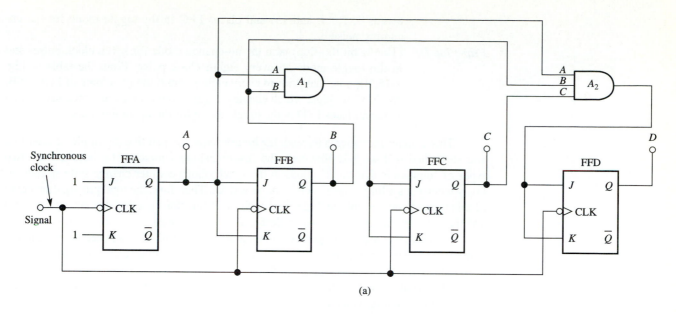

(a)

Count	D	C	B	A
0	0	0	0	0
1	0	0	0	1
2	0	0	1	0
3	0	0	1	1
4	0	1	0	0
5	0	1	0	1
6	0	1	1	0
7	0	1	1	1
8	1	0	0	0
9	1	0	0	1
10	1	0	1	0
11	1	0	1	1
12	1	1	0	0
13	1	1	0	1
14	1	1	1	0
15	1	1	1	1

(b)

FIGURE 8.16
Mod-16 synchronous counter: (a) Logic diagram. (b) Count sequence.

For example:

Flip-Flop A: (The 1s bit flip-flop) toggles every time a clock pulse arrives because the J and K inputs are at permanent 1 states.

Flip-Flop B: (The 2s bit flip-flop) toggles on every other clock pulse. On the arrival of an odd clock pulse, lows are at the J and K inputs from the Q output of FFA. Therefore, FFB is in the no-change mode. On the arrival of an even clock pulse, highs from FFA cause FFB to complement because it is in the toggle mode.

Flip-Flop C: (The 4s bit flip-flop) is in the no-change mode for three clock pulses and in the toggle mode on every fourth clock pulse. This can be understood from the table of Fig. 8-16(B). Prior to every fourth count, the Q outputs of FFA and FFB are high, which enables A_1 (this condition occurs on counts 3, 7, 11, and 15) and places FFC in the toggle mode for the upcoming pulse.

Flip-Flop D: (The 8s bit flip-flop) is in the no-change mode for seven clock pulses and in the toggle mode on every eighth clock pulse. From the table in Fig. 8-16(B), note that prior to every eighth count, the Q outputs of FFA, FFB, and FFC are high, which enables A_2 (this condition occurs on counts 7 and 15) and places FFD in the toggle mode for the upcoming pulse.

This counter increments through the binary sequence just like the ripple counter that was described in Fig. 8-1. However, the distinguishing characteristic between the two types of counters is that the flip-flops in the synchronous counter change state simultaneously when the trailing edge of a clock pulse arrives. Therefore, the total propagation delay of this type of counter only includes that of the J–K flip-flops (each delay simultaneously) and the delay of the AND gates (both delays at the same time).

As a result, a higher-frequency input signal can be applied to a synchronous counter compared to an asynchronous ripple counter, where the total time delay is an accumulation of the time delays of all individual flip-flops.

The disadvantage of the synchronous counter is that more logic circuitry and connections are required, which makes them more expensive.

8.12   SYNCHRONOUS UP/DOWN-COUNTER ICs

Synchronous counters are available in IC form. Two that have similar features are the 74192 decade counter (mod-10) and the 74193 binary counter (mod-16). Both contain four flip-flops and are capable of counting both up or down. Parallel load lines are included that allow the user to load any number within the counter's modulus capacity. The count capacity of these counters can be increased by cascading two or more of them.

A block diagram used for both counters is shown in Fig. 8-17(a). The function of each line is as follows:

MR (Master Reset): This is an active-high input line that resets the counter to 0000. The counter holds at 0000 as long as MR = 1. It overrides all other inputs.

$\overline{PL}$***, D_0–D_3 (Preset Inputs):*** The counter can be preset by placing a binary value on the parallel data lines (D_0-D_3) and then driving the parallel load ($\overline{PL}$) line low. The preset lines override any count in the counters except when MR is activated.

Q_0–Q_3 (Counter Outputs): The current count of the counter is always present at these output lines. Q_0 is the LSB and Q_3 is the MSB.

(CP)$_U$ and (CP)$_D$ (Clock Inputs): There are two positive-edge-triggered clock inputs for this circuit. The (CP)$_U$ lead is used during the up-count operation, and the (CP)$_D$ lead is used during the down-count operation.

$\overline{(TC)}_U$ *and* $\overline{(TC)}_D$ *(Terminal Count Outputs):* These output lines are normally high and are used when two or more counters are cascaded to produce a larger modulus number.

The $\overline{(TC)}_U$ line indicates that the maximum count is reached and the count is about to recycle to 0000. The $\overline{(TC)}_U$ line goes low when the highest count exists *and* the clock signal applied to the (CP)$_U$ line is low. For example, when the count of the mod-10 74192 IC is 9 and the clock signal at the (CP)$_U$ line is low, its $\overline{(TC)}_U$ line goes low. When the count of the mod-16 74193 IC is 15 and the clock signal at its (CP)$_U$ line is low, its $\overline{(TC)}_U$ lines goes low. The $\overline{(TC)}_U$ for each of these ICs remains low until the next low-to-high transition at the (CP)$_U$ line causes either counter to recycle to 0000.

When cascading either type of IC as up-counters, the $\overline{(TC)}_U$ pin of the first counter is connected to the (CP)$_U$ pin of the next counter. Every time the first counter reaches its

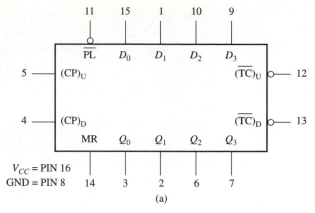

(a)

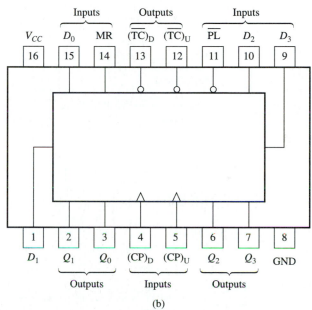

(b)

Operating mode	Inputs								Outputs					
	MR	$\overline{PL}$	$(CP)_U$	$(CP)_D$	D_0	D_1	D_2	D_3	Q_0	Q_1	Q_2	Q_3	$\overline{(TC)}_U$	$\overline{(TC)}_D$
Reset	H	X	X	L	X	X	X	X	L	L	L	L	H	L
	H	X	X	H	X	X	X	X	L	L	L	L	H	H
Parallel load	L	L	X	L	L	L	L	L	L	L	L	L	H	L
	L	L	X	H	L	L	L	L	L	L	L	L	H	H
	L	L	L	X	H	H	H	H	H	H	H	H	L	H
	L	L	H	X	H	H	H	H	H	H	H	H	H	H
Count up	L	H	↑	H	X	X	X	X	Count up				H	H
Count down	L	H	H	↑	X	X	X	X	Count down				H	H

H = High-voltage level; L = Low-voltage level; X = Don't care; ↑ = Low-to-high clock transition

(c)

FIGURE 8.17
The 74192 and 74193 synchronous up/down-counter ICs: (a) Block diagram. (b) Pin diagram.
(c) Function table.

maximum count, the $\overline{(TC)_U}$ line goes low; when it recycles to 0000, the $\overline{(TC)_U}$ line goes high and provides a positive-edge-triggered signal to increment the next counter.

The $\overline{(TC)_D}$ line indicates that the minimum count of 0000 is reached. When cascading either the 74192 or 74193 IC as down-counters, the $\overline{(TC)_D}$ pin of the first counter is connected to the $(CP)_D$ pin of the next counter. Every time the clock signal is low while the first counter is decremented to 0000, its $\overline{(TC)_D}$ pin goes low. When the low-to-high transition of the clock pulse causes the counter to recycle to its maximum count, the $\overline{(TC)_D}$ line goes high and decrements the next counter with the positive-edge transition.

The pin diagram of these counters is shown in Fig. 8-17(b) and their function table is shown in Fig. 8-17(c).

EXPERIMENT

Up/Down Counter

Objectives

- To wire a 74193 IC.
- To operate the 74193 IC as an up-counter.
- To operate the 74193 IC as a down-counter.
- To preset desired numbers into the 74193 IC.
- To operate any desired modulus count within the range of a 74193 IC.

Materials

1—74193 IC

1—7420 IC

4—LEDs

1—+5-Volt DC Power Supply

6—Logic Switches

1—Logic Pulse Push Button (Bounceless)

4—150-Ohm Resistors

Introduction

A binary counter is usually used in applications that require the count to increment from its lowest number to its highest number. However, it is occasionally desirable to have a counter decrement from its highest number to its lowest number. The 74913 IC counter is capable of performing both functions.

In the previous experiment, the 7493 counter was classified as *asynchronous*. The output of each flip-flop is connected to the clock input of the next flip-flop. This results in additive propagation delays that limit the operating speed of the counter. The 74193 IC counter is *synchronous*, which means that all of the flip-flops are connected to the same clock pulse. As a result, they all change states simultaneously. Because the 74193 flip-flops toggle at the same rate without the additive propagation delays that exist with asynchronous counters, they are capable of operating at higher frequencies.

The 74193 is capable of operating at different modulus counts. To perform the operation of counting up or down at different counts, parallel data entry input leads are used to preset the counter to any number from 0 to 15.

PART A—BASIC OPERATION

Procedure

Step 1.   Assemble the circuit shown in Fig. 8-18.

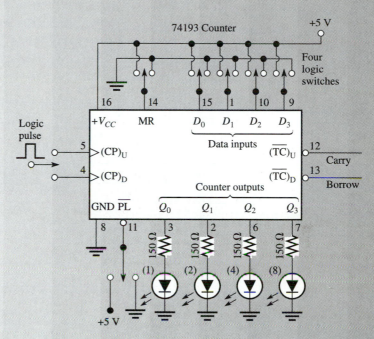

FIGURE 8.18

Count-Up Function

Step 2.   Set $\overline{PL}$ switch to the +5 volt logic high position. Clear the counter by applying a momentary high to the MR (Master Reset) input.

Procedure Information

The four output leads are labeled Q_0, Q_1, Q_2, and Q_3. The LEDs displaying the output should all be off to show that 0000 is in the counter.

Step 3.   Apply single clock pulses to the up-counter input $(CP)_U$. For each clock pulse, fill in the output portion of Table 8-1.

Parallel Load Function

Step 4.   Apply a momentary high to the MR input. Set all of the switches connected to the parallel data entry inputs high.

Procedure Question 1

What are the contents of the counter?

Step 5.   Apply a momentary low to the $\overline{PL}$ input.

TABLE 8.1

MR	$(CP)_U$	Q_3	Q_2	Q_1	Q_0
0	—	0	0	0	0
0	$1^{st}\uparrow$				
0	$2^{nd}\uparrow$				
0	$3^{rd}\uparrow$				
0	$4^{th}\uparrow$				
0	$5^{th}\uparrow$				
0	$6^{th}\uparrow$				
0	$7^{th}\uparrow$				
0	$8^{th}\uparrow$				
0	$9^{th}\uparrow$				
0	$10^{th}\uparrow$				
0	$11^{th}\uparrow$				
0	$12^{th}\uparrow$				
0	$13^{th}\uparrow$				
0	$14^{th}\uparrow$				
0	$15^{th}\uparrow$				
0	$16^{th}\uparrow$				

NOTE: $(CP)_D$ = LOGIC 1

$\overline{PL}$ = LOGIC 1

$\uparrow$ = LOGIC 0 TO LOGIC 1 TRANSITION

Procedure Question 2

What are the contents of the counter?

Count-Down Function

Step 6. Apply single clock pulses to the down-counter $(CP)_D$ input with a logic pulse push button. For each clock pulse, fill in the output portion of Table 8-2.

PART B—ALTERING THE MODULUS COUNT OF THE 74193 IC

Mod-6 Up-Counter Operation (0 to 5)

Step 7. Turn the power off.
Step 8. Connect the NAND gate to the counter as shown in Fig. 8-19.
Step 9. Set all of the parallel data entry switches low.
Step 10. Turn the power on.
Step 11. Clear the counter with the MR input logic switch.
Step 12. Apply single clock pulses to the up-counter input with a logic push button. For each clock pulse, fill in the output portion of Table 8-3.

TABLE 8.2

MR	$(CP)_D$	Q_3	Q_2	Q_1	Q_0
1	—	1	1	1	1
0	1st↑				
0	2nd↑				
0	3rd↑				
0	4th↑				
0	5th↑				
0	6th↑				
0	7th↑				
0	8th↑				
0	9th↑				
0	10th↑				
0	11th↑				
0	12th↑				
0	13th↑				
0	14th↑				
0	15th↑				
0	16th↑				

NOTE: $(CP)_U$ = LOGIC 1

$\overline{PL}$ = LOGIC 1

↑ = LOGIC 0 TO LOGIC 1 TRANSITION

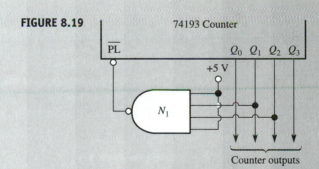

FIGURE 8.19

74193 Counter

Counter outputs

Background Information

When the counter increments to 6, all highs are applied to the NAND gate N_1 inputs. This causes N_1 to produce a low. With the $\overline{PL}$ input activated by a low, the number 0000 at the parallel data entry input is loaded into the counter. The count of 6 is never displayed because it only lasts for a few nanoseconds.

Mod-6 Down-Counter (5 to 0)

Step 13. Turn the power off.
Step 14. Connect the NAND gate to the counter as shown in Fig. 8-20.

TABLE 8.3

MR	$(CP)_U$	Q_3	Q_2	Q_1	Q_0
1	—	0	0	0	0
0	$1^{st}\uparrow$				
0	$2^{nd}\uparrow$				
0	$3^{rd}\uparrow$				
0	$4^{th}\uparrow$				
0	$5^{th}\uparrow$				
0	$6^{th}\uparrow$				
0	$7^{th}\uparrow$				

NOTE: $(CP)_D$ = LOGIC 1

$\uparrow$ = LOGIC 0 TO LOGIC 1 TRANSITION

FIGURE 8.20

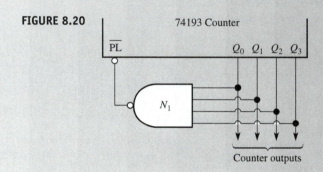

74193 Counter

Counter outputs

Step 15. Set the parallel data entry input switches as follows: 0101_2 ($D_0 = 1$, $D_1 = 0$, $D_2 = 1$, $D_3 = 0$).

Step 16. Turn the power on.

Step 17. Load 0101_2 into the counter by applying a momentary low to the $\overline{PL}$ input with a logic switch ($\overline{PL}$).

Step 18. Apply a clock pulse to the down-counter.

Procedure Question 3

The number in the counter becomes _____.

Background Information

When the down-counter input receives a clock pulse with the counter contents at 0000, the counter recycles to 1111. All highs at the Q outputs cause the N_1 to go low, which activates the $\overline{PL}$ input. A low at input $\overline{PL}$ causes the binary number 5 at the parallel data entry inputs to be loaded into the counter. The number 1111 is never displayed because it only lasts for a few nanoseconds.

Step 19. Apply single clock pulses to the down-counter input with a logic pulse push button. For each clock pulse, fill in the output portion of Table 8-4.

TABLE 8.4

$\overline{PL}$	$(CP)_D$	Q_3	Q_2	Q_1	Q_0
0	—	0	1	0	1
1	$1^{st}\uparrow$				
1	$2^{nd}\uparrow$				
1	$3^{rd}\uparrow$				
1	$4^{th}\uparrow$				
1	$5^{th}\uparrow$				
1	$6^{th}\uparrow$				
	$7^{th}\uparrow$				

NOTE: $(CP)_U$ = LOGIC 1

$\uparrow$ = LOGIC 0 TO LOGIC 1 TRANSITION

■ EXPERIMENT QUESTIONS

1. Why does the synchronous counter operate faster than the asynchronous counter?
2. To load a 7 into the counter, list what states are required at the following inputs:

   $D_0 =$

   $D_1 =$

   $\overline{PL} =$

   $D_2 =$

   $D_3 =$

3. Clock pulses applied to the 74193 IC pin _____ causes it to increment.
4. The 74193 IC changes counts on every _____ (negative, positive) edge signal applied to its clock inputs.
5. To cause the counter in Fig. 8-21 to recycle its count range from 0011_2 to 1010_2, connect the input leads of N_1 to the required Q outputs and place the necessary 0s or 1s at the parallel data entry inputs.

FIGURE 8.21

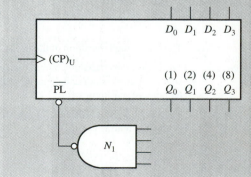

EXAMPLE 8.5

The input waveforms in Fig. 8-22 are applied to the 74192 IC decade up/down-counter. Draw the output waveforms that result.

Solution   See Fig. 8-22.

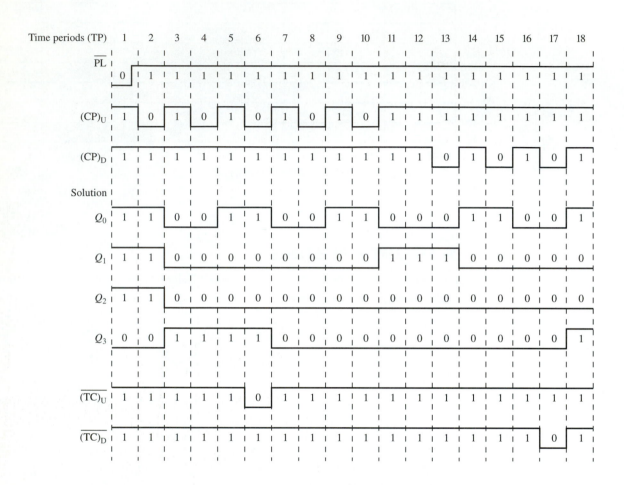

Time Period (TP)

1   With the $\overline{PL}$ input activated by a low, and a binary value of 0111_2 applied to the D_{0-3} parallel data lines (not shown), a 7_{10} is parallel loaded into the counter.

2   The signal applied to the $(CP)_U$ input changes from a high to low state. The contents of the counter remain unchanged.

3   The low-to-high transition of the $(CP)_U$ input causes the counter to increment from 7 to 8.

4   The high-to-low transition of the $(CP)_U$ input has no effect on the counter. Its contents remain at a count of 8.

5   The low-to high transition of the $(CP)_U$ input causes the counter to increment from 8 to 9.

6   The high-to-low transition of the $(CP)_U$ input does not increment the count. Whenever the contents of the counter is at its maximum count of 9 and the logic state of the clock signal is a low at $(CP)_U$, the $\overline{(TC)_U}$ output line goes low.

7   The low-to-high transition of $(CP)_U$ causes the counter to increment and recycle to a count of 0. The $\overline{(TC)_U}$ line returns to a high state.

FIGURE 8.22
Input and output waveforms for Example 8.5.

8   The high-to-low transition of the $(CP)_U$ input does not increment the count.

9   The low-to-high transition of the $(CP)_U$ line causes the counter to increment from 0 to 1.

10   The high-to-low transition of the $(CP)_U$ input does not increment the count.

11   The low-to-high transition of the $(CP)_U$ input causes the counter to increment.

12   All inputs from the previous time period do not change. The count of the counter remains unchanged.

13   The signal applied to the $(CP)_U$ input changes from a high to low state. The contents of the counter remain unchanged.

14   The low-to-high transition of the $(CP)_D$ input causes the counter to decrement from 2 to 1.

15   The high-to-low transition of the $(CP)_D$ input has no effect on the counter. The count remains at 1.

16   The low-to-high transition of the $(CP)_D$ input causes the counter to decrement from 1 to 0.

17   The high-to-low transition of the $(CP)_D$ input does not decrement the count. Whenever the contents of the counter is at its minimum count of 0 and the logic state of the clock signal is a low at $(CP)_D$, the $\overline{(TC)_D}$ output line goes low.

18   The low-to-high transition of the $(CP)_D$ input causes the counter to decrement and recycle to a count of 9 (1001_2). The $\overline{(TC)_D}$ line returns to a high state.

FIGURE 8.22
(Continued)

EXPERIMENT

Cascading Up/Down Counters

Objectives

- To cascade two 74193 ICs as 8-bit up-counters or down-counters.

Materials

     2—74193 ICs

     1—Logic Pulse Push Button (Bounceless)

     8—LEDs

     8—150-Ohm Resistors

     1—Logic Switch

     1—3-Hz Clock Pulse Generator

     1—+5-Volt DC Power Supply

Introduction

The maximum range of counts for a 74193 synchronous counter is 0000 to 1111. When a count larger than the maximum count of a single IC is required, two or more chips can be connected together. This is called *cascading*.

Cascading Up-Counters

To cascade two or more up-counters, the carry output $\overline{(TC)_U}$ lead of the first counter is connected to the up-counter clock input CP_U of the next counter. On the highest count of 1111 in the first counter, its carry output $\overline{(TC)_U}$ goes low. When the next clock pulse is applied to the first counter, it recycles to 0000 and its carry output lead goes high. The low-to-high transition of the $\overline{(TC)}_U$ signal is fed into the up-count clock input CP_U of the next counter to cause it to increment. Every time the first counter recycles, it causes the second counter to increment once.

Step 1. Assemble the circuit in Fig. 8-23.

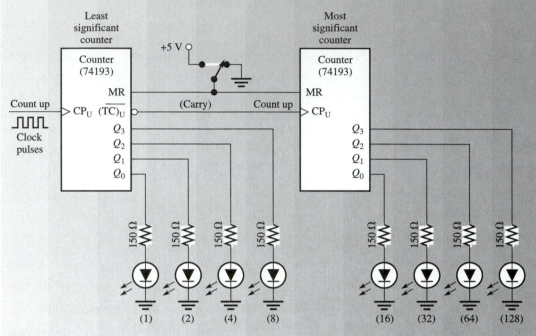

FIGURE 8.23

Step 2. Load 0000 0000 into the counter by applying a momentary high to the MR input.
Step 3. Apply a 3-Hz clock pulse frequency to the input of the cascaded up-counter. Periodically stop the clock and observe how the counters are incrementing.

Procedure Question 1

How many clock pulses are applied to the cascaded counters before the second counter increments once?

Procedure Question 2

What is the highest decimal equivalent count the two cascaded counters can make?

Procedure Question 3

After the cascaded counters are cleared, how many clock pulses are applied to cause them both to recycle at the same time?

Cascading Down-Counters

To cascade two or more 74193 ICs to operate as down-counters, the borrow output lead $\overline{(TC)}_D$ of the first counter is connected to the down-counter clock input (CP_D) of the next counter. When the lowest count of 0000 is present in the first counter, and the clock pulse applied to CP_D is low, the *borrow* output goes low. On the next clock pulse to the first counter, it recycles to 1111 and causes the borrow lead to go high. The low-to-high transition of the borrow lead is fed to the count-down clock input (CP_D) of the next counter to cause it to decrement. The second counter decrements every time the first counter recycles.

Step 4. Assemble the circuit shown in Fig. 8-24.

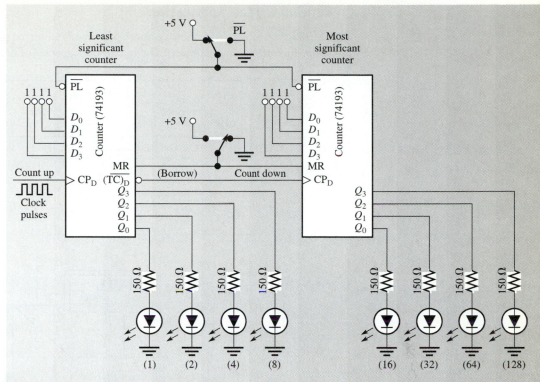

FIGURE 8.24

Step 5. Load 1111 1111 into the counter by applying a momentary low to the parallel load ($\overline{PL}$) input. *Note:* 1s are present at the data entry inputs D_0–D_3 when they are left disconnected.

Step 6. Apply a 3-Hz clock pulse frequency to the input of the cascaded down-counters. Periodically stop the clock and observe how the counters are decrementing.

Procedure Question 4

What two conditions must exist to cause the borrow lead to go low?

Procedure Question 5

Every time the second cascaded counter decrements, _____ clock pulses have to be applied to the first cascaded counter.

■ EXPERIMENT QUESTIONS

1. The term used to describe the connecting of two or more IC counters is _____.
2. When cascading two 74193 ICs as down-counters, the _____ (carry, borrow) output of the first counter is connected to the _____ (CP_U,CP_D) input of the second flip-flop.
3. The carry output goes low on count _____ of the 74193 IC when the CP_U input is _____, and the borrow output goes low on count _____ when the CP_D input is _____.

   (a) 0000

   (b) 1111

   (c) Low

   (d) High

Parking Garage Application

The diagram in Fig. 8-25 shows a booth for a semiautomated parking lot. It uses digital circuitry to perform its operations. The maximum capacity of the lot is 99 cars.

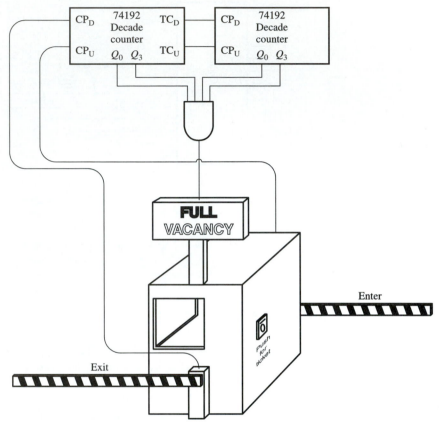

FIGURE 8.25
Semiautomated parking lot.

When a car approaches the gate on the right, a button is pushed to discharge a ticket stub. When the driver removes the ticket from the slot, the gate opens until the car passes. To exit from the parking area, the driver presents the ticket to the cashier at the window. After paying, the exit gate is opened until the car passes.

Two 74192 IC decade up/down counters are used to maintain a count of the cars that enter and exit the lot. They are cascaded so that one chip stores the 1s digit, and the other chip stores the 10s digit. Every time the entry gate opens, the counter increments because a pulse is applied to the CP_U input. Every time the exit gate opens, the counter decrements because a pulse is applied to the CP_D input.

Whenever the number of cars inside the lot becomes 99, the *FULL* light is illuminated. When the 99th car exits, the counter decrements to 98 and the *VACANCY* light illuminates.

8.13 FREQUENCY DIVIDER

Frequency divider:

A circuit that produces an output pulse after receiving a certain number of input pulses.

Asynchronous counters can also be used as frequency dividers. A **frequency divider** is a circuit that provides an output pulse after receiving a certain number of inputs. The binary ripple counter described earlier in this chapter is capable of frequency division. The circuit and its timing diagram in Fig. 8-1 illustrate this characteristic.

From the timing diagram, the output of each flip-flop is exactly one-half the frequency applied to its input. Because each flip-flop divides its incoming frequency in half, several flip-flops can be cascaded to reduce a high frequency to a low frequency. For example, if the incoming square wave frequency to the counter is 100 kHz, the outputs of each flip-flop are

$A = 50$ kHz

$B = 25$ kHz

$C = 12.5$ kHz

$D = 6.25$ kHz

Any asynchronous counter is capable of frequency division. However, those that are not pure binary do not necessarily reduce the frequency by a descending order of 2 at each flip-flop. For example, a 4 flip-flop mod-10 counter divides the incoming frequency by 10. A 3 flip-flop mod-6 counter divides the frequency by 6. Whatever the modulus number of a counter is, the incoming frequency is divided by the same number.

Frequency dividers are often used in timekeeping devices. One practical application is a digital watch. This device contains a small crystal that has a small DC voltage applied to it. When current flows through it, the crystal vibrates at a very precise frequency (i.e., 50,000 pulses per second). Divider circuits are used to reduce the high frequency down to a usable value of one pulse per second.

Frequency Divider Application

The square wave frequency generator used for electronic experiments by students illustrates frequency division. Figure 8-26 shows six cascaded mod-10 counters fed by a clock generator that produces a square wave frequency of 100 k to 2 MHz. The frequency is varied by turning the fine-tune knob, which is connected to a variable resistor. The clock signal feeds through the series of counters. Each one divides the signal by 10. For example, suppose the clock generator produces a frequency of 1 MHz. Counter 6, to which this signal is fed, reduces the frequency to 100 kHz. Counter 5 reduces the signal to 10 kHz, counter 4 to 1 kHz, counter 3 to 100 Hz, counter 2 to 10 Hz, and counter 1 to 1 cycle per second. The desired frequency range is obtained when the proper counter is connected to the output terminal by using one of the selector push buttons.

■ REVIEW QUESTIONS

14. A 6-bit ripple counter has a propagation delay of _____ ns if each flip-flop has a delay of 20 ns.

15. A _____ counter clocks all flip-flops simultaneously.

16. The _____ (synchronous/asynchronous) mod-16 counter operates at a faster speed than the _____ (synchronous/asynchronous) mod-16 counter.

17. Which of the following is *not* an operational characteristic of the 74193 counter?
    (a) The counter increments as well as decrements.
    (b) The counter is cleared by an active-low pulse.
    (c) The counter can be preset to a 4-bit number.
    (d) To cascade two IC counters, $\overline{(TC)}_U$ of one counter is fed to the $(CP)_U$ of another counter when incrementing.

18. To clear the 74193 counter, a positive pulse is applied to input _____ . To preset a number into the counter, data bits are applied to inputs _____ to _____ and a _____ pulse is applied to input PL.

19. A binary counter that contains three flip-flops has a frequency at the output of the third flip-flop that is (the same, $\frac{1}{2}$, $\frac{1}{4}$, $\frac{1}{8}$, or $\frac{1}{16}$) the input frequency.

20. A mod-8 counter divides the incoming frequency by _____ .

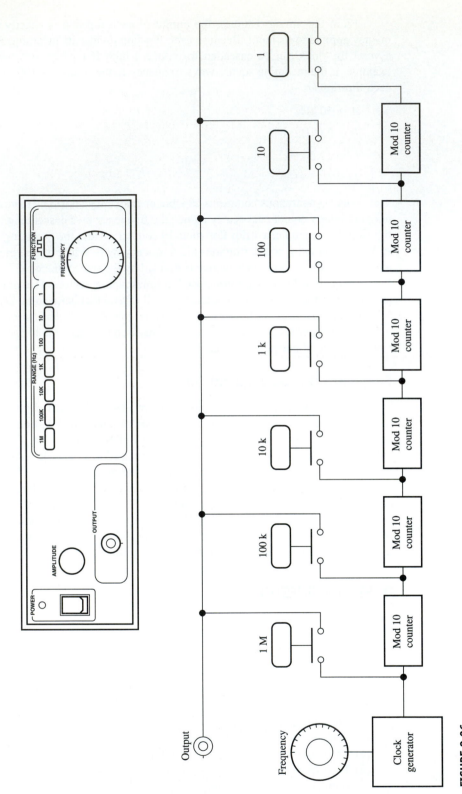

FIGURE 8.26
Frequency division.

8.14   TROUBLESHOOTING

When a malfunction develops in a counting circuit, there are a few different ways in which the fault can be found.

Frequently, counting circuits are connected to display devices. By observing how the display responds to the signal applied to it, the troubleshooter is sometimes able to use the information to determine the cause of a counter problem. For example, suppose a technician is testing the operation of a newly assembled device and observes that the display increments from 0 to 1, and then starts over again at 0. Because a 7490 decade counter IC drives the display, it is supposed to count from 0 to 9 before recycling. Further checking found that the jumper wire was not connected from pin 12 to 1, as shown in Fig. 8-27. Therefore, the counter operates as a mod-2 counter instead of a decade counter.

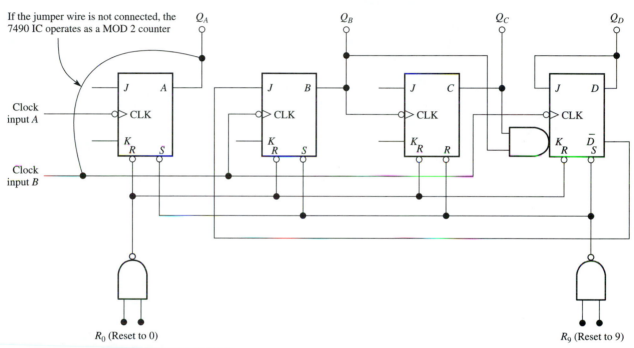

FIGURE 8.27
A 7490 IC configured to operate as a mod-10 counter.

If a display is not connected, another method used to troubleshoot a counter is the dynamic test. An oscilloscope or logic probe can be used as a troubleshooting device to find the problem. In the dynamic test, the incoming clock pulse to the counter is provided by the system of which it is a part. If an oscilloscope is used to troubleshoot a mod-16 counter, for example, the input and outputs of the flip-flops can be viewed simultaneously, as shown in Fig. 8-28. If each flip-flop divides the frequency applied to its clock input line, it is operating properly.

A logic probe is also a useful tool for the troubleshooter performing a dynamic test. It is first used to determine if the clock signal is present at the clock input pin. The probe can also be used to check the output states of each flip-flop, as shown in Fig. 8-29. If the probe indicates activity, it is likely that the counter is operating properly up to the location where the signal is detected. A flip-flop output that does not show a changing signal indicates a fault at or somewhere before the test-point location. The problem can be isolated by moving the probe toward the beginning portion of the counter until a signal is found.

If the dynamic test fails or at least indicates a possible fault, a counter is usually tested further by using the static-test troubleshooting technique.

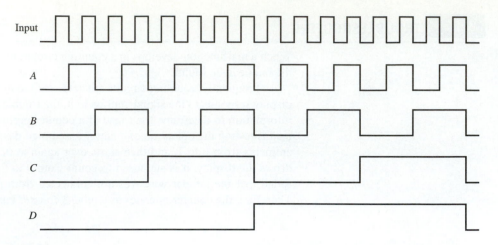

FIGURE 8.28
Input and output signals of each mod-16 up-counter flip-flop simultaneously displayed on an oscilloscope screen.

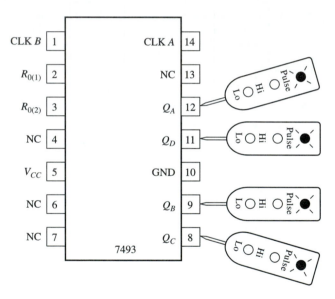

FIGURE 8.29
Logic probe used to troubleshoot an up-counter under a dynamic testing procedure.

Troubleshooting a 7490 Decade Counter

Static Test   The following procedure uses a logic clip and pulser to troubleshoot a 7490 IC configuration as a decade counter.

1. Attach the 16-pin logic clip to the 14-pin IC, as shown in Fig. 8-30(b) so that the end LEDs are not connected.
2. Apply the pulser probe to the reset pin of the IC as shown in Fig. 8-30(a) and press the pulse button once so that a zero pulse causes the counter to clear. The display should be as shown in Fig. 8-30(b), which indicates that the Q outputs of each flip-flop are low.
3. Apply the pulser probe to the clock input pin of the IC and press the button once. After this first clock pulse, the LED for Q_A should light, as shown in Fig. 8-30(c), indicating a count of 1.
4. By continuing to apply pulses to the clock input of the counter, its count should increment through its entire cycle before recycling back to 0 again.
5. If the circuit does not reach its maximum count before recycling, determine if an external problem is causing the malfunction before replacing the IC.

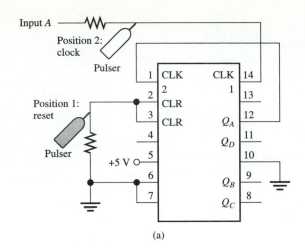

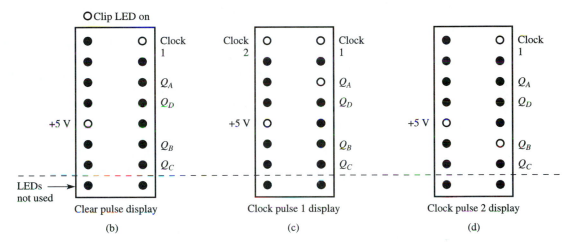

FIGURE 8.30
Checking a 7490 decade counter with a logic probe and pulse. (a) IC pins and pulse-signal injection points. (b–d) The probe's expected display.

Using a Logic Analyzer If the maintenance department can afford to purchase a logic analyzer, it is an excellent troubleshooting tool when trying to detect a fault with a counter. For example, Fig. 8-31 shows how the sequence of counts of a decade counter can be displayed as it increments and recycles. The leftmost bit is the MSB (weight of 8), and the next bits are 4, 2, and 1. The analyzer shows that the counts of 8 and 9 do not occur. Further tests reveal that output D of the counter is shorted to ground.

FIGURE 8.31
Logic analyzer displaying the output of a faulty decade counter.

0	0	0	0
0	0	0	1
0	0	1	0
0	0	1	1
0	1	0	0
0	1	0	1
0	1	1	0
0	1	1	1
0	0	0	0
0	0	0	1
0	0	0	0
0	0	0	1
0	0	1	0
0	0	1	1

■ SUMMARY

- The basic building block of a sequential circuit is a flip-flop.
- Sequential circuits are classified as counters and registers.
- A counter is a device that contains a count that is determined by the number of pulses applied to its input.
- There are up-counters and down-counters. An up-counter counts up from zero to its maximum count. A down-counter counts down from its maximum count to zero.
- The modulus, or mod, of a counter is the number of times a counter increments or decrements before the counter repeats itself.
- Counters are classified as asynchronous or synchronous. A synchronous counter has all flip-flops clocked at the same time.
- Counters are capable of frequency division.

■ PROBLEMS

1. Digital circuits such as counters and registers that use flip-flops are classified as (combination or sequential) circuits. (Introduction to Sequential Circuits)

2. The _____ in a counter is representative of the exact number of pulses applied to the input. (8-1)

3. Other than the flip-flop located at the ripple counter input, where do the rest of the flip-flops receive their clock pulses? (8-1)

4. A decade counter manipulates bits that represent the _____ number system. (8-6)

5. The maximum number of counts a counter is capable of making is defined by the word _____. The highest number a mod-6 counter counts to is _____. (8-5)

6. A 5-bit binary counter counts up to a decimal output of _____. It has _____ different numbers that it can count. (8-5)

7. A 6-bit binary down-counter counts from which of the following to 0? (8-5)
   (a) 64
   (b) 128
   (c) 63
   (d) 127

8. It is customary to designate the Q output of FFA as the (LSB or MSB) of a counter. (8-2)

9. A ripple down-counter is a(n) (asynchronous or synchronous) device. (Synchronous Section)

10. Explain why an asynchronous counter cannot operate at high frequencies. (Synchronous Section)

11. Clock inputs are connected in (series or parallel) in a synchronous counter. (Synchronous Section)

12. The larger the number of flip-flops in a synchronous counter, the more its operating speed decreases. True or false? (Synchronous Section)

13. The _____ inputs are used to load a specific number other than 0000 into a counter. (8-4)

14. What are three practical applications of counters? (8-1)

15. A mod-6 up-counter is used to count the (least or most) significant digit of minutes and seconds. (8-7)

16. What is a practical application for a mod-10 down-counter? (8-8)

17. What is the frequency at the Q output of the last flip-flop of a 4 flip-flop asynchronous counter when 100 kHz is applied to its clock input? (8-13)

18. If the input applied to the first flip-flop of a decade counter is 5000 Hz, the output frequency of the last flip-flop is _____ Hz. (8-6, 8-13)

19. What is the primary application of counters used as frequency dividers? (8-13)

20. Using the logic diagrams of Fig. 8-32, draw the connections that make (a) a ripple up-counter, and (b) a ripple down-counter. (8-2, 8-3)

21. Before a counter sequence begins, the preset and clear inputs to the counter in Fig. 8-3 must be brought to a (low or high). (8-4)

22. A 7490 IC will operate as a mod- _____ counter if a jumper wire is not connected from pin _____ to _____. (8-11)

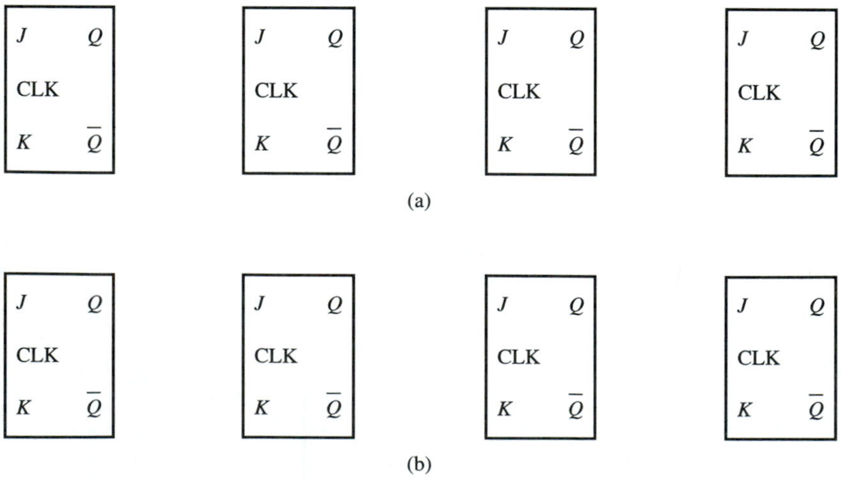

(a)

(b)

FIGURE 8.32
Logic diagrams for Problem 20.

23. At what number does the counter of Fig. 8-33 stop counting? (8-2)

24. Assuming that the counter of Fig. 8-34 is initially cleared, at what number does it stop incrementing? (8-2)

25. Draw the timing diagram for the counter circuit of Fig. 8-35. Assume that the counter starts at 111. (8-3)

26. Fill in the table in Fig. 8-36 for its accompanying counter. (8-7).

27. Answer the following for the 7493 counter. (8-10)
    (a) Name the different counters that make up this IC.
    (b) This is a(n) (up or down) counter.
    (c) This is a(n) (synchronous or asynchronous) counter.
    (d) What conditions must be at the reset inputs to clear the counter?

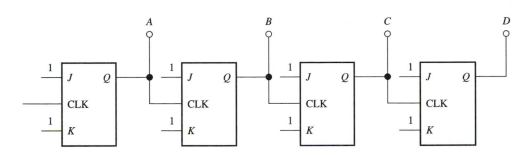

FIGURE 8.33
Counter for Problem 23.

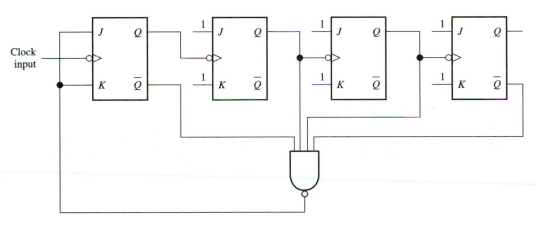

FIGURE 8.34
Counter for Problem 24.

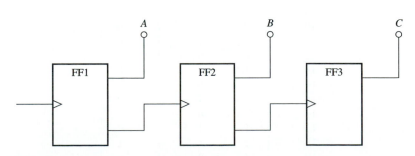

FIGURE 8.35
Counter for Problem 25.

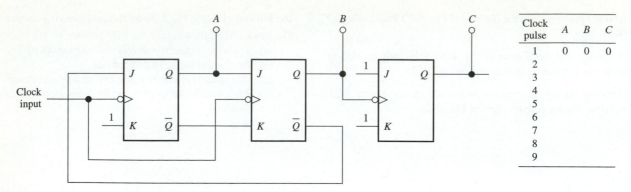

FIGURE 8.36
Counter and count sequence for Problem 26.

28. Based on the data bits parallel loaded during event number 1 and the signals applied to the input pins during events 2-20, complete Table 8-5 by filling in the appropriate output binary bits for a 74193 up/down-counter. (8-12)

Troubleshooting

29. What fault condition causes the 7490 IC decade counter to develop the waveform patterns of Fig. 8-37 on a multichannel oscilloscope? (8-11, 8-14)
   (a) Q_B internally shorted to ground.
   (b) Internal fault of the IC.

(c) Line from Q_A to CLK_B is open.
(d) All of the above.

30. The bit patterns of a mod-16 ripple up-counter is displayed on a logic analyzer, as shown in Fig. 8-38. It should count from 0 to 15 in pure binary. What is the cause of the malfunction? (8-2, 8-14)
   (a) The Q output of FFD is shorted to ground.
   (b) The FFD clock input is open.
   (c) The FFD J input is shorted to ground.
   (d) All of the above.

TABLE 8.5
Operation of a 74193 up/down counter for Problem 28.

EVENT NUMBER	INPUTS				OUTPUTS					
	$(CP)_U$	$(CP)_D$	$\overline{PL}$	MR	Q_0	Q_1	Q_2	Q_3	$(TC)_U$	$(TC)_D$
1	1	1	0	0	1	0	1	1	1	1
2	0	1	1	0						
3	1	1	1	0						
4	0	1	1	0						
5	1	1	1	0						
6	0	1	1	0						
7	1	1	1	0						
8	0	1	1	0						
9	1	1	1	0						
10	0	1	1	0						
11	1	1	1	0						
12	1	0	1	0						
13	1	1	1	0						
14	1	0	1	0						
15	1	1	1	0						
16	1	0	1	0						
17	1	1	1	0						
18	1	1	1	1						
19	1	0	1	0						
20	1	1	1	0						

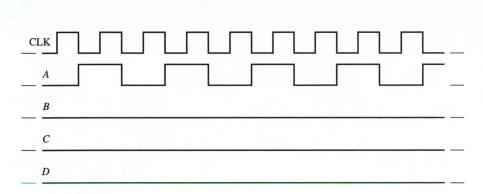

0	0	0	0
0	0	0	1
0	0	1	0
0	0	1	1
0	1	0	0
0	1	0	1
0	1	1	0
0	1	1	1
0	0	0	0
0	0	0	1
0	0	1	0
0	0	1	1
0	1	0	0
0	1	0	1
0	1	1	0

FIGURE 8.37
Waveforms for Problem 29.

FIGURE 8.38
Bit patterns for Problem 30.

■ ANSWERS TO REVIEW QUESTIONS

1. combination, sequential   **2.** counters, registers

3. asynchronous, synchronous   **4.** asynchronous

5. clock   **6.** $Q, \overline{Q}$ **7.** four, 00, 11

8. mod-8, down   **9.** four   **10.** 11111   **11.** recycles

12. *B; B, C,* and *D*   **13.** 2, 5, Q_A, clock input   **14.** 120

15. synchronous   **16.** synchronous, asynchronous

17. (b)   **18.** MR, D_0, D_3, negative   **19.** $\frac{1}{8}$ **20.** Eight

CHAPTER 9

REGISTERS

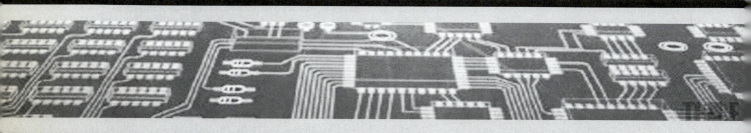

When you complete this chapter you will be able to:

1. List the classifications of registers and describe their characteristics.
2. Identify the types of register configurations from a schematic diagram.
3. List practical applications of each type of register.
4. List the correct outputs of a register when given specific input signals.
5. Multiply or divide a given binary number by using a shift register
6. Determine the operation of registers that are in IC form.
7. Perform basic troubleshooting of registers.

9.1 INTRODUCTION

As the computer processes data, there is a continual flow of data within and between its major sections. It is necessary to provide temporary storage locations for the data as well as transfer information from one location to another. These operations are performed by registers.

A register is a group or array of flip-flops driven by a common clock signal. It performs four main functions in digital electronics:

1. Data storage
2. Conversion of data from one form to another
3. Data manipulation
4. Counting

These functions are performed by two main types of registers, the *storage register* and the *shift register*.

9.2   STORAGE REGISTERS

Computer word:

Binary bits divided into organized groups that are stored into registers.

Data word:

Binary bits divided into organized groups that are usually 4, 8, 16, or 32 bits in length.

Storage registers:

Circuits consisting primarily of flip-flops that are used to temporarily store digital information.

A flip-flop can store only 1 bit of information. The bit of information can be a high or low (1 or 0). Flip-flops are often arranged into groups to retain multibit information, such as binary numbers, binary-coded numbers, or binary-coded alphanumeric characters. These flip-flops are called *storage registers*. The number of flip-flops in each register determines the storage capacity of 1s and 0s it can retain, which are often referred to as **computer word, data word,** or just *word*. In various computing devices, word lengths are commonly 4, 8, 16, 32, or 64 bits. By further arranging registers into organized groups, it is possible to manageably store millions of bits of information. For simplification, 4-bit registers will be described in this chapter.

Storage registers are primarily used to temporarily store binary information. They provide a place to hold data until it is processed. For example, when two binary numbers are added together in a calculator, each of them must be stored in a register before the addition can take place. This is similar to a person writing two numbers down before adding them. The answer is then displayed for as long as the user wishes, because the answer is held in a register. The data in each register are removed by clearing the register, writing over the bits that are already there, or by turning off the power supply voltage.

9.3   SHIFT REGISTERS

Shift register:

Circuits consisting primarily of flip-flops that are used to store and move data within the register.

Data manipulation:

The process of moving binary data by a shift register within itself.

SISO (serial-in serial-out):

A type of shift register that enters and removes data in serial form.

PIPO (parallel-in parallel-out):

A type of shift register that enters and removes data in parallel form.

PISO (parallel-in serial-out):

A type of shift register that enters data in parallel form and removes data in serial form.

SIPO (serial-in parallel-out):

A type of shift register that enters data in serial form and removes data in parallel form.

The primary function of the storage register just described is that it stores commands and data until they are transferred to another section of the computing device. Another type of register is called a **shift register.** It, too, may also serve as a storage register when proper controls exist. However, a shift register differs from a storage register in that it has the ability to move, or shift, data bits within itself. This process is known as *bit*, or **data, manipulation** and is a common function in most computer systems.

Registers are further classified according to how data are placed into and taken from the circuit. Information is entered or removed one bit at a time (serial form), all bits at once (parallel form), or a combination of both. There are four basic types of registers: **serial-in serial-out (SISO), parallel-in parallel-out (PIPO), parallel-in serial-out (PISO),** and **serial-in parallel-out (SIPO).** These four methods are illustrated in Fig. 9-1. Each block represents an individual flip-flop, and the arrows indicate the directions and types of data movement. The SIPO register converts serial data into parallel data, and the PISO converts parallel data into serial data.

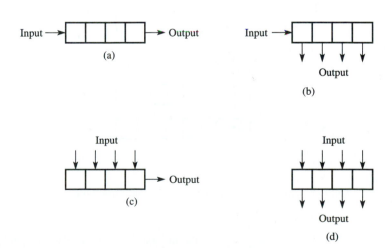

FIGURE 9.1
General types of shift registers: (a) SISO. (b) SIPO. (c) PISO. (d) PIPO.

9.4 SERIAL-IN SERIAL-OUT (SISO) SHIFT REGISTER

A serial-in serial-out (SISO) shift register is one that enters and removes data in serial form. Figure 9-2 shows a 4-bit SISO shift register that consists of four J–K flip-flops all connected to a common clock. Data are placed into the input of the first flip-flop (FFA), then is transferred from flip-flop to flip-flop whenever a clock pulse arrives. The process of shifting data into and out of a register in serial fashion is illustrated in Fig. 9-3. It shows the data that are applied to the serial input of the register before each of four clock pulses and the data found in the register after each clock pulse. The four rectangular boxes represent the Q output of each flip-flop. The register is first reset, making the Q output of each flip-flop a low. Data are then transferred into the first flip-flop (FFA) one bit at a time. Before each new bit enters the input, all of the bits presently in the register are simultaneously moved one position to the right to make room for the new bit. Each new bit that enters FFA writes over the previous bit that was stored.

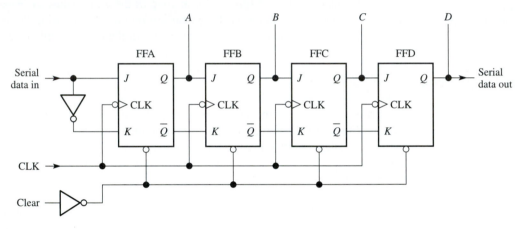

FIGURE 9.2
Four-bit SISO shift register.

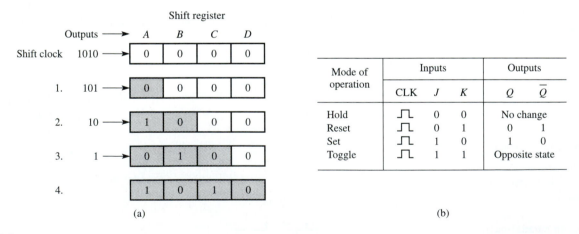

(a)   (b)

FIGURE 9.3
(a) Data movement in a SISO shift register. (b) J–K flip-flop truth table.

To understand why the data are shifted in this manner, it is necessary to examine how each flip-flop responds to the data. Notice that an inverter is connected between the J and K inputs of FFA. This ensures that the signal of the incoming data to FFA is complemented. The J and K inputs of FFB-FFD are also complemented, because they are connected to the Q and $\overline{Q}$ complementing outputs of the previous flip-flops.

The importance of complementing each flip-flop is revealed by examining the truth table of a synchronous J–K flip-flop in Fig. 9-3(b). Note that whatever logic level (1 state or 0 state) a data bit is at when it is applied to the J input of a J–K flip-flop before a clock pulse arrives, it is transferred to the Q output after the clock pulse arrives. The configuration of the universal J–K flip-flop in this circuit makes it operate like a D flip-flop.

The SISO is the slowest shift register. The circuit in Fig. 9-2 requires four clock pulses to enter the data and four more clock pulses to read out the data stored in the register. The number of flip-flops used in a SISO shift register is not limited to four. Many of them can be connected in series. One type of SISO shift register in IC form contains 4096 flip-flops connected in series.

Recirculating shift register:

The process of preserving data in a SISO shift register by reloading the data that leaves the output back into the input.

Recirculating Shift Register

It is sometimes desirable to preserve the data after it is loaded into a SISO register. This is accomplished by first disconnecting the serial-data-in line of Fig. 9-2. Then the Q output of FFD is connected to the J input of FFA, and the $\overline{Q}$ output of FFD to the K input of FFA. The result is a recirculating shift register whose output is fed back to the input and reloaded into the register. This type of register configuration is used by digital circuits that perform the arithmetic functions that are described in Chapter 10.

EXAMPLE 9.1

Suppose that 0110 is initially loaded into a recirculating 4 flip-flop shift register that transfers data to the right. List the contents after each of four clock pulses.

Solution After 1st clock pulse, 0011
After 2nd clock pulse, 1001
After 3rd clock pulse, 1100
After 4th clock pulse, 0110

Two other types of recirculating registers are the *ring counter* and the *Johnson counter.*

Ring Counter

Recirculating shift registers are also used to create a sequence of equally spaced clock pulses. Because the path in which the data travels forms a circle, this device is called a *ring counter,* also a *sequencer.* The block diagram in Fig. 9-4(a) illustrates the data movement when the register is preset with the number 0100_2. After each clock pulse, data is shifted one location to the right. Data at FFD is recirculated to FFA. After the fourth clock pulse, the data completely recirculates. This cycle repeats continuously.

Figure 9-4(b) shows a waveform diagram produced at the Q outputs of the flip-flops. Each flip-flop will be high for one clock period, then low for the next three, and then repeated. When one flip-flop is set and the other three cleared, the input frequency is divided by 4 at any defined output. If, for example, six flip-flops are used instead, the input frequency would be divided by 6 if only one flip-flop is set.

Johnson Counter

Another version of the ring counter is the *Johnson counter.* The configuration of this circuit is the same as the ring counter except that the connections between the outputs of the last flip-flop and the input lines of the first flip-flop are crossed. For example, the Q output of FFD is connected to the K input of FFA of Figure 9–2 (with the inverter removed), and the

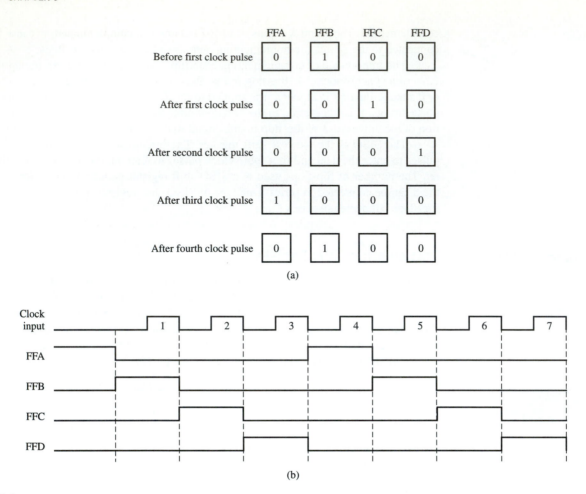

FIGURE 9.4
Ring counter: (a) Shifting sequence. (b) Output waveforms.

$\overline{Q}$ output of FFA is connected to the J input of FFA. The result is that each data bit that leaves the last output is inverted as it is recycled into the first flip-flop. For example, if the data word 0100 is loaded into the register before the first clock pulse is applied, the contents will be 1011 after four clock pulses. The block diagram of the Johnson counter in Fig. 9-5(a) illustrates the sequence.

Figure 9-5(b) shows the waveform diagram produced at the Q output of each flip-flop during eight clock pulses. During time periods X and Y, all outputs are low. Between these time periods, eight sets of output waveforms occur. This gives the Johnson counter, before the same output pattern is repeated, twice as many output states as it has flip-flops.

SISO Shift Resister Application

Figure 9-6(a) illustrates a conveyer system that uses opposed optical sensors to determine if caps are securely placed onto spraypaint cans. If a cap is crooked or too high, the can will be removed from the belt at a rejection point downstream from the inspection location.

When an unsecured cap breaks the light beam, the sensor produces a momentary logic 1 pulse. This pulse is entered as data into a serial shift register. The data movement in the register is synchronized with the conveyer by clock pulses. The clock pulses are obtained from a gear that is coupled to a rotating member of a conveyer drive. As each gear tooth passes an inductive sensor, a clock pulse is produced and fed into the register.

The register in Fig. 9-6(b) illustrates how the clock pulses cause inspection data to move within the 6-bit register. Suppose the sensor detects a tilted cap. A logic 1 is loaded into the first flip-flop of the register. As each clock pulse is applied to the register, the logic 1 bit is shifted toward the right. After six clock pulses, the 1 bit is shifted to the last flip-flop and energizes a pneumatic air valve, which blows the can off the belt.

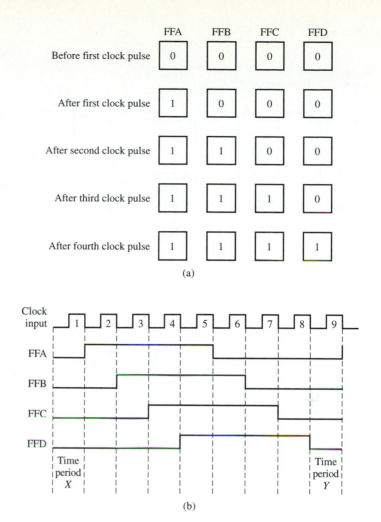

	FFA	FFB	FFC	FFD
Before first clock pulse	0	0	0	0
After first clock pulse	1	0	0	0
After second clock pulse	1	1	0	0
After third clock pulse	1	1	1	0
After fourth clock pulse	1	1	1	1

(a)

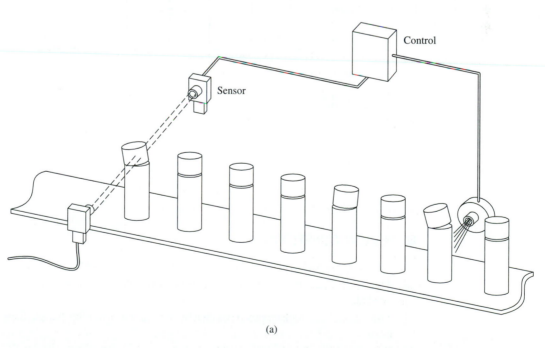

(b)

FIGURE 9.5
Johnson counter: (a) Shifting sequence. (b) Output waveforms.

(a)

FIGURE 9.6
A conveyer system application: (a) Paint cans inspected on a conveyer belt. (b) Data movement within a shift register.

(continues)

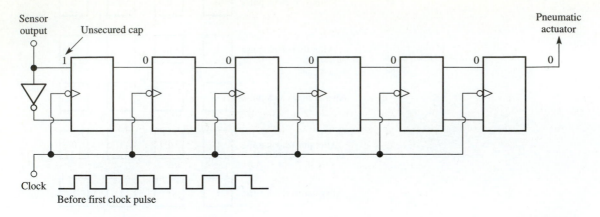

Before first clock pulse

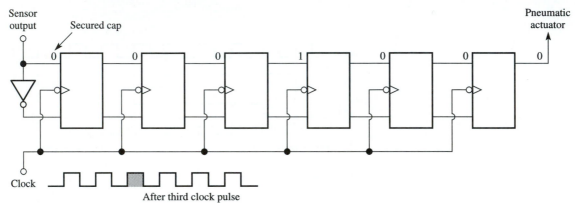

After third clock pulse

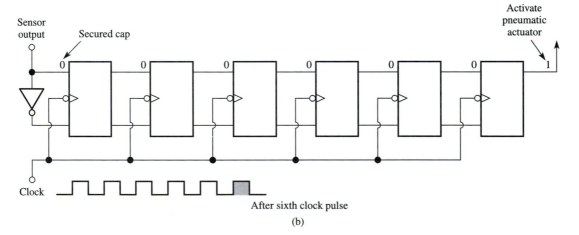

After sixth clock pulse

(b)

FIGURE 9.6
(Continued)

■ REVIEW QUESTIONS

1. The two types of register classification are _____ and _____.

2. The _____ register is primarily used whenever temporary storage of information is needed.

3. The _____ register processes data by moving bits from flip-flops to their adjacent flip-flops.

4. Registers that are classified by how data are entered and removed are _____, _____, _____, and _____.

5. How many clock pulses are needed to load 4-bit data into and shift data out of a SISO shift register consisting of four flip-flops?
   (a) 1
   (b) 4
   (c) 8
   (d) 16

6. As data are shifted right out of the SISO shift register in Fig. 9-2, they can be recirculated back in by connecting the _____ output of FFD to the J input of FFA and the _____ output of FFD to the _____ input of FFA.

7. When one flip-flop in a ring counter that consists of five flip-flops is set and the remaining flip-flops are cleared, the input frequency is divided by _____.
   (a) 1
   (b) 2
   (c) 5

8. As data is recirculated in a _____ counter, each bit is inverted.
   (a) Ring
   (b) Johnson

9.5 PARALLEL-IN PARALLEL-OUT (PIPO) SHIFT REGISTER

The parallel-in parallel-out (PIPO) shift register is shown in Fig. 9-7. It consists of four J–K flip-flops configured as D flip-flops. The clock inputs to the flip-flops are connected in parallel and are negative edge triggered. The inputs to the register are connected to a 4-bit bus. Each bus line is labeled with the same number as the flip-flop to which it supplies data. When 4-bit data is stored in the register, each bit is present at the Q output of each flip-flop.

Suppose the register is used in a computer. When the computer is ready to store information in the register, it sends a clock pulse. All four flip-flops are clocked at the same time. The four flip-flops latch the data supplied by the data-in lines. This data is stored into the register until one of two conditions occurs: either the power fails and the data vanishes, or another clock pulse is delivered to it and new data overrides the data previously stored in the register.

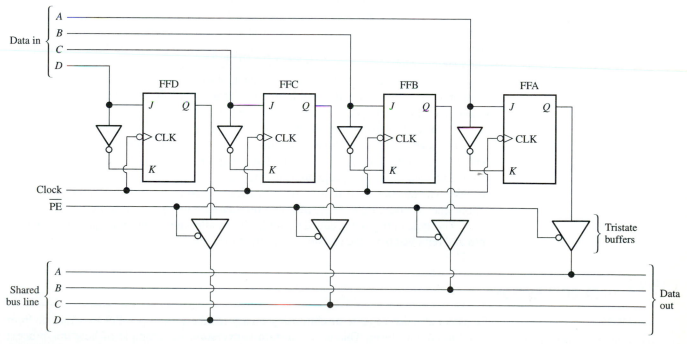

FIGURE 9.7
PIPO shift register.

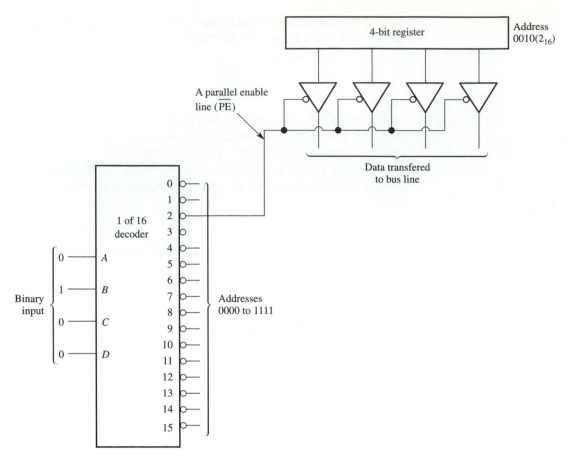

FIGURE 9.8
A decoder is used to transfer data from a register with a particular address.

Any data in the register is stored until it is needed at another location. The data is transferred over a 4-bit bus line that is also shared by other registers. A bus line is designed to pass information from only one register at any given moment. The four tristate buffers connected between the Q outputs and the buffer lines in Fig. 9-7 are used for this purpose. The diagram shows a common enable line, labeled $\overline{PE}$, connected to all four buffers. As long as $\overline{PE}$ remains high, the buffers are in the high-impedance state and the data remains in the register. Only when the enable line is activated will the data from each output be transferred through the buffer to the bus lines in a parallel fashion.

Each register that shares the bus lines has a set of four buffers. Only one set can be activated at any given time. Figure 9-8 shows a simplified diagram of how a register is selected to pass information to a bus line. Each register is assigned a numerical address. A command from a software program instruction, for example, is applied to the 1–16 address decoder. The number applied to the decoder causes its corresponding output to go low. The output line is connected to the buffers of a register with the specified address.

This type of register is called a parallel-in, parallel-out device because data is entered into and removed from the register in parallel form.

9.6   DATA CONVERSION

Data conversion:

The process of converting the movement of data from parallel to serial or from serial to parallel.

Registers are also capable of converting binary numbers from parallel to serial, or from serial to parallel form. This is called **data conversion.** An example of how this process can be used in a practical application is shown by examining the operation of a computer.

All computers use multibit data words. When moved from one location to another inside the computer, the bits in a word are transferred simultaneously in parallel form by data busses connected between PIPO registers. The number of wires in the data bus and the number of flip-flops in the register are equal to the number of bits in the words used by a particular computer.

When data must be sent externally from one computer to another computer a long distance away—say, Los Angeles to New York—it cannot be transferred in parallel form. If it was, large cables with many conductors would be required, which would mean high costs and a high probability of broken wires.

By converting parallel data to serial form for transmission, it is possible to send data long distances over a telephone line consisting of one serial wire. Once the serial data arrives at its destination, it is converted back to the parallel form required by the receiving computer. Parallel-in serial-out (PISO) and serial-in parallel-out (SIPO) shift registers are used to perform this conversion process, as shown in Fig. 9-9.

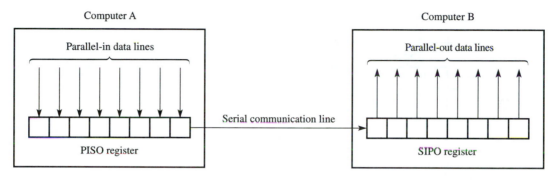

FIGURE 9.9
Using registers to transfer data over serial lines between computers that use parallel data.

An 8-bit data word is parallel loaded into a PISO shift register in computer A. Seven clock pulses are required to shift data out of the register across the serial communication line to the SIPO register in computer B, which concurrently loads the 8 bits. Eight clock pulses are required to shift the 8-bit data word into the register. After the data is completely loaded into the receiving register, it is parallel transferred to another location in computer B.

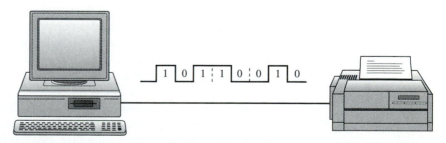

FIGURE 9.10
Serial transfer of binary data from a computer to a printer.

This is a simplified explanation of how data are sent in serial form between computers, or from a computer to peripheral equipment over a single transmission wire. A device called a *modem* performs this function. The technique is also used by the RS232 standard to send ASCII data from a computer to a printer, as shown in Fig. 9-10.

9.7 PARALLEL-IN SERIAL-OUT (PISO) SHIFT REGISTER

The parallel-in serial-out register converts parallel data into serial form. This type of circuit is shown in Fig. 9-11. It consists of four J–K flip-flops. Each one is configured to operate as a D flip-flop. A common clock input is parallel connected to each flip-flop.

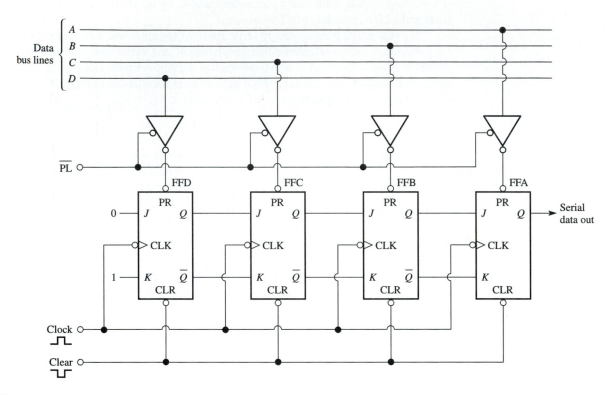

FIGURE 9.11
PISO shift register.

Data are supplied to the register from one location by using a four-line bus, which is shared by other registers. A set of four tristate buffer inverters is connected between the bus and the preset (PR) leads of the flip-flop. The enable line for the buffers is labeled $\overline{PL}$. If line $\overline{PL}$ is high, the buffers are in their high-impedance state. Before loading data into the register, all the flip-flops are cleared by applying a momentary low at the CLR inputs. The register output is 0000. When line $\overline{PL}$ is activated by a low, the data is parallel loaded into the register through the buffers. If a 1 is present on the bus line, it is inverted to a logic 0 and applied to the active-low $\overline{PR}$ line of the flip-flop. The flip-flop sets and a logic 1 is present at its Q output. If a 0 is present on the bus line, it is inverted to a logic 1 and is applied to the active-low $\overline{PR}$ line of the flip-flop. This flip-flop remains cleared and a logic 0 is present at its Q output. Making line $\overline{PL}$ high again disables the buffer, and the 4-bit data is latched in the register.

Suppose that a 0101 (5_{10}) is loaded into the register. Data are serial fed out of the register at the Q output lead of FFA. With each succeeding clock pulse, data are shifted right one bit at a time and out of the register. After the third clock pulse, the last bit is supplied to the serial output line. The LSB is output first and the MSB is output last. Any more serial outputs will be 0s until new parallel data are loaded into the register.

9.8   SERIAL-IN PARALLEL-OUT (SIPO) SHIFT REGISTER

Serial data is converted back to parallel form by a serial-in parallel-out shift register. A 4-bit SIPO shift register is shown in Fig. 9-12. It consists of four *J–K* flip-flops. Each one is configured as a *D* flip-flop. A common clock input and a common CLR input are parallel connected to each flip-flop. To load data into the SIPO register, data are applied serially to the input lead of FFD.

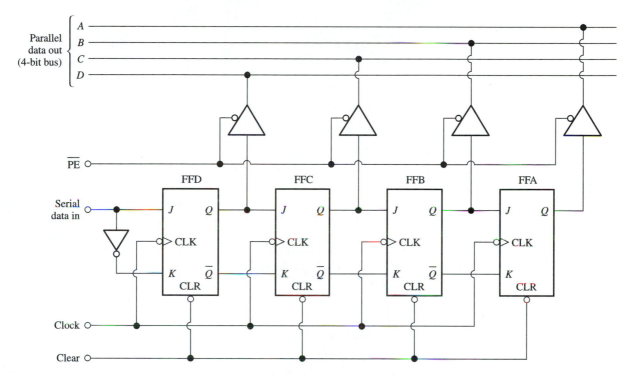

FIGURE 9.12
SIPO shift register.

Before the register is loaded with data, it is cleared by a momentary low clear pulse. Suppose a 4-bit data word, 0101 (5_{10}), is transmitted over the serial line. The LSB, which is a 1, arrives first at the *J* input of FFD. Its complement, a low, is applied to the *K* input, causing the flip-flop to be in the set mode. After the first clock pulse arrives, the 1 latched into the flip-flop, is present at its *Q* output and is applied to the *J* input of FFC. The next bit, which is a 0, is applied from the serial input line of the register to the *J* input line of FFD, causing it to be in the reset mode. After the second clock pulse, it shifts to the *Q* output of FFD. The bit that was previously stored in FFD before the second clock pulse shifts to the output of FFC. With each clock pulse that follows, the data bits shift one position to the right and a new bit enters the register. After four clock pulses, the register is fully loaded with the 4-bit word.

A series of four tristate buffers are connected between the *Q* outputs of the register and the bus lines to which they are sent. A common active-low enable line, labeled PE, is parallel connected to the buffers. The data from the register cannot be transferred to the bus line unless the PE line is activated. After the momentary enable pulse, the contents remain in the register until power is lost, the register is cleared, or new data is transferred in.

Bidirectional shift register:

A serial shift register capable of moving data to the right or to the left.

EXPERIMENT

Shift Registers

Objectives

- To construct two 7474 ICs to operate as a 4-bit serial load register.
- To construct two 7476 ICs to operate as a 4-bit serial/parallel load register.
- To demonstrate the four different ways that data is loaded into and read out of a 4-bit register.

Materials

2—7474 ICs

2—7476 ICs

4—LEDs

4—150-Ω Resistors

5—Logic Switches or SPDT Switches

1—Logic Pulse Push Button

1—+5-V DC Power Supply

Introduction

A shift register consists of a series of flip-flops configured with the outputs of one flip-flop connected to the inputs of the next. The number of flip-flops in a register depends on the number of bits to be stored or processed. The register operates as a memory device when it is used to store information until it is needed. The register is also capable of processing information by moving data from one flip-flop to an adjacent flip-flop.

Shift registers are classified according to three factors:

1. Their storage capacity (number of flip-flops).
2. Their direction of data movement (shift-right/shift-left).
3. Their method of handling data:
   (a) Serial-in serial-out (SISO)
   (b) Serial-in parallel-out (SIPO)
   (c) Parallel-in serial-out (PISO)
   (d) Parallel-in parallel-out (PIPO)

   Serial-in means that data is shifted into the register one bit at a time.

   Serial-out means that data is shifted out of the register one bit at a time.

   Parallel-in means that several bits of data are shifted into the register all at the same time.

   Parallel-out means that several bits of data are shifted out of the register all at the same time.

The primary objective of this experiment is to demonstrate the four different methods of handling data.

PART A—SERIAL-LOAD REGISTER

Background Information

The wiring diagram for a 4-bit serial-load register is shown in Fig. 9-13. The circuit uses two 7474 ICs consisting of two *D* flip-flops each. A common clock input is connected to each flip-flop in the register. When a synchronous (synchronized clock signal sent to all

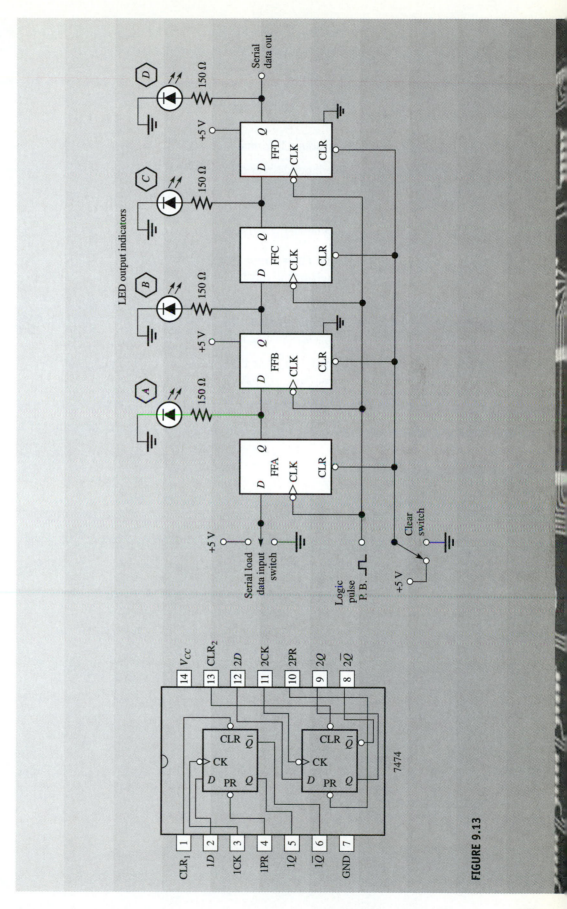

FIGURE 9.13

flip-flops simultaneously) clock pulse is applied, all the bits in the register shift together one place to the right. The circuit can be used as a SISO or SIPO register.

Procedure

Step 1. Assemble the circuit shown in Fig. 9-13.
Step 2. Turn the power on.

Procedure Information

The input devices on the left of the register are used to operate the register. They are used to perform the following functions:

Clear Switch   By setting the switch to 0, the data in the register is cleared out so that 0000 is in the register. The switch must be at a 1 when data is in the register.

Logic Pulse Input   The logic pulse applies a single clock pulse to all of the flip-flops simultaneously. All data in the register shifts one place to the right after each clock pulse.

Data Switch   This switch applies binary data to the input of the register. When the data switch is high, a 1 enters the register at FFA after the clock pulse. When the data switch is low, a 0 enters the register after a clock pulse.

Step 3. Operate the register according to Table 9-1.

TABLE 9.1
Serial load register

	Inputs			Outputs			
Line	Clear	Data input line	Clock pulse	LED indicators			
				A	B	C	D
A	0			0	0	0	0
B	1	0		0	0	0	0
C	1	1	⎍	1	0	0	0
D	1	0	⎍				
E	1	1	⎍				
F	1	1	⎍				
G	0	1					
H	1	1	⎍				
I	1	1	⎍				
J	1	0	⎍				
K	1	0	⎍				
L	1	0	⎍				
M	1	1	⎍				
N	1	0	⎍				
O	0	0					
P	1	1	⎍				
Q	1	0	⎍				
R	0	1	⎍				

Example:

> *Line A* shows that data is cleared from the register by setting the CLR switch to 0.
>
> *Line B* shows the CLR switch setting is back to 1 and the data switch is at 0.
>
> *Line C* shows the data switch at 1 and the contents in the register after a clock pulse is applied.
>
> Observe and record the results in the output section of the table.

Step 4. Go through the sequence of inputs on lines *D* through *R* and record the results in the output section of the table.

Procedure Question 1

What happens to a data bit at FFD after each clock pulse?

PART B—PARALLEL-LOAD REGISTER

Background Information

The wiring diagram for a 4-bit parallel-load register is shown in Fig. 9-14. The circuit uses two 7476 ICs consisting of two *J–K* flip-flops each. The circuit can be used as a PISO or PIPO register.

This circuit is available on Electronics Workbench by downloading file *Fig. 9-14*.

Step 5. Assemble the circuit in Fig. 9-14.
Step 6. Turn the power on.

Procedure Information

Instead of using a single data switch for serial transfer into the register, four switches located at the left of the diagram are used to parallel load the data. When a switch is high, the flip-flop to which it is connected remains in its present state. When a switch is low, a 1 is loaded into the flip-flop. The register is capable of retaining its data by feeding flip-flop FFA with the data from the outputs of FFD. A register that retains its data is also called a *recirculating* shift register.

Step 7. Operate the register according to Table 9-2.

Example:

> *Line A* shows that the data is cleared from the register by setting the CLR switch to 0.
>
> *Line B* shows that the CLR switch setting is back at 1 and a binary 1000 is parallel loaded into the register. This is accomplished by placing switch A at position 0. The 0 at the active-low preset input causes FFA to set and store a 1.
>
> *Line C* shows all of the parallel-load data switches back in their inactive 1 states. As a clock pulse is applied, all of the data in the register is shifted one position to the right.

Procedure Question 2

What happens to the data bit at FFD after each clock pulse?

Step 8. Go through the sequence of inputs remaining in the table and record the results in the output portion.
Step 9. Turn off the power and disassemble the circuit.

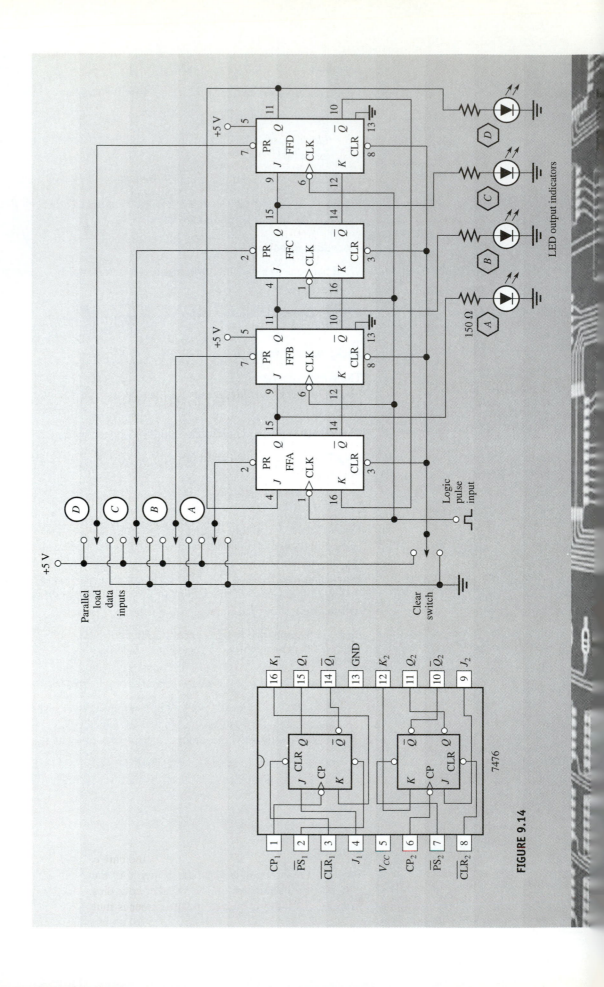

FIGURE 9.14

TABLE 9.2
Parallel load register

Line	Clear	Parallel data lines				Clock pulse	LED indicators			
		A	B	C	D		A	B	C	D
A	0						0	0	0	0
B	1	0	1	1	1		1	0	0	0
C	1	1	1	1	1	⊓	0	1	0	0
D	1	1	1	1	1	⊓				
E	1	1	1	1	1	⊓				
F	1	1	1	1	1	⊓				
G	0	1	1	1	1					
H	1	1	0	0	1					
I	1	1	1	1	1	⊓				
J	1	1	1	1	1	⊓				
K	1	1	1	1	1	⊓				
L	0	1	1	1	1	⊓				
M	1	0	1	1	0					
N	1	1	1	1	1					
O	1	1	1	1	1	⊓				
P	1	1	1	1	1	⊓				
Q	1	0	1	1	1					
R	1	1	1	1	1	⊓				
S	1	1	1	1	1	⊓				
T	1	1	1	1	1	⊓				

■ **EXPERIMENT QUESTIONS**

1. A register that stores 6 bits has how many flip-flops?
2. Registers are classified according to what three factors?
3. The data that is shifted to the right goes from _____ (Q_A, Q_D) to _____ (Q_A, Q_D).
4. The shift register in Fig. 9-14 is a _____ (synchronous, asynchronous) device.
5. The faster operating shift register is the _____ (SISO, PISO) type.
6. How many clock pulses are required to transfer 1011 into a 4-bit SISO shift register?

9.9   BIDIRECTIONAL SHIFT REGISTER

The movement of data in a shift register is not limited to being unidirectional. A **bidirectional shift register** is one in which the data can be shifted to the right or left. Therefore, it is also capable of performing the multiplication and division of binary numbers by powers of 2. Figure 9-15 illustrates how these two mathematical functions operate. Figure 9-15(a) shows that a binary number is multiplied by 2 every time it is shifted one position to the left. Likewise, Fig. 9-15(b) shows that a binary number is divided by 2

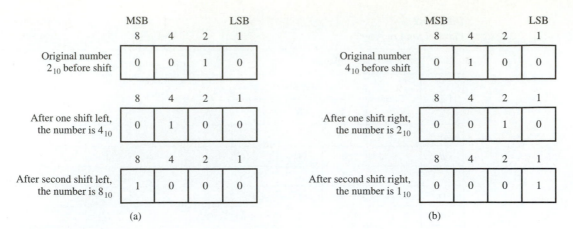

FIGURE 9.15
Multiplication and division functions: (a) Shift-left multiplication. (b) Shift-right division.

every time it is shifted one position to the right. Shift-left and shift-right instructions are used by microprocessors to perform multiplication and division operations utilizing this method.

A 4-bit register that is capable of performing the bidirectional function is shown in Fig. 9-16. A high on the direction-control input enables data to be shifted to the right, and a low input allows data to be shifted left. When a high is applied to the right/$\overline{\text{left}}$ control, gates G_1 through G_4 are enabled and the logic state at the Q output of each flip-flop is passed through to the inputs of the flip-flop located at the immediate right. The data are effectively shifted one place to the right when a clock pulse occurs. When a low is applied to the right/$\overline{\text{left}}$ control, gates G_5 through G_8 are enabled, and the logic states of the Q output of each flip-flop are passed through to the inputs of the flip-flops located at the immediate left. The data are effectively shifted one place to the left after a clock pulse arrives.

■ REVIEW QUESTIONS

9. When 0110 is applied to the input of a decoder used to activate address buffers, what is the address of the register from which data will be transferred to a bus line?
    (a) 0
    (b) 1
    (c) 6
    (d) 9

10. The shift register classification PISO means _____ _____.

11. The two types of shift registers commonly used for long-distance transmission of binary data are the _____ and _____.

12. How many clock pulses are needed to load an 8-bit SIPO register?
    (a) 1
    (b) 4
    (c) 8
    (d) 16

13. The _____ register is the fastest of all registers and is the one primarily used as a storage register.

14. In a bidirectional shift register, if data are shifted to the left, which of the following arithmetic functions is performed?
    (a) Addition
    (b) Subtraction
    (c) Multiplication
    (d) Division
    (e) None of the above

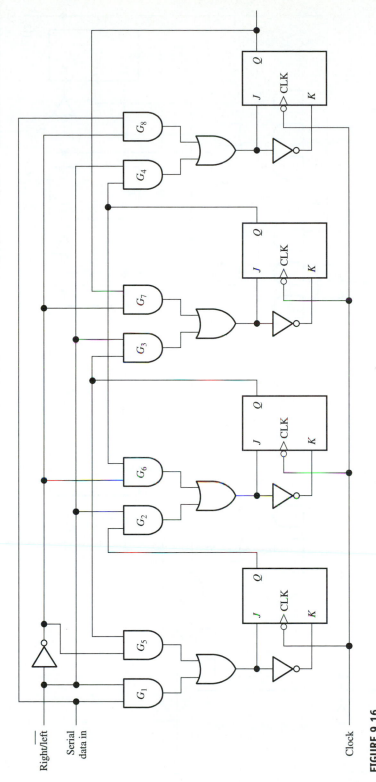

FIGURE 9.16
Bidirectional shift register.

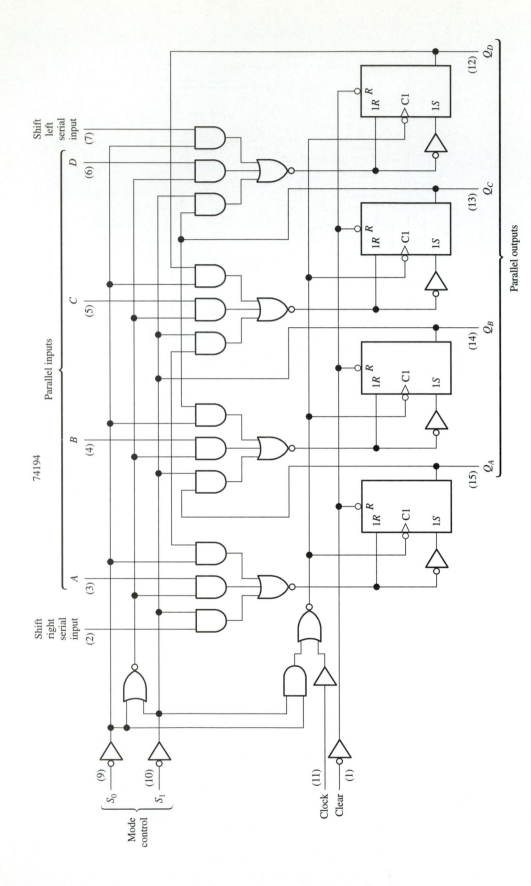

FIGURE 9.17
Schematic diagram of the internal circuitry of the 74194 universal shift register IC.

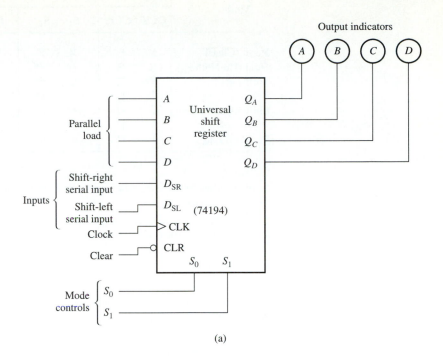

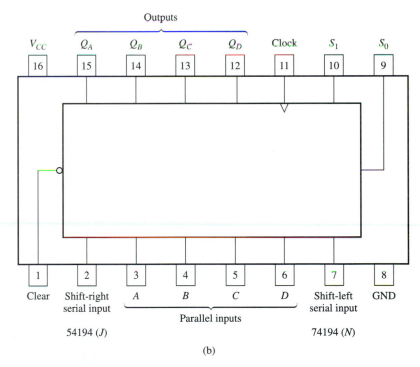

FIGURE 9.18
The 74194 universal shift register IC: (a) Logic symbol. (b) Pin diagram. (c) Timing diagram.

(continues)

9.10 THE 74194 UNIVERSAL SHIFT REGISTER

Universal shift register:

A register capable of functioning in any of the SISO, SIPO, PISO, and PIPO modes of operation.

The 74194 IC is called a **universal shift register** because it is capable of operating in any mode. To operate universally, it contains serial inputs, serial outputs, parallel inputs, parallel outputs, and must be able to serially shift data to the right or to the left, hold the data, or be reset. The schematic diagram of the 74194 IC is shown in Fig. 9-17. It consists of four flip-flops, and more than one 74194 IC can be cascaded to make a shift register of 8 bits or larger.

Operating Mode	Inputs						Outputs			
	CLK	CLR	S_1	S_0	D_{SR}	D_{SL}	Q_A	Q_B	Q_C	Q_D
Reset (Clear)	X	L	X	X	X	X	L	L	L	L
Hold (Do Nothing)	X	H	ℓ^*	ℓ^*	X	X	q_a	q_b	q_c	q_d
Shift-Left	↑	H	h	ℓ^*	X	ℓ	q_b	q_c	q_d	L
	↑	H	h	ℓ^*	X	h	q_b	q_c	q_d	H
Shift-Right	↑	H	ℓ^*	h	ℓ	X	L	q_a	q_b	q_c
	↑	H	ℓ^*	h	X	H	H	q_a	q_b	q_c
Parallel-Load	↑	H	h	h	X	X	d_a	d_b	d_c	d_d

H= High-voltage level

h= High-voltage level one setup time prior to the L-to-H clock transition

L= Low-voltage level

ℓ = Low-voltage level one setup time prior to the L-to-H clock transition

$d_n(q_n)$ = (Lowercase letters indicate the state of the referenced input, or output, one setup time prior to the L-to-H clock transition)

X = Don't care

↑ = L-to-H clock transition

*The H-to-L transition of the S_0 and S_1 inputs on the 74194 should only take place while CLK is high for conventional operation.

(c)

FIGURE 9.18 (Continued)

TABLE 9.3

Operating modes of the 74194

OPERATING MODE	MODE CONTROL INPUT LINES
Parallel load	$S_0 = 1, S_1 = 1$
Shift right	$S_0 = 1, S_1 = 0$
Shift left	$S_0 = 0, S_1 = 1$
Inhibit	$S_0 = 0, S_1 = 0$

A block diagram used for the 74194 IC register is shown in Fig. 9-18(a). How these lines affect the operation of this register are as follows:

CLR (Clear): This is an active-low input line that resets the register to 0000. The register holds at 0000 as long as CLR = 0. It overrides all other inputs.

Q_A–Q_D (Register Outputs): The binary contents of the register always present at these output lines.

D_{SR} (Data Shift-Right Input): Shifting data to the right is defined by the manufacturer as moving data serially from Q_A to Q_D. Shift-right data enter input lines D_{SR}, and as they leave Q_D, they are lost.

D_{SL} (Data Shift-Left Input): Shifting data to the left is defined by the manufacturer as moving data serially from Q_D to Q_A. Shift-left data enter input line D_{SL}, and as they leave Q_A, they are lost.

Parallel Inputs: Data can be parallel loaded into the register at these inputs, which are labeled *A–D*.

S_0 and S_1 (Mode-Control Inputs): The operating mode that this register functions at is determined by the states applied to mode-control inputs S_0 and S_1 shown in Table 9-3.

Parallel Load: The parallel-load function is performed by making S_0 and S_1 high, placing data at *A–D,* and applying a positive-edge-triggered signal to the clock input (CLK).

Shift Right: Data are shifted to the right by making $S_0 = 1$ and $S_1 = 0$ and applying a clock signal to CLK. Shift-right data can be recirculated by connecting output Q_D to input D_{SR}.

Shift Left: Data can be shifted to the left by making $S_0 = 0$ and $S_1 = 1$ and applying a clock pulse to CLK. Shift-left data can be recirculated by connecting output Q_A to input D_{SL}.

Hold: Data in the register can be frozen by making S_0 and S_1 low.

The function table for the 74194 register is in Fig. 9-18(c).

EXAMPLE 9.2

Demonstrate how the 74194 IC serial loads and shifts data when the mode control switches are in various positions.

Solution   Examine the serial load shift register table in Fig. 9-19.

Line	Mode control S_0	S_1	Clear	Serial data inputs D_{SR}	D_{SL}	Clock pulse	Outputs A	B	C	D
A			0				0	0	0	0
B	1	0	1	1	0	⊓	1	0	0	0
C	1	0	1	0	0	⊓	0	1	0	0
D	1	0	1	1	1	⊓	1	0	1	0
E	0	1	1	0	0	⊓	0	1	0	0
F	0	1	1	0	1	⊓	1	0	0	1
G	0	1	1	0	1		1	0	0	1
H	0	1	1	0	0	⊓	0	0	1	0
I	0	0	1	0	1	⊓	0	0	1	0
J	0	1	1	0	1		0	0	1	0
K	1	0	1	1	0	⊓	1	0	0	1
L	0	0	1	1	0	⊓	1	0	0	1
M	1	0	0	1	0	⊓	0	0	0	0
N	0	1	1	0	1	⊓	0	0	0	1

Line	Operation
A	The register is cleared.
B	In the shift-right mode, a 1 enters the shift-right input after a clock pulse.
C	In the shift-right mode, a 0 enters the shift-right input after a clock pulse.
D	In the shift-right mode, a 1 enters the shift-right input after a clock pulse.
E	In the shift-left mode, a 0 enters the shift-left input after a clock pulse.
F	In the shift-left mode, a 1 enters the shift-left input after a clock pulse.
G	No data movement because there is no clock pulse.
H	In the shift-left mode, a 0 enters the shift-left input after a clock pulse.
I	In the hold mode, there is no data movement.
J	No data movement because there is no clock pulse.
K	In the shift-right mode, a 1 enters the shift-right input after a clock pulse.
L	In the hold mode, there is no data movement.
M	The register is cleared.
N	In the shift-left mode, a 1 enters the shift-left input after a clock pulse.

(c)

FIGURE 9.19
Serial load shift register for Example 9.2.

EXAMPLE 9.3

Demonstrate how the 74194 IC operates as a bidirectional recirculating shift register after data is parallel loaded.

Solution   Examine the parallel load shift register table in Fig. 9-20.

- The register becomes 0000 with a 0 applied to the clear input. (Lines A and M)
- Data shifts one position to the right and is recirculated from the D flip-flop to the A flip-flop when $S_0 = 1$ and $S_1 = 0$ and a clock pulse is applied. (Lines C and D)
- Data shifts one position to the left and is recirculated from the A flip-flop to the D flip-flop when $S_0 = 0$ and $S_1 = 1$ and a clock pulse is applied. (Lines, E, F, H, I, J, and N)
- Data is parallel loaded into the register whenever $S_0 = 1$ and $S_1 = 1$ and a clock pulse is applied. (Lines B, G, and K)

	Inputs										Outputs			
Line	Mode control		Clear	Parallel data inputs				Clock pulse		Outputs				
	S_0	S_1		A	B	C	D			A	B	C	D	
A			0							0	0	0	0	
B	1	1	1	0	0	1	0	⊓		0	0	1	0	
C	1	0	1					⊓		0	0	0	1	
D	1	0	1					⊓		1	0	0	0	
E	0	1	1					⊓		0	0	0	1	
F	0	1	1					⊓		0	0	1	0	
G	1	1	1	1	0	0	0	⊓		1	0	0	0	
H	0	1	1					⊓		0	0	0	1	
I	0	1	1					⊓		0	0	1	0	
J	0	1	1							0	0	1	0	
K	1	1	1	0	1	1	0	⊓		0	1	1	0	
L	0	0	1					⊓		0	1	1	0	
M	1	0	0					⊓		0	0	0	0	
N	0	1	1					⊓		0	0	0	0	

FIGURE 9.20
Parallel load shift register for Example 9.3.

EXPERIMENT

74194 Universal Shift Register

Objectives

- To construct and operate the 74194 IC to manipulate data the following ways:

  Serial load   Shift right

  Parallel load   Shift left

Materials

  1—74194 IC   4—LEDs

  7—Logic Switches   4—150-Ω Resistors

  1—Logic Clock Pulse Push Button   1—75-Ω Resistor
  (Bounceless)
                                       1—+5-V DC Power Supply

Introduction

The 74194 IC is a 4-bit register that is capable of transferring data in the SISO, SIPO, PISO, and PIPO modes of operation. It also is capable of shifting data to the right or to the left. Because it is so versatile, it is often called the *universal shift register*.

Background Information

Before proceeding with the experiment, carefully read the following information, which describes the function of each type of input and output terminal on the 74194 IC. Refer to Fig. 9-21(a).

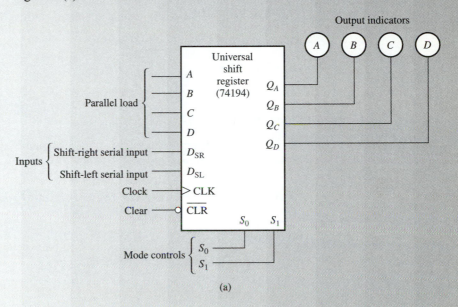

(a)

Mode control		Data enters
Parallel load	$S_0 = 1, S_1 = 1$	Parallel-load inputs
Shift right	$S_0 = 1, S_1 = 0$	Serial-load input (shift right)
Shift left	$S_0 = 0, S_1 = 1$	Serial-load input (shift left)
Inhibit	$S_0 = 0, S_1 = 0$	Data movement frozen

(b)

FIGURE 9.21
(a) Block diagram of the 74194 TTL IC. (b) Function table of the 74194 TTL IC.

Data Inputs

Parallel-Load Inputs: Four bits of data can be parallel loaded directly into the register at these terminals.

Shift-Right Serial Input: The input where data can be serially loaded into the register one bit at a time to the right (Q_A toward Q_D).

Shift-Left Serial Input: The input where data can be serially loaded into the register one bit at a time to the left (Q_D toward Q_A).

Clock

This input enables data to be loaded into the register in either a serial or a parallel fashion when it receives a positive-edge-triggered clock pulse. It also causes the shifting of data within the register.

Mode Control Inputs

The mode control inputs are labeled S_0 and S_1. The combination of logic states applied to them determines how the shift register operates.

$S_0 = 1, S_1 = 1$: Data at inputs A, B, C, and D is parallel loaded into the shift register when a pulse is applied to the clock input.

$S_0 = 1, S_1 = 0$: Data synchronously shifts right and new data enters the shift-right serial input when pulses are applied to the clock input.

$S_0 = 0, S_1 = 1$: Data synchronously shifts left and new data enters the shift-left serial input when pulses are applied to the clock input.

$S_0 = 0, S_1 = 0$: The clock input is inhibited. Nothing happens.

The table in part (b) of Fig. 9-21 summarizes the functions of the mode control inputs.

Clear Input $\overline{CLR}$

The contents of the register are cleared to 0000 when a low is applied to the CLR input.

Output Leads Q_A to Q_D

The contents of the register can be displayed or transferred out by the four output terminals of the register. Q_A represents the LSB and Q_D represents the MSB.

PART A—SERIAL LOAD SHIFT-RIGHT/LEFT

Procedure

Step 1. Assemble the circuit shown in Fig. 9-22.

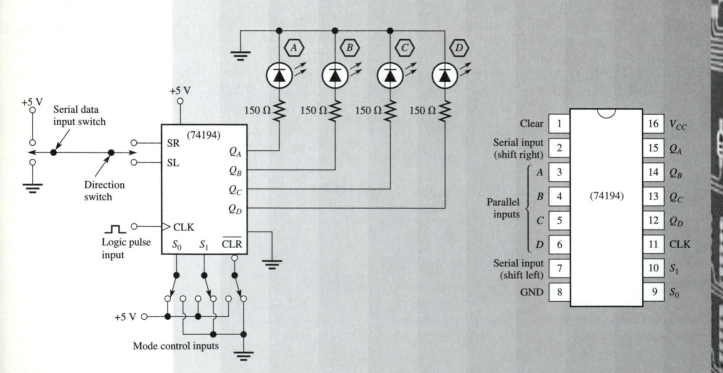

FIGURE 9.22
A 4-bit serial-load shift register using the 74194 TTL IC.

Procedure Information

Figure 9-22 shows the 74194 IC wired as a serial-load shift-right register.

Serial Load Shift-Right

The manufacturer defines shifting right as shifting from Q_A to Q_D. To operate this way, mode control S_0 must be 1 and S_1 must be 0. Data is entered at the shift-right serial input when pulses are applied at the CLK terminal. Data is shifted out of Q_D.

Step 2. Set the mode control switches to the shift-right positions ($S_0 = 1$, $S_1 = 0$). Set the *direction* switch in the (SR) position to enter data into the shift-right input.

Step 3. Turn the power on. As shown on line A of Table 9-4, clear data from the register. This is done by placing the $\overline{\text{CLR}}$ switch to 0, then back to 1. The output section of the table becomes 0000.

TABLE 9.4
Serial-load shift register

	Inputs			Outputs			
Line	Clear	Data input line	Clock pulse	LED indicators			
				A	B	C	D
A	0			0	0	0	0
B	1	1		0	0	0	0
C	1	1	⊓				
D	1	0	⊓				
E	1	0	⊓				
F	1	1	⊓				
G	1	0	⊓				
H	1	1					
I	1	0	⊓				
J	0						
K	1	1	⊓				
L	1	1	⊓				
M	1	0	⊓				
N	1	1	⊓				
O	1	0	⊓				
P	1	0					
Q	1	0	⊓				
R	1	1	⊓				

Step 4. Operate the shift register according to lines B through I in the table's input section. Record the results in the output section.

Serial Load Shift-Left

The manufacturer defines shifting left as shifting from Q_D to Q_A. To operate this way, mode control S_0 must be 0 and S_1 must be 1. Data is entered at the shift-left serial input when pulses are applied at the CLK terminal. Data is shifted out of Q_A.

Step 5. Set the mode control switches to the shift-left positions ($S_0 = 0$, $S_1 = 1$). Set the *direction* switch to the (SL) position to enter data into the shift-left input.

Step 6. Operate the shift register according to lines J through Q and record the results in the output section.

Inhibit Mode: The movement of data within the shift register is frozen when the mode control switches are both 0.

Step 7. Set the mode control switches S_0 and S_1 to 0 so that the register is in the inhibit mode.

Step 8. Apply several clock pulses. Observe that the data in the register shown on line Q in Table 9-4 does not move. Record the results on line R of the output section in Table 9-4.

PART B—PARALLEL LOAD SHIFT-RIGHT/LEFT

Step 9. Assemble the circuit shown in Fig. 9-23.

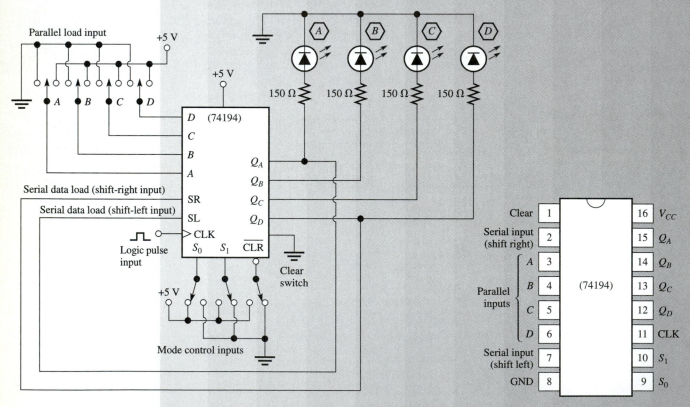

FIGURE 9.23
A 4-bit parallel load recirculating shift-right/left register using the 74194 TTL IC.

Procedure Information

Figure 9-23 shows the 74194 IC wired as a parallel-load shift register.

Parallel Load

Data can be parallel loaded into the shift register at the four active-high inputs called the *parallel load lines*. To perform this operation, 1s must be applied to both mode control switches before a clock pulse is applied.

Step 10. Clear the register with a momentary low at the $\overline{CLR}$ input. Line *A* of Table 9-5 shows 0000. Set mode control switches S_0 and S_1 to 1. Turn the power on. Parallel load data into the register by setting the input switches according to line *B* of Table 9-5 and then applying a clock pulse.

TABLE 9.5
Parallel-load shift register

Line	Inputs								Outputs			
	Mode control		Clear	Parallel data inputs				Clock pulse	A	B	C	D
	S_0	S_1		A	B	C	D					
A			0						0	0	0	0
B	1	1	1	0	0	1	0	⎍				
C	1	0	1					⎍				
D	1	0	1					⎍				
E	0	1	1					⎍				
F	0	1	1					⎍				
G	1	1	1	1	0	0	0	⎍				
H	0	1	1					⎍				
I	0	1	1					⎍				
J	0	1	1									
K	1	1	1	0	1	1	0	⎍				
L	0	0	1					⎍				
M	1	0	0					⎍				
N	0	1	1					⎍				

Recirculating Data

Once data is parallel loaded into the register, the mode control switches can be set so that the data is shifted to the right or left. By connecting the Q_D output line to the shift-right input (SR), data that shifts out of the register can be recirculated into the register in a left-to-right direction. Likewise, data shifting towards the left can be recirculated by connecting Q_A to the shift-left serial input (SL).

Step 11. Operate the shift register according to lines *C* through *N* of the input section in Table 9-5. Observe and record the results in the output columns.

■ **EXPERIMENT QUESTIONS**

1. Why is the 74194 IC called a universal shift register?
2. To parallel load data into the 74194 IC, what requirements need to be met?
3. What requirements need to be met when serial loading data into a 74194 IC in a right-to-left direction?

Cascading Registers

Figure 9-24 shows how two 74194 ICs can be wired as an 8-bit, bidirectional, recirculating shift register. The Q_D output from each register is connected to the SR input of the other register, and the Q_A output from each register is connected to the SL input of the other register. Several inputs of one register are connected to the input with the same name at the other register. They include the clock, clear, S_0, and S_1 inputs.

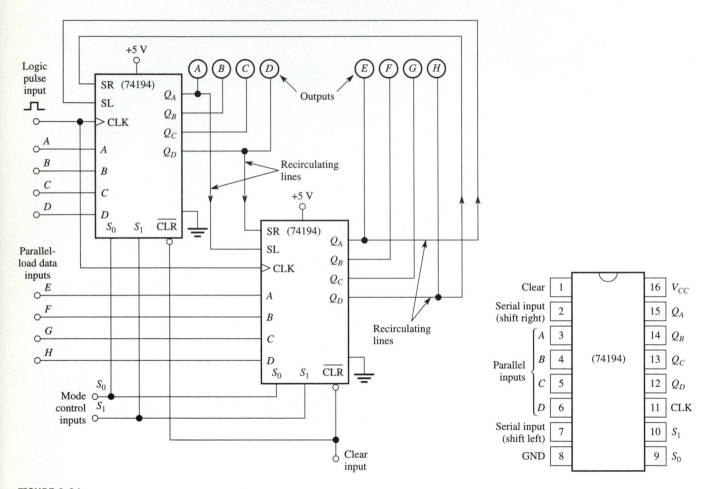

FIGURE 9.24
An 8-bit parallel-load recirculating shift-right/left register using two cascaded 74194 TTL ICs.
*Note: Use LEDs for outputs A through H. Place a 68 Ω current limiting resistor between each output lead and the LED.

EXPERIMENT

CASCADING THE 74194 Universal Shift Register

Objectives

- To construct and operate two cascaded 74194 ICs as an 8-bit shift register.
- To demonstrate how shift registers can multiply and divide numbers by 2.

Materials

2—74194 ICs

11—Logic Switches

1—Logic Clock Pulse Push Button (Bounceless)

8—LEDs

8—150-Ω Resistors

1—75-Ω Resistor

1—+5-V DC Power Supply

Introduction

The 4-bit capacity of the 74194 IC shift register can be increased by cascading two or more ICs. Figure 9-24 shows how two registers are connected to form an 8-bit shift register. This circuit has all of the operating capabilities of the single 74194 IC shift register.

Step 1. Assemble the circuit shown in Fig. 9-24.

Step 2. Turn the power on. Clear the register.

Step 3. Parallel load 00000100_2 into the register (00100000 shown on the diagram). The decimal value stored into the register is 4.

Background Information—Multiplication and Division

Numbers in the register can be multiplied by shifting the contents to the right and divided by shifting them to the left. Output A of Fig. 9-24 is the LSB of a pure binary number, and output H is the MSB.

Step 4. Set the mode control switches in the following positions: $S_0 = 1$ and $S_1 = 0$. Apply a clock pulse. The display should show a 00010000 (which is a binary number, 00001000_2, or decimal 8).

Step 5. Apply one more clock pulse. The display should show a 00001000 (which is a binary number, 00010000_2, or decimal 16).

Procedure Question 1

How much is the number multiplied by every time a clock pulse is applied?

Step 6. Clear the contents. Then parallel load 00001100_2 into the register (00110000 shown on the diagram). The decimal value stored in the register is 12.

Step 7. Set the mode control switches in the following positions: $S_0 = 0$ and $S_1 = 1$. Apply two clock pulses. The display should show a 11000000 (which is a binary number, 00000011_2, or decimal 3).

Procedure Question 2

How much is the number divided by each time the contents are shifted to the left by a clock pulse?

Step 8. Parallel load 00010000 into the register. Set the mode control switches so that the data shifts to the right. Apply seven clock pulses and observe how the data is recirculated as it transfers from output H to output A.

Step 9. Set the mode control switches so that the data shifts to the left. Apply several clock pulses and observe how the data is recirculated as it transfers from output A to output H.

■ EXPERIMENT QUESTIONS

1. How many 74194 ICs are cascaded to store a 16-bit binary number?

2. Data can be recirculated to the right by a 74194 IC when a jumper wire is connected from the Q_____ output to the _____ (SL, SR) input.

3. The contents in a register is divided by _____ every time data is shifted one adjacent position to the _____ (right, left).

■ REVIEW QUESTIONS

15. What are the four modes of operation of the 74194 IC?

16. To cause the shift-left operation by the 74194 IC, S_0 is (high or low), S_1 is (high or low), and the clock goes from _____ to _____.

17. When inputs S_0 and S_1 are both high, the 74194 IC operates in the _____ mode.
    (a) Hold
    (b) Load

9.11 TROUBLESHOOTING A SHIFT REGISTER

One method of constructing a 4-bit shift register is by wiring together two dual J–K negative-edge-triggered flip-flop IC packages. Figure 9-25 shows how they can be configured to form a universal shift register that functions in the SISO, SIPO, PISO, and PIPO modes of operation.

If the register does not appear to be functioning properly during operation, the troubleshooter performs the following tests to find the problem. See Fig. 9-25.

1. *Action:* Apply a momentary 0 state to the clear inputs.
   Result: The register shows 0000.
   Conclusion: The clear function operates correctly.

2. *Action:* Place parallel load 1111 into the register by applying a momentary 0 to the preset inputs of each flip-flop.
   Result: The register shows 1111.
   Conclusion: The preset function operates correctly.

3. *Action:* Apply a momentary 0 state to the clear inputs.
   Result: The register shows 0000.
   This clears the register so that data can be serially loaded into the register.

4. *Action:*
   ■ Apply a 1 to the J input of FFA.
   ■ Use a logic pulser to apply a single pulse to the clock input.
   Result: The output reads 1000.
   Conclusion: Flip-flop A serially loads 1s correctly.

5. *Action:*
   ■ Apply a 0 to the J input of FFA.
   ■ Apply a single pulse to the clock input.
   Result: The output reads 0100.
   Conclusion: FFA and FFB load 1s and 0s correctly (FFB loaded 0s correctly, after the clock pulse in Step 4).

6. *Action:*
   ■ Apply a 1 to the J input of FFA.
   ■ Apply a single pulse to the clock input.
   Result: The register reads 1010.
   Conclusion: FFA, FFB, and FFC serially load 1s and 0s correctly.

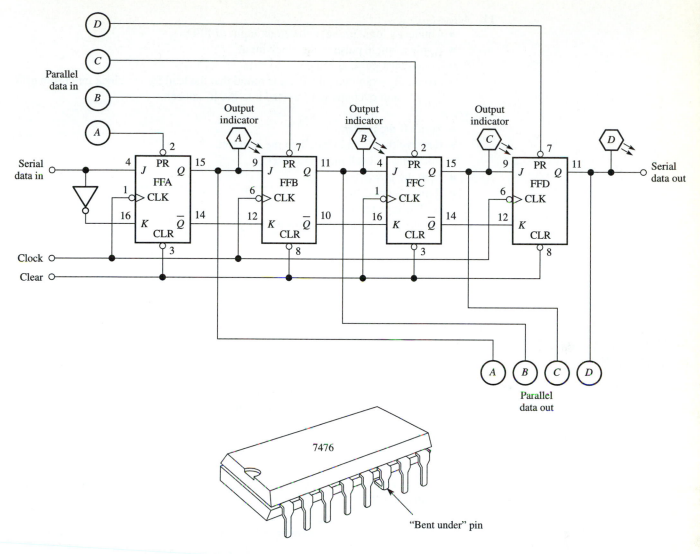

FIGURE 9.25
Troubleshooting a shift register. The bent pin caused an open clock input.

7. *Action:*
   ■ Apply 0 to the *J* input of FFA.
   ■ Apply a single pulse to the clock input.
   Result: The register reads 0100.
   Conclusion: The register should read 0101, so a problem associated with FFD exists because it did not load a 1 properly.
8. *Action:* Parallel load 1s into FFA-C by applying a momentary low to their preset inputs.
   Result: The register reads 1110.
9. *Action:* Connect a logic probe to the *J* input of FFD.
   Result: *J* = 1 at FFD.
   Conclusion: The *J* input to FFD is correct.
10. *Action:*
    ■ Apply a 1 to the *J* input of FFA.
    ■ Apply a single pulse to the clock inputs.
    Result: The register reads 1110.

11. *Action:*
    ■ Connect a logic probe to the clock input of FFD.
    ■ Apply a single pulse to the clock input.
Result: No clock pulse at FFD.
Conclusion: By examining the IC, it is found that the lead for the clock input (pin 6) is bent over and not inserted into the IC socket.

12. *Action:*
    ■ Turn off the power.
    ■ Remove the IC and straighten the bent pin.
    ■ Insert the IC back into the socket.
    ■ Turn the power on.
    ■ Repeat Steps 3 to 7.
Results: The register reads 0101.
Conclusion: The register functions properly.

■ SUMMARY

■ Registers are classified as storage registers and shift registers.

■ Storage registers are primarily used to temporarily store binary data until they are needed elsewhere in the circuit.

■ Shift registers are capable of moving data into, within, and out of the register to perform storage, data conversion, and mathematical operations.

■ Data are moved into and out of registers in either a serial or parallel fashion. Serial data are stored or retrieved one bit at a time. Parallel data are stored or retrieved more than one bit at a time.

■ Shifting data serially to the right or left allows the multiplication or division of binary numbers by a power of 2.

■ PROBLEMS

1. How many bits can be stored in a register containing four flip-flops? (Introduction to Sequential Circuits)
   (a) 1
   (b) 2
   (c) 4
   (d) 8
   (e) 16

2. A computer that operates using 16-bit words uses storage registers consisting of _____ flip-flops. (9-2)

3. A _____ register has the ability to move data bits within itself. (9-3)

4. Shift registers can also operate as storage registers. True or false? (9-3)

5. The _____ shift register is the slowest of all registers. (9-4)

6. A binary bit applied to the $J–K$ flip-flop's J input of a serial register is transferred to the (Q or $\overline{Q}$) output after a pulse arrives at its clock input (9-4)

7. Which of the following are performed by registers? (9-1)
   (a) Data storage.
   (b) Data conversion
   (c) Manipulation
   (d) Counting
   (e) All of the above.

8. A _____ shift register uses serial-input and parallel-output formatting. (9-6)

9. What types of registers do not lose their contents when data is read out? (9-5, 9-8)

10. What are the number of clock pulses required for the data transfer of 4 bits into and out of the following 4-bit shift registers? (9-4, 5, 7, 8)
    (a) SISO
    (b) PISO
    (c) PIPO
    (d) SIPO

11. Converting binary numbers from serial to parallel and parallel to serial is called _____ _____. (9-6)

12. The bits 0110 are placed into a 4-bit SISO shift-right register. What are the contents after one clock pulse if the J input of the flip-flop on the far left is a 0? What are the contents after one more clock pulse if the J input is changed to a 1 immediately after the first clock pulse? (9-4)

13. If the binary number 00001000 is loaded into a bidirectional shift register, what will be its decimal equivalent value after being shifted two places to the left? (9-9)

14. The _____ and _____ inputs are used to quickly preload specific data into a shift register. (9-5)

15. Draw the connections between the appropriate J and K inputs and the Q and $\overline{Q}$ outputs of the $J–K$ flip-flops of Fig. 9-26 to make this register shift the data around in a continuous loop (9-4)

16. If 0110 is the content in a recirculating shift-right register, what will be the content after three clock pulses? (9-4)

17. What is stored in register *B* of Fig. 9-27 after three clock pulses? (9-4)

18. After a bit pattern of 1011 is parallel loaded into the register of Fig. 9-28, what bit pattern is read at the output after each of three clock pulses? (9-7)

19. The 74194 IC needs (no, one, two, or four) clock pulses to parallel load data into the register. (9-10)

20. Based on the data bits applied to the inputs, fill in the appropriate output bits in Table 9-6 for a 74194 universal shift register. (9-10)

Troubleshooting

21. Assume that the register of Fig. 9-19 is initially cleared. If the *J* input of FFC were shorted to ground, what will be the

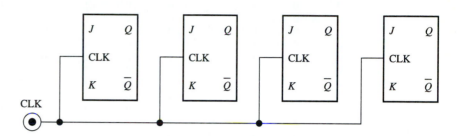

FIGURE 9.26
J–K flip-flops for Problem 15.

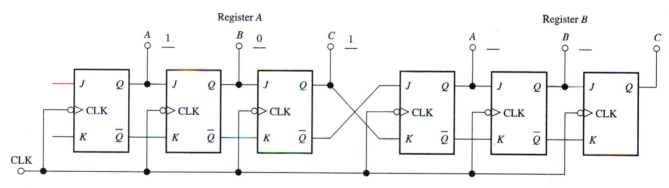

FIGURE 9.27
J–K flip-flops for Problem 17.

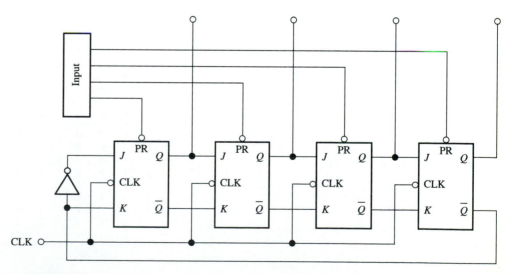

FIGURE 9.28
Flip-flops for Problem 18.

TABLE 9.6
74194 Universal shift register operation for Problem 20

Event number	CLK	S_0	S_1	$\overline{CLR}$	D_{SR}	D_{SL}	A	B	C	D	Q_A	Q_B	Q_C	Q_D
							Parallel data				Outputs			

Event number	CLK	S_0	S_1	$\overline{CLR}$	D_{SR}	D_{SL}	A	B	C	D	Q_A	Q_B	Q_C	Q_D
1	‾‾	1	1	1	0	0	0	1	0	1	0	0	0	0
2	⊓	1	1	1	0	0	0	1	0	1				
3	‾‾	1	0	1	0	0	0	1	0	1				
4	⊓	1	0	1	0	0	0	1	0	1				
5	‾‾	1	0	1	1	0	0	1	0	1				
6	⊓	1	0	1	1	0	0	1	0	1				
7	‾‾	1	0	0	1	0	0	1	0	1				
8	⊓	1	0	1	1	0	0	1	0	1				
9	‾‾	0	1	1	0	1	0	1	0	1				
10	⊓	0	1	1	0	1	0	1	0	1				
11	‾‾	0	1	1	0	1	0	1	0	1				
12	⊓	0	1	1	0	1	0	1	0	1				
13	‾‾	0	1	1	0	0	0	1	0	1				
14	⊓	0	1	1	0	0	0	1	0	1				
15	‾‾	0	0	1	0	0	0	1	0	1				
16	⊓	0	0	1	0	0	0	1	0	1				
17	‾‾	1	1	1	0	0	1	1	0	0				
18	⊓	1	1	1	0	0	1	1	0	0				

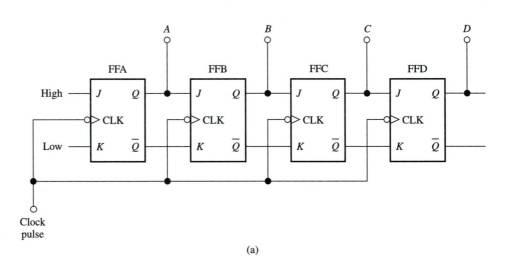

(a)

Clock pulse	Expected waveform				Observed waveform			
	A	B	C	D	A	B	C	D
0	0	0	0	0	0	0	0	0
1	1	0	0	0	1	0	0	0
2	1	1	0	0	1	1	0	0
3	1	1	1	0	1	1	1	0
4	1	1	1	1	1	1	0	1
5	1	1	1	1	1	1	1	0
6	1	1	1	1	1	1	0	1
7	1	1	1	1	1	1	1	0
8	1	1	1	1	1	1	0	1

(b)

FIGURE 9.29
Flip-flops and logic analyzer pattern for Problem 22.

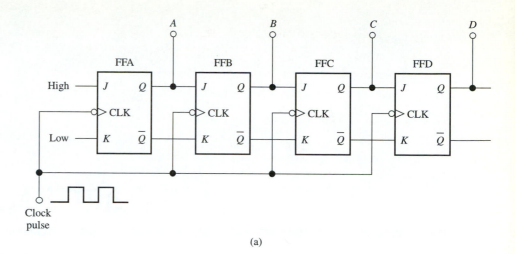

(a)

Clock pulse	Expected waveform				Observed waveform			
	A	B	C	D	A	B	C	D
0	0	0	0	0	0	0	0	0
1	1	0	0	0	1	0	0	0
2	1	1	0	0	1	1	0	0
3	1	1	1	0	1	1	0	0
4	1	1	1	1	1	1	0	0
5	1	1	1	1	1	1	0	0
6	1	1	1	1	1	1	0	0
7	1	1	1	1	1	1	0	0
8	1	1	1	1	1	1	0	0

(b)

FIGURE 9.30
Flip-flops and logic analyzer pattern for Problem 23.

content of the register after four clock pulses if 1111 is serial loaded into it? (9-11)

22. The technician tests the serial shift register in Fig. 9-29 (a) with a logic analyzer and observes the pattern in Fig. 9-29 (b). Which of the following conditions causes this symptom?
(a) The FFB Q output is shorted to ground.
(b) The FFB K input is shorted to ground.
(c) The FFC K input is open.
(d) All of the above.

23. The technician tests the serial shift register in Fig. 9-30 (a) with a logic analyzer and observes the pattern in Fig. 9-30 (b). Which of the following conditions causes this symptom?
(a) The FFB Q output is shorted to ground.
(b) The FFC J input is shorted to ground.
(c) The FFC clock input is open.
(d) The FFC Q output is open.
(e) All of the above.

24. Suppose that a number can be parallel loaded into the register in Figure 9-30 so that FFA = 1, FFB = 0, FFC = 1, and FFD = 0. If the clock input line is open at FFC, a _____ (0,1) state will be observed at output pin D after the second clock pulse.

■ ANSWERS TO REVIEW QUESTIONS

1. storage, shift   2. storage   3. shift   4. SISO, SIPO, PISO, PIPO   5. (c)   6. $Q, \overline{Q}, K$ 7. (c)
8. (b)   9. (c)   10. parallel-in serial-out   11. PISO, SIPO   12. (c)   13. PIPO   14. (c)
15. (1) parallel load, (2) shift right, (3) shift left, and (4) inhibit   16. low, high, low, high   17. (b)

CHAPTER 10

ARITHMETIC CIRCUITS

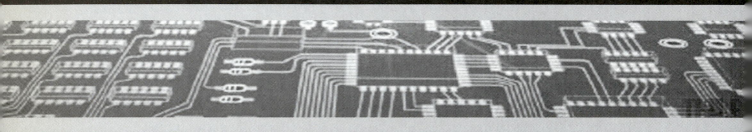

When you complete this chapter, you will be able to:

1. Explain the operation of a binary serial adder circuit.
2. Explain the operation of a binary serial subtractor circuit.
3. Identify the difference between the serial adder and serial subtractor circuits.
4. Explain the operation of serial adder/subtractor vs. a parallel adder/subtractor circuit.
5. Explain the operation of a multiplier circuit.
6. Explain the operation of a divider circuit.
7. Identify the differences between the multiplier and divider circuits.
8. Troubleshoot a serial adder circuit.

10.1 INTRODUCTION

In Chapter 5, which explained the operation of combination circuits, examples were provided to demonstrate how logic gates connected together in certain configurations can make complex logic decisions. Chapters 8 and 9 described how binary numbers can be loaded, transferred, stored, and counted by sequential circuits. This chapter examines arithmetic circuits that utilize various combination and sequential circuits that work together to perform the four basic arithmetic functions: addition, subtraction, multiplication, and division of binary numbers.

When an electronic calculator performs the four basic arithmetic functions, it does so by *adding*. The key device that performs this adding process is the **full-adder** circuit. It is capable of adding binary numbers, one column at a time. The operation of the full adder is described in detail in Section 9 of Chapter 5. How the bits are manipulated before entering the full adder determines which one of the four functions is processed.

The reader who is not familiar with how to add and subtract binary numbers should read Sections 15 and 16 of Chapter 2 before continuing in this chapter.

Full-adder:

A combination logic circuit capable of adding two binary bits and a carry-in bit and that generates a sum and carry-out.

■ REVIEW QUESTIONS

1. The full-adder circuit adds _____ column(s) at a time.
2. The four arithmetic functions are performed by a circuit that does which of the following?
   (a) Adds
   (b) Subtracts
   (c) Multiplies
   (d) Divides
3. What determines which of the four arithmetic functions are performed by the full adder?

10.2 SERIAL ADDER CIRCUIT

Serial adder circuit:

A network of logic gates and flip-flops capable of adding two multibit binary numbers one column at a time.

Accumulator:

A shift register that receives the answer from a full adder at the completion of an arithmetic operation.

Figure 10-1 shows a block diagram of a **serial adder circuit** that adds two 4-bit binary numbers. Each number is loaded into separate shift registers that contain four flip-flops, called registers A and B. The squares in each register represent a flip-flop. Both registers are of the synchronous type and are synchronized together because they share the same clock pulses. Also synchronized with registers A and B is a third shift register called an **accumulator.** The function of the accumulator is to receive the answer from the full adder.

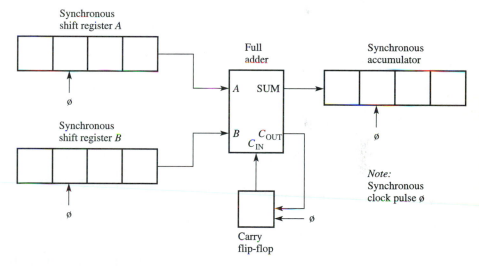

FIGURE 10.1
Block diagram of a serial adder.

Carry flip-flop:

A flip-flop that accepts a carry-out bit from the full adder and then places it back in the full adder for when the next column is added.

The numbers are shifted into and out of the full adder one column at a time as each clock pulse arrives. A synchronous flip-flop that also shares the same clock pulses with the registers is the **carry flip-flop.** It accepts a carry-out bit from the full adder and places it back into the full adder through the carry-in line. The carry bit is then added to the bits of the next column when they are shifted into the FA inputs A and B. It takes four clock pulses for this circuit to add two 4-bit numbers.

Figure 10-2 shows a more detailed version of this addition circuit. Notice that registers A and B, the accumulator, and the carry flip-flop use J–K flip-flops. An exclusive-OR gate is also used. One of its inputs is connected to a switch that is in the 0-state *add* position. This enables the data that enters the other input from register B to pass straight through to the full adder.

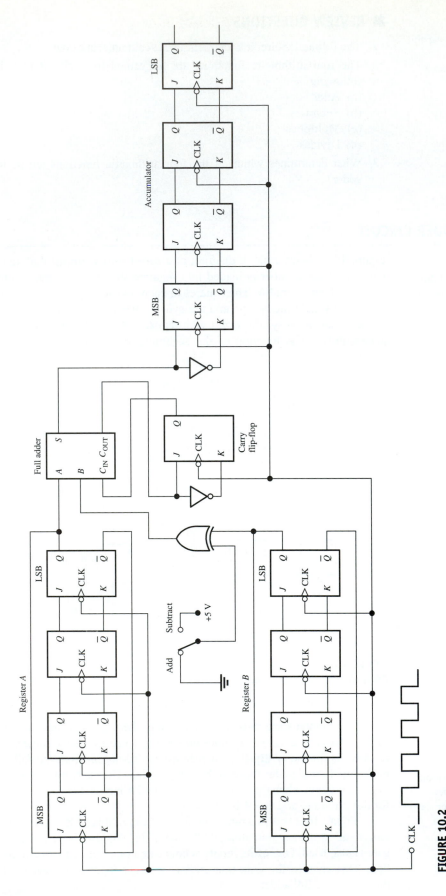

FIGURE 10.2
Schematic diagram of a serial adder.

Before the First Clock Pulse Arrives

Figure 10-3 shows that the number 1001_2 is loaded into register A and 0011_2 is loaded into register B. Because a carry is never added in a LSB column, the carry flip-flop must be cleared. Therefore, a 0 state is at the carry flip-flop Q output. The full adder immediately adds the numbers applied to its inputs, which include 1 from register A, 1 from register B, and 0 at the carry-in line from the carry flip-flop. It generates a sum of 0 and a carry out of 1, which goes to the carry flip-flop (for when the next column is added).

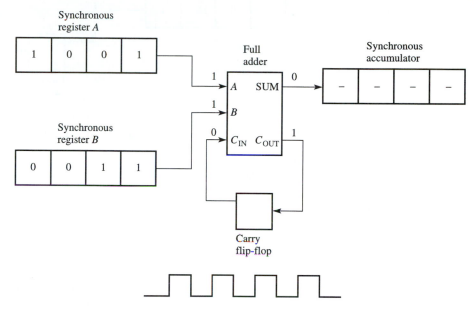

FIGURE 10.3
Before the first clock pulse arrives.

After the First Clock Pulse but Before the Second Clock Pulse Arrives

Figure 10-4 shows what happens after the first of four clock pulses arrives. The numbers in registers A and B are shifted one place to the right, and the first bit of the answer is shifted into the MSB flip-flop of the accumulator. This enables the next column to be added by the full adder. A 0 is at the A input, 1 is at the B input, and 1 is carried over from the previous column by the carry flip-flop and is entered at the carry-in line. As a result, the full adder generates a 0 at the sum output and a 1 at the carry-out line.

After the Second Clock Pulse but Before the Third Clock Pulse Arrives

Figure 10-5 shows the status of the addition circuit after the second clock pulse arrives. Again, the numbers are shifted one place to the right in both registers, and in the accumulator, the first bit of the partial answer is shifted one place to the right to make room for the second bit of the answer.

After the Third Clock Pulse but Before the Fourth Clock Pulse Arrives

Figure 10-6 shows the status of the adder circuit after the third clock pulse arrives, which shifts the contents of the registers another place to the right. The accumulator now has three bits of the answer shifted into it.

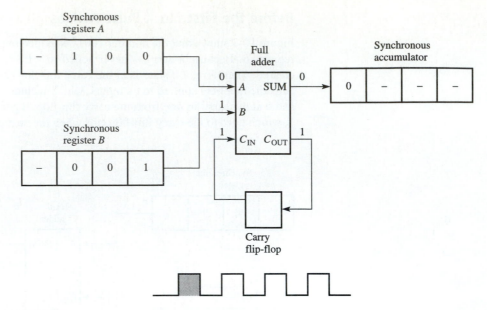

FIGURE 10.4
After the first clock pulse but before the second clock pulse arrives.

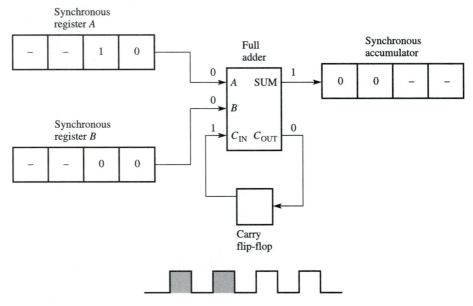

FIGURE 10.5
After the second clock pulse but before the third clock pulse arrives.

After the Fourth Clock Pulse Arrives

Figure 10-7 shows the circuit after the fourth and final clock pulse. All of the bits from each register have been shifted into the full-adder circuit. As a result, the two binary numbers of 1001_2 and 0011_2 have been added by the full adder to produce a sum of 1100_2, which is placed in the accumulator.

Suppose that 1111_2 was placed in register A and 0001_2 was placed in register B. Because the sum total would be 10000_2, the accumulator consisting of four flip-flops could not display the answer. However, the answer would be available because the carry flip-flop can be used as the fifth accumulator flip-flop.

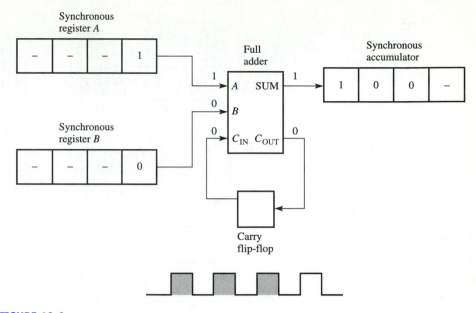

FIGURE 10.6
After the third clock pulse but before the fourth clock pulse arrives.

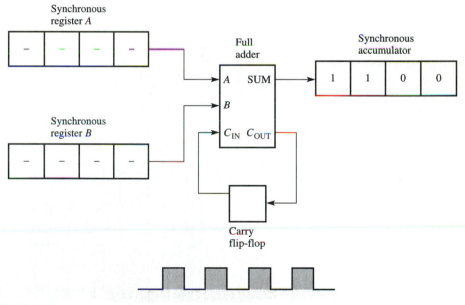

FIGURE 10.7
After the fourth clock pulse arrives.

■ REVIEW QUESTIONS

4. What does the symbol ϕ represent?

5. Which of the following devices in a serial adder circuit are activated by the same synchronous clock pulse?
   (a) Registers A and B
   (b) Accumulator
   (c) Full adder
   (d) Carry flip-flop
   (e) Exclusive-OR gate

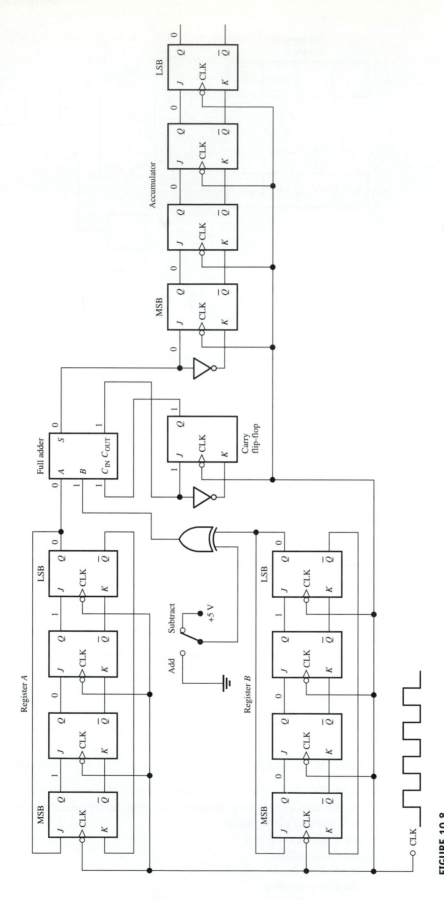

FIGURE 10.8
Before the first clock pulse arrives.

6. How many clock pulses take place before the full adder adds the LSB column?
   (a) 0
   (b) 1
   (c) 4
   (d) 5

7. When one of the inputs of an exclusive-OR gate is at a (0 or 1), data applied to the other pass straight through the gate as if it were a single wire.

8. Before the serial adder circuit is able to add two binary numbers, what two circuit requirements must be met?

9. If the sum of two 4-bit numbers is 5 bits long, where is the fifth bit stored?

10.3 SERIAL SUBTRACTOR CIRCUIT

Serial subtractor circuit:

A network of logic gates and flip-flops that subtracts two multibit binary numbers one column at a time using the 2's complement method.

The adder circuit just described can also be used for subtracting binary numbers. As shown in Fig. 10-8, this circuit becomes a subtractor circuit after two requirements are met. First, the add/subtract switch must be in the *subtract* position, which makes a connection to a +5-volt 1 state. Also, before any clock pulses arrive, the Q output of the carry flip-flop must be set to a 1 state. The reason for these two requirements is explained at the end of this section.

Figures 10-8 to 10-12 illustrate the sequence of how the number 0010_2 is subtracted from 1010_2.

Before the First Clock Pulse Arrives

Figure 10-8 shows that the number 1010_2 is placed in register A, 0010_2 is placed in register B, and the carry flip-flop is preset so that a 1 state is at its Q output. The 0 from register A is applied directly to the A input of the full adder. However, the 0 from register B is complemented as it passes through the exclusive-OR gate to input B of the full adder. This occurs because the add/subtract input line of the exclusive-OR gate is at a 1 state (subtract position), and any number going into the other input line is always inverted at the gate output line. Also, the 1 from the Q output of the carry flip-flop is applied to the full-adder carry-in input.

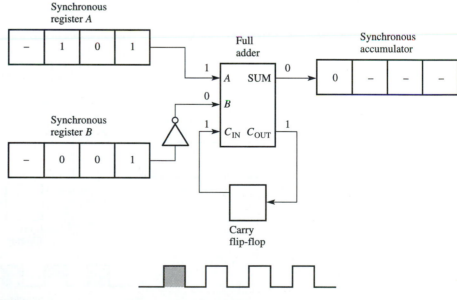

FIGURE 10.9
After the first clock pulse but before the second clock pulse arrives.

Before any clock pulses arrive, the full adder initially generates a sum of 0 with a carry-out of 1 from the numbers in the LSB column of registers *A, B,* and the carry flip-flop.

Figures 10-9 to 10-12 show the subtractor-circuit contents after each of the four successive clock pulses arrives. The full adder adds the numbers of each column as they are shifted in after each clock pulse. After the fourth and final clock pulse of the add cycle, the answer of 1000_2 is shifted into the accumulator.

The primary difference between this circuit operating as an adder or subtractor circuit is how the numbers are manipulated before they are applied to the full adder. One difference is that as a subtractor, a carry-in is added into the LSB column. When the device acts as an adder circuit, this does not occur. The other difference is that when the circuit

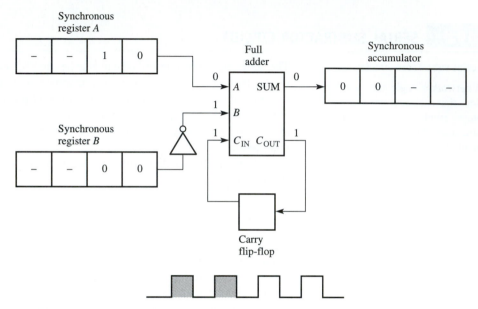

FIGURE 10.10
After the second clock pulse but before the third clock pulse arrives.

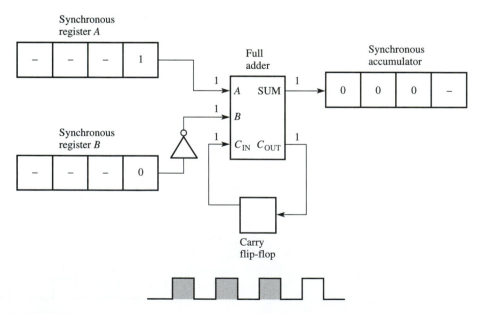

FIGURE 10.11
After the third clock pulse but before the fourth clock pulse arrives.

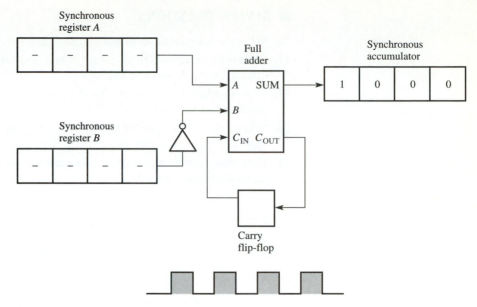

FIGURE 10.12
After the fourth clock pulse arrives.

operates as a subtractor, the exclusive-OR inverts the numbers from register B before entering the B input of the full adder; when operating as an adder circuit, the bits from register B pass straight through the exclusive-OR gate to the B input of the full adder.

The reasons why a 1 from the carry-in line is added to the LSB column and why the inverting of the bits from register B takes place when the circuit operates as a subtractor are explained by examining the 2's complement arithmetic method in Fig. 10-13. From the 2's complement method, note that the subtractor circuit manipulates numbers exactly the same way. The carry-in number at the LSB column is from the preset carry flip-flop (Step 3). The bits in the subtrahend are inverted by the exclusive-OR gate as they are transferred from register B to the full adder (Step 2). The numbers are shifted into the full adder, one column at a time (Steps 1 and 2), as it generates the sum and carry-out from the bits of the minuend, subtrahend, and the carry-in bit (Step 4).

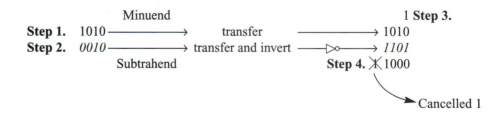

Step 1. Transfer the minuend 1010.
Step 2. Transfer the subtrahend 0010 while inverting each bit.
Step 3. Place a 1 into the carry-in location of the least significant bit column.
Step 4. Add the minuend, inverted subtrahend, and carry-in number for the answer. Ignore any 1 that is carried into the (fifth) column that exceeds the number of bits that make up the original numbers being added.

FIGURE 10.13
2's complement arithmetic method.

■ REVIEW QUESTIONS

10. What method is used to subtract binary bits?

11. Before the serial subtractor circuit is able to subtract two binary numbers, what two circuit requirements must be met?

12. How should a 1 in the carry flip-flop be treated after the answer is produced by the serial subtractor circuit?

13. When one of the inputs of an exclusive-OR gate is at a (0 or 1), data from the other input are inverted as they pass through the gate to its output.

14. How many synchronous clock pulses are required to subtract two 4-bit numbers in a serial subtractor circuit?

EXPERIMENT

Adder/Subtractor Circuit

Objectives

■ To understand the operation of a binary serial adder circuit.
■ To understand the operation of a binary serial subtractor circuit.
■ Using ICs and discrete components, construct and operate a serial adder/subtractor circuit.

Materials

1—+5-V DC Power Supply

4—7476 *J–K* Flip-Flop ICs

1—7404 Inverter IC

1—7486 Exclusive-OR IC

1—7483 4-bit Adder IC

1—SPDT Switch

1—Logic Pulse Push Button (Bounceless)

7—LEDs

7—68-Ω Resistors

1— Logic Data Manual

Introduction

One of the primary operations of an electronic calculator and computer is the arithmetic function. This experiment demonstrates how binary numbers can be added or subtracted using various types of digital circuits examined in earlier experiments. The diagram of an adder/subtractor circuit is shown in Fig. 10-14. The circuit consists of two synchronous shift registers. Each is capable of storing 3-bit binary numbers. One number is loaded into register *B,* which is configured to operate as a ring counter. Its binary contents are recirculated back into the register and become the original value after three clock pulses. The other binary number is loaded into register *A.*

The binary numbers in registers *A* and *B* are shifted into the full adder one column at a time. The contents of register *A* are shifted into input *A* of the full adder. The number in register *B* is fed into input *B* of the full adder through the XOR gate. Any carry bits generated from one column are transferred to the next column by the carry-out and carry-in leads of the full adder and the carry flip-flop. After the completed add cycle, consisting of three clock pulses, the sum of the two numbers is serial loaded into register *A.* When the answer is held in register *A,* it becomes classified as an accumulator.

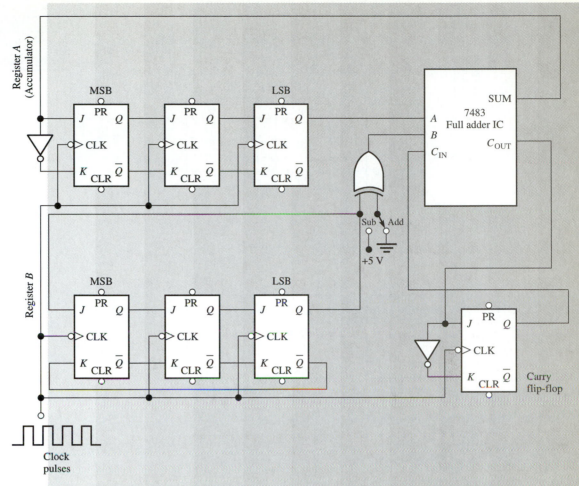

FIGURE 10.14
Adder/subtractor circuit

PART A—ADDITION OPERATION

Procedure

Step 1. Assemble the circuit shown in Fig. 10-15. Refer to the pin diagrams in a logic data manual to determine which pins of ICs are used when connecting the flip-flops, exclusive OR gate, and inverters.

This circuit is available on Electronics Workbench by downloading file *Fig. 10-15.*

Background Information

One of the input leads for the exclusive-OR gate is connected to a switch. When the switch is in the ground (add) position, the binary bits applied to the other XOR input are passed straight through to the output as if there were a straight wire connection.

Step 2. Position the add/subtract switch in the 0-V ground (add) position to perform the add function.

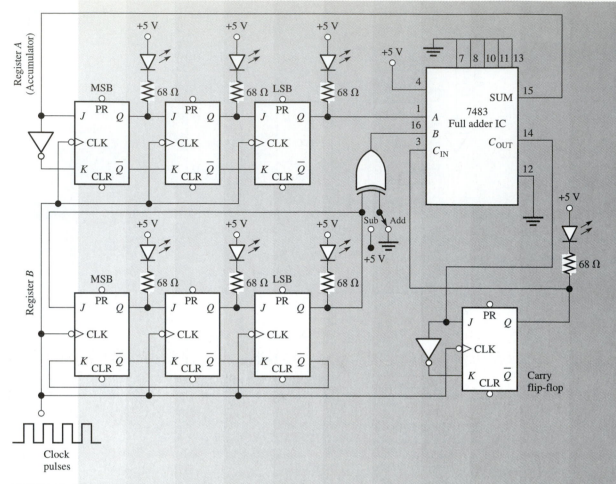

FIGURE 10.15

Procedure Information

Each time numbers are loaded into the registers, it is necessary to preset or clear each flip-flop. This function can be accomplished by connecting one end of a long wire to ground. The other end of the wire is momentarily connected to either the active-low preset (PR) or clear (CLR) lead of each flip-flop.

Step 3. Clear the carry flip-flop.

Step 4. Load 011_2 into register A and 001_2 into register B. The LEDs should verify that the proper binary bit is inserted. If the LED is on, the flip-flop is preset and stores a logic 1. If the LED is off, the flip-flop is cleared and stores a logic 0.

Step 5. Use the push button to apply three clock pulses. After the three clock pulses, the accumulator reads binary number _____.

Note: **If the circuit does not operate properly, it may be necessary to tie the CLR and PR leads of each flip-flop to +5 V after the numbers are loaded in.**

Step 6. Clear the carry flip-flop. Load 101_2 into register A and 010_2 into register B. After the three clock pulses, the accumulator reads binary number _____.

If a 4-bit sum is generated, the fourth bit is stored in the carry flip-flop.

Step 7. Clear the carry flip-flop. Load 101_2 into register A and 110_2 into register B. After the third clock pulse, the accumulator reads binary number _____.

PART B—SUBTRACTION OPERATION

Background Information

When the subtraction function of two binary numbers is performed, it is accomplished by using the 2's complement method. The following example using the pencil-and-paper method shows the steps used to perform 2's complement subtraction:

Example:

Minuend:	$101 \longrightarrow$ transfer $\longrightarrow 101$		**Step A**
Subtrahend:	$010 \longrightarrow$ transfer $\longrightarrow +101$		**Step B**
		$\overline{X}010$	**Step C**
		$\underline{+1}$	**Step D**
	Answer $\longrightarrow 011$		**Step E**

Steps:

    A. Transfer the minuend.

    B. Transfer and invert the subtrahend.

    C. Add the two binary numbers (ignore the carry).

    D. Insert a 1 to be added to the sum from Step C.

    E. Add the sum and 1 to obtain the answer.

Procedure Information

When the switch connected to the XOR gate lead is in the $+5$-V position, the bits applied to the other gate input are inverted as they pass through to the output. The 1 inserted in Background Information Step D can be just as well added to the minuend and the inverted subtrahend immediately instead of being added to their sum. The subtraction circuit adds this 1 immediately by presetting the carry flip-flop before any clock pulses are applied.

Step 8.   Load 100_2 into register A and 001_2 into register B.

Step 9.   Preset a 1 into the carry flip-flop.

Step 10.   Place the add/subtract switch into the $+5$-V (sub) position.

Step 11.   Use the logic push button to apply three clock pulses. After the third clock pulse, the accumulator reads binary number _____.

Step 12.   Load 111_2 into register A and 011_2 into register B, and repeat Steps 9 through 11. After the third clock pulse, the accumulator reads binary number _____.

■ EXPERIMENT QUESTIONS

   1. What is the function of the inverter connected across the carry flip-flops J and K inputs?

   2. After the three-clock-pulse add cycle, what happens to the contents in register B?

   3. How many J–K flip-flops are needed to add or subtract two 8-bit binary numbers?

   4. When are the sum and carry values for the LSB generated at the full-adder outputs?

      (a) Before the first clock pulse.

      (b) After the first clock pulse.

      (c) Before the third clock pulse.

      (d) After the third clock pulse.

      (e) After the fourth clock pulse.

   5. Why is the carry flip-flop cleared before the addition operation?

   6. How does the subtrahend become inverted?

   7. Observe Step D of the subtraction operation example, which shows how subtraction is performed using the 2's complement pencil-and-paper method. How is the 1 inserted into the circuit?

10.4 PARALLEL ADDER/SUBTRACTOR CIRCUIT

Serial addition:

The process of adding multibit binary numbers one column at a time.

The addition process described earlier is performed by a method called **serial addition.** One 4-bit number is loaded into serial register A and the other is loaded into the other 4-bit serial register B. These two numbers are simultaneously transferred bit-by-bit into the full adder, which individually adds two bits at a time. The sum of each pair of bits is transferred into another 4-bit register called the *accumulator*. The process of adding these two numbers requires four separate clock pulses to complete the entire operation.

Electronic devices such as hand-held calculators add numbers using the serial addition method. However, some types of electronic equipment, such as bank computers, which process large volumes of numbers, require a much faster method of adding called **parallel addition.**

Parallel addition:

The process of adding multibit binary numbers simultaneously.

Figure 10-16 shows a parallel adder/subtractor that operates as a parallel adder. It is capable of adding the same two numbers just as the serial adder described in Figs. 10-3 to 10-7. It consists of three parallel-loaded shift registers, two of which store the two numbers

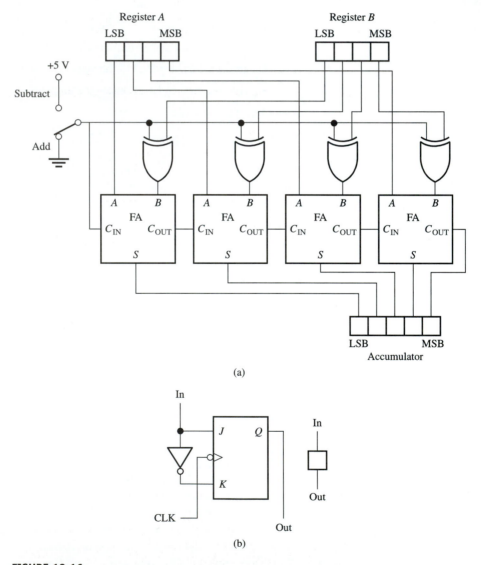

FIGURE 10.16

(a) Parallel adder/subtractor circuit. (b) This diagram shows that each block in the shift registers and accumulator is made up of a negative-edge-triggered J–K flip-flop with complemented inputs.

to be added and a third that stores the answer. Four separate full adders are used. Each one adds a pair of bits from each column and any carry bits generated by the full adder in the preceding column. When the add/subtract switch is in the add position, a 0 is applied to the carry-in lines of the LSB full adder. Zeros are also applied to one input of each exclusive-OR gate. This allows each of the bits from register B to pass straight through to the full adders.

The reason why the parallel adder operates faster is that it requires one clock pulse to perform the entire addition process. The bits from each number are transferred from registers A and B through the chain of full adders to the accumulator simultaneously.

The disadvantage of a parallel adder/subtractor is that it requires more circuitry than serial adders. Therefore, it is more expensive.

Parallel adder/subtractor circuit:

A network of logic gates and flip-flops capable of adding or subtracting two multibit binary numbers quickly by using one clock pulse.

The circuit in Fig. 10-16 is also capable of operating as a **parallel subtractor.** To do so, a 1 must first be applied to the carry-in line of the LSB full adder. 1s must also be connected to one input of each exclusive-OR gate so that the bits are inverted as they transfer from register B to the full adder. Placing the add/subtract switch in the subtract position provides the necessary 1s for these two requirements to be met.

Similar to the operation of the parallel adder, the pairs of bits from each column are transferred through the full adders simultaneously. The answer is displayed in the accumulator after one clock pulse.

■ REVIEW QUESTIONS

15. What is the primary advantage of a parallel adder or subtractor circuit over a serial circuit?

16. What is the primary disadvantage of a parallel adder or subtractor circuit over a serial circuit?

17. How many clock pulses are required to activate a 4-bit parallel adder or subtractor circuit?

18. How are carry bits transferred from one column to the next significant column in a parallel adder circuit?

19. To satisfy one of the two 2's complement requirements, how is the 1 inserted into the LSB column of the parallel subtractor circuit?

10.5 MULTIPLICATION CIRCUIT

When multiplying two numbers by each other, different methods can be used to find the answer. For example, when multiplying 3 times 4, the answer can be found in the following ways:

	Method A	Method B	Method C
	4	4	3
	×3	+4	+3
	12	+4	+3
		12	+3
			12

Method A: This is the memorization method of solving the problem.

Method B: This shows how the answer is obtained by adding 4 three times consecutively.

Method C: This shows how the answer is obtained by adding 3 four times consecutively.

Multiplication circuit:

A network of logic gates and flip-flops capable of multiplying binary numbers.

The **multiplication circuit** described in this section utilizes an operation similar to Method B. Figures 10-17 (a) to (c) are three block diagrams that pictorially show how 4 is added to itself three times.

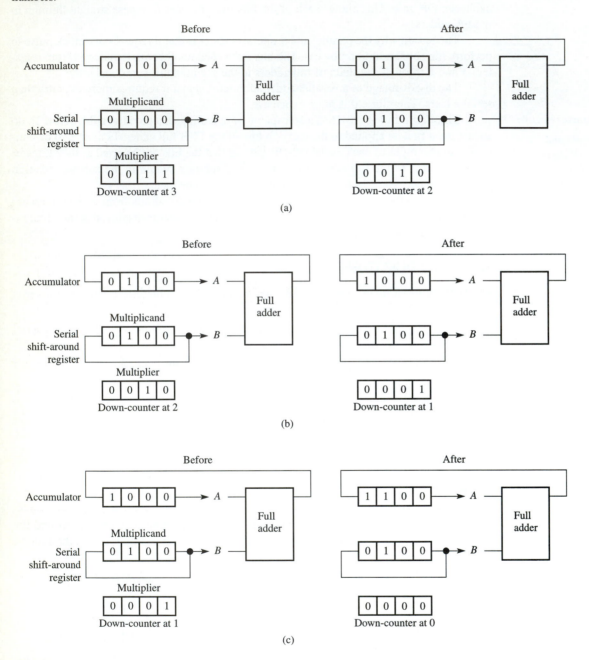

FIGURE 10.17
Block diagram of a multiplier circuit: (a) Add cycle 1. (b) Add cycle 2. (c) Add cycle 3.

The full adder is again used as the key device in the circuit as the bits are manipulated in various ways before they enter the FA inputs. Data that are transferred into the full adder and throughout the rest of the circuitry are activated synchronously by pulses from a clock source. Every four of these clock pulses produces an *add cycle*. By using a down-counter as the multiplier, the 3 is initially stored into it. The multiplicand of 4 is initially stored into a serial shift-around (ring counter) register. The accumulator is a serial register that is initially cleared. As the bits from the multiplicand shift register and the accumulator are synchronously shifted four places to the right into the full adder, the sum is stored in the accumulator. The 4 is added to the contents in the accumulator during each add cycle until the 3 in the multiplier decrements to 0.

Add Cycle 1

After one add cycle is performed, the sum of the initial multiplicand register number 4 and the initial accumulator number 0 ends up in the accumulator as 4. The down-counter (multiplier) is decremented from 3 to 2.

Add Cycle 2

After the second add cycle is performed, the sum of the numbers that were in the multiplicand register and accumulator $(4 + 4)$ before the second add cycle ends up in the accumulator as 8. The down-counter is decremented to 1.

Add Cycle 3

After the third and final add cycle, the sum of the numbers that were in the multiplicand register and accumulator $(4 + 8)$ before the third add cycle ends up in the accumulator as 12. The 0000 at the down-counter indicates that the multiplication process is complete. Therefore, the product of 3×4 (which is 12) is displayed in the accumulator.

These block diagrams show how a multiplication circuit operates. However, the circuit has limited capabilities. Suppose that the number of bits that make up the product exceeds the number of bits in the multiplier or multiplicand. For example, a multiplicand of 8_{10} would require a 4-bit register and a multiplier of 8_{10} would require a 4-bit register. If 8_{10} (1000_2) were multiplied by 8_{10} (1000_2), an accumulator register consisting of more than four flip-flops would be required to store the product, 64_{10} (01000000_2).

Therefore, it is necessary to modify the circuit by adding four flip-flops to the accumulator. A block diagram of such a circuit is shown in Fig. 10-18(a). Figures 10-18(a) to (c) graphically show how this circuit operates.

Circuit Operation

- 0_{10} is loaded into the accumulator, 8_{10} is loaded into the multiplicand, and 8_{10} is loaded into the multiplier. Like the multiplication circuit previously discussed, this circuit also operates by add cycles. The number of these cycles is determined by the number placed in the multiplier. However, instead of four clock pulses generating a cycle, eight clock pulses are required.
- During the first four clock pulses, a timer circuit generates a high to enable A_1 to pass the four multiplicand bits into the full adder.
- Figure 10-18(b) shows the start of the second half of the first add cycle. The multiplicand bits were shifted around one time, and the sum was shifted into the four most significant bits of the accumulator during the first four clock pulses. During the second four clock pulses, the timer output goes to a low to disable A_1. The purpose of disabling A_1 is to prevent the bits from the multiplicand from passing into the full adder as it shifts around while the sum in the accumulator is shifted into the proper position of the four least significant bits for the next cycle to occur.
- Figure 10-18(c) shows that the multiplier decremented after the first add cycle. The timer circuit goes to a high to enable A_1 so that the four bits from the accumulator and multiplicand can be shifted together into the full adder and generate their sum during the first four clock pulses of the second add cycle. During the second four clock pulses of the second add cycle, the timer output goes low to disable A_1.
- The number of times the add cycles are repeated is determined by the number left in the multiplier.

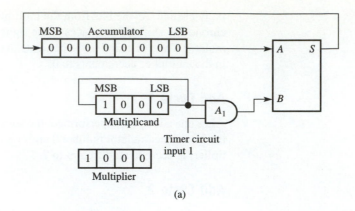

(a)

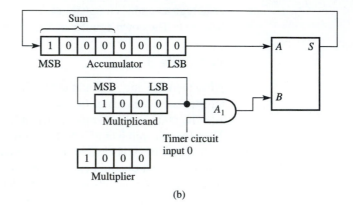

(b)

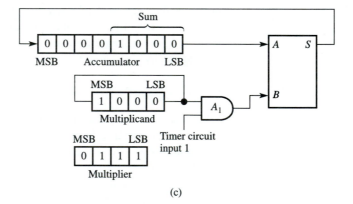

(c)

FIGURE 10.18
(a) Before the first add cycle. (b) Start of the second half of the first add cycle. (c) The multiplier after the first add cycle.

Figure 10-19 shows a more detailed version of the circuit described in Figs. 10-18(a) to (c). The accumulator, multiplicand, gate A_1, and full adder of this circuit operate exactly like the ones in the block diagrams. The other sections of Fig. 10-19 operate as follows:

Timer Circuit

Consisting of FF17-FF19, the timer is a mod-8 counter. The primary function of this circuit is to provide an output that is at a high during the first four counts of the add cycle and a low during the last four counts of the add cycle. The output of this circuit is generated at the $\overline{Q}$ output of FF19.

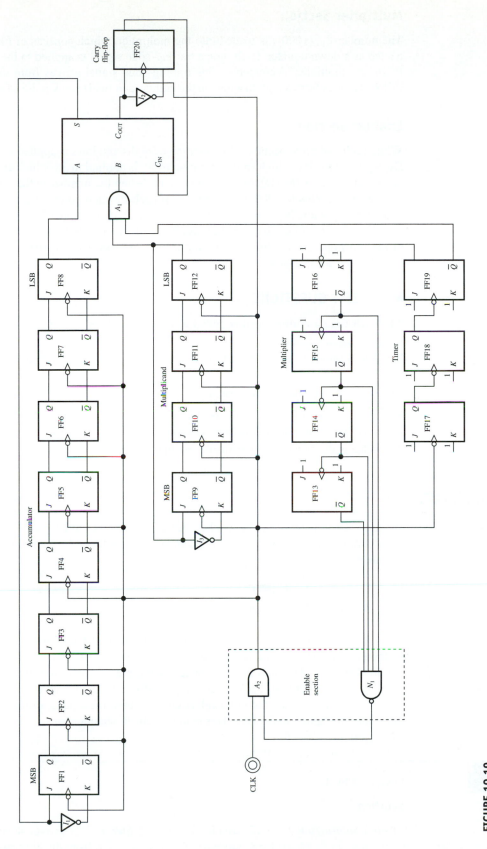

FIGURE 10.19
Schematic diagram of a multiplier circuit.

Multiplier Section

The number 8_{10} (1000_2) is loaded into the multiplier, which consists of FF13-FF16 connected as a down-counter. Each time a trailing-edge signal is applied to the clock input of FF16, the multiplier decrements. This trailing-edge signal arrives from the Q output of FF19 as it goes from a high to a low on the eighth (and final) clock pulse of each add cycle.

Enabler Section

While each add cycle occurs, a combination of highs and lows is applied to NAND gate 1. During these cycles, a high from the output of N_1 is applied to A_2, which enables the main clock pulse to pass through to the entire circuit. When the number in the multiplier down-counter decrements to 0000, the $\overline{Q}$ outputs of each flip-flop go to a 1 state. With all highs applied to N_1, a low is generated at its output. This 0-state output of N_1 disables A_2, which causes the entire circuit to freeze because the main clock pulses can no longer pass to any of the flip-flops. At this time, the multiplication process is completed and the product is stored in the accumulator.

■ REVIEW QUESTIONS

20. Each time a multiplicand is added to the accumulator contents, this action is called an _____ _____.
21. The number of times the add cycles are repeated is determined by the number in the _____.
22. Where is the final answer stored?
23. What is the purpose of A_1 in Fig. 10-19?
24. What is the purpose of the timer circuit?
25. What is the function of the enabler section?

10.6 DIVIDER CIRCUIT

Divider circuit:

A network of logic gates and flip-flops capable of dividing binary numbers.

When dividing one number by another, different methods can be used to find the answer. For example, when dividing 12 by 4, the answer can be found by the following memorization method:

$$\text{Divisor } 4)\overline{12} \quad \begin{matrix} 3 & \text{Quotient} \\ & \text{Dividend} \end{matrix}$$

Subtract and count method:

A procedure that obtains an answer for division by subtracting numbers.

Another way of solving the problem is through the use of the pencil-and-paper method called **subtract and count.** This procedure begins by subtracting the divisor from the dividend. The divisor is then subtracted again, but this time from the answer obtained from the first subtraction. The divisor is repeatedly subtracted from each new answer *until* the divisor is larger than the dividend. A count is then made to determine the number of times the subtraction took place. This count results in the quotient.

EXAMPLE 10.1

Divide 12 by 4.

Solution

First Subtraction:

```
   12  Dividend
  − 4  Divisor
    8  First Answer
    ↓  "8 is transferred"
```

Step 1: The divisor 4 is subtracted from the dividend 12 to produce an answer of 8.

Second Subtraction:	8	*Step 2:*	The divisor 4 is subtracted
	-4 Divisor		from 8 (which is the answer
	4 Second Answer		obtained from the previous
	↓ "4 is transferred"		subtraction). An answer of
			4 is produced.

Third Subtraction:	4	*Step 3:*	The divisor 4 is subtracted
	-4 Divisor		from 4 (which is the answer
	0 Third Answer		obtained from the previous
	↓ "0 is transferred"		subtraction). An answer of
			0 is produced.

Fourth Subtraction:	0	*Step 4:*	Since the divisor 4 is greater
	-4 Divisor		than dividend 0, the sub-
	—̶4̶ Does not occur		traction process stops.
		Step 5:	To determine the quotient
			answer, count the number of
			times the subtraction func-
			tion took place. The answer
			is 3.

Divisors do not always divide evenly into dividends. As a result, a remainder is developed. When dividing such numbers using the subtract-and-count method, the subtraction steps are repeated until the divisor is greater than the dividend. The quotient answer is again obtained by counting the number of subtraction steps, and the remainder is the number left at the end of the final subtraction step.

EXAMPLE 10.2

Divide 8 by 3.

Solution

First Subtraction:	8 Dividend	*Step 1:*	The divisor 3 is subtracted
	-3 Divisor		from the dividend 8 to pro-
	5		duce an answer of 5.
	↓		

Second Subtraction:	5	*Step 2:*	The divisor 3 is subtracted
	-3 Divisor		from 5 (which is the answer
	2 Remainder		obtained from the previous
			subtraction step). An answer
			of 2 is produced.

Third Subtraction:	2	*Step 3:*	Since the divisor 3 is larger
	-3 Divisor		than the new dividend 2, the
	Does not occur		subtraction process stops.
		Step 4:	To determine the quotient an-
			swer, count the number of
			times the subtraction function
			took place. The answer is 2.
		Step 5:	Any remainder is determined
	Remainder		by observing the answer that
			was produced by the final
			subtraction step. The remain-
			der of this problem is 2.

Digital circuits utilize a similar technique to divide numbers. However, instead of using decimal integers, digital circuits manipulate binary bits.

The block diagram of Fig. 10-20 shows a circuit that performs the subtract-and-count division function similarly to the pencil-and-paper method just explained. However, it uses the 2's complement method to perform each subtraction step.

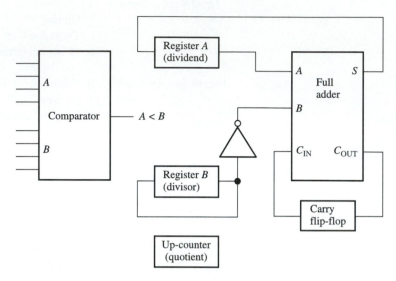

FIGURE 10.20
Block diagram of a divider circuit.

Register *A*

This circuit is where the dividend is loaded before the first subtraction step takes place. It also serves as an accumulator for the answer that is produced after each subtraction step. When the subtraction steps finish, the number in this accumulator shows the remainder. The contents of this accumulator shift one bit at a time into input *A* of the full adder.

Register *B*

This circuit is where the divisor is loaded. It is a shift-around serial register (ring counter), and its contents are continuously recycled from its output back into its input.

Inverter

To enable the full adder to subtract, the 2's complement method is used. Therefore, the divisor contents in register *B* are complemented by the inverter before entering input *B* of the full adder.

Carry Flip-Flop

This flip-flop transfers any carry bits from one column to the next. By initially loading a 1 into it before any subtraction steps are performed, it also fulfills the 2's complement subtraction requirements.

Up-Counter

After each subtraction step, this counter increments. When the subtraction steps stop, the number in the counter equals the quotient answer.

Comparator

This device compares the dividend number in accumulator/register A to the divisor number in register B. When the divisor is greater than the dividend, output $A < B$ goes high and disables the circuit.

Figures 10-21(A) to (D) illustrate each of the subtraction steps that takes place to divide 12 by 4. Each block in the registers and up-counter represents a flip-flop. The process of shifting data through the full adder during each subtraction step is called an *add cycle*. The dividend 12_{10} (1100_2) is loaded into register A, the divisor 4_{10} (0100_2) is loaded into register B, the up-counter is cleared, and the carry flip-flop is preset to a 1.

At the end of each add cycle, the divisor remains the same, the dividend becomes smaller by the same numerical value as the divisor, and the up-counter increments by 1. The subtraction steps continue until the divisor becomes greater that the dividend. Because

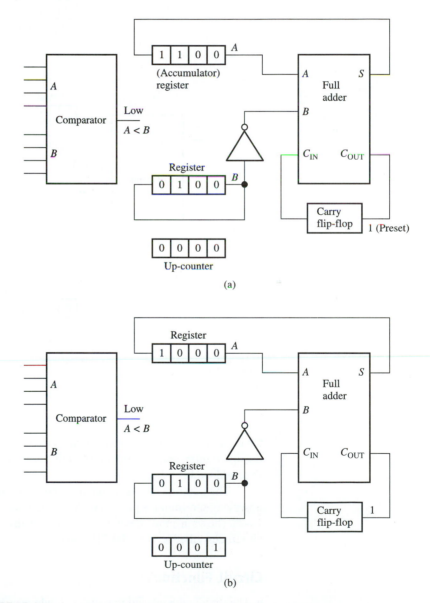

FIGURE 10.21

Divider circuit: (a) Before the first add cycle arrives. (b) After the first add cycle but before the second add cycle arrives. (c) After the second add cycle but before the third add cycle arrives. (d) After the third and final add cycle, the answer quotient (3) and remainder (0) are displayed.

(continues)

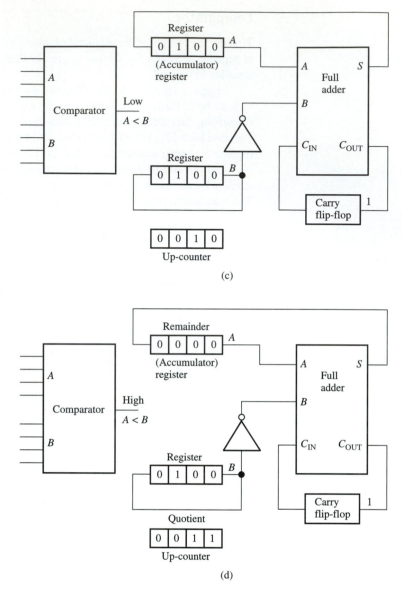

FIGURE 10.21
(Continued)

the quotient answer is the same number as the add cycles that took place, it can be found in the up-counter, and any remainder is the number in accumulator/register A at the end of the final add cycle.

The quotient of 3 is found in the counter contents, and the remainder of 0 is found by observing the contents present in the accumulator after the third and final add cycle. Figure 10-22 shows a more detailed version of a divider circuit that operates similarly to the circuit described in Figs. 10-21(a) to (d).

Circuit Functions

- The dividend is placed in register A, which consists of FF1-FF4.
- The divisor is loaded into register B, which consists of FF5-FF8. As the add cycles take place, each bit from register B is applied to the full adder through inverter I_3, which complements the divisor value.

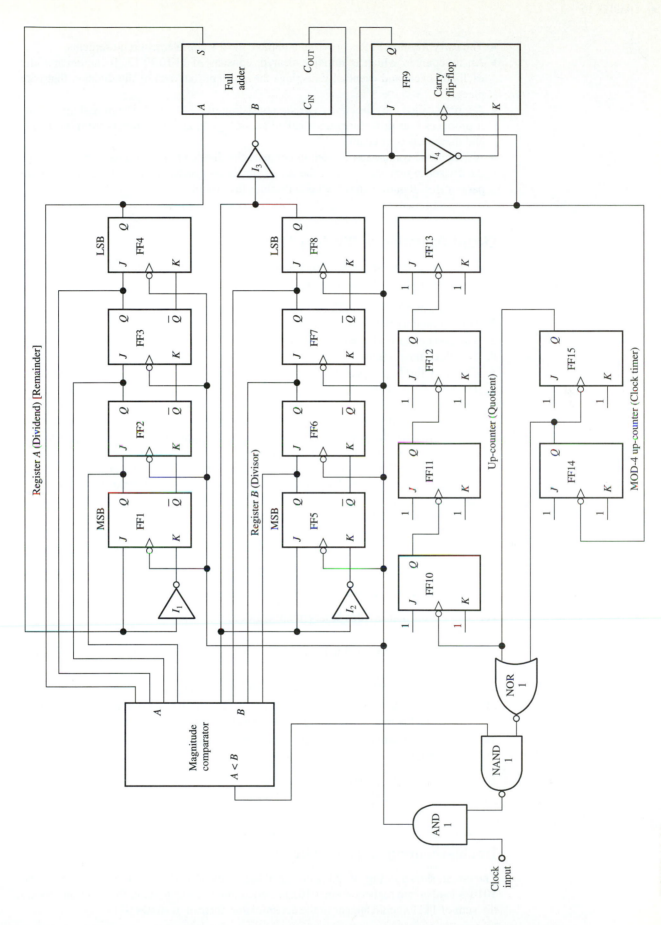

FIGURE 10.22
Schematic diagram of a divider circuit.

359

- The carry flip-flop FF9 is preset to complete the 2's complement requirements.
- An up-counter, which is initially cleared, consists of FF10-FF13. It increments after each add cycle and eventually displays the quotient (answer) of the division that takes place.
- A circuit called the *clock timer circuit,* consisting of FF14-FF15, is a mod-4 up-counter. It produces four different counts, called an *add cycle,* as it increments from 00_2 to 11_2, and then starts over again.
- A magnitude comparator is used to compare the dividend in accumulator/register A and the divisor in register B. When the divisor becomes greater than the dividend, the comparator $A < B$ output goes high and disables the circuit.

Circuit Operation of Dividing 12 by 4

Each add cycle consists of four counts, which is the number of clock cycles required to feed the bits from the divisor and dividend into the full adder.

1. Dividend number 1100_2 (12_{10}) is loaded into register A.
2. Divisor number 0100_2 (4_{10}) is loaded into register B.
3. The up-counter is cleared to 000_2.
4. The clock timer circuit is cleared to a 00_2.
5. The carry flip-flop is preset.
6. The add cycles continue until two conditions are met:
   - The clock timer circuit has just completed a cycle, so that its Q outputs are 00_2.
   - The $A < B$ output of the magnitude comparator is 1.
7. At this time, the division process is stopped. All 0s applied to NOR gate 1 produce a 1 at its output. This 1 state and the 1 from the $A < B$ output of the comparator cause the output of NAND gate 1 to go low. The 0 from the NAND 1 output applied to AND gate 1 disables the main clock signal from passing through to the flip-flops throughout the circuit.
8. The up-counter displays the quotient (answer to the division problem), and register A displays the remainder.

■ REVIEW QUESTIONS

26. The circuit that divides binary numbers does so by the _____-and-_____ method.
27. In Fig. 10-22, the dividend is placed in _____ and the divisor is placed in _____.
28. After each add cycle, the dividend number becomes (smaller/larger).
29. When the division circuit has finished its process of dividing one number by another, the quotient is found in the _____ and any remainder is found in the _____.
30. In Fig. 10-22, what circuit condition determines when the division operation has been completed?

10.7 TROUBLESHOOTING

Troubleshooting an Adder Circuit

The circuit shown in Fig. 10-23 is designed to add two 4-bit binary numbers. The number 1010 is loaded into register A and 0101 is loaded into register B. After the addition process, the sum of 1111 should appear in the accumulator. Instead, it shows 0111.

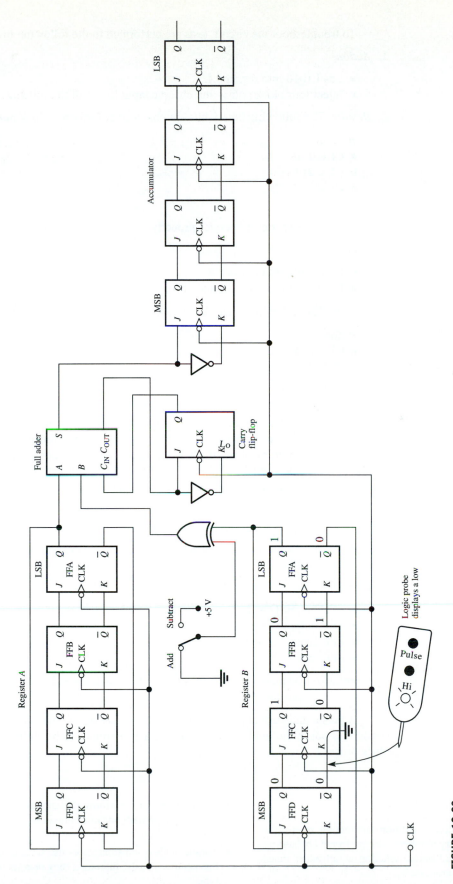

FIGURE 10.23

To troubleshoot the circuit, tests are performed in the following order:

1. *Action:*

   - Load 1010 into register *A*.
   - Inject four clock pulses and observe input line *A* of the full adder.

   Results: The following binary bits are observed before each clock pulse.

   Before:
   - CLK 1: 0
   - CLK 2: 1
   - CLK 3: 0
   - CLK 4: 1

   Conclusion: Register *A* operates properly.

2. *Action:*

   - Load 0101 into register *B*.
   - Inject four clock pulses and observe input line *B* of the full adder.

   Results: The following binary bits are observed before each clock pulse.

   Before:
   - CLK 1: 1
   - CLK 2: 0
   - CLK 3: 1
   - CLK 4: 1

   Conclusion: Register *B* and the exclusive-OR gate appear to operate properly for the first three clock pulses. However, before the fourth clock, a 1 appears at input *B* instead of a 0. Because the data that reach the full adder after three clock pulses originate from FFC, the defect is probably either the flip-flop itself or the lines that feed the data to its inputs.

3. *Action:*

   - Load 0101 into register *B*.
   - Use a logic probe to find what logic states are present at the input and output data lines of FFC.

   Results:

   - FFC output Q = 1
   - FFC output $\bar{Q}$ = 0
   - FFC input J = 0
   - FFC input K = 0

   Conclusion: Because input *K* of FFC should be a 1 instead of a 0, it appears that it is shorted.

   A pulser and current tracer confirm that FFC has an internal short at its *K* input. As a result, it is put in the hold mode. Therefore, a 1 continues to be shifted to FFB after each clock pulse.

■ SUMMARY

- All four arithmetic functions—addition, subtraction, multiplication, and division of binary numbers—are performed by adding. How bits are manipulated before entering a full adder determines which function is processed.

- Serial adders perform the addition function by a full adder adding binary bits one column at a time.

- Serial subtractors perform the subtraction function by a full adder adding binary bits one column at a time using the 2's complement method.

- Parallel adders and subtractors use several full adders to add all bits simultaneously; therefore, they operate faster than the serial type.

- Binary bits are multiplied by a full adder, which adds a number (multiplicand) to itself the number of times specified by the number (multiplier) in a down-counter. The product answer is displayed by an accumulator.

■ Binary bits are divided by a full adder that adds using the 2's complement method of subtracting a number (divisor) from another number (dividend). The number of times the 2's complement addition takes place is recorded by an up-counter; this number is the quotient answer.

■ PROBLEMS

Problems 1 to 7 refer to Fig. 10-8.

1. How does the single-pole double-throw switch affect the operation of the circuit? (10-3)

2. When the circuit is used as a subtractor, why is it necessary to preset the carry flip-flop before the clock pulses are applied? (10-3)

3. The binary bit that is applied to the K input of any J–K flip-flop in the registers before a clock pulse is at the $\overline{Q}$ output after the clock pulse. True or false? (10-3)

4. How many clock pulses are required to generate the answer? (10-2, 10-3)

5. With the switch in the add position, if the 4-bit number in register A added to the 4-bit number in register B generates a 5-bit answer, what happens to the fifth bit? (10-2)

6. With the switch in the subtract position, if 0011 is loaded into register A and 0001 is loaded into register B, the full adder generates a 0 at the sum output and a 1 at the carry output under which of the following conditions? (10-3)
   (a) Before the first clock pulse.
   (b) After the first clock pulse.
   (c) After the second clock pulse.
   (d) After the third clock pulse.
   (e) All of the above.

7. Explain why the carry-out bit of the full adder is fed to the carry flip-flop instead of being fed directly back to the carry-in input of the FA. (10-2)

8. The parallel adder of Fig. 10-16 requires one clock pulse to generate an answer. When does this occur? (10-4)

Problems 9 to 12 refer to Fig. 10-19.

9. If a 6 is loaded into the multiplicand register and a 5 is loaded into the multiplier register, how many add cycles are required to generate the answer? (10-5)

10. What function does the $\overline{Q}$ output of FF19 provide? (10-5)

11. Could I_1, I_2, and I_3 be removed from the circuit? (10-5)

12. How many clock pulses are required to multiply the following numbers? (10-5)

	Multiplicand		*Multiplier*	
(a)	2	×	2	=
(b)	3	×	2	=
(c)	2	×	3	=

Problems 13 to 17 refer to Fig. 10-22.

13. When the dividend number is greater than the divisor number, the divider circuit is (able or unable) to continue through its next subtraction step. (10-6)

14. What is the function of the Q output of FF15? (10-6)

15. Is it possible to eliminate I_3? (10-6)

16. To disable the circuit, why is it necessary to have the clock timer content be 00 at the same time that output $A < B$ is high? (10-6)

17. What is the largest number that can be divided in this circuit? (10-6)

Troubleshooting

Problems 18 and 19 refer to Fig. 10-24.

18. The parallel adder/subtractor circuit first adds 7 + 2 (0111 [Reg A] + 0010 [Reg B]) and generates a correct answer of 9 (1001). It then subtracts numbers 7 − 2 (0111 [Reg A] − 0010 [Reg B]) and generates a wrong answer of 7 (0111). Which of the following is the most likely fault? (10-4, 10-7)

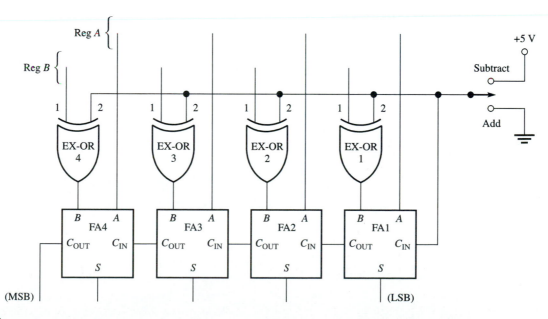

FIGURE 10.24

(a) The FA1 C_{OUT} is shorted to V_{CC}.

(b) The exclusive-OR 2 output is open and observes a high.

(c) The FA2 output is shorted to V_{CC}.

(d) All of the above.

19. The parallel/subtractor circuit is first set up to add $8 + 1$ (1000 [Reg A] + 0001 [Reg B]) and generates a correct answer of 9 (1001). After being set up to subtract $8 - 1$ (1000 [Reg A] − 0001 [Reg B]), it generates an incorrect answer of 9 (1001). Which of the following is the most likely fault? (10-4, 10-7)

(a) Input 2 of exclusive-OR gate 3 is shorted to ground.

(b) Input 2 of exclusive-OR gate 1 is shorted to V_{CC}.

(c) The FA1 C_{OUT} is shorted to V_{CC}.

(d) All of the above.

■ ANSWERS TO REVIEW QUESTIONS

1. one   **2.** (a)   **3.** How the bits are manipulated before entering the full adder.   **4.** Clock pulse

5. (a), (b), and (d)   **6.** (a)   **7.** 0

8. (1) Clear the carry flip-flop, and (2) place the add/subtract switch at a 0 state   **9.** In the carry flip-flop

10. 2's complement method   **11.** (1) Preset the carry flip-flop, and (2) place the add/subtract switch at a 1 state

12. It should be ignored.   **13.** 1   **14.** 4

15. Faster operating speed   **16.** Higher cost

17. One clock pulse   **18.** By straight wires connected from the carry-out full-adder outputs to the carry-in inputs of the FA located in the next significant column.   **19.** By connecting the carry-in input of the FA that adds the LSB column to a +5-V source.   **20.** add cycle   **21.** multiplier

22. In the accumulator   **23.** It passes the bits of the multiplicand to the full adder during the first four clock pulses of the add cycle. It then blocks the multiplicand bits from passing to the full adder during the second four clock pulses of the add cycle when the accumulator bits shift into their proper position before the next add cycle begins.   **24.** (1) Decrements the multiplier circuit. (2) Alternately enables A_1 to pass multiplicand bits to the full adder for four clock pulses, then disables A_1 from passing multiplicand bits to the full adder during the next four clock pulses.   **25.** When the multiplier decrements to 0000, it causes the output of N_1 to go low, which disables A_2 from passing clock pulses throughout the circuit. The circuit is then frozen.   **26.** subtract, count   **27.** register A, register B **28.** smaller

29. up-counter, accumulator/register A **30.** (1) The clock timer circuit contains 00. (2) Output $A < B$ of the comparator is high.

CHAPTER **11**

CONVERSION DEVICES
AND CIRCUITS

When you complete this chapter, you will be able to:

1. Identify the schematic diagrams of the comparator, inverting amplifier, and summing amplifier.
2. Calculate the output voltages of the comparator, inverting amplifier, and summing amplifier given the component values and input voltages.
3. Explain the operation of the binary-weighted digital-to-analog converter using discrete components.
4. List the disadvantages of the binary-weighted digital-to-analog converter and describe the modifications that have been developed to overcome them.
5. Determine the resolution of a digital-to-analog converter given the maximum analog voltage and number of binary input lines.
6. Make the external connections to a digital-to-analog converter in IC form.
7. List the sections and describe the operation of an analog-to-digital converter.
8. Make the external connections to an analog-to-digital converter in IC form.

9. Describe the problems caused by switch bouncing of SPST and SPDT switches and how to eliminate their effects.
10. Describe the wave-shaping capability and operating characteristics of a Schmitt trigger in IC form.
11. Describe the characteristics of the bistable, astable, and monostable multivibrators.
12. Construct an astable multivibrator by using a reference chart to determine which component values are needed to produce a desired frequency.
13. Construct a monostable multivibrator by using a reference chart to determine which component values are needed to produce a desired unstable state for a useful length of time.
14. Explain how the Gray code is used and describe the differences between it and the binary code.
15. Convert the binary code to Gray and the Gray code to binary by using the pencil-and-paper method.
16. Describe how logic gates are used to convert binary to Gray and Gray to binary codes.

1.1 INTRODUCTION

The material covered so far in this book has been about digital electronics. Many examples have been provided to explain the various functions these circuits are capable of performing. Most references were made to either calculator or computer operations in which the data being processed and the internal circuitry are digital in nature. However, digital equipment is not limited to only processing and controlling digital data. Digital devices are also used to control analog information, such as fuel levels, meter readings, temperatures, weights, velocity, and position.

If a digital system is used to interact with real-world information, it must be capable of converting analog signals to digital data. Likewise, after processing the data, it must convert digital data back to analog signals. These interaction functions are performed by conversion devices and circuits. Because there are many conversion devices and circuits that have been developed for electronic equipment over the years, they are too numerous to address in this chapter. The ones that are included have been selected because they are used in the functional circuits described in Chapter 13.

1.2 OPERATIONAL AMPLIFIERS

Operational amplifier:

An analog electronic circuit that has a high input impedance, low output impedance, and a high voltage gain.

One of the most important devices used in the conversion process is the **operational amplifier (op-amp).** The original design of the op-amp utilized a configuration of several cascaded amplifiers made up of discrete components. It was used in analog computers to perform mathematical operations. Since then, the sophisticated fabrication techniques used to construct ICs have allowed present-day op-amps to be manufactured inside eight-pin miniature integrated circuit packages.

There are three important characteristics of op-amps that make them ideal amplifiers:

1. high input impedance
2. high voltage gain
3. low output impedance

Figure 11-1 shows the standard schematic symbol of the popular 741 op-amp. Represented by a triangle, the op-amp has two input terminals located at the base on the left and has a single output located at the apex of the triangle. There are also two separate power supply lines. The one located at the top base is connected to a positive potential, and the other located at the bottom base is connected to a negative potential. These two power supplies allow the output voltage to swing to either a positive or negative voltage with respect to ground.

FIGURE 11.1
Standard symbol of an operational amplifier.

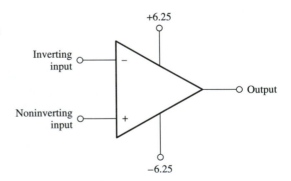

One of the inputs has a minus sign. This is called the *inverting input* because any DC or AC signal applied to its input is 180 degrees out of phase at the output. The other input has a plus sign and is called the *noninverting input*. Any DC or AC signal applied at this input is in phase at the output.

There are several other types of op-amps available, but the 741 is used most frequently. When external components are connected to the input and output leads, the op-amp is capable of performing several functions. How the components are connected determines which function the op-amp performs. However, only the types of operations that apply to the conversion applications in this chapter are described. These op-amp applications include:

- comparators
- inverting amplifiers
- summing amplifiers

11.3 COMPARATORS

Figure 11-2 shows an op-amp configuration that operates as a voltage comparator. This device compares the voltage applied at one input to the voltage applied at the other input. Any minute difference between the voltages drives the op-amp output into a +5-V or −5-V saturation. The polarity of the output is determined by the polarity of the voltages applied at the inputs. When the voltage applied to the inverting input is more positive than the voltage at the noninverting terminals, the output swings to a −5-V saturation potential. Likewise, when the voltage applied to the inverting input is more negative than the voltage at the noninverting input, the output swings to a +5-V saturation potential. However, when the input voltages are the same amplitude and polarity, the output is zero. The following equations provide a summary of the operation for the voltage comparator.

FIGURE 11.2
Op-amp comparator.

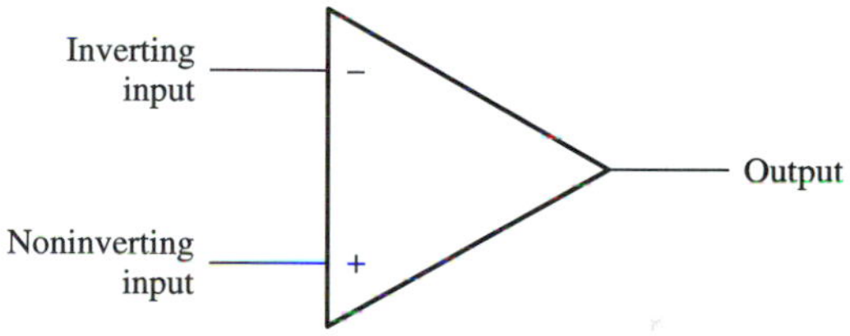

Inverting input voltage < noninverting input voltage = positive output voltage

Inverting input voltage > noninverting input voltage = negative output voltage

Inverting input voltage = noninverting input voltage = zero output voltage

Table 11-1 provides some examples of how the op-amp, operating as a comparator, responds to several different input voltages.

TABLE 11.1
Operation of an op-amp comparator

INVERTING INPUT TERMINAL (VOLTS)	NONINVERTING INPUT TERMINAL (VOLTS)	OUTPUT SATURATION VOLTAGE (VOLTS)
+1	−1	−5
+1	+2	+5
+2	+1	−5
0	0	0
−1	+1	+5
0	−1	−5
0	+1	+5
+3	+3	0

The reason an op-amp is capable of producing an output voltage that is either positive or negative is that it uses separate power supplies for each polarity. Some comparator applications require that the voltage produced at the output be of only one polarity. For example, suppose that only a positive output voltage was desired. By connecting a ground to the bottom power supply terminal, the output does not go negative when the inverting input is more positive than the noninverting input. Instead, it produces a 0-V ground potential. Because the positive power supply is still connected to the top terminal, the output goes into positive saturation when the noninverting input is more positive than the inverting input. The saturation voltage is approximately 80% of the supply voltage value. Likewise, when voltages supplied to both inputs are the same, the output produces 0 V.

EXAMPLE 11.1

Construct a table to show the output of an op-amp that has +6.25 V connected to the top power supply lead and ground connected to the bottom power supply lead when the inputs are applied to the op-amp of Fig. 11-3(a).

Solution   See Fig. 11-3(b).

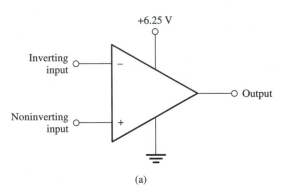

Inverted input terminal (volts)	Noninverted input terminal (volts)	Output saturation voltage (volts)
+2	−1	0
+1	+2	+5
+2	−2	0
0	−1	0
0	+1	+5
−2	−2	0

(a)   (b)

FIGURE 11.3
Op-amp and operational table for Example 11.1.

TABLE 11.2
The op-amp comparator operation for Review Question 4

INVERTING INPUT		NONINVERTING INPUT	OUTPUT (VOLTS)
	( )		+5
	( )		0
	( )		−5
	( )		+5
	( )		−5
	( )		0

■ REVIEW QUESTIONS

1. Conversion devices and circuits are used to enable digital equipment to interact with _____ equipment.

2. The phase difference between the output of an op-amp and a signal applied to its (−) input is _____ degrees. The phase difference between the output and a signal at the (+) input is _____ degrees.

3. What factors determine the type of function an op-amp performs?

4. Fill in the symbols (>), (<), or (=) in the appropriate areas in Table 11-2 on the previous page to complete the formula that explains the operation of an op-amp comparator circuit.

11.4 INVERTING AMPLIFIERS

A typical op-amp can have a voltage gain of approximately 200,000. However, the output cannot exceed a voltage level that is above approximately 80% of the supply voltage. For example, the maximum output voltages of the op-amp in Fig. 11-1 are +5 V and −5 V because the power supply potentials are +6.25 V and −6.25 V. Therefore, it only takes a 25-μV input to result in a positive or negative 5-V output voltage, depending on the input-signal polarity and the terminal to which it is applied.

Feedback:

A technique used to control the gain of an operational amplifier.

However, the op-amp is used for many applications that require a voltage gain of less than 200,000. A technique called **feedback** is used to control the gain of this device, and it is accomplished by connecting a resistor from the output terminal to an input lead. A negative-feedback circuit is shown in Fig. 11-4. Its operation is as follows:

- Both input terminals have high impedances; therefore, they do not allow current to flow into or out of them.
- The potential at the negative input lead is called 0-V *virtual ground* (in effect, but not actual). The positive input lead is connected to an actual 0-V ground potential.
- Because point VG is 0 V, there is a voltage drop of 2 V across the 2-kΩ resistor, R_{IN}. 1 mA flows through it.
- The 1 mA cannot flow into the op-amp. Therefore, it flows up through the 10-kΩ feedback resistor, R_F, developing a 10-V drop across it.
- Because V_{OUT} is measured with respect to the virtual ground, its voltage is −10 V.

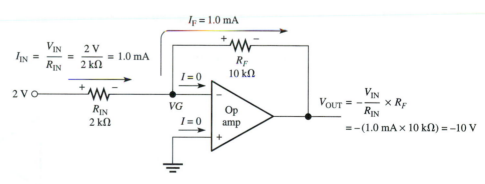

FIGURE 11.4
Inverting op-amp.

Gain:

The amount an analog input voltage is increased by an amplifier at the output terminal.

The **gain** of the op-amp is determined by

$$V_{GAIN} = \frac{V_{OUT}}{V_{IN}}$$

TABLE 11.3

Input and output voltages of an inverting op-amp with a gain of 10

V_{IN}	V_{OUT} (VOLTS)
+0.2	−2
−0.4	+4
0	0
+0.32	−3.2

The gain of the inverting op-amp in Fig. 11-4 is 5 because a 2-V signal is applied to the input and an inverted −10-V signal is at the output. A negative input voltage applied to this amplifier produces a positive output. The gain is influenced by the resistance ratio of R_F compared to R_{IN}. The larger R_F becomes compared to R_{IN}, the larger the gain.

The output voltage can also be determined by

$$V_{OUT} = \frac{R_F}{R_{IN}} \times V_{IN}$$

Table 11-3 provides some examples of how the inverting amplifier with a gain of 10 responds to several input voltages.

11.5 VOLTAGE-SUMMING AMPLIFIER

Voltage-summing amplifier:

A type of operational amplifier configuration capable of adding the algebraic sum of several voltages applied to one of its input lines.

When two or more inputs are tied together and then applied to an input lead of an op-amp, a summing amplifier is developed. This type of amplifier is capable of adding the algebraic sum of DC or AC signals. The circuit of Fig. 11-5 is that of an inverting summing amplifier. It consists of a 20-kΩ feedback resistor, R_F, three parallel 20-kΩ summing resistors tied together and connected to the inverting input lead, and +2-V, +1-V, and +3-V signals applied to the inputs. The following calculations show how to determine the voltage at the output terminal:

$$I_{R1} = \frac{V_{R1}}{R_1} = \frac{2\text{ V}}{20\text{ k}\Omega} = 0.1\text{ mA}$$

$$I_{R2} = \frac{V_{R2}}{R_2} = \frac{1\text{ V}}{20\text{ k}\Omega} = 0.05\text{ mA}$$

$$I_{R3} = \frac{V_{R3}}{R_3} = \frac{3\text{ V}}{20\text{ k}\Omega} = 0.15\text{ mA}$$

$$I_{R_F} = 0.1\text{ mA} + 0.05\text{ mA} + 0.15\text{ mA}$$

$$= 0.3\text{ mA or } -0.3\text{ mA (inverted)}$$

$$V_{OUT} = I_{R_F} \times R_F$$

$$= -0.3\text{ mA} \times 20\text{ k}\Omega = -6\text{ V}$$

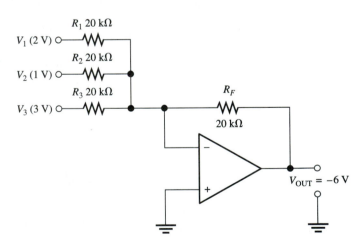

FIGURE 11.5
Inverting summing amplifier.

The gain of each input is found and then summed to obtain the resulting output.

Table 11-4 provides some examples of how the summing amplifier in Fig. 11-5 responds to several input voltages.

TABLE 11.4
The operation of an inverting summing amplifier

INPUT VOLTAGES			ALGEBRAIC SUM OF OUTPUT VOLTAGES
V_1	V_2	V_3	
+1	+1	+1	−3
+1	−1	−1	+1
+2	−1	−1	0
−3	−1	+3	+1
+1	+2	−1	−2

■ REVIEW QUESTIONS

5. The saturation output voltages of an op-amp cannot exceed _____% of the supply voltage.

6. A typical op-amp without a feedback resistor has a gain of _____, but with a feedback resistor, the gain is reduced to a controllable level.

7. An inverting amplifier has its _____ lead tied to ground.

8. An inverting amplifier with a signal applied to its input has a gain of _____ if $R_{IN} = 50\,k\Omega$ and $R_F = 100\,k\Omega$.

9. An inverting summing amplifier adds the _____ sum at its _____ input terminal, and ground is connected to the (+) input terminal.

EXPERIMENT

Operational Amplifiers

Objectives

- To construct and operate a comparator circuit using an operational amplifier.
- To construct and operate an inverting operational amplifier.
- To construct and operate a summing operational amplifier.

Materials

1—Dual Power Supply (+15 V and −15 V)

1—741 Linear IC

1—2-kΩ Resistor

3—10-kΩ Resistors

4—20-kΩ Resistors

1—25-kΩ Resistor

2—10-kΩ Potentiometers

1—Voltmeter

Introduction

The previous experiments in this textbook have examined the operation of digital circuits. These devices function by using signals that are at one of two distinct voltage levels. The other classification of electronic circuits is **analog.** These devices function by operating with signals that vary anywhere between the low and high extreme voltages supplied by the power supply. Because these devices produce an output that is directly proportional to the input, they are known as **linear circuits.** One type of linear circuit commonly used with digital circuits is the operational amplifier. Also known as an op-amp, it is capable of performing many types of functions. Two functions, comparing and amplifying, will be examined. Figure 11-6 shows the symbol of an op-amp. It has two inputs. One of them is called an *inverting* input and is labeled with a minus sign. The other is called a *noninverting* input and is labeled with a positive sign. The output lead is located at the apex of the triangle.

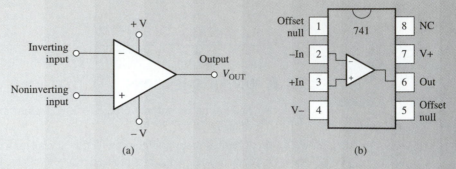

(a)   (b)

FIGURE 11.6
Operational amplifier: (a) Standard schematic symbol. (b) Pin diagram.

Background Information—Comparator

The op-amp in Fig. 11-7(a) is capable of comparing the voltage applied to one input to the voltage applied to the other input. When the voltage at the noninverting terminal is greater than the voltage at the inverting terminal, the output will go to an approximate positive 5-V saturation potential. When the voltage at input $(-)$ is greater than the voltage at input $(+)$, the output will go to an approximate negative 5-V saturation level. When the voltage at both input terminals is the same, the output will go to 0 V.

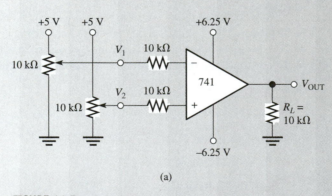

	Inputs		V_{OUT} (V)
	V_1 $(-)$	V_2 $(+)$	
	+4 V	+1 V	
	+2 V	+3 V	
	+1 V	0 V	
	+4 V	+4 V	
	0 V	+1 V	
	+3 V	+2 V	

(a)   (b)

FIGURE 11.7
Op-amp comparator: (a) Schematic diagram. (b) Data table.

Procedure

Step 1. Assemble the circuit shown in Fig. 11-7(a).
Step 2. Fill in the output portion of the table in part (b) by applying the voltages listed in the input section.

Background Information—Inverting Operational Amplifier

The op-amp in Fig. 11-8(a) shows an inverting op-amp. An input resistor (R_{IN}) is connected between the input terminal and the $(-)$ op-amp lead, and a feedback resistor (R_F) is connected between the output and the $(-)$ input. The name of the op-amp circuit is derived from the way in which it operates. When a voltage is applied to the (−) input lead, a voltage of the opposite polarity develops at the output. The gain of the inverting op-amp is determined by the resistance ratio of R_F compared to R_{IN}. The larger R_F becomes compared to R_{IN}, the larger the gain.

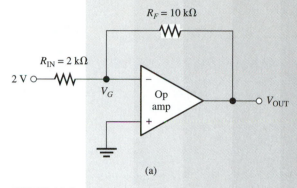

V_{IN}	V_{OUT}
+0.2 V	
−0.3 V	
0 V	
+0.32 V	

V_{IN}	V_{OUT}
+0.3 V	
−0.25 V	
−0.2 V	
+0.4 V	

(a)   (b)   (c)

FIGURE 11.8
Inverting op-amp.

The output voltage can be determined by the following formula:

$$V_{OUT} = \frac{V_{IN}}{R_{IN}} \times R_F$$

where, R_F = feedback resistor
R_{IN} = input resistor
V_{IN} = input voltage

Step 3. Assemble the circuit in Fig. 11-8(a).
Step 4. Fill in the output portion of the table in part (b) by applying the voltages listed in the input section. Use the formula to verify that the measured voltages are correct.
Step 5. Change R_F in Fig. 11-8(A) to 25 kΩ and R_{IN} to 10 kΩ.
Step 6. Fill in the output portion of the table in part (c) by applying the voltages listed in the input section. Use the formula to verify that the measured voltages are correct.

Background Information—Summing Amplifier

The circuit in Fig. 11-9(a) shows an op-amp circuit with more than one input tied at its inverting input lead. Called a *summing* amplifier, it is capable of adding the algebraic sum of all the input voltages applied. The sum of these voltages is inverted to the opposite polarity. Each of the summing input resistors are tied together, and a 20-kΩ feedback resistor is used. The following calculations show how to determine the voltage at the output terminal:

$$I_{R1} = \frac{V_{R1}}{R_1} = \frac{2\text{ V}}{20\text{ k}\Omega} = 0.1\text{ mA}$$

$$I_{R2} = \frac{V_{R2}}{R_2} = \frac{1\text{ V}}{20\text{ k}\Omega} = 0.05\text{ mA}$$

$$I_{R3} = \frac{V_{R3}}{R_3} = \frac{3\text{ V}}{20\text{ k}\Omega} = 0.15\text{ mA}$$

$$I_{RF} = 0.1\text{ mA} + 0.05\text{ mA} + 0.15\text{ mA} = 0.3\text{ mA}$$

$$V_{OUT} = I_{RF} \times R_F$$

$$= 0.3\text{ mA} \times 20\text{ k}\Omega = -6\text{ V}$$

FIGURE 11.9
Summing amplifier.

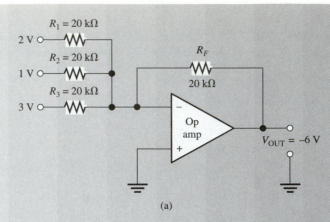

(a)

Input voltage			Output voltage	
V_1	V_2	V_3	Measured	Calculated
+1 V	+1 V	+1 V		
−1 V	−1 V	+1 V		
−1 V	−1 V	+2 V		
+3 V	−1 V	−3 V		
+2 V	+1 V	−1 V		

(b)

Step 7. Assemble the circuit shown in Fig. 11-9(a).

Step 8. Fill in the output portion of the table in part (b) labeled *measured* by applying the voltages listed in the input section.

Step 9. Using the formula to determine the output of a summing amplifier, verify that the measured values are correct and place the answers in the column labled "Calculated."

■ EXPERIMENT QUESTIONS

1. A/n _____ (analog, digital) signal can vary at any value between the low and high voltage range supplied by the power supply.

2. A _____ (linear, digital) circuit produces signals that are analog.

3. When the voltage applied to the noninverting terminal of an op-amp comparator is _____ (less, greater) than the voltage at the inverting input, the output will be driven into positive saturation.

4. If −2 V is applied to an inverting op-amp that has an input resistance of 2 kΩ and a feedback resistance of 5 kΩ, the voltage at the output is _____ volts, and the polarity is _____ (−,+).

5. The summing op-amp in Fig. 11-9 has the following voltages applied to the inputs: +2 V, +3 V, and −1 V. What is the output voltage?

11.6 DIGITAL-TO-ANALOG CONVERTERS

Digital-to-analog converter (DAC):
A circuit that converts a digital value into a proportional analog output voltage.

Digital-to-analog converters (DACs or D/A converters) are used to convert digital signals representing binary numbers into proportional analog voltages. Although these devices are now available in IC packages, they are analyzed in a discrete form to better describe how they function.

A 4-bit input DAC is shown in Fig. 11-10. It consists of a summing amplifier with its feedback resistor (R_F), four summing resistors, and four switches that are used to provide a 4-bit binary input. A switch in the open position represents a 0 state. In the closed position, it represents a 1 state. The placement of each switch corresponds to the same 8-4-2-1 weighted values of a 4-bit binary number. Resistors R_1 through R_4 are also selected with a weight proportional to the next. The 12.5-kΩ resistor R_4 is connected at the MSB input line. The values of the remaining resistors are selected by making each progressive resistor twice the size of the preceding one. The analog voltage is always at the op-amp output. The circuit is designed to operate so that a 4-bit binary number represented by the four switches is converted into voltages. Because 16 different combinations of switch positions are possible (0–15), 16 different analog voltage levels proportional to the digital number applied are produced. The circuit in Fig. 11-10 is designed so that it develops an analog output voltage that is equivalent to the binary number applied. For example, when all switches are in the open position to represent a binary input of 0000, the output is 0 V. If SW1 is moved to the closed position (binary 0001), the op-amp output will be −1 V. If SW1 and SW3 are in the closed position (binary 0101), the op-amp output will be −5 V. If all four switches

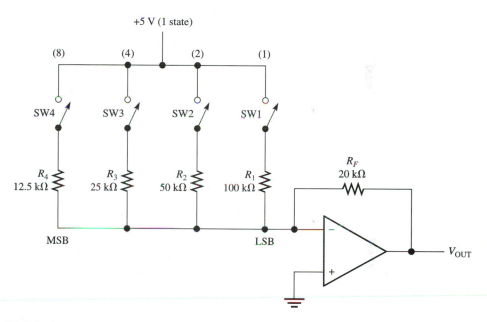

FIGURE 11.10
Binary-weighted D/A converter.

EXAMPLE 11.2

What is the analog output voltage of the DAC in Fig. 11-10 when a binary 1001 is applied?

Solution

$$I_{R1} = \frac{V_{R1}}{R_1} = \frac{5 \text{ V}}{100 \text{ k}\Omega} = 0.05 \text{ mA}$$

$$I_{R4} = \frac{V_{R4}}{R_4} = \frac{5 \text{ V}}{12.5 \text{ k}\Omega} = 0.4 \text{ mA}$$

$$I_{RF} = 0.05 \text{ mA} + 0.4 \text{ mA} = 0.45 \text{ mA}$$

$$V_{OUT} = I_{RF} \times R_F$$

$$= 0.45 \text{ mA} \times 20 \text{ k}\Omega = 9 \text{ V}$$

are in the closed position (binary 1111), the analog output voltage will be -15 V. The analog output voltage for each combination of switch settings can be determined by the same formula used in Section 11-5, "Voltage-Summing Amplifiers."

Figure 11-11(a) provides all possible digital inputs and the corresponding output voltages for the circuit of Fig. 11-10. Figure 11-11(b) provides the same information in a graphical format.

SW4 (8)	SW3 (4)	SW2 (2)	SW1 (1)	V_{OUT} (−V)
0	0	0	0	0
0	0	0	1	1
0	0	1	0	2
0	0	1	1	3
0	1	0	0	4
0	1	0	1	5
0	1	1	0	6
0	1	1	1	7
1	0	0	0	8
1	0	0	1	9
1	0	1	0	10
1	0	1	1	11
1	1	0	0	12
1	1	0	1	13
1	1	1	0	14
1	1	1	1	15

(a)

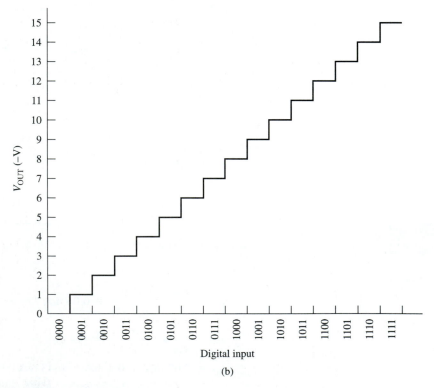

(b)

FIGURE 11.11
Analog output vs. digital input for the circuit of Fig. 11-10.

EXPERIMENT

Binary-Weighted D/A Converter

Objective

- To construct and operate a binary-weighted DAC.

Materials

> 1—Dual Power Supply (+18 V and −18 V)
>
> 1—39-kΩ Resistor
>
> 1—741 Linear IC
>
> 1—16-kΩ Resistor
>
> 1—10-kΩ Resistor
>
> 1—82-kΩ Resistors
>
> 1—20-kΩ Resistor
>
> 1—39-kΩ Resistor

Introduction

Figure 11-12 shows a DAC consisting of a summing amplifier with its feedback resistor, four summing resistors, and four logic switches that are used to provide a 4-bit binary input. The placement of the switch corresponds to the same 8–4–2–1 weighted values of a 4-bit binary number. The summing resistor values are also selected with a weight proportional to the next. The smallest-valued summing resistor is connected to the MSB input line. The values of the remaining resistors are twice the resistance size of the previous one.

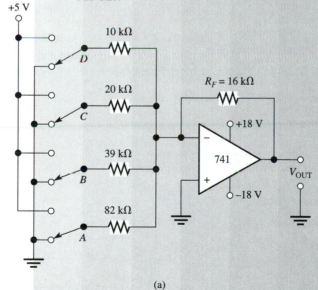

DCBA	V_{OUT}	
Inputs	Measured	Calculated
0000		
0001		
0011		
0101		
1000		
1011		
1100		
1111		

(a) (b)

FIGURE 11.12

Binary-weighted D/A converter: (a) Schematic diagram. (b) Function table.

Procedure

Step 1. Assemble the circuit shown in Figure 11-12(a).

Step 2. Fill in the output portion of the table in part (b) labeled *measured* by applying the binary numbers listed in the input section.

Step 3. Using the formula to determine the output of a summing amplifier, calculate the output voltage of the DAC for the numbers applied to the inputs. Place the answers in the output columns labeled *Calculated*, and compare them to the measured values.

■ EXPERIMENT QUESTIONS

1. The number of binary inputs applied to the binary weighted DAC can be expanded from 4 to 8 by _____.
   (a) Using two op-amps.
   (b) Connecting four more resistors in parallel to the input resistor network of the summing op-amp.
2. The resistor value at the LSB input of a 4-input binary weighted DAC is _____ times greater than the MSB resistor.

For some applications the binary-weighted DAC is sufficient. However, it does have its limitations, which result in potential problems:

Problem 1

If the circuit is expanded to convert binary numbers larger than 4 bits, a larger number of different resistor values is required. Also, some resistor values become very high. Because most DACs are now constructed on an IC chip, the manufacturing process necessary to fabricate resistors having many different and high values would be cost-prohibitive.

Solution 1

This problem can be overcome by the *R–2R* ladder resistor network. Figure 11-13 shows this type of resistor configuration. This network gets its name because two resistor values are used, one twice (2*R*) the other resistor (*R*). For example, if each of the *R* resistor values equals 10 kΩ, then each of the 2*R* resistors would have twice that value, 20 kΩ. The

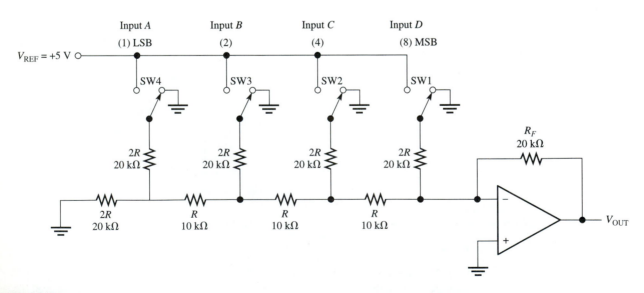

FIGURE 11.13
The **R–2R** *ladder DAC.*

function of the R–$2R$ resistor network is the same as the binary-weighted circuit in that it weighs the inputs so that each one has twice the value as the previous one. For example, the current supplied by the resistor network that flows through the feedback resistor of the op-amp should be twice as much when input B is activated as when input A is activated. Activating input C should cause the current supplied by the resistor network to be four times greater than when input A is activated, and D eight times greater than when A is activated.

The R–$2R$ network is commonly used in integrated circuit DACs. Because there are resistors of only two values, the range and high-resistance problems are eliminated. Furthermore, the R resistors have almost identical values because they are formed together. The same applies to the $2R$ resistors. Therefore, the accuracy of the network is maximized.

Problem 2

An actual DAC does not use switches to represent its low-level 0-V or high-level +5-V inputs. Instead, these inputs can come from a TTL counter that can provide a low-level voltage that ranges from 0 to +0.8 V or a high-level voltage that ranges from +2.8 to +5 V. Input voltages that vary this much do not allow the op-amp to produce the precise analog voltages it is designed to generate.

Solution 2

Input voltages must be precise for this circuit to operate properly. Therefore, actual DACs contain transistorized switches that are located at each input, as shown in Fig. 11-14. Each data input drives a base of a transistor through a current-limiting resistor. When an input is high, it causes the transistor to saturate and acts like a closed switch that connects an accurate +5-V reference to a $2R$ resistor. When an input is low, the transistor is cut off and a 0-V potential is at the R–$2R$ junction under the transistor.

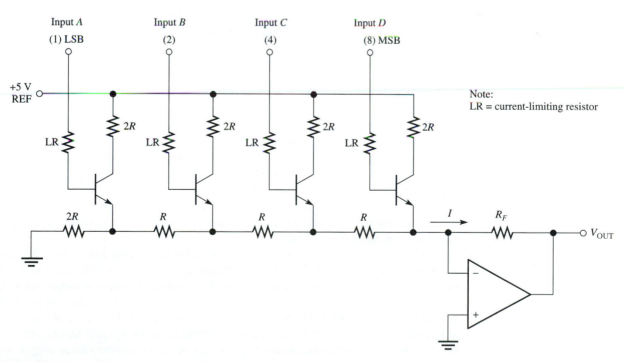

FIGURE 11.14
Transistor switches for a DAC.

The DACs described to this point are capable of converting 4-bit binary numbers, the smallest being 0000_2 and the largest 1111_2. Therefore, this DAC divides the reference analog output into 15 equal divisions.

DACs in IC form are also available with 8, 12, and 16 binary inputs. As the number of inputs increases, the reference analog voltage is divided into smaller divisions. For example, 8-bit DACs divide the analog output voltage into 255 equal parts, 12-bit converters into 4095 equal parts, and 16-bit converters into 65,535 equal divisions.

The number of equal divisions a DAC divides the reference voltage into is called **resolution.** The resolution of a DAC can be determined by the formula $V_{REF}/2^n - 1$.

Resolution:

The number of equal divisions a digital-to-analog converter divides the reference voltage into.

- The 2 in the formula represents the binary number system.
- The n is the exponent that specifies to what power 2 is multiplied. It is determined by how many binary inputs are used at the input of the DAC. By multiplying 2 to the nth power, the maximum binary (equivalent decimal) number is determined.
- A 1 is subtracted from the maximum binary number to determine the number of equal steps (resolution) between the maximum binary number and minimum binary number.

EXAMPLE 11.3	Find the resolution of a DAC with a reference voltage of 30 V and four inputs.

Solution

- Determine that the reference voltage is 30 V.
- Because the number of digital inputs is 4, this number is n.
- Multiply $2^4 = 16$.
- Subtract $16 - 1 = 15$.
- Divide 30 V/15 = 2-V resolution.

Normally, resolution is expressed in terms of the number of binary input bits that are converted. A DAC with high resolution requires an accurate reference voltage because any variation can cause an error.

Resolution is obviously an important factor to consider when purchasing a DAC. Also important are its accuracy and operating speed.

11.7 INTEGRATED CIRCUIT DIGITAL-TO-ANALOG CONVERTER

One popular DAC is the 8-bit DAC0808. Its block diagram and pin diagram are shown in Fig. 11-15. The DAC contains transistor switches (called *current switches*), an R–$2R$ ladder resistor network, and a reference-current amplifier. Even though some types of DACs contain an internal op-amp, this one does not. The internal components supply proportional currents to its output lead.

Figure 11-16 shows the DAC0808 connected to an external 741 op-amp. The current range is dictated by the 10-V 5-kΩ combination connected to pin 14. The 2 mA flowing through resistor R_F is the maximum amount of current that can flow at output pin 4 (I_{OUT}). When the digital input is $0000\ 0000_2$, the minimum current of 0 mA flows at pin 4. When the digital input is $1111\ 1111_2$, the maximum current of 2 mA flows at pin 4. By using a 5-kΩ feedback resistor (R_F), the analog output voltage at the op-amp output ranges from 0–10 V. The 10 V is produced when I_{OUT} is 2 mA. If a different analog output-voltage range is desired, this can be accomplished by adjusting the gain of the op-amp by changing resistor R_F to a different value.

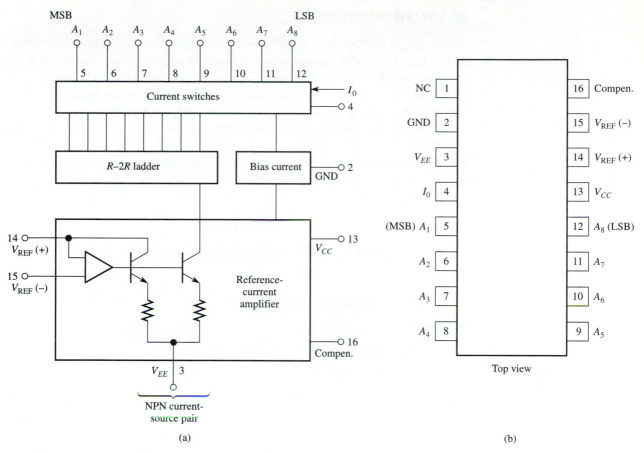

FIGURE 11.15
The DAC0808 DAC: (a) Block diagram. (b) Pin diagram.

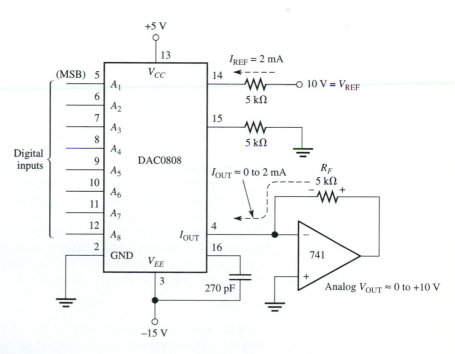

FIGURE 11.16
The DAC0808 and the 741 op-amp connected to form a DAC.

■ REVIEW QUESTIONS

10. A DAC is a device that converts _____ inputs into _____ analog voltages.

11. A binary-weighted resistor uses the (smallest/largest) valued resistor at the MSB input and the (smallest/largest) _____ valued resistor for its LSB input.

12. A 5-bit input DAC produces how many different analog output-voltage levels?

13. An _____ ladder resistor network is made up of resistors that are of only _____ different values and are usually used when the number of inputs to the DAC exceeds _____.

14. The number of equal divisions a DAC divides the reference voltage into is called _____.

EXPERIMENT

Integrated Circuit Digital-to-Analog Converter

Objective

■ To construct and operate a DAC in IC form.

Materials

     1—Dual Power Supply (+5 V and 0 to +10 V)

     1—Dual Power Supply (+18 V and −18 V)

     1—Signal Generator or Astable Multivibrator (10 kHz)

     1—Oscilloscope

     1—741 Linear IC

     1—DAC0808 IC

     2—7493 ICs

     1—10-kΩ Resistor

     1—1-kΩ Resistor

     3—5-kΩ Resistors

     1—2.5-kΩ Resistor

     1—270 pF Capacitor

Introduction

One popular DAC in IC form is the 8-bit DAC0808 shown in Figure 11–17. An inverting op-amp is externally connected to the DAC IC package.

    The maximum current level of 2 mA at the output of the op-amp is dictated by the 10-V supply and 5-kΩ resistor connected to pin 14. When the digital input is 0000 0000, the minimum current of 0 mA flows through output pin 4. When the digital input is 1111 1111, the maximum current of 2 mA flows through output pin 4. Using a 5-kΩ feedback resistor (R_F), the analog output voltage at the op-amp output can range from 0 V to 10 V. If a different voltage range is desired, the V_{REF} voltage applied to pin 14 can be changed, the resistor at pin 14 can be changed, or the R_F resistor value can be used to change the gain.

Procedure

Step 1. Assemble the circuit shown in Fig. 11-17.

Step 2. Turn on the power and apply a clock frequency of 10 kHz to the cascaded counters. Observe the analog output with an oscilloscope as the digital input continuously increases, recycles to zero, and increments again.

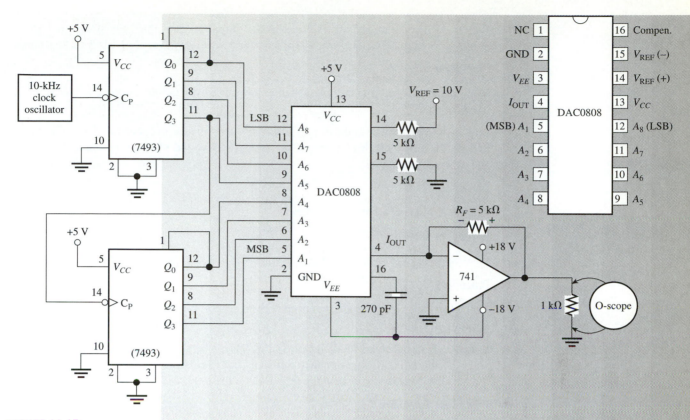

FIGURE 11.17

Step 3. Reduce the reference voltage applied to pin 14 to 5 V and observe the analog output with the oscilloscope.

Procedure Question 1

Compared to when 10 V were applied to the reference input, what happens to the analog output voltage when 5 V are applied?

    (a) It is one-half.

    (b) It doubles.

    (c) No change occurs.

Step 4. Turn the power off and replace the 5-kΩ R_F resistor connected across the op-amp with a 2.5-kΩ resistor.

Step 5. Turn the power back on and apply 10 V to the reference input.

Procedure Question 2

Compared to when 10 V were applied to the reference input and a 5-kΩ feedback resistor was used, what happens to the analog output voltage when a 2.5-kΩ resistor is used?

    (a) It is one-half.

    (b) It doubles.

    (c) No change occurs.

Procedure Information

With eight digital inputs, there are 256 (2^8) different analog voltages produced. To make it more manageable to read the different voltage levels on the oscilloscope screen, it will be necessary to use only four digital input leads.

Step 6. Turn the power off. Disconnect the output leads at the 7493 IC located on the bottom of the diagram and connect pins A_1 to A_4 of the DAC0808 IC to ground.

Procedure Question 3

How many different analog voltages are produced at the output of the DAC if only four inputs are used?

Step 7. Apply power to the circuit and verify your answer to Procedure Question 3.

Step 8. Remove the lead at pin 12 of the DAC0808 and observe the waveform on the oscilloscope.

Procedure Question 4

Is a 0 or 1 recognized at the open input?

Procedure Question 5

How many voltage levels are produced with an open at the LSB input? Explain your answer.

Procedure Question 6

How many voltage levels are produced with an open at pin 9 instead of pin 12 of the DAC? Explain your answer.

Step 9. Reconnect pin 12 of the DAC0808. Disconnect the ground connection at pin 8 of the DAC0808 and replace it with a wire from pin 12 of the bottom IC shown in the diagram. Observe the waveform at the output of the DAC.

Procedure Question 7

How many voltage levels are produced at the output of the DAC and why?

■ EXPERIMENT QUESTIONS

1. How many different analog voltage levels can be produced by the DAC0808 IC?
2. An open at an input of a DAC0808 is recognized as a logic _____ (0, 1).
3. List three ways in which the voltage range of the DAC0808 IC can be varied.

11.8   ANALOG-TO-DIGITAL CONVERTER

Analog-to-digital converter (ADC):

A circuit that converts an analog voltage applied to its input into a proportional digital output.

The **analog-to-digital converter (ADC or A/D converter)** illustrated in Fig. 11-18 is capable of converting analog input voltages ranging from 0 to +15 V into proportional digital numbers that are stored in a binary counter. The circuit consists primarily of a binary up-counter, a comparator, a 20-kHz astable multivibrator, a DAC, and a reset switch.

Functions

The Reset Switch   This switch is in the normally high state. When depressed, a grounded 0-state potential is applied to the active-low clear pins at each flip-flop, causing the counter to clear.

Comparator:

A type of operational amplifier circuit that produces an output voltage when both inputs are not at the same potential.

The Comparator   The analog input voltage to be converted is applied to the positive input of the comparator. The analog output voltage of the DAC is applied to the negative input of the comparator. Whenever the voltage applied to the positive lead is more positive than the voltage applied to the negative lead, the comparator output is driven to a positive-voltage saturation level. Whenever the voltage at the positive lead is equal or less positive than the voltage at the negative lead, the comparator output is 0 V.

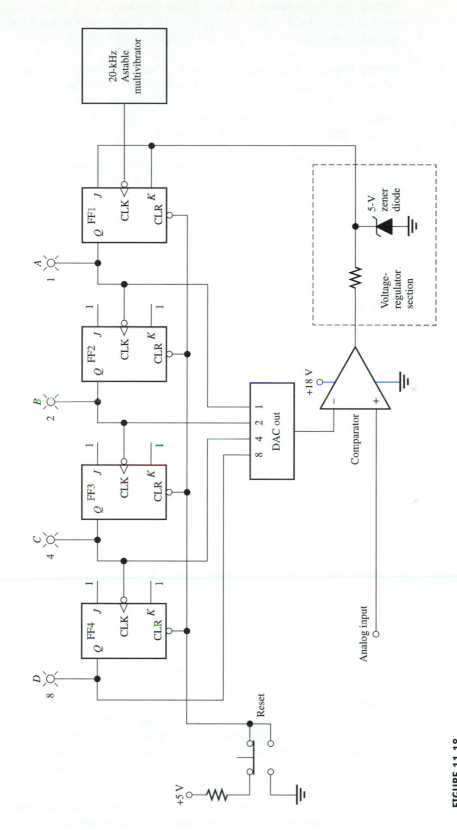

FIGURE 11.18
Analog-to-digital converter.

Voltage-Regulator Network The regulator consists of a 5-V zener diode and resistor. It is used to ensure that a voltage greater than 5 V is not applied to the *J* and *K* inputs of FF1 when the comparator output goes to a positive-voltage saturation level.

Binary Counter The counter is capable of counting from 0000_2 (0_{10}) to 1111_2 (15_{10}). When a high is applied to the *J* and *K* inputs of FF1, the flip-flop is in the toggle mode. Therefore, the 20-kHz clock applied to the clock input of FF1 causes the counter to increment. When the *J* and *K* inputs go low, FF1 is in the hold mode. Therefore, the up-counter is frozen and does not continue to increment.

DAC The *Q* outputs of the binary-counter flip-flops are connected to the digital inputs of the DAC. The analog output voltage generated by the DAC is proportional to the digital input number registered in the counter.

Operation

- Assume that the reset switch produces a negative pulse that causes the counter to clear (reset). The number registered in the counter becomes 0000, which is applied to the DAC's digital inputs.
- The 20-kHz signal is always connected to the clock input of the binary counter. If 0 V is applied to the analog input of the analog-to-digital converter circuit and 0 V is generated at the output of the DAC, the comparator output will be low. Therefore, the 20-kHz clock input is unable to cause the counter to increment because FF1 is in the hold mode, with a low at both the *J* and *K* inputs.
- Suppose that a positive potential is applied to the analog input of the ADC. Also, assume that the counters have just been reset by a pulse. The output of the DAC is 0 V. Because the voltages applied to the positive lead of the comparator are more positive than the 0 V applied to the negative lead, the comparator output is driven into positive-voltage saturation. With a high at the *J* and *K* inputs of FF1, it is in the toggle mode, allowing the counter to increment. It counts until the binary number registered in the counter causes the DAC to generate an analog output voltage that is equal to the analog voltage applied to the A/D circuit input.
- When both analog voltages at each comparator input are equal, its output goes to 0 V and provides a low to the *J* and *K* inputs of FF1. This puts FF1 into the hold mode, which causes the counter to display a digital binary number that is proportional to the analog input voltage applied to the analog-to-digital converter circuit.
- n Suppose that the analog-to-digital converter is recording 14.00 V applied to its analog input terminal. If the analog voltage is increased, a more positive voltage will be applied to the positive lead of the comparator than the 14.00 V (that comes from the DAC output) applied to the negative lead of the comparator. The comparator is driven into positive-voltage saturation and allows the counter to increment. When the counter causes the DAC to increase to the same voltage that is at the analog input, the comparator goes to 0 V and causes the counter to freeze. The new digital output number is displayed by the counter.
- The counter in the analog-to-digital converter is a binary up-counter. It counts up to a number that is equal to the voltage that is applied to the analog input of the analog-to-digital converter. Suppose that the converter is recording a +4.00-V input when the analog input voltage decreases to +2.00 V. Since the up-counter cannot decrement, it must be reset to 0000 by the reset switch. It can then count up until it equals the 2.00 V reading.

Counter ramp:

A type of analog-to-digital converter that produces an analog output waveform that resembles a staircase as digital numbers applied to its inputs increase.

Table 11-5 is the function table for the analog-to-digital converter described in Fig. 11-18. It lists the digital binary count that is generated by a given analog input. Note that the binary numbers are proportional to the analog voltages. The counter increments at every 1-V increase of the analog voltage. This type of A/D converter is called a **counter ramp.**

TABLE 11.5

Function table for a 4-bit ADC with a maximum of 15 V applied

| ANALOG INPUT VOLTAGE (VOLTS) | BINARY (DIGITAL) OUTPUT | | | |
| | 8 | 4 | 2 | 1 |
	D	C	B	A
0	0	0	0	0
1	0	0	0	1
2	0	0	1	0
3	0	0	1	1
4	0	1	0	0
5	0	1	0	1
6	0	1	1	0
7	0	1	1	1
8	1	0	0	0
9	1	0	0	1
10	1	0	1	0
11	1	0	1	1
12	1	1	0	0
13	1	1	0	1
14	1	1	1	0
15	1	1	1	1

EXPERIMENT

Counter-Ramp A/D Converter

Objective

■ To convert analog signals to digital signals using discrete components.

Materials

1—DC Dual Power Supply -18V, $+18$ V

1—DC Dual Power Supply 0–15 V DC $+5$ V

1—Signal Generator (4 Hz)

1—SPDT Switch

2—741 Linear ICs

1—50-kΩ Rheostat

1—7408 IC

1—7493 IC

4—LEDs

4—330-Ω Resistors

2—10-kΩ Resistors

1—18-kΩ Resistor

1—39-kΩ Resistor

1—75-kΩ Resistor

1—150-kΩ Resistor

1—1-kΩ Resistor

1—SPST Switch

1—1N5230 Zener Diode

Introduction

The analog-to-digital converter illustrated in Fig. 11-19 is capable of converting analog input voltages ranging from 0 to 15 V into proportional 4-bit binary numbers. The equivalent digital value is stored in the up-counter. The ADC uses a digital-to-analog converter (summing op-amp), an AND gate, a 4-bit up-counter, a voltage regulator, and an op-amp voltage comparator.

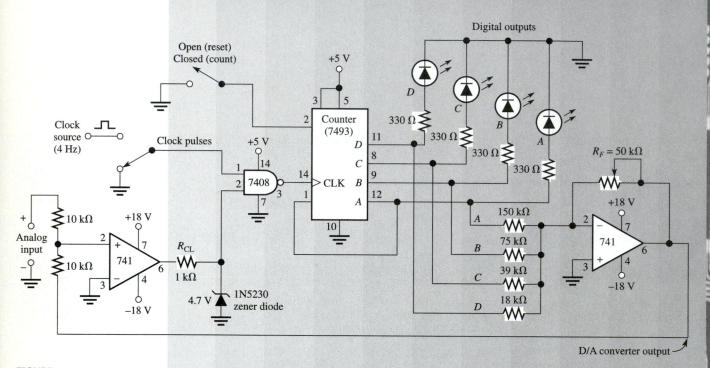

FIGURE 11.19
Counter ramp analog-to-digital converter.

The Operation

Step A. An analog voltage of 0 V is applied to the input of the ADC.

Step B. The 7493 up-counter is cleared by using the reset switch.

Step C. The contents of 0000 in the counter is applied to the DAC inputs, which produces 0 V at its output.

Step D. The analog input of 0 V and the DAC output of 0 V are joined at the noninverting input of the voltage comparator. Because 0 V is applied to both inputs of the voltage comparator, 0 volts is produced at its output.

Step E. The 0 V from the voltage comparator output disables the 7408 AND gate and prevents it from passing any clock signals through to its output.

Step F. Suppose that an analog voltage of +7 V is applied to the ADC. The analog voltage of +7 V is compared with the DAC voltage of 0 V at the noninverting junction of the voltage comparator.

Step G. Because the voltage at the noninverting input of the voltage comparator is greater than the 0-V ground potential at the inverting input, the comparator output goes high. The AND gate then becomes enabled and passes clock pulses into the counter input.

> *Note:* **Any voltage at the voltage comparator output that exceeds 4.7 V is dropped across the resistor (R_{LC}).**

Step H. The counter increments until it reaches the count of 0111_2 (7_{10}), at which time the DAC produces an analog output of -7 V.

Step I. The analog $+7$ V applied to the ADC cancels the -7 V of the DAC, producing 0 V at the voltage comparator output.

Step J. The 7408 AND gate becomes disabled and prevents any more clock pulses that would increment the counter.

Step K. The numerical contents of the 7493 up-counter is frozen, and the count of 7 is displayed at the four LEDs that represent the digital output.

Procedure

Step 1. Assemble the ADC circuit shown in Fig. 11-19.

Step 2. Clear the counter by momentarily opening the switch connected to pin 2 of the 7493 IC.

Step 3. Apply 10 V to the analog input of the ADC.

Step 4. Apply the clock pulses until the counter stops incrementing. (If the counter does not stop at the count of 10, it needs to be calibrated. If the count is too high, increase the resistance of rheostat R_F. If the count is too low, decrease the R_F resistance. After clearing the counter, repeat the R_F adjustments until the count of 10 is obtained.)

Step 5. After clearing the counter and disconnecting the clock pulse source, apply the analog input voltage listed on the first line of Table 11-6.

TABLE 11.6

INPUT	OUTPUT (BINARY)			
Measured Analog Voltage	*D* (8)	*C* (4)	*B* (2)	*A* (1)
2 V				
6 V				
8 V				
11 V				
13 V				
15 V				

Step 6. Connect a voltmeter or oscilloscope across the zener diode located at the output of the voltage comparator.

Procedure Question 1

In Step 6, what is the voltage? (0 V, 4.7 V)

Procedure Question 2

In Step 6, the 7408 gate is _____ (enabled, disabled)

Step 7. Reconnect the clock source and apply clock pulses until the counter stops incrementing. Record the binary number in the digital output section of the table.

Procedure Question 3

In Step 7, what is the voltage of the voltage comparator after the up-counter has stopped incrementing? (0 V, 4.7 V)

Procedure Question 4

In Step 7, the 7408 gate is _____ (enabled, disabled)

Step 8. Repeat Steps 5 and 7 for each of the remaining analog input voltages listed in Table 11-6.

■ EXPERIMENT QUESTIONS

1. List the four primary circuits that are required to construct a counter-ramp ADC.
2. If an analog voltage of 7 V is applied to the ADC in Figure 11-19, how many clock pulses are required to make the conversion?

The counter-ramp ADC has one major flaw in its operation.

Problem

The operation of a counter-ramp ADC is adequate where the conversion-speed requirements are not fast. However, for faster-speed applications, it is too slow. For example, an 8-bit converter consists of eight flip-flops in the up-counter. The counter increments from 0 as the D/A comparator output is compared with the analog input on every successive count. If it were necessary for the counter to increment to its maximum number before the DAC output equaled the analog voltage, it would take 255 clock pulses for this to occur.

Successive-approximation register (SAR):

A type of analog-to-digital converter that requires the same number of clock pulses to operate as the number of digital bits being converted.

Solution

Analog-to-digital converters that operate at high speeds employ a circuit called a **successive-approximation register (SAR)** in place of the up-counter; it uses a method to narrow in on the unknown analog voltage much faster.

    Figure 11-20(a) shows a simplified block diagram of an ADC that uses the SAR instead of the counter. Its eight output lines D_0–D_7 cause the DAC to produce a different voltage, as shown in Fig. 11-20(b). These voltages will result if 10 V are supplied to the V_{REF} input line. Its operation is as follows:

1. When the START button is pressed, the SAR is reset on the negative edge of the pulse applied to the $\overline{WR}$ input.
2. The conversion is begun on the leading edge of the conversion pulse after the START button is released.
3. When the positive transition of the first clock pulse occurs, the SAR produces a high at its MSB output, D_7. This causes the DAC to produce an analog voltage that is one-half its maximum value.
4. If the DAC output is higher than the unknown analog voltage (analog V_{IN}), the SAR output returns low. If the DAC output is lower than the analog input voltage, the SAR leaves bit 7 high.
5. The second clock pulse causes the next lower bit, D_6, to produce a high. If it causes the DAC output to be higher than the analog input, it returns to a low. If not, the SAR leaves D_6 high.

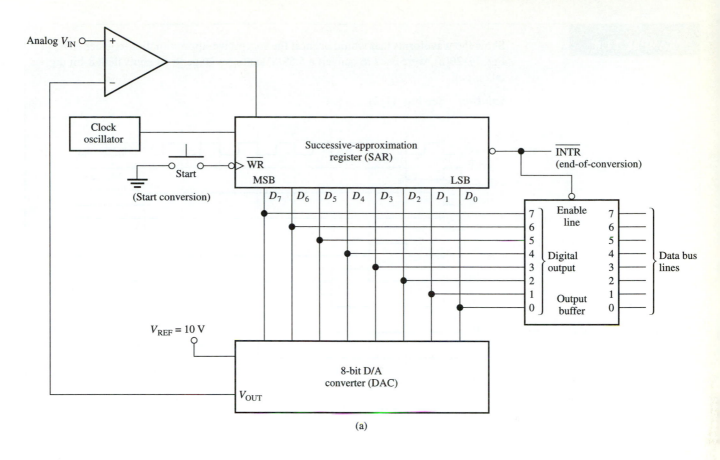

(a)

DAC input	DAC V_{OUT}
D_7	5.0000
D_6	2.5000
D_5	1.2500
D_4	0.6250
D_3	0.3125
D_2	0.15625
D_1	0.078125
D_0	0.0390625

(b)

FIGURE 11.20
(a) Simplified SAR ADC. (b) Voltage-level contributions by each successive-approximation-register (SAR) bit.

6. This process continues with the remaining six bits, D_5 to D_0.
7. At the end of the process, the SAR contains an 8-bit binary output that causes the DAC to produce an analog output voltage equal to the unknown analog input. This occurs at the end of the eighth clock pulse. The 8-bit binary number contained by the SAR that represents the analog input is present at the eight output lines.
8. At the moment the eight-step conversion process is complete, the end-of-conversion $\overline{INTR}$ lines goes low. Because the ADC outputs are often shared with other devices on a common data bus line, an 8-bit tristate buffer is often connected to the digital outputs. When low, the $\overline{INTR}$ signal is used to enable the buffer to pass the digital count of the ADC to the bus lines. When the $\overline{INTR}$ output is high, the buffer outputs go into a high-impedance state, which allows another device to use the data bus lines.

EXAMPLE 11.4

Show the waveforms that would occur if the successive-approximation-register ADC in Fig. 11-20(a), were used to convert a 5.59-V analog voltage to an equivalent 8-bit digital output.

Solution See Fig. 11-21.

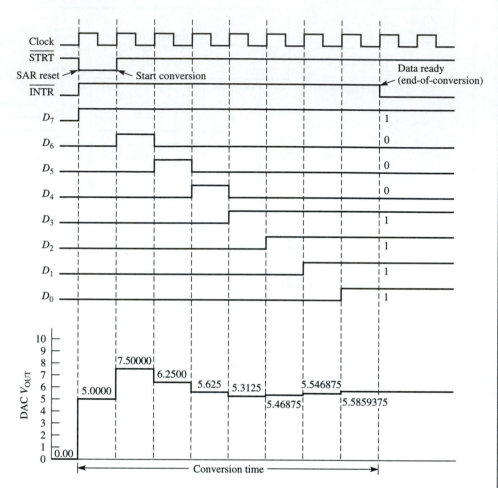

FIGURE 11.21
Timing diagram for a successive-approximation-register (SAR) ADC for Example 11.4.

The reason why the SAR is faster is because an 8-bit SAR only requires eight clock pulses to perform the process compared to an 8-bit counter-ramp ADC, which requires 255.

11.9 **INTEGRATED CIRCUIT ANALOG-TO-DIGITAL CONVERTER**

Figure 11-22 shows the block diagram of the ADC0804 analog-to-digital converter IC. The circuit shown is capable of converting an analog voltage into a proportional 8-bit digital output. The analog voltage range to be converted is determined by applying the desired maximum voltage to V_{DC}. For fine tuning, half of the V_{DC} voltage is applied to input $V_{REF/2}$. A slight voltage change at $V_{REF/2}$ will then bring the ADC into calibration if necessary. By

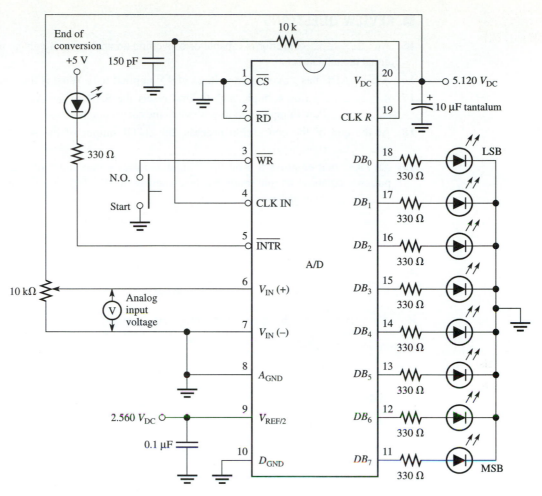

FIGURE 11.22
The block diagram of the ADC0804 analog-to-digital converter.

applying 5.12 volts to V_{DC} and 2.56 V to $V_{REF/2}$, the circuit is capable of converting an analog voltage connected across $V_{IN}(+)$ and $V_{IN}(-)$ ranging from 0 to 5.12 V. With eight output leads, there are 255 different analog voltage levels that are converted into digital outputs. Therefore, the resolution of this device is 0.39% ($\frac{1}{255} = 0.0039 = .39\%$). With 5.12 V as the maximum input voltage, each 0.02-V (5.12 V × 0.0039) increase causes the binary count to increase by 1.

The ADC0804 IC contains an internal clock. To operate, a resistor and capacitor are connected to the CLKR and CLK IN inputs. The ADC0804 IC also contains an 8-bit successive approximation register for its conversion process. The SAR is reset on a negative edge of a pulse applied to the $\overline{WR}$ input lead by the closure of the START push button. When the push button is released, the pulse applied to the $\overline{WR}$ input returns high and the conversion process begins. At the end of this process, which takes eight clock pulses, output $\overline{INTR}$ goes low. The eight outputs that represent the analog input voltage will be present at the active-high output lines DB0 to DB7. To continue updating the applied analog input voltage, the $\overline{INTR}$ pin is connected to the $\overline{WR}$ input line. By doing so, 5,000 to 10,000 conversions can be made per second.

The ADC0804 IC is a CMOS device that is designed to interface directly with some of the more popular 8-bit microprocessors. Therefore, some of its pins, such as $\overline{RD}$, $\overline{WR}$, $\overline{CS}$, and $\overline{INTR}$, correspond to leads of the similarly labeled microprocessors.

■ REVIEW QUESTIONS

15. An _____ converter is capable of converting an analog input voltage into a proportional digital number.

16. A 6-bit ADC has a maximum voltage of 9 V applied to it. What is the resolution?

17. A _____ (low-to-high, high-to-low) resets the SAR in the ADC0804 IC, and a _____ (low-to-high, high-to-low) starts the conversion process.

18. At the end of the conversion process, the $\overline{INTR}$ output of the ADC0804 IC goes _____ (low, high).

19. An ADC that employs a 6-bit up-counter requires a maximum of _____ clock pulses to complete its conversion process, whereas an ADC that uses a 6-bit SAR requires _____ clock pulses.

EXPERIMENT

IC Analog-to-Digital Converter

Objective

■ To convert analog signals to digital signals using a successive-approximation IC DAC.

Materials

1—Power Supply, 0 to 5.12 V DC

1—ADC0804 IC

1—10-kΩ Potentiometer

8—LEDs

8—330-Ω Resistors

1—10-kΩ Resistor

1—10 μF Tantalum Capacitor

1—150 ρF Capacitor

Introduction

Eight-bit ADCs in IC form require only eight clock pulses and use a register that performs a function called *successive approximation.* Figure 11-23 shows the block diagram of the ADC0804 analog-to-digital converter IC. It has an internal successive-approximation register (SAR). The circuit is capable of converting an analog voltage into a proportional 8-bit digital output. The voltage range to be converted is determined by applying the maximum desired voltage to V_{DC} at pin 20. The analog input is applied across $V_{IN}(+)$ and $V_{IN}(-)$. Because there are 256 (2^8) different binary digital numbers produced at the eight output leads, the analog voltage changes by .39% ($1/255 = 0.0039 = .39\%$) before one binary bit changes. This is called *resolution*.

The IC contains an internal clock. Its pulses are applied to the IC when an external resistor and capacitor are connected to the CLK R and CLK IN inputs. The conversion begins every time a low is applied to $\overline{WR}$. At the end of the conversion process, output line $\overline{INTR}$ goes low. Also, the eight outputs that represent the analog input voltage will be present at the active-high output lines DB_0 to DB_7. Connecting output line $\overline{INTR}$ to input $\overline{WR}$ will cause the IC to continually update itself by repeating the conversion process.

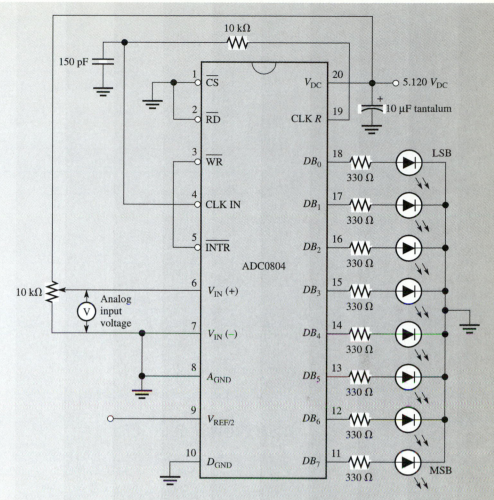

FIGURE 11.23
ADC circuit.

Procedure Information

The output voltage range of the ADC used in this experiment is 0 to 5.12 V. A variable analog voltage is applied to the input by a potentiometer. LEDs at the output leads show the converted binary number.

Step 1. Assemble the circuit shown in Fig. 11–23.

Step 2. Turn on the power supply.

Step 3. With one end of a jumper wire connected to ground, momentarily touch the other end to input $\overline{WR}$.

Step 4. Apply the analog voltage listed on the top line of the input section of Table 11-7.

Step 5. Fill in the output section to show the logic states of each digital output lead. Place a 1 in the column to indicate if the corresponding LED turns on, and 0s to indicate which LEDs do not light.

Step 6. Add all of the resolution values of each column that has a 1. Place the sum in the column labeled *Resolution Total*. Does the number equal the applied analog input voltage?

Step 7. Repeat Steps 5 and 6 for each of the remaining input analog voltages in Table 11-7.

TABLE 11.7
A/D converter test results

INPUT	OUTPUT								
	RESOLUTION VALUES								
Measured Analog Voltage	DB_7 2.56 V	DB_6 1.28 V	DB_5 0.64 V	DB_4 0.32 V	DB_3 0.16 V	DB_2 0.08 V	DB_1 0.04 V	DB_0 0.02 V	Resolution Total
0 V									
0.4 V									
1.0 V									
1.6 V									
2.3 V									
3.5 V									
4.6 V									
5.12 V									

Procedure Question 1

How should the sum of resolution values and the analog voltages compare to each other?

■ EXPERIMENT QUESTIONS

1. How many clock pulses are required by the ADC0804 IC to complete one analog-to-digital conversion?
2. The resolution of the ADC0804 IC is _____ %.
3. At the end of each analog-to-digital conversion, the $\overline{INTR}$ lead goes _____ (low, high).
4. An analog voltage of 1 V applied to the circuit in Fig. 11-23 will produce a digital output of _____ .

DIGITAL SIGNAL-GENERATING DEVICES

Digital components are capable of either recognizing or producing 1- or 0-state voltages that are in the form of square waves. The characteristic of a square wave is that it changes from one state to another very rapidly; the time duration that it is in one state or another is often irrelevant.

Digital signals can be produced by different types of devices:

logic gate devices

mechanical switches

wave-shaping devices

multivibrators

optical couplers

11.10 LOGIC-GATE DEVICES

Logic gate devices are described in Chapters 3 to 5. Based on what digital signals are applied to their inputs and what type of logic device it is, a certain 1- or 0-state digital signal is produced by its output.

11.11 MECHANICAL SWITCHES

Switch bounce:

Occurs when a mechanical switch is closed; it bounces several times before its contacts come to rest.

Mechanical switches are often used to provide input signals to digital circuits. Unfortunately, however, they have an undesirable characteristic called **switch bounce,** which occurs when a mechanical switch is opened or closed. For example, when a switch is closed, the movable part makes contact with a stationary fixed terminal. The electrical and mechanical connection is made. However, due to a spring action of the movable part of the switch, it bounces several times before coming to rest on the stationary terminal. The bouncing action can last as long as 40 milliseconds.

If a mechanical switch is used as a digital input device, the extra pulses cause a false signal. For example, a single-pole single-throw (SPST) switch is used to provide an input pulse to a 4-bit binary counter, as shown in Fig. 11-24(a). Suppose that the counter is cleared. When the switch closes, it generates several low pulses instead of one because of the bouncing action, which is graphically shown in Fig. 11-24(b). Instead of the counter incrementing once, it counts the number of times the switch bounces up and down.

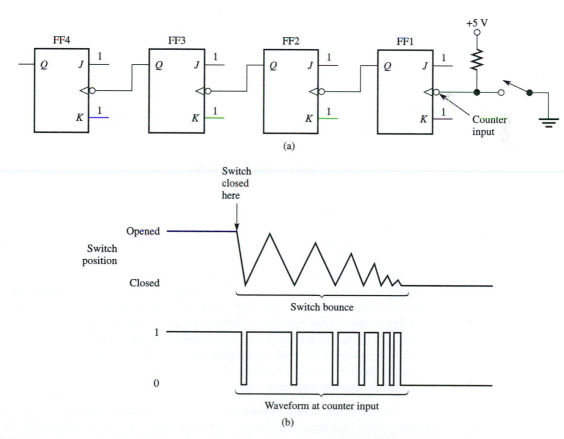

FIGURE 11.24
(a) Switch used as a clock input to a 4-bit binary counter. (b) Extraneous pulses caused by switch–contact bouncing.

Schmitt trigger:

A device that produces sharply defined square waves from distorted signals or sine waves.

There are several ways of eliminating the effects of switch bounce. To debounce a single-pole single-throw switch or push button, a device called a **Schmitt trigger** can be used. Figure 11-25 shows how a Schmitt trigger is used to debounce a switch.

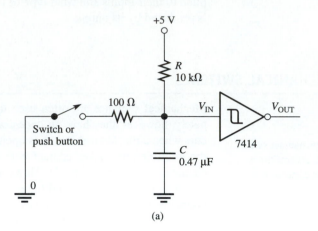

(a)

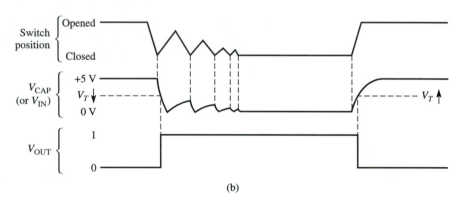

(b)

FIGURE 11.25
Debouncing a single-pole single-throw (SPST) switch or push button.

The Operation of Debouncing a SPST Switch

- When the switch is open, it keeps the capacitor charged to +5 V.
- The +5-V high input to the Schmitt trigger inverter is inverted to a (V_{OUT}) low.
- When the switch is closed, the capacitor discharges quickly to zero through the 100-Ω resistor. When the charge goes below the trigger voltage, $V_{T\downarrow}$, it causes the Schmitt trigger output to change to a high.
- As the switch bounces, the capacitor charges slowly when the contact is open and discharges rapidly when the contact closes. Because the RC time constant of the 10-kΩ resistor and the 0.47-μF capacitor is long, the capacitor does not have enough time to charge to the trigger voltage $V_{T\uparrow}$ before the switch bounces back to a momentary closure. Therefore, the output of the Schmitt trigger remains high.
- When the switch is opened, the capacitor charges. The moment it reaches $V_{T\uparrow}$, the high input at the Schmitt trigger causes it to switch to a low output.
- The capacitor continues to charge to +5 V.

The operation of this debouncing circuit results in producing a single pulse at the output even though the switch bounces several times. This is accomplished by the resistors and capacitor. The purpose of the Schmitt trigger can be observed by examining the waveforms in Fig. 11-25(b). When the switch closes, the discharging capacitor produces a gradual voltage increase. The Schmitt trigger increases the transition speed of the changing voltage in either direction as it crosses V_T (↓ or ↑) both ways.

The Operation of Debouncing a SPDT Switch

When a single-pole double-throw (SPDT) switch is used, it is necessary to use a different circuit configuration to perform the debouncing operation. Figure 11-26(a) shows a typical SPDT switch used in a circuit. A 0 V ground potential connected to the movable part of the switch is used as the input signal. The switch has three positions:

1. Position A
2. Position B
3. Between positions A and B while making the transition from one to another.

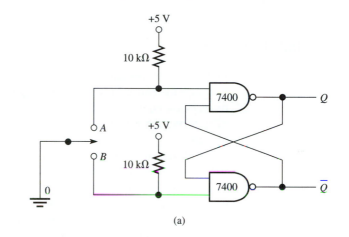

(a)

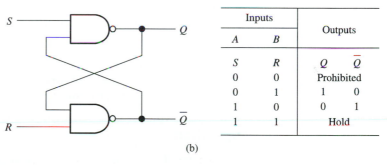

(b)

FIGURE 11.26
(a) Cross-NAND method of debouncing a single-pole double-throw (SPDT) switch. (b) Cross-NAND latch and its truth table.

To perform the debouncing operation of a single-pole double-throw switch, the cross-NAND circuit shown in Fig. 11-26(a) can be used. The Q and $\overline{Q}$ outputs of the flip-flop circuit become the bounceless output terminal of the switch. It operates like the R–S latch described in Chapter 7. The truth table for an R–S latch is provided in Fig. 11-26(b) to help illustrate the operation.

■ Assume that the switch is connected to the input A terminal.
   —A 0-state ground is felt at input A.
   —A 1 state is felt at input B because the +5-V potential is applied through the 10-kΩ resistor.
   —According to the truth table, the Q output is high and $\overline{Q}$ is low.

- When the switch is being moved from position *A* to *B,* both inputs observe 1 states applied through both 10-kΩ resistors. According to the truth table, the *R–S* flip-flop is in the hold mode, so the *Q* and $\overline{Q}$ output states remain.
- The movable part of the switch makes initial contact with terminal *B.*
  - –A 0-state ground is felt at input *B.*
  - –A 1 state is felt through the 10-kΩ resistor at the unconnected *A* input.
  - –According to the truth table, the *Q* output changes to a low and $\overline{Q}$ goes high.

The action of the NAND latch flip-flop outputs changing states is initiated by the first contact closure pulse. However, the switch contacts bounce several times before coming to rest. The debouncing operation takes place because every time the switch bounces off terminal *B,* the flip-flop is in the hold mode because both *A* and *B* inputs are disconnected from ground. Because of the regenerative behavior of the NAND latch circuit, the switch action once initiated continues even with the loss of switch closure.

11.12 WAVE-SHAPING DEVICES

Logic gate devices and flip-flops require that the waveforms applied to their inputs change very rapidly. However, when digital signals are transmitted long distances, they are sometimes distorted or pick up noise. Therefore, it is sometimes necessary to use some means of restoring the waveforms back to a pure square wave before they are applied to digital circuity.

Waveform reconditioning is performed by a Schmitt trigger, which develops sharply defined square waves by utilizing positive feedback internally to speed up level transitions. It also utilizes an effect called *hysteresis,* which means that the switching threshold on a positive-going input signal is at a higher voltage level than the switching threshold on a negative-going input signal. Schmitt triggers can also be used to transform the following waveforms into rectangular-shaped signals:

- A low-voltage AC wave.
- Signals with slow rise times such as those that are produced from charging and discharging capacitors and temperature-sensing transducers.

They are also used for voltage-level restoration. Figure 11-27 illustrates the switching action of a Schmitt trigger inverter and also shows how the hysteresis characteristics recondition a distorted square wave.

Operation

Time Period 1   A logic 0 is recognized at the input, and a 1 state is generated at the inverting output.

Time Period 2   A logic 1 at the input is recognized if the input voltage exceeds the 1.7-V positive-going threshold level that causes the output to snap to a logic-0 value. Note the ragged spike on the input signal caused from noise drops below 1.7 V into the hysteresis region during period 2. The output does not change unless the input drops below the 0.9-V negative-going threshold level.

Time Period 3   A logic 0 at the input is recognized if the voltage drops below the 0.9-V negative-going threshold level, which causes the output to snap to a logic 1 value. Note that a noise spike on the input signal rises above 0.9 V into the hysteresis region during time period 3. The output does not change unless the input reaches the 1.7-V positive-going threshold level.

The logic symbol for a Schmitt trigger inverter is shown in Fig. 11-28(A). It includes a miniature hysteresis waveform inside the symbol to indicate that it is a Schmitt trigger

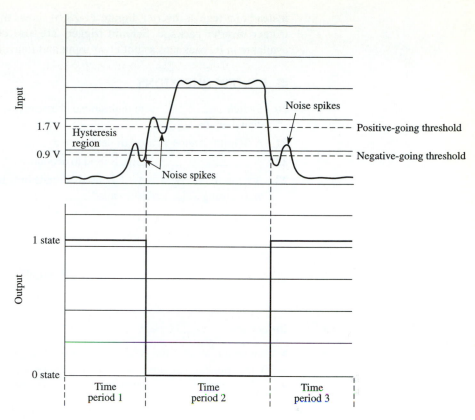

FIGURE 11.27
Schmitt trigger switching action.

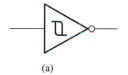

(a)

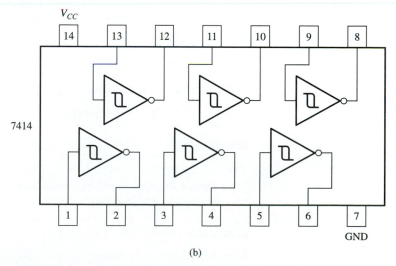

(b)

FIGURE 11.28
Schmitt trigger: (a) Logic symbol. (b) Pin diagram.

instead of a regular inverter. Figure 11-28(b) shows the pin diagram of a TTL hex Schmitt trigger inverter package. Schmitt triggers are also constructed as NAND gates and are available in IC packages as both two-input and four-input gate devices.

■ REVIEW QUESTIONS

20. Switch _____ is an undesirable characteristic of most mechanical switches used for digital inputs.

21. A Schmitt trigger is used to debounce a _____ switch.

22. A _____ _____ is used to debounce a SPDT switch.

23. A _____ _____ is used to transform nonrectangular waveforms into square waves using a characteristic called _____ .

EXPERIMENT

Schmitt Trigger

Objectives

■ To wire an inverter and NAND gate Schmitt trigger.
■ To determine the positive and negative threshold voltages.
■ To demonstrate how a sine wave signal is converted to a square wave.

Materials

  1—7414 IC

  1—74132 IC

  1—7476 IC

  1—LED

  1—1N4006 Diode

  1—Signal Generator (Sine Wave)

  1—150-Ω Resistor

  1—470-Ω Resistor

  1—1-kΩ Resistor

  1—1-kΩ Potentiometer

  1—Dual Trace Oscilloscope

  1—+5-V DC Power Supply

Introduction

The Schmitt trigger is available in IC form. It performs the dual function of operating as a logic device and as a waveform conditioner. The 7414 Schmitt trigger IC operates as an inverter, and the 74132 Schmitt trigger operates as a NAND gate.

Procedure

Step 1. Assemble the circuit shown in Figure 11-29(a).

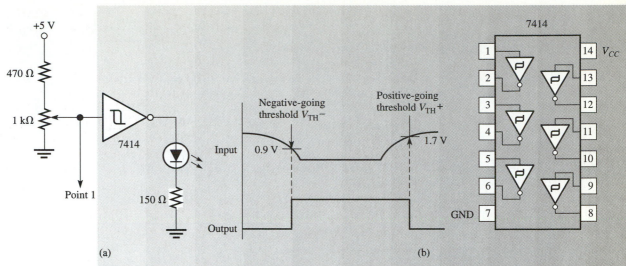

FIGURE 11.29

Background Information

Part (a) of Fig. 11-29 shows the waveform of a signal applied to the input of the Schmitt trigger inverter. As the voltage gradually decreases to 0.9 V, the *negative-going threshold voltage* ($V_{TH}-$) is reached and causes the inverter output to abruptly switch from low to high. When the input voltage is increased to 1.7 V, the *positive-going threshold voltage* ($V_{TH}+$) of the Schmitt trigger is reached. The output of the inverter rapidly switches from high to low.

Note that the Schmitt trigger switches at one voltage level when going high and at a different voltage level when going low. This characteristic is caused by a storage of energy in the device. The asymmetrical response is referred to as *hysteresis*. The Schmitt trigger has a rectangle with extended lines inside its symbol. This shape indicates the hysteresis properties of the device.

Step 2. Turn on the oscilloscope. Adjust the vertical range switch of the scope to measure a DC voltage range between 0 and 5 V.

Step 3. Connect the channel A lead to *Point 1* of the circuit in Figure 11-29(a).

Step 4. Apply the power to the circuit. Adjust the potentiometer so that the wiper arm is in the position that provides the highest voltage at *Point 1*. The LED should be off.

Step 5. Slowly adjust the potentiometer so that the voltage is reduced. The negative-going threshold input voltage is reached when the LED turns on. Observe the oscilloscope and record the $V_{TH}-$ voltages in Table 11-8.

TABLE 11.8

Point 1	
$V_{TH} - =$	_____ V_{DC}
$V_{TH} + =$	_____ V_{DC}

Step 6. Slowly adjust the potentiometer so that the voltage at *Point 1* is increased. The positive-going threshold input voltage is reached when the LED turns off. Observe the oscilloscope and record the $V_{TH}+$ voltage in Table 11-8.

Step 7. Assemble the circuit shown in Fig. 11-30(a).

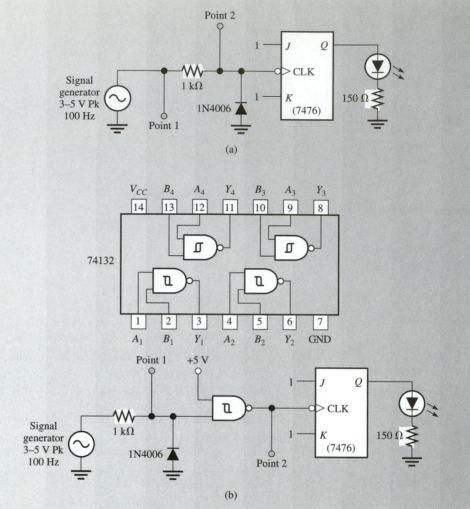

FIGURE 11.30

Background Information

A very useful type of instrument used for testing electronic circuitry is a digital frequency counter. This device is often used to count the number of sine waves that occur every second. A frequency counter uses cascaded flip-flops to form an up-counter to count and display the number of sine waves. Before being applied to the clock input of the first flip-flop in the counter, the sine wave is conditioned by a Schmitt trigger so that it becomes a square wave.

Step 8. Turn on the power.

Step 9. Connect channel A of a dual trace oscilloscope to *Point 1* and channel B to *Point 2* of the circuit.

Step 10. Adjust the signal generator to produce a sine wave with a frequency of 100 Hz and a peak amplitude of 3 to 5 V. Adjust the frequency range switch on the oscilloscope to read a frequency of 100 Hz.

Step 11. Display the input waveforms on channels A and B of the oscilloscope. Draw the waveforms in the top portion of the chart provided in Table 11-9, and indicate whether the flip-flop is toggling.

TABLE 11.9

Waveform	At Point 1	At Point 2	Is the Flip-Flop Toggling? (Yes, No)
Circuit (A)			
Circuit (B)			

Procedure Question 1

Should the flip-flop toggle? Why?

Background Information

The sine wave is partially conditioned by the diode, which removes the negative alternation. The positive alternation is left to be applied to the flip-flop.

Step 12. Assemble the circuit shown in Fig. 11-30(b).

Step 13. Repeat Steps 9 and 10. Display the waveforms on channels A and B of the oscilloscope. Draw the waveforms in the bottom portion of Table 11-9, and indicate whether the flip-flop is toggling.

■ EXPERIMENT QUESTIONS

1. A Schmitt trigger has a dual function. What are the functions?

2. Which types of the following electrical signals will a Schmitt trigger convert to a square-shaped waveform?
   (a) Sine waves
   (b) Sawtooth waves
   (c) Distorted square waves
   (d) All of the above

3. Rectangular-wave-shaped signals are especially necessary for _____ (level, edge)-triggered flip-flops.

4. The negative-going threshold voltage of a Schmitt trigger inverter occurs when its output goes _____ (low, high)-to-_____ (low, high).

5. A rectangle with extended lines inside a symbol for a Schmitt trigger indicates the _____ properties of the device.

6. Explain why a Schmitt trigger is needed by a digital frequency counter.

11.13 MULTIVIBRATORS

Multivibrator:

A device that has two complementary outputs that produce rectangular output pulses.

Multivibrators are devices that have two complementary output leads, called Q and $\overline{Q}$, from which rectangular output pulses are produced. There are three main types of multivibrator devices:

1. Bistable
2. Astable
3. Monostable

Bistable Multivibrators

Bistable multivibrator:

Also known as a flip-flop, the circuit produces one logic level signal until an input signal causes it to produce the opposite logic level output signal.

A **bistable multivibrator** is another name for the flip-flop described in Chapter 7. A large percentage of square wave signals received by flip-flops is provided by the outputs of other bistable multivibrators. For example, most flip-flops in counters receive their clock signals from adjacent flip-flops. The other two types of multivibrators are produced by a linear integrated circuit specifically designed for timing applications.

The 555 IC Chip

One of the most popular and versatile linear integrated circuit chips is the 555 monolithic IC. It was originally developed by the Signetics Corporation as an 8-pin DIP device. It is also available in a dual version, the 556 IC, which contains two independent 555 ICs on a 14-pin DIP IC package. A pin diagram of a 555 chip is shown in Fig. 11-31. When a minimal number of external resistors and capacitors are connected to various pins of the 555 IC, it operates as an astable or monostable multivibrator. It can also be used as a switch component device called a *debouncing* switch, which is used to provide single pulses for digital circuitry.

FIGURE 11.31
555 IC package.

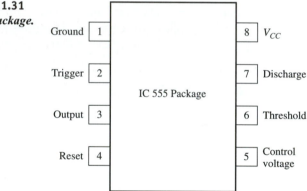

Figure 11-32 shows a schematic diagram of the 555 IC. It consists of the following sections:

1. A voltage-dividing network—R_1, R_2, and R_3
2. Two voltage comparators
3. An R–S flip-flop
4. An NPN transistor
5. An output buffer

Voltage-Divider Network: Resistors R_1, R_2, and R_3 are all 5 kΩ. They form a voltage divider that biases the inverting (−) input of comparator A at 2/3 the power supply voltage (3.33 V), and the noninverting (+) input of comparator B at 1/3 the power supply voltage (1.65 V).

Voltage Comparators: Each comparator has one of its inputs connected to an external pin. The noninverting input of comparator A is connected to external pin 6, called the *threshold terminal.* The inverting input of comparator B is connected to external pin 2, called the *trigger terminal.* The output of comparator A is low if the voltage at the threshold terminal is lower than 3.33 V. The output of comparator B is low if the voltage at the trigger terminal is greater than 1.65 V. The logic levels at the comparator control the flip-flop.

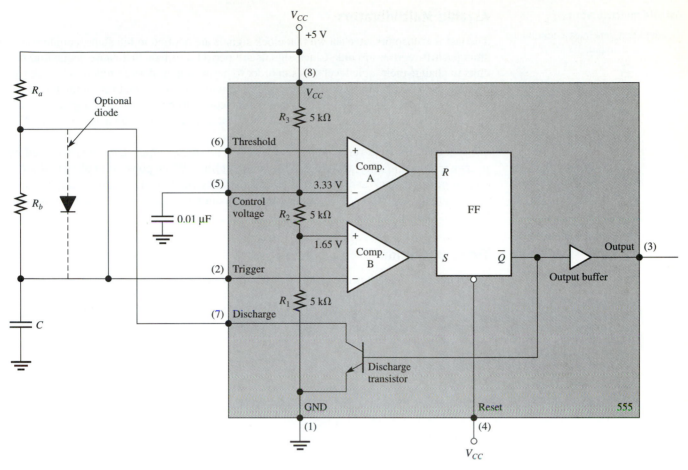

FIGURE 11.32
Schematic diagram of an astable 555 timer.

R–S Flip-Flop: The output of comparator A is connected to the *R* input of the flip-flop, and the output of comparator B is connected to the *S* input. The outputs of the two comparators are never on simultaneously. Only output $\overline{Q}$ of the *R–S* flip-flop is used. The $\overline{Q}$ lead is connected to the base of the transistor and the input of the output buffer. When the output of comparator A goes high, it causes the flip-flop to reset, generating a high at the $\overline{Q}$ output. When the output of comparator B goes high, it causes the flip-flop to set, generating a low at output $\overline{Q}$.

Transistor: The NPN transistor operates like a switch. When the $\overline{Q}$ output of the flip-flop is a logic high, the transistor turns on, acting like a closed switch. When the $\overline{Q}$ output is low, the transistor turns off.

Output Buffer: The function of the output buffer is to produce a high current output to provide a sufficient signal for external circuitry. The buffer goes low when $\overline{Q}$ is high, and goes high when $\overline{Q}$ is low, because it is an inverting amplifier.

The 555 IC has been specifically designed for timing applications. It produces a square wave with the following characteristics:

1. High-output-current capability
2. Can drive both TTL and CMOS
3. Excellent accuracy and temperature stability
4. Generates a wide range of frequencies
5. Stable under temperature and power supply fluctuation conditions

Astable multivibrator:

A circuit that generates a continuous square wave output.

Astable Multivibrators

The last few chapters have shown that clock signals are applied to flip-flops, counters, and storage/shift-register circuits. Clock signals are periodic signals that cause sequential circuits to change their logic level states in order to perform their desired function. Clock signals are the heartbeat in computer devices and provide the pulses that time and control the proper synchronizing of all events throughout the computing system. Clock signals that are applied to the input leads of flip-flops and sequential circuits are frequently generated by astable multivibrators.

The astable (free-running) multivibrator has no stable output states. It is triggered by its own internal circuitry; therefore, it has no input lines. When power is applied to this device, it switches back and forth at a desired rate between two states, producing a square wave signal at the output. Devices that produce continuous square wave outputs are also known as **timing circuits.**

Timing circuit:

A device that produces a continuous square wave output in a digital-type circuit.

Timer Operation

To operate as a timer, it is necessary to connect three external components to the pins of the 555 IC. As shown in Fig. 11-32, these include two resistors (R_a and R_b) and one capacitor (C). The figure also shows two jumper wires and a coupling capacitor, which shunts any unwanted noise to ground.

Assume that

The capacitor is discharged.

Comparator A output is low.

Comparator B output is high.

Flip-flop $\overline{Q}$ output is low.

Transistor is off.

Therefore,

When power is applied to the circuit, current flows through the RC network of R_a, R_b, and C. When the capacitor charges to 1.66 V, this potential is felt at the trigger input (2) and causes the comparator B output to go low.

When the capacitor charges to 3.34 V, it is felt at the threshold input (6) and comparator A goes high.

With a low at flip-flop input S, and a high at input R, the $\overline{Q}$ output goes high.

A high at $\overline{Q}$ causes the output line of the output buffer to go low.

A high at $\overline{Q}$ turns the transistor on, which allows the capacitor to discharge through the transistor and R_b.

When the charge on the capacitor goes less than 3.33 V, the threshold potential causes the comparator A output to go low.

When the discharging capacitor goes less than the 1.65 V, the trigger input causes comparator B to go high.

When comparator A output is low and comparator B output is high, the flip-flop $\overline{Q}$ output goes low.

A $\overline{Q}$ output causes the output line of the output buffer to go high.

A low turns the transistor off, which opens the discharge path of the capacitor and starts the charging phase of the next cycle.

The rate at which the ICs internal components turn on and off is determined by the values of the external components connected to the IC.

The frequency of the output can be determined by

$$f = \frac{1.44}{(R_a + 2R_b)C}$$

Figure 11-33 includes a graph for quick reference to determine which combinations of resistance and capacitance values generate a desired frequency.

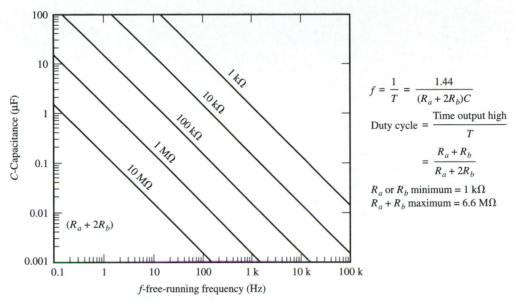

$$f = \frac{1}{T} = \frac{1.44}{(R_a + 2R_b)C}$$

$$\text{Duty cycle} = \frac{\text{Time output high}}{T}$$

$$= \frac{R_a + R_b}{R_a + 2R_b}$$

R_a or R_b minimum = 1 kΩ
$R_a + R_b$ maximum = 6.6 MΩ

1. Select the desired frequency at the bottom of the chart.
2. Follow this frequency line upward until it intersects an available capacitor value.
3. Read the combined resistance value for $R_a + 2R_b$ from the nearest diagonal line.

FIGURE 11.33
Graph for determining **R** *and* **C** *values for a desired frequency using the 555 astable multivibrator.*

A few rules must be observed when selecting values for the 555 astable multivibrator:

1. The maximum $R_a + 2R_b$ resistance is about 6.6 Megaohms.
2. The minimum R_a or R_b resistance value is 1 kΩ.
3. Resistor R_b can be replaced with a rheostat and a fixed 1-kΩ resistor in series to provide variable-frequency capabilities.
4. The value of the timing capacitor can vary from a few hundred picofarads to more than 1000 μF.

Miscellaneous Rule: Place a 0.01-μF capacitor from pin 5 to ground.

EXAMPLE 11.5

Assume that there are some 100 Ω, 1-kΩ, and 5-kΩ resistors and a 1-μF capacitor available. Use Fig. 11-33 to determine which combined R + 2R resistance value can be connected to the 555 IC to generate an approximate 100-Hz square wave frequency.

Solution

1. Find the 100-Hz frequency at the bottom of the chart.
2. Follow the 100-Hz vertical line upward to where it intersects with the μF capacitor.
3. Read the values for $R_a + 2R_b$ from the nearest diagonal line.

Answer: The 10-kΩ diagonal line. Use a 5 kΩ resistor for R_a, and a 5 kΩ resistor for R_b.

Duty cycle:

The ratio of time a square wave signal is high to the total time period of one cycle.

Initially, the external capacitor charges through R_a and R_b and then discharges through R_b. These charging and discharging times affect what is called a **duty cycle.**

Duty Cycle: The duty cycle of a 555 timer astable multivibrator is the ratio of time the output terminal is high to the total time period of one cycle. The duty cycle is set precisely by the ratio of these two resistors.

The charging time (output buffer is high) is T_1. The discharging time (output buffer is low) is T_2. The total period of time for one cycle is T. These values are calculated as follows:

$$T_1 = 0.693(R_a + R_b)C$$
$$T_2 = 0.693(R_b)C$$
$$T = T_1 + T_2 = 0.693(R_a + 2R_b)C$$

The duty cycle is

$$\text{DC} = \frac{T_2}{T} \quad \text{or} \quad \text{DC} = \frac{R_b}{R_a + 2R_b}$$

Because the capacitor charges up through R_a and R_b and then discharges only through R_b, the duty cycle is always less than 50%, as shown in the top waveform of Fig. 11-34. However, in most situations, a square wave with a duty cycle less than 50% fulfills the same requirements as a square wave with a 50% symmetrical duty cycle. For example, Fig. 11-34 shows two 1-Hz square waves with different duty cycles. If either one is providing a timing pulse to a clock circuit that is triggered by a negative-edge pulse, they both provide the same required signal. However, if it is desirable to have a square wave with a duty cycle of 50% or greater, a diode can be placed across resistor R_b with the cathode lead connected to *discharge* pin 7 and the anode lead to the *threshold* pin 6. Depending on the resistance ratio of R_a to R_b, this configuration allows the duty cycle to operate over a range of 5% to 95%.

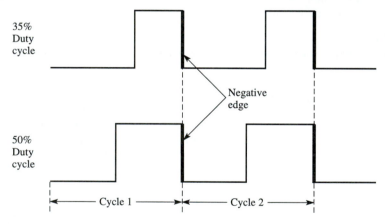

FIGURE 11.34
Two waveforms with different duty cycles providing negative-edge transitions at the same frequency.

EXPERIMENT

555 Clock Timer (Astable Multivibrator)

Objectives

■ To wire and operate an astable multivibrator.
■ To determine the timer frequency by mathematical calculations, by using a graph, and by using test instruments.

Materials

    1—+5-V DC Power Supply

    1—555 Linear IC

    2—10-kΩ Resistors

    1—2.2-kΩ Resistor

    1—22-kΩ Resistor

    1—47-kΩ Resistor

    1—1-μF Capacitor

    1—0.1-μF Capacitor

    2—0.01-μF Capacitor

Introduction

A circuit that produces a continuous rectangular signal is called an *astable* (*free-running*) *multivibrator.* An astable multivibrator, using a 555 timer chip, is shown in Fig. 11-35. With only two resistors and a capacitor, the 555 IC can produce a highly accurate rectangular signal.

FIGURE 11.35

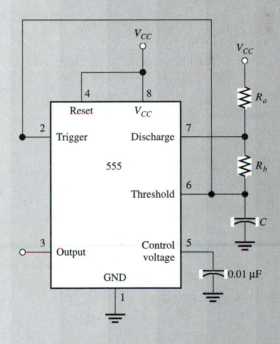

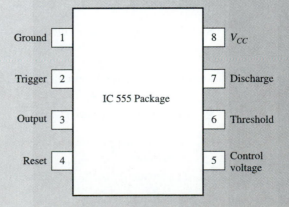

Procedure

Step 1. Assemble the circuit shown in Fig. 11-35. Use the component values listed on the first line of Table 11-10.

Background Information

The frequency of the timer depends on how fast the internal components of the 555 IC turn on and off. This rate is determined by the values of the components connected externally to the IC. The frequency of the output can be calculated by the following formula:

$$f = \frac{1.44}{(R_a + 2R_b)C}$$

Note: **The capacitor connected to pin 5 of the IC does not affect the frequency of the circuit. It is used as a precaution to pass any unwanted noise to ground.**

Step 2. Calculate f_{OUT} (frequency out) and record your calculation in Table 11-10.

TABLE 11.10

R_a (Ω)	R_b (Ω)	C_1 (μF)	$f = \frac{1.44}{(R_a + 2R_b)C}$ calculated (Hz)	f_{out} measured (Hz)	Chart Values (Hz)
47 k	10 k	1			
22 k	10 k	1			
22 k	10 k	0.1			
22 k	2.2 k	0.1			
10 k	10 k	0.01			

Step 3. Turn on the power. Using the oscilloscope, measure f_{out} at output pin 3 and record it in the data table.

Step 4. Repeat Steps 2 and 3 for the remaining values of R_a, R_b, and C_1 given in the table.

Procedure Information

Figure 11-33 is a quick reference chart that provides a way of determining which combination of external resistance and capacitance values generate a desired frequency. To use the chart,

1. Select the desired frequency on the bottom horizontal line on the chart.
2. Follow this frequency line upward until it intersects an available capacitor value.
3. Read the combined resistance value for R_a and $2R_b$ from the nearest diagonal line.

Step 5. Using the chart, find the astable multivibrator frequency for each of the R_a, R_b, and C_1 values listed in Table 11-10. Place the answer in the right column labeled *Chart Values.*

Step 6. Dismantle the circuit.

■ EXPERIMENT QUESTIONS

1. A circuit that produces a continuous rectangular signal is called a(n) _____ multivibrator.
2. The frequency of the 555 IC timer depends on what factors?

3. What activates the 555 IC timer?

4. What are clock signals used for?

5. Using the formula $f = 1.44/(R_a + 2R_b)C$, determine the frequency of a 555 IC timer when the component values are as follows:

$$R_a = 2 \text{ k}\Omega \qquad R_b = 15 \text{ k}\Omega \qquad C = 0.1 \text{ }\mu\text{F} \qquad f = \underline{\hspace{2cm}}$$

6. Suppose a frequency of 1100 Hz is desired and a .001 μF capacitor is available. Using the chart in Fig. 11-33, determine the required combined values of R_a and R_b.

Monostable Multivibrator

Monostable multivibrator:

Also known as a one-short, a circuit which produces a temporary logic level voltage after an activating signal is applied to its input.

The **monostable multivibrator** (also known as a one-shot multivibrator) is characterized as only having one stable state. Its primary output, Q, is normally 0. When a triggering signal is applied to its input, the Q output changes from its normal stable state to a logic 1 (unstable state) for a specified length of time before it automatically returns to its original stable state. The trigger signal is obtained from either a mechanical switch or from another circuit. The period of time it remains in its unstable state is determined by a combination of internal circuit delays of an IC and an external RC timing circuit. Therefore, it is independent of an input trigger-pulse width. Depending on the component values of the external resistor and capacitor selected, the output pulse generated can either be a longer pulse (stretching) or a shorter pulse (shortening) than the input pulse.

Monostable multivibrators are primarily used for producing timing-delay (one-shot) signals. They can also be used as pulse stretchers and bounceless switches and can reshape a *ragged* input pulse.

The 555 IC makes an ideal one-shot multivibrator, which is one reason for its design.

One-Shot Operation. To operate as a one-shot multivibrator, it is necessary to connect two external components to the pins of the 555 IC. These include one resistor (R_a) and one capacitor (C). Resistor R_a is connected from $+V_{CC}$ to the discharge (pin 7), and capacitor C is connected from the threshold (pin 6) to ground. Pins 6 and 7 are connected together by a jumper wire, and a noise-eliminating coupling capacitor is placed between pin 5 and ground. See Fig. 11-36. Its operation is as follows:

Assume that

- The capacitor is discharged.
- Comparator A output is low.
- Comparator B output is low.
- Flip-flop $\overline{Q}$ output is high.
- The transistor is on.
- The output buffer is low.
- A +5-V high is applied to the trigger input.

Therefore,

- While the trigger signal is brought from a high to a temporary 0-V potential by a push button closure, the comparator B output goes high. The comparator B output returns to a low when the push button is released.
- A low applied to the flip-flop's R input from comparator A and a temporary high applied to the flip-flop's S input from comparator B cause the flip-flop's $\overline{Q}$ output to go low.
- A low at the $\overline{Q}$ output of the flip-flop causes the buffer output to go high.
- A low at the $\overline{Q}$ output turns off the discharge transistor, which enables the capacitor to begin charging up toward $+V_{CC}$.

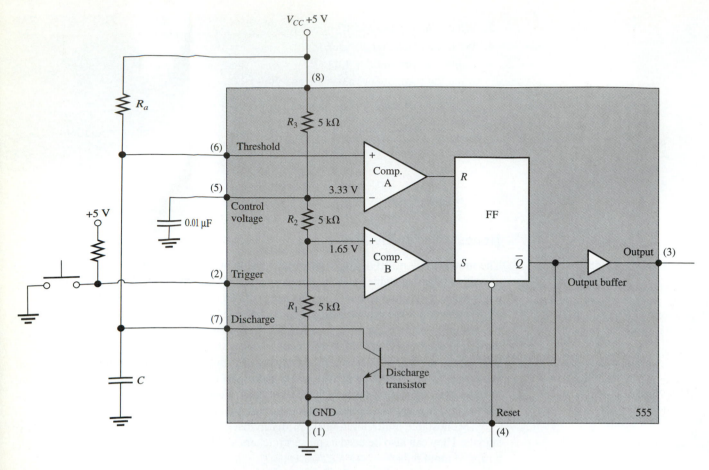

FIGURE 11.36
Schematic diagram of a 555 timer with the external timing components to form a monostable multivibrator.

- When the capacitor charges to 3.34 V, the comparator A output goes high.
- A high at the output of comparator A and a low at the output of comparator B causes the RS flip-flop to reset and develop a high at its $\overline{Q}$ lead.
- A high at the $\overline{Q}$ output causes the output buffer to go back to a normal low state, and the one-shot time duration is complete.
- The high $\overline{Q}$ output turns on the discharge transistor, which provides a discharge path for the capacitor.
- When the capacitor is discharged, the one-shot awaits another negative-going pulse at the trigger input.

The capacitor reaches a 3.34-V charge after 1.1 time constants. This time period determines the width of the output pulse of the one-shot. The time duration of the pulse is expressed in the following formula:

$$T = 1.1 \, RC$$

where T is the time in seconds, R is the resistance in ohms, and C is the capacitance in farads.

The one-shot pulse duration can range from microseconds to several minutes. Figure 11-37 provides a graph for quick reference to determine which combination of resistance and capacitance values generates a desired pulse duration (ranging from 10 μs to 100 s).

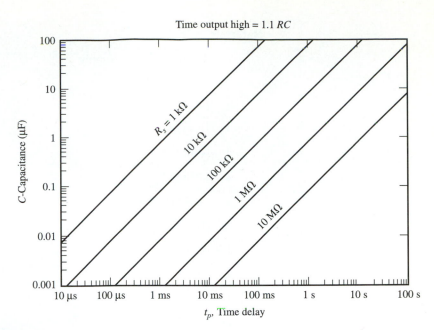

FIGURE 11.37
Graph for a 555 monostable multivibrator **R** *and* **C** *values.*

| EXAMPLE 11.6 | Assuming that a 10-kΩ resistor and 0.01-, 0.1-, and 1-μF capacitors are available, use Fig. 11-37 to determine which external components can be connected to the 555 IC to produce an approximate 1-ms one-shot pulse.

Solution Find 1 ms on the horizontal axis. Follow the 1-ms vertical line upward to where it intersects with a combination of an available capacitor (horizontal line) and resistor (diagonal line). The answer is a 0.1-μF capacitor and a 10-kΩ resistor. |

■ REVIEW QUESTIONS

24. The _____ multivibrator is also known as a flip-flop.
25. The _____ multivibrator has no stable state and, therefore, produces a continuous _____ when power is applied.
26. The _____ _____ of a square wave is the length of time an alternation is low versus the total period of the wave.
27. The _____ multivibrator temporarily changes to an unstable state before returning to its original state by itself.
28. The _____ linear IC is used with external components in the construction of the astable and monostable multivibrators.

EXPERIMENT

555 Clock Pulse (Monostable One-Shot Multivibrator)

Objectives

- To wire and operate a monostable multivibrator.
- To determine the pulse width of a monostable multivibrator by mathematical calculations, a graphical chart, or test instruments.

Materials

1—+5-V DC Power Supply

1—555 Linear IC

1—1-kΩ Resistor

1—10-kΩ Resistor

1—100-kΩ Resistor

1—470-kΩ Resistor

1—1-MΩ Resistor

1—.01-μF Capactitor

1—10-μF Capacitor

1—50-μF Capacitor

1—100-μF Capacitor

1—N.O. Push Button

1—Oscilloscope

Introduction

One type of signal often used by digital circuitry is a single clock pulse. A circuit that produces this type of signal is called a *monostable (one-shot) multivibrator*. Figure 11-38 shows a monostable multivibrator 555 linear IC. The output lead is normally resting at a low state. When a triggering signal is applied to its input, the output changes from its normal stable state to an unstable 1 for a specified length of time. It then automatically returns to its stable state. The trigger signal is obtained from either a mechanical switch or from another circuit.

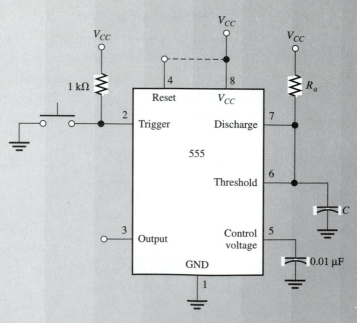

FIGURE 11.38
The 555 IC connected as a monostable multivibrator.

Procedure

Step 1. Assemble the circuit shown in Fig. 11-38. Use the R_a and C_1 component values listed on the first line of Table 11-11.

TABLE 11.11

R_a (Ω)	C_1 (μF)	$T = 1.1\,R_aC_1$ Calculated	T Measured	Chart Value
1 M	10			
470 k	10			
100 k	50			
10 k	100			
470 k	50			

Step 2. Connect channel A of an oscilloscope to input pin 2, and channel B to output pin 3.

Step 3. Quickly press and release the push button. Watch the oscilloscope to measure the length of time the one-shot is unstable and record it on the appropriate line of Table 11-11.

Background Information

The period of time the one-shot remains in its unstable state is determined by the values of the resistor (R_a) and the capacitor (C_1) externally connected to the 555 IC. The unstable state produced at the output terminal occurs over a time period of 1.1 time constants. Therefore, the time duration of the one-shot pulse can be found by using the following formula:

$$T = 1.1\,RC$$

where:

T is in seconds
R is in ohms
C is in farads

Step 4. Using the formula $T = 1.1RC$, calculate the length of time the output is in the unstable state and record it on the appropriate line of Table 11-11.

Step 5. Repeat Steps 3 and 4 for the values of R_a and C_1 given on each line of Table 11-11.

Procedure Information

Figure 11-37 is a quick reference chart that provides a way of determining which combination of external resistance and capacitance values generate a desired pulse width. To use the chart:

1. Select the desired time delay on the bottom horizontal line of the chart.
2. Follow this time delay line upward until it intersects an available capacitor value.
3. Obtain the required resistor value from the nearest intersecting diagonal line.

Step 6. Using the chart, find the one-shot time delay for each of the R_a and C_1 values listed in Table 11-11. Place the answer in the column labeled *Chart Value*.

11.14 OPTICAL ENCODERS

Digital circuits are often used to control the physical positioning of mechanical equipment. For example, a digital circuit may be used to turn a rotary device, such as a mechanical shaft, a given number of degrees. When the desired position is reached, data must be fed to a conversion device that makes the rotation stop. Because circular rotation is often divided into angular sectors, an encoder converts the analog value that represents the sectors into binary data. One type of encoding device used to convert angular position into binary data is an **optical encoder.**

The position-detecting section of an optical encoder is shown in Fig. 11-39(A). It consists of three major elements: a light source, an optical disc, and a light sensor.

The light source can be an incandescent lamp or a light-emitting diode (LED). The optical disc is usually made of plastic or glass with opaque and translucent areas. The light sensor is usually a phototransistor with a load resistor, as shown in Fig. 11-39(B). When the phototransistor base is struck by light, it turns on and acts like a closed switch. Therefore, 0 V is dropped across the phototransistor and +5 V across the load resistor. When light does not shine on the base, the phototransistor turns off and acts like an open switch. Therefore, the entire +5-V supply is dropped across it and 0 V across the load resistor.

The disc is mounted on the motor shaft between the light source and the four light sensors. As the disc rotates, light from the source passes through the translucent areas and is blocked by the opaque areas. Therefore, the absence or presence of light at the transistor base generates a logic +5-V 1 state or a 0-V 0 state, respectively, across the output. Optical encoders are often used because they are immune to noise problems.

Figure 11-40 shows an encoding disc with four tracks of opaque and translucent areas that represent the binary number system. By using 4-bit numbers, the binary count of 0000 through 1111 is possible. If each number represents a sector of equal size, the disc is divided into sixteen $22\frac{1}{2}°$ sections.

Using a pure binary encoder sometimes presents a problem, however. If all the bits do not change at exactly the same instant when the encoder crosses over from one sector to the next, a false number may be temporarily detected. For example, when going from the seventh sector to the eighth, all four bits on the encoder must change from 0111_2 to 1000_2 exactly at the same time. However, if the most significant bit changes slightly earlier than the other three bits, the encoder output would temporarily read 1111_2. This represents an 180° error of the encoder position. Because it is extremely difficult to construct a disc with enough precision to prevent such slight differences in the bit-switching times, another number code, called a **Gray code,** is frequently used.

Optical encoder:

A circuit that uses opto couplers to sense the positional location of a linear or rotary device.

Gray code:

A number system that uses multibit 0s and 1s and only one of the bits changes when incrementing or decrementing the count.

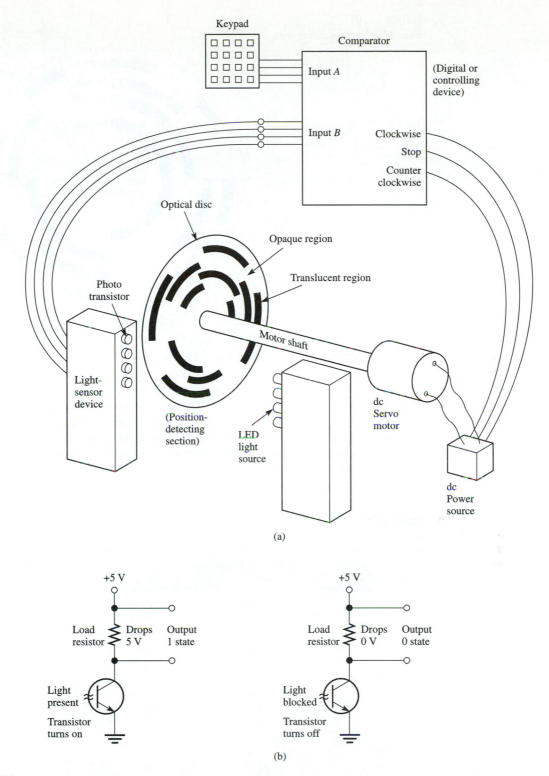

(a)

(b)

FIGURE 11.39
(a) Position-detecting device using an optical encoder. (b) How the phototransistor turns on when light is present and off when light is blocked.

FIGURE 11.40
Four-bit binary encoding disc.

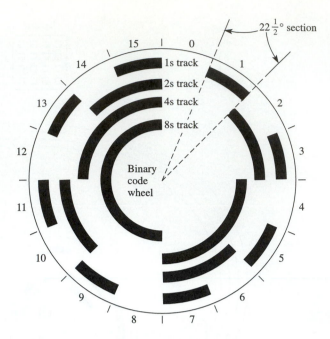

The Gray code is a system that also uses 1s and 0s. Table 11-12 shows a comparison between the standard binary number system and the Gray code. When counting is done in the standard binary system, it is not uncommon to have more than one bit change at a time. The Gray code, however, is designed so that only one bit changes at a time. This characteristic makes the Gray code system an ideal choice for encoder discs. Even with a disc that lacks precision, if the bit changes too soon or too late, the switching-time error is insignificant. A 4-bit Gray code encoder disc divided into 16 sectors is shown in Fig. 11-41. To operate as a 4-bit optical encoding device, four separate light sources with four separate phototransistors are used for each of the four tracks.

TABLE 11.12
Gray code

DECIMAL	GRAY COLUMNS ABCD	BINARY $2^3 2^2 2^1 2^0$
0	0000	0000
1	0001	0001
2	0011	0010
3	0010	0011
4	0110	0100
5	0111	0101
6	0101	0110
7	0100	0111
8	1100	1000
9	1101	1001
10	1111	1010
11	1110	1011
12	1010	1100
13	1011	1101
14	1001	1110
15	1000	1111

When the Gray code is used by digital systems, it is usually necessary to convert the Gray code numbers to pure binary numbers.

FIGURE 11.41
Four-bit Gray code encoding disc.

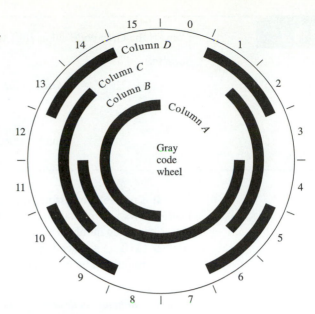

Gray-Code-to-Binary-Code Conversion

One method used to convert a Gray code number to its equivalent binary value is shown in Fig. 11-42. This table is used to convert 4-bit Gray numbers to 4-bit standard binary numbers. Similar tables can be used to make the conversion of numbers that contain a smaller or larger number of bits than 4.

Binary number system	2^3	2^2	2^1	2^0
Gray number system	Column A	Column B	Column C	Column D
Gray section	1	1	1	0
Binary section	1	0	1	1

FIGURE 11.42
Gray-code-to-binary-code conversion table.

Steps

1. Enter the Gray code number to be converted into the Gray section of the table. The number 1110 is entered. From Table 11-12, which compares the standard binary system and the Gray code, note that the most significant bit is the same for both.
2. Because the MSB of the Gray code number is 1, place another 1 in column A of the binary section of the table.
3. Exclusive-OR the bit in column A of the binary section with the bit in column B of the Gray section. Place the result in column B of the binary section of the table.
4. Exclusive-OR the bit in column B of the binary section with the bit in column C of the Gray section. Place the result in column C of the binary section of the table.
5. Exclusive-OR the bit in column C of the binary section with the bit in column D of the Gray section. Place the result in column D of the binary section of the table.

The equivalent standard binary number 1011_2 is generated from the Gray code number of 1110_{Gray}.

EXAMPLE 11.7

Convert the Gray code value 1001 to an equivalent binary number.

Solution

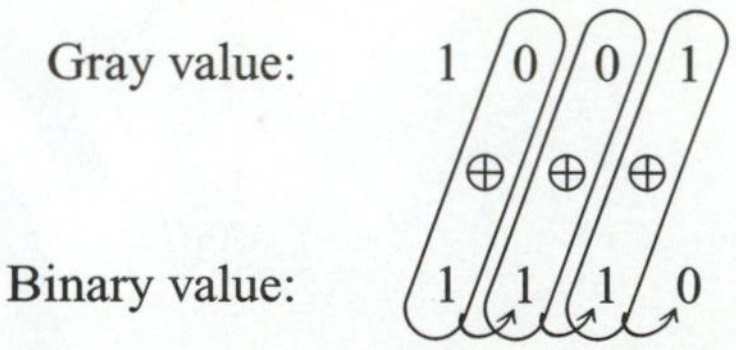

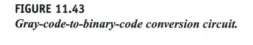

Gray value: 1 0 0 1

⊕ ⊕ ⊕

Binary value: 1 1 1 0

It is possible to perform the same conversion process by using logic circuitry. Figure 11-43 shows a combination circuit that consists of three exclusive-OR gates that convert 4-bit Gray code numbers to equivalent 4-bit standard binary numbers. As the Gray code number 1110_{Gray} is placed into the register on the left, the figure illustrates how the 1s and 0s are manipulated by each gate to generate 1011_2 at the output.

FIGURE 11.43
Gray-code-to-binary-code conversion circuit.

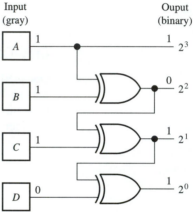

Input (gray)		Ouput (binary)

Binary-Code-to-Gray-Code Conversion

Some applications require that binary numbers be converted to equivalent Gray code values. Figure 11-44 shows how this process is performed.

Binary number system	2^3	2^2	2^1	2^0
Gray number system	Column A	Column B	Column C	Column D
Binary section	1 ⊕ 0	⊕ 1	⊕ 1	
Gray section	1	1	1	0

FIGURE 11.44
Binary-code-to-Gray-code conversion table.

Steps

1. Enter binary 1011 number into the binary section of the table.
2. Since the MSB of the standard binary numbers is 1, the same value is placed in column A of the Gray section.
3. Exclusive-OR the binary bits of columns A and B and place the result in column B of the Gray code section.

4. Exclusive-OR the binary bits of columns B and C and place the result in column C of the Gray code section.

5. Exclusive-OR the binary bits of columns C and D and place the result in column D of the Gray code section.

The equivalent Gray code number 1110 is generated from the standard binary value of 1011.

EXAMPLE 11.8

Convert the binary number 1010 to its Gray code equivalent.

Solution

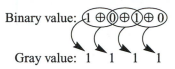

Binary value: 1 ⊕ 0 ⊕ 1 ⊕ 0

Gray value: 1   1   1   1

The binary-to-Gray conversion function can be performed by a combination circuit consisting of three exclusive-OR gates, as shown in Fig. 11-45. The placement of 1s and 0s at the inputs and outputs of the gates shows how the binary number 1011 is converted to its equivalent Gray code number 1110 by this circuit.

FIGURE 11.45
Binary-code-to-Gray-code conversion circuit.

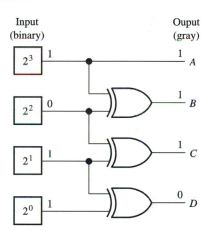

11.15 TROUBLESHOOTING A D/A CONVERTER

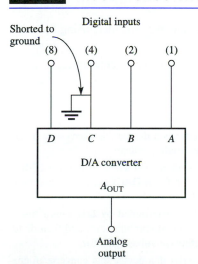

FIGURE 11.46
Block diagram of a D/A converter.

Figure 11-46 shows the block diagram of a DAC. It consists of four digital inputs that receive their data from a 4-bit mod-16 binary up-counter. It is assumed that the DAC is not operating properly.

The first test that should be performed is a dynamic test.

Action: Unless the counter continuously increments and recycles in the circuit it operates in, a signal generator should be connected to the counter input so that it does. By connecting an oscilloscope to the DAC output, the analog signal is observed. The waveform should look like the "staircase" sample shown in Fig. 11-47(a), where the output increases step-by-step as the binary input increments from 0 to 15.

Results: The output waveform shown in Fig. 11-47 (b) appears on the scope.

Conclusion: The staircase does not increment from 0 to 15 as it should. Instead, the waveform reveals that as the binary count increments, the positional value of 4_{10} (0100_2) never goes to 1. Therefore, the

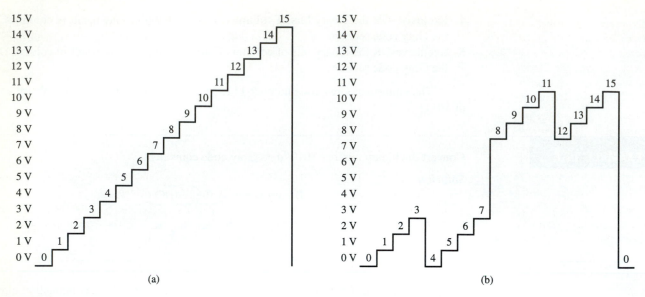

FIGURE 11.47
(a) Proper output of a DAC. (b) The DAC output when input C is shorted to ground.

fault appears to be associated with input *C* of the DAC. Further examination shows that input *C* is externally shorted to ground.

If the dynamic test does not reveal a problem, the static test should be performed next. This test is used to check if the analog output falls within the specified voltage range of the DAC. Some of the probable causes of such a problem are the result of the following:

- Component values change due to temperature, aging, and other factors.
- A faulty voltage reference.

■ REVIEW QUESTIONS

29. _____ _____ are often connected to a mechanical shaft to provide digital data on angular positioning of a physical device.

30. The advantage of the _____ code over binary numbers is that only one bit changes at a time when its numbers increment or decrement.

31. (Binary, Gray) optical encoders require more accuracy than encoders using the other code.

32. Logic circuits that convert binary numbers to Gray code or vice versa use _____ gates.

■ SUMMARY

- When the voltages applied to the inputs of a comparator are different, the output goes into either positive or negative saturation. The saturation of an op-amp is approximately 80% of the supply voltage.

- An inverting amplifier inverts the input signal to its opposite polarity.

- An inverting summing amplifier gives an inverted algebraic summation of the voltages applied to its input.

- A digital-to-analog converter is a device that converts a binary number into a proportional analog voltage.

- A binary-weighted DAC is often used when the number of digital inputs is 4 or less and the *R–2R* DACs are ideal when more than 4 bits are used.

- The resolution of a DAC is determined by first subtracting 1 from the number of binary input combinations, and then dividing the answer into the reference voltage number.

- An analog-to-digital converter is a device that converts an analog voltage into a proportional digital binary number.

- To prevent mechanical switches that are used as digital input devices from providing extra pulses that cause a false signal, a Schmitt trigger or a NAND latch is used to allow only the first switch closure to be recognized.

- A Schmitt trigger is used to shape or recondition distorted or slow-rising waveforms into pure square waves.

- The three types of multivibrators are the bistable, astable, and monostable. The bistable multivibrator is a common flip-flop; the astable multivibrator produces a continuous clock pulse when power is applied; the monostable multivibrator produces a temporary output voltage level and then returns to its original state.

- Optical encoders provide digital data on angular positioning.

- The Gray code uses a code consisting of several 1s and 0s to represent an equivalent decimal number; only one bit changes at a time when the numbers increment or decrement.

■ PROBLEMS

1. Why do operational amplifiers have two power supplies? (11-2)

2. An inverting operational amplifier has a positive power supply of + 12.5 V and a ground at its negative power supply terminal. Assuming that it has a gain of 30, draw the waveform at its output if it has a 2-V peak-to-peak sine wave applied to its input. (11-4)

3. Fill in the parentheses of each of the following equations with a $<$, $>$, or $=$ symbol to describe how an op-amp comparator operates. (11-3)

    (a) Inverting input voltage ( ) noninverting input voltage $=$ positive output voltage

    (b) Inverting input voltage ( ) noninverting input voltage $=$ zero output voltage

    (c) Inverting input voltage ( ) noninverting input voltage $=$ negative output voltage

4. An inverting op-amp circuit has $R_f = 5\ k\Omega$ and $R_{IN} = 1\ k\Omega$. What is the gain of this circuit? (11-4)

5. If the input voltage of the circuit in Problem 4 is 2 V, what is the output voltage? (11-4)

6. Negative feedback is used to control _____. (11-4)

7. What is the output voltage of the circuit in Fig. 11-48? (11-5)

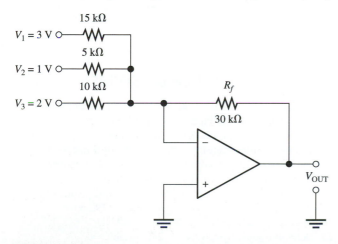

FIGURE 11.48
Circuit for Problem 7.

8. An inverting amplifier has an input voltage of +0.7 V. If $R_{IN} = 8k\Omega$ and $R_f = 24k\Omega$, which of the following is the output voltage? (11-4)
    (a) −1.4 V
    (b) +1.4 V
    (c) −2.1 V
    (d) −3.4 V

9. When its maximum output voltage is +5 V, the analog output of a four-input DAC is _____ volts when the binary digital input is 1001. (11-6)

10. DAC resolution is determined by the number of digital _____ lines available. (11-6)

11. The LSB input line of a binary-weighted DAC uses a resistor that is _____ the value of the resistor connected to the input line next to it. (11-6)

12. What is the resolution of a DAC with a maximum voltage of 15 V and five input lines? (11-6)

13. Draw the timing diagram of a three-input DAC if a mod-8 down-counter is connected to it. (11-6)

14. What are the problems associated with a binary-weighted DAC and how are they resolved? (11-6)

15. What are three ways to enable the IC DAC to operate within a voltage range of 0 to 5 V? Refer to Fig. 11-16. (11-7)

16. What are the circuits and component devices that make up a discrete ADC? (11-8)

17. In an ADC, what stores a digital value that is proportional to the analog input voltage applied to it? (11-8)

18. What component or circuit enables and disables the clock pulse to increment the up-counter of an ADC? (11-8)

19. How does the voltage at the inverting input compare with the voltage at the noninverting input of the comparator when the digital value displayed by an ADC is equal to the analog value? (11-8)

20. What is the resolution of a 5-bit ADC that has a maximum analog input voltage of +10 V? (11-8)

21. An 8-bit "counter-ramp" ADC requires a maximum of _____ clock pulses to generate its digital output, whereas an 8-bit "successive-approximation register" ADC requires a maximum number of _____ clock pulses. (11-8)

22. From Fig. 11-22, if the output reads 10000000, the analog voltage applied to the input is _____ volts. (11-9)

23. The cross-NAND gate circuit debounces _____ switches. (11-11)

24. A _____ _____ is used to both reshape distorted square waves and to transform slow-rising and -falling waveforms into rectangular-shaped signals. (11-12)

25. Describe the differences among bistable, astable, and monostable multivibrators. (11-13)

26. What determines the output frequency of a 555 IC when it is configured as an astable multivibrator? (11-13)

27. Using Fig. 11-33, select the capacitor value to produce an astable multivibrator frequency of 1100 Hz if the sum of resistors R_a and R_b equals 10 $k\Omega$. (11-13)

28. The duty cycle of a square wave is the ratio of time the output terminal is (low, high) to the total time period of one cycle. (11-13)

29. Determine the frequency of the 555 IC astable multivibrator consisting of components with the following values: R_a = 1kΩ, R_b = 500 Ω, and C = 5 μF (11-13)

30. How can a 555 astable multivibrator be constructed to produce a square wave that ranges from a 5% to 95% duty cycle? (11-13)

31. A 555 monostable multivibrator is activated by a (negative, positive) edge-triggered signal. (11-13)

32. Determine the output pulse width of the 555 IC monostable multivibrator when R_a = 5 kΩ and C = 5 μF. (11-13)

33. Using Fig. 11-37, select the component values for R_a and C to produce a monostable multivibrator output of 10 seconds. (11-13)

34. Why are optical encoders often used to provide digital data on the physical positioning of mechanical devices? (11-14)

35. A 5-bit Gray code wheel can be divided into _____ sectors of equal size that are _____ degrees each. (11-14)

36. Convert the Gray code number 1000 to an equivalent binary number. (11-14)

37. Convert the binary number 0101 to an equivalent Gray code number. (11-14)

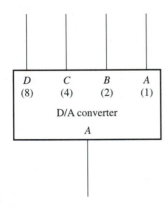

FIGURE 11.49
Circuit for Problems 38 and 39.

Troubleshooting

Refer to Fig. 11-49 for Problems 38 and 39.

38. Which 4-bit input of a binary-weighted DAC network would be shorted to ground to cause the analog output waveform shown in Fig. 11-50? (11-6, 11-15)

39. The DAC in Fig. 11-49 is designed to produce an analog voltage that is equal to the binary number applied to its input. Suppose that when a binary number of 7_{10} (0111) is applied to the input, the analog voltage is 6 V. Which of the following is the most likely cause? (11-6, 11-15)

(a) The reference voltage is high.

(b) Input B is shorted to V_{CC}.

(c) Input A is open.

(d) All of the above.

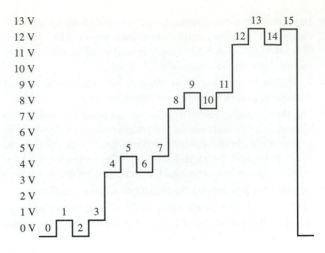

FIGURE 11.50
Waveform for Problem 38.

TABLE 11.13
Answer to Review Question 4

			OUTPUT (VOLTS)
I N V E R T I N G I N P U T	(<)	N O N I N V E R T I N G I N P U T	+5
	(=)		0
	(>)		−5
	(<)		+5
	(>)		−5
	(=)		0

■ ANSWERS TO REVIEW QUESTIONS

1. analog   2. 180, 0   3. How external components are connected to the input and output leads
4. See Table 11-13   5. 80   6. 200,000
7. positive (noninverting)   8. 2   9. algebraic, (−)
10. binary or digital, proportional   11. smallest, largest
12. 32   13. *R-2R*, two, four   14. resolution
15. analog-to-digital   16. 9V/64 − 1 = 0.1428 volts
17. high-to-low, low-to-high   18. low   19. 63, 6
20. bounce   21. SPST   22. cross-NAND
23. Schmitt trigger, hysteresis   24. bistable
25. astable, square wave   26. duty cycle
27. monostable   28. 555   29. Optical encoders
30. Gray   31. Binary   32. exclusive-OR

CHAPTER 12

MEMORY DEVICES

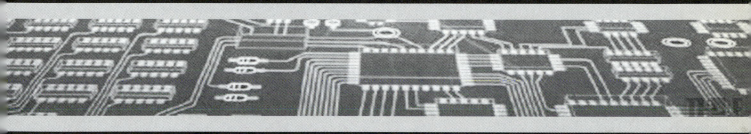

When you complete this chapter, you will be able to:

1. List and describe the four important characteristics of memory devices.
2. Explain the purpose of working memories.
3. Describe how to address memories by using the one-dimensional, two-dimensional, and stacking methods.
4. List the various types of ROMs and describe the operation of each.
5. List the various types of RAMs and describe the operation of each.
6. Describe the operation of the 7489 IC device.
7. Describe how memory capacity is determined.
8. List the types of mass storage memories and explain why they are used.
9. Troubleshoot a RAM memory device.

12.1 INTRODUCTION

One of the major advantages of digital equipment compared to analog equipment is its capability to receive, hold, and retrieve information. For example, a pocket calculator receives information resulting from a keypad entry, holds the information when the *memory store* button is pressed, and retrieves the information when the *memory recall* button is pressed.

The hold function is performed by **memory circuits,** which are also called **storage devices.** These circuits come in different forms and are found in all parts of digital equipment.

Memory devices are currently one of the most rapidly changing and expanding areas of digital electronics. The memory technology of today is the result of the modern computer. Because every computer must continually store and retrieve digital data, the memory element is an integral part of all computers. Therefore, new memory components are continually being developed or improved.

Memory circuits:

Circuits found in digital equipment that are used to store data.

Storage device:

Memory circuits used by digital equipment to store data.

Because each type of memory device has its own unique characteristics, some are better suited for certain applications than others. Therefore, the important characteristics of each type described in this chapter are identified to provide a basis for comparison. These include:

Packing Density: A term used to indicate memory capacity within a certain area. The higher the packing density is, the more bits are stored within a certain area.

Speed: Indicates how fast data can be placed in memory or retrieved from memory.

Power Consumption: Indicates how many watts the memory device draws from the power source.

Cost: The price for storing memory, which is usually measured on a per-bit basis.

Each characteristic is considered independently, and its importance depends on the type of application for which a system is used. For example, if a system were used for a spacecraft, small size and low power consumption would be a high priority, whereas high cost would be a low priority.

12.2   MEMORY ORGANIZATION

Memory cell:

A basic element of a memory that stores one bit of information.

Byte:

A binary word that is divided into eight bits.

The single requirement of a memory bit is that it have two distinct electrical states that can be translated into the 1- and 0-state logic levels with which digital circuitry operates. Each bit is stored by some basic element of memory called a **memory cell.** These bits alone, however, seldom provide adequate information. The memory, therefore, has bits that are grouped together into word configurations. In some cases, the bits are grouped together into 4-bit words called *nibbles* or 8-bit words called **bytes.** However, word sizes in modern computers can be 16, 32, or even 64 bits long, depending on the size of the computer. Binary words can represent a numerical value (data), a memory address, or a computer instruction.

Memory devices are arranged so that they store binary words in an organized array of addressable locations. Table 12-1 shows how a 4-bit word is stored at one of 16 different memory addresses. The 4-bit data word located at address 6 is 1001.

TABLE 12.1
Organization of memory consisting of 16 addresses for 4-bit words

ADDRESS	BIT D	BIT C	BIT B	BIT A
Word 0				
Word 1				
Word 2				
Word 3				
Word 4				
Word 5				
Word 6	1	0	0	1
Word 7				
Word 8				
Word 9				
Word 10				
Word 11				
Word 12				
Word 13				
Word 14				
Word 15				

Figure 12-1 is a block diagram of a memory device from which data can be retrieved. It is comprised of three major sections: the address decoder, the memory storage elements, and the output circuits. The address decoder accepts a 4-bit binary number applied to it. Because there are four input bits, a total of 2^4, or 16, different inputs can be decoded. The second section is the memory itself and is made of elements that store the 4-bit words. There are 16 words that can be stored. The address number applied to the input of the decoder selects which one of the 16 words are to be read. The third section is the output, which consists of tristate buffers that permit the addressed data word bits to reach the final output lines when an enable terminal is high. If the enable line is low, the buffer outputs float and prevent the data from passing through them.

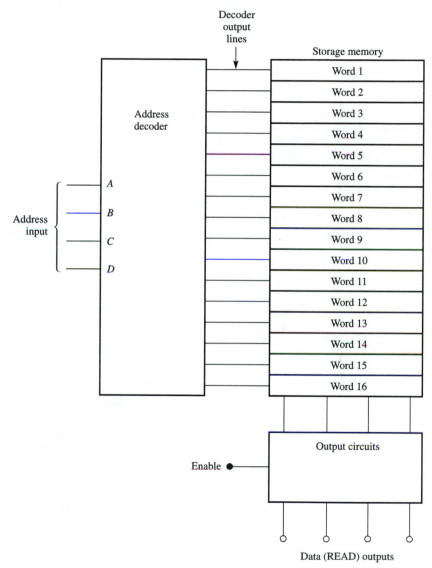

FIGURE 12.1
Block diagram of a memory device from which data can be retrieved.

One-dimensional memory:

A type of memory configuration that uses one decoder to address a memory device.

The type of memory configuration just described is **one-dimensional memory** because it uses one decoder for addressing. This type of design is adequate for small memories. As the size of a memory device increases, it becomes increasingly difficult to construct ICs with many wires to address the memory words.

If two decoders are used to address, one for horizontal lines (x-axis) and the other for vertical lines (y-axis), as shown in Fig. 12-2(a), the number of address wires is reduced. This is called **two-dimensional memory** because it uses two decoders. The memory organization becomes more complex as memories become even larger. Memory cells are placed on planes that are stacked one on top of the other. This is called a **stacked memory** and is shown in Fig. 12-2(b). How two-dimensional and stacked-memory devices are organized is not described in this chapter. However, the reader should be aware of their existence and can learn about them in advanced digital or computer books.

Two-dimensional memory:

A type of memory configuration that uses two decoders to address a memory device.

Stacked memory:

The memory storage architecture where memory cells are placed on planes that are stacked on top of each other.

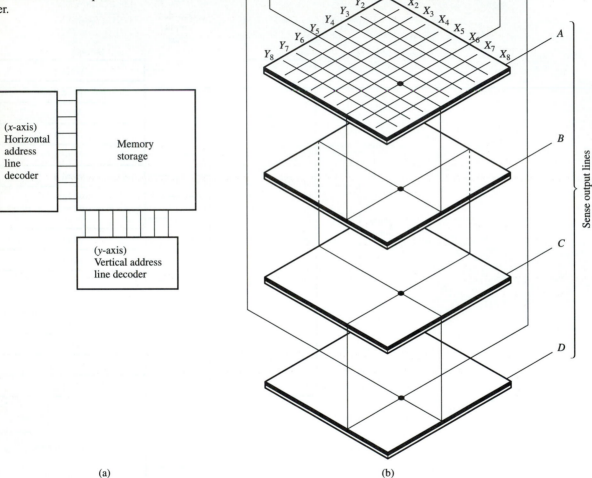

(a)

(b)

FIGURE 12.2

Memory organization of high-density memory devices: (a) Two-dimensional addressing. (b) Stacked memory planes.

■ REVIEW QUESTIONS

1. One of the major advantages of digital equipment over analog equipment is its ability to _____, _____ and _____ information.
2. Memory devices are also called _____ devices.
3. When selecting a memory device, what four characteristics should be considered before making a decision?
4. Each binary bit is stored in a _____ _____.
5. A one-dimensional memory device is comprised of what three major sections?
6. Two-dimensional memory devices use _____ address decoders.

WORKING MEMORY

Program:

A list of instructions stored in a memory device that tell a computing device what operations to perform.

The microprocessor of a computer or programmable controller is the major workhorse of these devices. The microprocessor is primarily made of logic circuits that perform arithmetic functions and make various decisions. To operate as it does, it must be provided with a detailed list of instructions, called a **program,** before it can do useful work. The program, along with input and output data, is stored in the internal working memory of a computer.

During the evolution of computers, devices such as relays, vacuum tubes, transistors, and magnetic cores (called donuts) have made up the main working memory. However, the advancement of semiconductor integrated-circuit technology with its mass storage capability, fast operating speed, and low power consumption has made it the current candidate to perform this function. This technology has caused the entire computer circuit architecture to change by introducing microprogramming, in which a given part of a program is permanently stored in memory. Constructed from both the bipolar and MOS processes, semiconductor memories have several designs, which are divided into two major groups, ROM and RAM.

12.3 READ-ONLY MEMORY (ROM)

Read-only memory (ROM):

A semiconductor memory device from which data can only be read and in which data is permanently stored.

Read-only memory (ROM) is the type from which data can be repeatedly read out of but not written into. Moreover, stored information is not lost when the power supply voltage is removed from this type of memory device. Therefore, it is said to be *nonvolatile.* The data placed into the ROM memory (programmed) is called **firmware.** To read data out of ROM, the user supplies an address of the word that is to be read out. A sample of the word is then sent to the output lines. The ROM memory is primarily used when a particular memory output is called frequently.

Firmware:

Programs that are permanently stored in a ROM memory device.

There are three general variations of read-only memory: mask-programmable, programmable, and reprogrammable.

Mask-Programmable ROM

Mask-programmable ROM:

A type of read-only memory device that can only be programmed by the IC manufacturer.

This type of read-only memory device can only be programmed by the IC manufacturer. A customer who purchases a mask-programmed ROM must supply the manufacturer with the software program containing information that is to be stored in the ROM. This program is then converted to 1s and 0s. A photographic negative mask is developed that controls how interconnections are made during the IC manufacturing process which represents the 1s and 0s as high- and low-voltage levels, respectively.

The initial cost of developing a mask is very high. However, when a large number of chips is produced from the same mask, the unit cost drops significantly. Therefore, when large quantities are needed, masked ROMs are used. Usually, a minimum order (such as 1000) is required by the manufacturer.

EXAMPLE 12.1

Suppose that it is necessary to frequently obtain equivalent Gray code numbers of decimal numbers 0–15. Program a 4-bit 16-memory ROM so that it contains the equivalent Gray code number for each corresponding binary memory location number.

Solution Enter the binary number equivalent to the decimal number needed into the address input of the memory to obtain the desired equivalent Gray code number. See Table 12-2.

TABLE 12.2
A Binary-to-Gray code converter

DECIMAL NUMBER	MEMORY ADDRESS BINARY NUMBER	GRAY CODE NUMBER
0	0000	0000
1	0001	0001
2	0010	0011
3	0011	0010
4	0100	0110
5	0101	0111
6	0110	0101
7	0111	0100
8	1000	1100
9	1001	1101
10	1010	1111
11	1011	1110
12	1100	1010
13	1101	1011
14	1110	1001
15	1111	1000

Because the program cannot be modified once the ROM IC is produced, it is often desirable to verify that the firmware program performs exactly as required. The verification procedure is accomplished by writing the firmware program into a computer. The computer is then used in place of the desired ROM chip to simulate how it should operate. If any modifications are required, the program is easily changed on the computer. After all the bugs are taken out and the program has been verified, the firmware program can then be mass produced onto ROM chips.

Masked-programmed ROMs are constructed in a matrix configuration that uses either diodes, bipolar transistors, or MOSFETs as the memory cells.

Diode matrix:

A type of read-only memory device that uses diodes to store each data bit.

Diode-Matrix ROM A **diode-matrix ROM** is an array of intersecting conductors, as shown in Fig. 12-3. A diode that connects two intersecting conductors represents a logic 1, whereas the absence of a diode at an intersection is a logic 0.

FIGURE 12.3
Diode-matrix cells representing a logic 1 and 0.

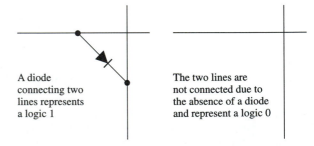

A diode connecting two lines represents a logic 1

The two lines are not connected due to the absence of a diode and represent a logic 0

Figure 12-4(a) shows a more detailed configuration of a diode-matrix ROM. There are eight memory locations that contain 4-bit data words. The word is read out of a memory address by first applying the octal code into the decoder that corresponds to the desired memory location and then applying a high to the tristate buffer enable line. For example, to read the data out of the memory location 2, an octal 010 is applied to the input of the decoder, which causes decoder output line 2 to go high. This forward biases the diodes that intersect horizontal line 2 with vertical lines 8, 2, and 1. A high is applied to the enable input line. This causes electron current to flow through the output read resistors R_4, R_2, and

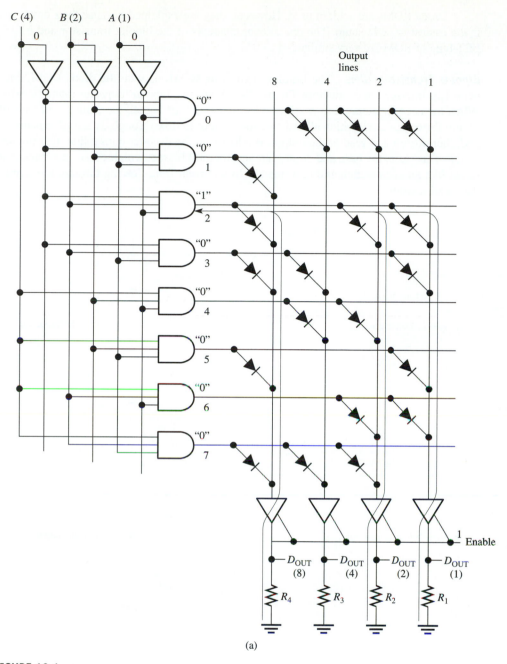

FIGURE 12.4
A diode-matrix read-only memory (ROM): (a) Schematic diagram. (b) Truth table.

Address line	8	4	2	1
0	0	1	1	1
1	1	0	0	0
2	1	0	1	1
3	1	1	0	1
4	0	1	1	0
5	1	0	0	1
6	0	0	1	1
7	1	1	1	0

(b)

R_1, where the data are read across. Because there are no diodes that intersect horizontal line 2 with vertical line 4, no electron current flows through resistor R_3. Therefore, 1011 is read across the output read resistors, which is consistent with the truth table in Fig. 12-4(b).

EXAMPLE 12.2

If 101 were applied to the address input lines of Figure 12-4(a), what resistors would electron current flow through?

Solution The decoder output line 5 would go high and forward bias the diodes connected to vertical lines 8 and 1. Therefore, electron current would flow through resistors R_4 and R_1, which would cause the data output to read 1001.

Diode ROMs are seldom used. However, they are explained because their operation is the easiest to understand. The operational concepts of the bipolar transistor and MOS-FET-type of ROM are very similar.

Bipolar transistor ROM:

A type of read-only memory device that uses bipolar transistors to store each data bit.

Bipolar Transistor ROMs The **bipolar transistor ROM,** as with the diode ROM, connects two intersecting conductors. Figure 12-5 shows two bipolar transistor memory cells. When the decoder output line is selected by an active-high signal, the emitter and base of Q_1 are forward biased, which allows electron current to flow through data read resistor R_1 and, therefore, a 1 is read at the output. If a low is at the decoder output line, it is not selected. Therefore, the base and emitter are not forward biased, which causes the transistor to act like an open switch and prevents electron current from flowing through the output data read resistor.

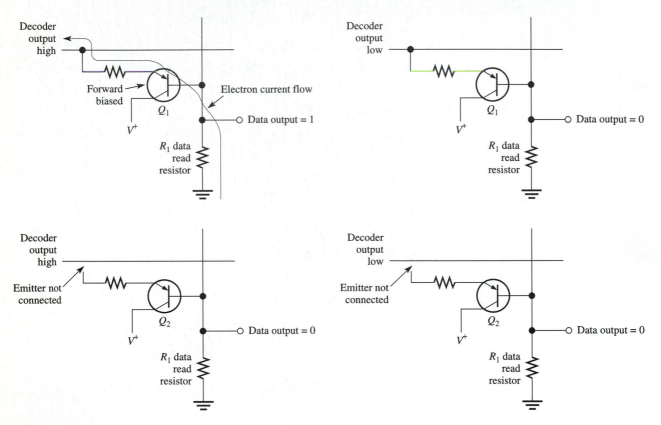

FIGURE 12.5
Bipolar transistor ROM memory cells.

Note that the emitter of Q_2 is not connected to the output line. When the decoder output line is either a high or a low, the transistor always acts like an open switch. Therefore, electron current never flows through the output read resistor and a 0 is always read across it. Any transistor with an open at the emitter is the same as the diode matrix without a diode. The open is formed during the IC masking steps by the manufacturer.

The advantages of a transistor ROM over a diode ROM are that it is faster and uses less power. As a result of less power usage, more bipolar memory cells can be placed in the same space, which improves the packing density of the chip.

MOS ROM:

A read-only memory device that uses miniature field-effect transistors to store binary data.

MOS ROMS As with the diode ROM and bipolar transistor ROM, the MOSFET (MOS for short) connects two intersecting conductors. Figure 12-6 shows three memory cells. The gate and drain of Q_1 are connected to a horizontal output line, and the substrate and source terminal are connected to a vertical line. When the decoder output line is selected by a high

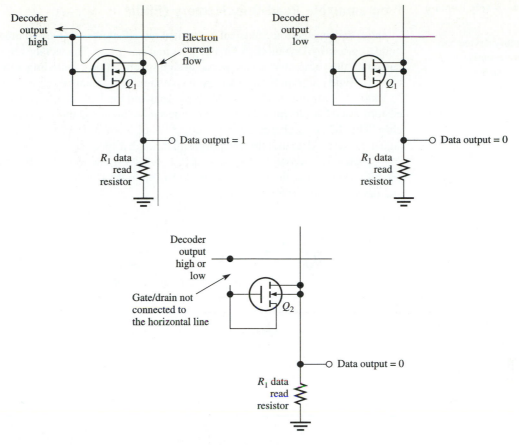

FIGURE 12.6
MOSFET ROM memory cells.

signal, the resistance of the FET is about zero ohms, which allows the corresponding lines to intersect. Therefore, electron current flows through the output read resistor and the FET to the positive high-voltage potential. This causes a 1 to be read across the output read resistor. Likewise, a low at the gate and drain causes the FET to open, which does not allow the corresponding lines to intersect. Therefore, no electron current flows and a 0 is read at the output. Note that the gate and drain of Q_2 are not connected to the horizontal output line. This produces the same result as the diode matrix without the diode. This open is formed during a masking step of the IC by the manufacturer.

The advantage of MOS ROMs is that they consume only 1/100 the power needed for transistor ROMs. Also, their packing density is better because 20 to 100 more cells can be placed in the same area as one transistor ROM cell.

■ REVIEW QUESTIONS

7. A microprocessor is primarily made of _____ circuits that perform _____ functions and make various _____.

8. Semiconductor memories are divided into two major groups, _____ and _____.

9. A memory that does not lose its contents when the power supply voltage is turned off is called _____.

10. Mask-programmable ROMs are programmed by whom?

11. The three types of mask-programmable ROMs are _____, _____, and _____ matrix.

Programmable Read-Only Memory (PROM)

Suppose that an electronic musical instrument manufacturer received an order to design and build a customized synthesizer for a popular musical group. To meet the requirements of the customer, the equipment needs special design specifications, including a ROM chip with a one-of-a-kind program for its memory. As mentioned, IC manufacturers require a minimum of 1000 units for nonstandard ROM chips. To avoid the high one-time cost of producing a customized ROM, IC manufacturers provide a more practical alternative, called the **PROM chip.** This IC is purchased unprogrammed. Once a software program is verified, the designer can place data into this memory chip through a special programming device.

Figure 12-7 shows two memory cells of a PROM memory chip. Initially, every memory bit consists of a diode in series with a fusible link. The chip is programmed by a special device called a *PROM programmer.*

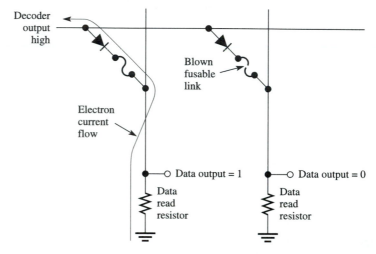

FIGURE 12.7
PROM memory cells.

The programming procedure required to place a word into a desired memory location involves addressing the memory address. The word to be permanently read is then applied to the PROM output. Only the fusible links connected to the PROM output lines that have a 0 state vaporize after a momentary high-current programming pulse is applied. A permanent 0 state is then read at locations where a fuse is blown. Likewise, a 1 state is read where any fuses remain. Once this process is performed, the program is permanent and cannot be changed.

Reprogrammable Read-Only Memory

There are three types of ROM ICs that can be reprogrammed. These include:

■ **Erasable Programmable Read-Only Memory (EPROM):** This type of ROM can be programmed after being entirely erased with the use of an ultraviolet light source. The complete erasure of a program is accomplished by shining an ultraviolet light through a window located at the top of the IC package for approximately 8 minutes or less.
■ **Electronically Alterable Read-Only Memory (EAROM):** These ROM devices are very similar to EPROMs. However, instead of using an ultraviolet source to erase them, an erasing voltage is applied to one of the EAROM chips to remove the old program.
■ **Electronically Erasable Programmable Read-Only Memory (EEPROM):** Like ROMs, EPROMs, or EAROMs, these are nonvolatile memory devices. They provide permanent storage of a software program but can easily be altered through the use of a CRT or manual programming unit.

Electronically alterable read-only memory (EAROM):

A type of memory device that can be reprogrammed after its previous contents are erased by a temporary applied voltage.

Electronically erasable programmable read-only memory (EEPROM):

A type of memory device that can be reprogrammed through the use of a computer or special programming device.

All of these programmable ROM devices are much slower and larger than masked-programmed ROMs.

ROM Applications

These types of memory are used in applications in which a certain output or the same fixed program is required repeatedly. These include:

- A microprogram stored in computers.
- Trigonometric and other mathematical lookup tables. An example is a multiplication reference table used by calculators. During the multiplication process of two numbers between 0 and 9, the calculator circuits enter the multiplicand and multiplier as two addresses into ROM and read out the product from this storage table.
- A ROM can be programmed to accomplish the same functions as gates and combination logic circuitry. Figure 12-8 shows how a ROM memory can be configured to perform the same function as an octal-to-binary encoder. Internally, the memory device does not perform the same logic functions as the gates inside the encoder. Instead, when one of the numbers to be encoded is applied to the memory address input, the same output number that the encoder supplies appears at the memory output data lines because it is stored at that location.

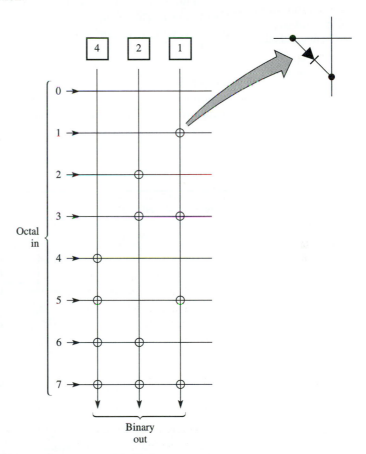

FIGURE 12.8
A diode-matrix ROM programmed as an octal-to-binary encoder, a circuit usually constructed with OR gates.

■ REVIEW QUESTIONS

12. A fusible link in a PROM matrix represents a (0, 1) state, and a blown fuse represents a (0, 1) state.

13. The three types of reprogrammable ROMs are _____, _____, and _____.

14. To enhance their programmability, programmable ROMs are much slower and larger than masked programmable ROMs. True or false?

15. ROM memory devices are used in applications where the stored information is used (once, repeatedly).

12.4 RANDOM-ACCESS MEMORY (RAM)

Software:

Programs that are temporarily stored in a RAM memory device.

Volatile:

Describes stored information in a memory device that is lost when power is removed from the circuit.

Random-access memory (RAM) is the type of memory from which data can be read or into which it can be placed. A RAM, like a ROM, consists of a matrix of conductors. However, instead of diodes, bipolar transistors, or MOSFETs, flip-flop latches are placed at the intersections of the grid. Because flip-flops can change states, data stored in RAM can be altered or erased very easily. For this reason, the data programmed into memory is called **software.** RAMs are **volatile,** which means that removing the operating power even momentarily causes the stored information to be lost because the internal flip-flops can assume either state at random. The reason why RAMs are called *random* is that data can be read out of or written into any of the memory locations without any restrictions in sequence.

Figure 12-9 is a block diagram illustrating how data is written into and read out of a RAM device. This memory device can store 4-bit words into 16 individual memory locations. To write data into a location, a high is applied to the *write enable line.* To select which of the 16 memory locations data are to be written into, a binary number equivalent to the desired location is placed at the address lines. The data to be stored into the selected location is applied to the *data input lines.* To read data from memory requires that a 1 state be applied to the *read enable line.* The specific memory location out of which data is read is determined by the binary number applied to the address line. The data appears at the *data output lines,* and as the information is read, the flip-flops retain their data.

FIGURE 12.9
Block diagram of a 4-bit 16-memory-location RAM.

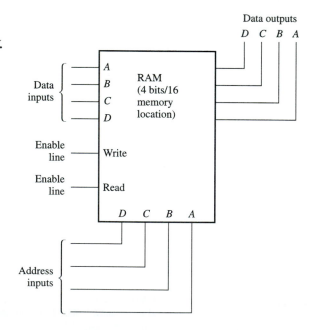

As with the ROM-type ICs, the RAM memory devices include internal address decoders and output buffers. Also, internal amplifiers are used to sense the low-voltage levels of data stored at each memory location so that the information can be increased to a usable level at the output of the IC.

There are two general categories of read/write (R/W) memory, *static* and *dynamic.*

Static RAM

Static RAM uses the flip-flop for its basic storage cell element. Recall that a flip-flop stays latched in one of two stages as long as power is supplied to it. The flip-flops in these devices are made either from cross-coupled TTL multiemitter bipolar transistors, as shown in Fig. 12-10(a), or cross-coupled MOSFETs, as shown in Fig. 12-10(b), that latch the data into each cell. As a result of MOSFETs using less power and taking up less space, the bit density is greater than that of TTL RAM. However, the TTL-type is faster.

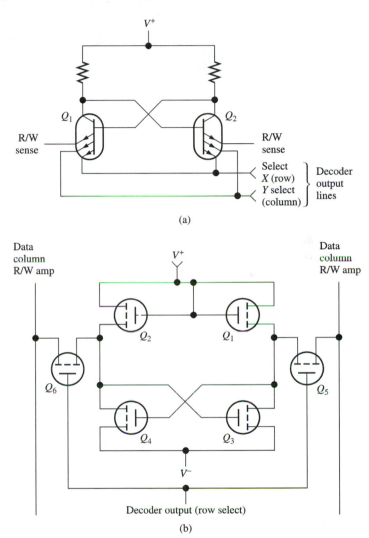

FIGURE 12.10
Static RAM memory cell. (a) Bipolar. (b) MOSFET.

Dynamic RAM

A **dynamic RAM** memory cell uses a capacitive element for storing the data bit. Binary information is stored as a charge. If a charge is present at a capacitive element, it represents a logic 1, whereas the absence of a charge represents a 0 state. Shown in Fig. 12-11, these dynamic memory cells use the capacitance between the gate and substrate of a MOS transistor to store the charge.

Because a capacitor-type device does not retain a charge indefinitely, it must be recharged. Dynamic RAM memory cells are only capable of holding their charge for a few milliseconds. Therefore, it is necessary to refresh these charges continuously, and this is accomplished by additional external circuitry that reads the stored data and then writes them back into the same memory location once every few milliseconds. The advantage of

FIGURE 12.11
Dynamic RAM memory cell.

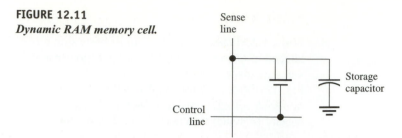

this type of RAM memory is that it has about four times greater density compared to static RAM, reducing the cost per bit. It is also faster than static RAM memory. However, this cost savings is offset by the "refresher" circuitry that is required. Dynamic RAM memory is often used in minicomputers and large computer applications where large amounts of memory are needed.

12.5   THE OPERATION OF A RAM IC

A common type of read/write RAM device is the 7489 IC. It consists of 64 memory cells and is capable of storing sixteen 4-bit words. A block diagram showing the input and output leads is shown in Fig. 12-12.

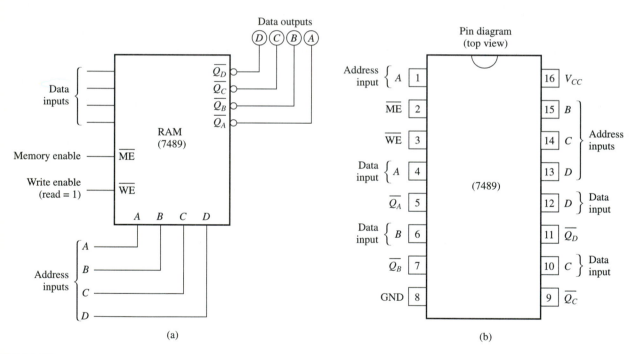

FIGURE 12.12
The 7489 RAM IC: (a) Block diagram. (b) Pin diagram.

Suppose it is desirable to write the 4-bit word 0011 into memory location 9. The following steps are required:

1. Place the binary number equivalent to 9 at the address input leads by applying $D = 1$, $C = 0$, $B = 0$, and $A = 1$.
2. Enter the desired 4-bit word 0011 into the memory by applying $D_{IN} = 0$, $C_{IN} = 0$, $B_{IN} = 1$, and $A_{IN} = 1$ to the input data lines.
3. Apply a 0 to the write enable line.
4. Apply a 0 to the memory enable line.

To read data out of memory, a similar procedure is necessary. Suppose it is desirable to read the word 1111 out of memory location 7. The following steps are required:

1. Place the binary number equivalent to 7 at the address input leads by applying $D = 0$, $C = 1$, $B = 1$, and $A = 1$.
2. Place the write enable line into the read mode by applying a 1 to it.
3. Apply a 0 to the memory enable line.
4. The memory contents 0000 appear at the data output lines; it is the complement of the stored number. To make the data at the output lines the same as that in memory, inverters can be attached to the outputs of the 7489 IC.

EXAMPLE 12.3

Draw the waveforms at the input and output leads of the 7489 RAM IC to show what happens when 0111 is first written into memory location 4 and then 0110 is read out of memory location 12.

Solution   See Fig. 12-13.

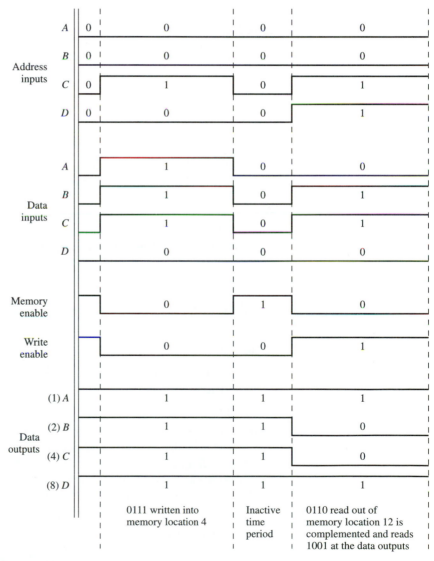

FIGURE 12.13
Waveforms for Example 12.3.

EXPERIMENT

Integrated Circuit Random-Access Memory (RAM)

Objectives

- To wire a 7489 IC memory device.
- To learn memory addressing.
- To write data into RAM.
- To read data from RAM.

Materials

1—7489 IC

10—SPDT Logic Switches

4—LED Indicators

4—330-Ω Resistors

1—+5-V DC Power Supply

Introduction

Many types of digital equipment require the storage of binary information called *data*. One type of memory device capable of storing 64 bits of information is the 7489 TTL MSI semiconductor chip. It is organized to store 4-bit words into 16 different rows called *memory locations*. Each of these rows is assigned an *address* to identify the location into which the word is stored or from which it is removed. Data can be stored into each memory location by a process called the *write* operation. Stored data can later be retrieved from a memory location by a process called the *read* operation. When the memory contents are read, the data is not erased. This type of reading is classified as *nondestructive*.

The 7489 IC is randomly accessible, which means that data can be written into or read out of any memory location, in any order. That is why it is called a *random access memory (RAM)* device. RAM devices are also classified as being *volatile,* which means that the data in each memory location is present while power is applied to the device. When power is lost, the data is lost. When power is applied again, a random pattern of bits will appear in the memory.

Procedure

To understand the experiment, the following sections provide information on how the 7489 IC operates. Refer to Fig. 12-14.

Input Lines

Data Input Lines   The 4-bit data words are entered into the 7489 memory device through the four data input lines. Logic switches can be used to enter the desired bit patterns of each word.

Address Input Lines   The 64-bit capacity of the memory device is divided into 16 locations where 4-bit data words can be stored. Which location each data word is written into or retrieved from is determined by four input terminals. These are called *address input lines*. Because any binary count of 0000_2 (00_{10}) to 1111_2 (15_{10}) can be applied, any one of the 16 memory locations becomes accessible. Logic switches can be used to apply these binary numbers to the address input terminals.

Mode Control Inputs   Two separate control lines provide a means for storing and retrieving data:

The memory enable ($\overline{\text{ME}}$) is used to enable the memory device to perform the reading or writing operation. It must be low to

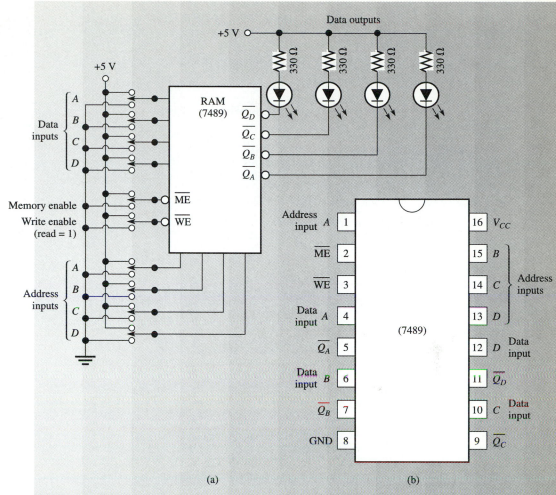

FIGURE 12.14
The 7489 64-bit Ram IC: (A) Block diagram. (B) Pin diagram.

perform either function. When a high is applied to the $\overline{ME}$ line, data cannot be written into the chip, and 1s are present at each of its data output terminals.

The write enable ($\overline{WE}$) line is used to control whether data can be written into memory, or read out of it. When the $\overline{WE}$ line is low, the data applied to the data input terminals is stored into the chip. A high at the $\overline{WE}$ line allows data to be read out of the memory device.

Output Lines

The contents of the memory are read out of the device from four output lines. Each line has an inverter bubble, which indicates that the contents are inverted before being read. Each line is an open collector, which means there may be some external load connection required for correct operation. If a LED is used to indicate the logic state of an output, a series current-limiting resistor is used in conjunction with the LED to perform the function of the pull-up resistor.

Step 1. Assemble the circuit shown in Figure 12-14(a).

PART A—WRITE OPERATION

Step 2. Write the contents of 6 into memory location 12.

(a) Set the switch at the $\overline{ME}$ input low.

(b) Set the data input lines as follows:
$A = 0$
$B = 1$
$C = 1$
$D = 0$

(c) Set the address lines as follows:
$A = 0$
$B = 0$
$C = 1$
$D = 1$

(d) Set the switch at the $\overline{WE}$ input low.

PART B—READ OPERATION

Step 3. Read the contents located at memory address 12.

(a) Set the switch at the $\overline{ME}$ input low.

(b) Set the address lines to 1100_2.

(c) Set the switch at the $\overline{WE}$ input High. Using a logic probe or voltmeter, record the logic states at the following outputs:
$\overline{Q_A} =$
$\overline{Q_B} =$
$\overline{Q_C} =$
$\overline{Q_D} =$

Procedure Question 1

Why do the IC output pins $\overline{Q_A}$ to $\overline{Q_B}$ read HLLH instead of LHHL?

Step 4. Turn the power off.
Step 5. Turn the power back on and read the data at memory location 12.

Procedure Question 2

Why have the contents changed at memory location 12?

Background Information

A special type of binary code called a *Gray code* is used for industrial applications that require precise positioning. It is used instead of the pure binary code because only one of its digits changes each time its value increments or decrements. Table 12-3 compares decimal, binary, and *Gray code* equivalent values.

Step 6. Load the *Gray code* contents into the memory device as shown in Table 12-4. As each number is written into the sequence of memory locations, place the $\overline{WE}$ and $\overline{ME}$ switches into the positions indicated by the table.

TABLE 12.3

Decimal Equivalent	Address (Binary)				Enable Inputs		(Gray Code) Data Inputs			
	D	C	B	A	$\overline{ME}$	$\overline{WE}$	D	C	B	A
0	0	0	0	0	0	0	0	0	0	0
1	0	0	0	1	0	0	0	0	0	1
2	0	0	1	0	0	0	0	0	1	1
3	0	0	1	1	1	0	0	0	1	0
4	0	1	0	0	0	0	0	1	1	0
5	0	1	0	1	0	0	0	1	1	1
6	0	1	1	0	0	0	0	1	0	1
7	0	1	1	1	0	1	0	1	0	0
8	1	0	0	0	0	0	1	1	0	0
9	1	0	0	1	0	0	1	1	0	1
10	1	0	1	0	0	0	1	1	1	1
11	1	0	1	1	0	0	1	1	1	0
12	1	1	0	0	0	0	1	0	1	0
13	1	1	0	1	0	1	1	0	1	1
14	1	1	1	0	0	0	1	0	0	1
15	1	1	1	1	0	0	1	0	0	0

TABLE 12.4

Reading data from a 7489 IC

Decimal Equivalent	Address (Binary)				Enable Inputs		(Gray Code) Data Inputs			
	D	C	B	A	$\overline{ME}$	$\overline{WE}$	D	C	B	A
0	0	0	0	0	0	1				
1	0	0	0	1	0	1				
2	0	0	1	0	0	1				
3	0	0	1	1	0	1				
4	0	1	0	0	0	1				
5	0	1	0	1	0	1				
6	0	1	1	0	0	1				
7	0	1	1	1	0	1				
8	1	0	0	0	0	1				
9	1	0	0	1	0	1				
10	1	0	1	0	0	1				
11	1	0	1	1	0	1				
12	1	1	0	0	0	1				
13	1	1	0	1	0	1				
14	1	1	1	0	0	1				
15	1	1	1	1	0	1				

Step 7. Read the Gray code contents from the memory device. Place the $\overline{\text{ME}}$ switch low and the $\overline{\text{WE}}$ switch high. Record the contents of each memory location in the output section of Table 12-4.

Procedure Question 3

Is any of the data that was read out of the memory different than the data that was written into it? If so, why?

Step 8. Correctly load the Gray code at the following memory locations. Make sure that the $\overline{\text{ME}}$ and $\overline{\text{WE}}$ inputs are both low.

Memory Locations	Gray Code Numbers
3	0010
7	0100
13	1011

Step 9. Read the new contents of memory locations 3, 7, and 13. Replace the incorrect numbers at these memory locations in Table 12-4 with the correct Gray code numbers.

Background Information

Placing the Gray code numbers into the memory device demonstrates that the memory device can function as a binary-to-Gray code decoder. By placing any binary number ranging from 0000 to 1111 into the address input lines, its equivalent Gray code output value can be read out of the memory location with the same number.

■ EXPERIMENT QUESTIONS

1. The 7489 IC is a _____ (RAM, ROM) _____ (TTL, CMOS) device.
2. What does the term *random access* mean?
3. The 7489 IC can store a total of _____ bits of binary data; the bits are arranged so that _____ 4-bit words are stored at _____ different memory locations.
4. To write into or read from memory location 10, what logic states should be applied to the following address input lines?
   $A =$
   $B =$
   $C =$
   $D =$
5. If $A = 0$, $B = 1$, $C = 1$, and $D = 1$ is stored into a memory location, list what binary data bits will be read at the output pins of the device.
   $\overline{Q_D} =$
   $\overline{Q_C} =$
   $\overline{Q_B} =$
   $\overline{Q_A} =$
6. What are the logic states required at inputs $\overline{\text{ME}}$ and $\overline{\text{WE}}$ to write data into memory?
   $\overline{\text{ME}} =$ _____ $\overline{\text{WE}} =$ _____

7. If input $\overline{ME}$ is high when data is read out of the 7489 IC, what data bits will appear at the outputs?

$$\overline{Q_D} =$$

$$\overline{Q_C} =$$

$$\overline{Q_B} =$$

$$\overline{Q_A} =$$

8. To read 1101_2 from memory location 3, what logic states will be found at the following pins?

Address: $D = \underline{\hspace{1cm}}$ $C = \underline{\hspace{1cm}}$ $B = \underline{\hspace{1cm}}$ $A = \underline{\hspace{1cm}}$

Output: $\overline{Q_D} = \underline{\hspace{1cm}}$ $\overline{Q_C} = \underline{\hspace{1cm}}$ $\overline{Q_B} = \underline{\hspace{1cm}}$ $\overline{Q_A} = \underline{\hspace{1cm}}$

RAM Applications

RAMs are sometimes used to store the same kind of information stored in ROMs. However, most frequently, they are used in applications where the data can be altered very easily. More specifically, they are used for:

- The main working memories of computers that require temporary data storage. These types of RAM memories consist of IC chips that are placed on printed circuit cards that are then stacked. The memory capacity for these RAM memory cards ranges from 250,000 to more than 10 million bytes.
- Storing software programs in computers, microprocessors, robots, and programmable controllers. RAMs are popular for this application because data can be written into and out of very easily.
- As with ROMs, RAMs can be programmed to output specific control signals in place of those that are generated by combination circuits.

■ REVIEW QUESTIONS

16. The (ROM or RAM) memory is the type (from which data can be read, into which it can be placed).

17. A memory device that loses its contents when power is removed is called _____.

18. The basic storage cell element of a static RAM device is a _____, which uses either a _____ or a _____ as the basic component.

19. The basic storage cell element of a dynamic RAM device is a _____ element, which stores a charge. To keep its data, the dynamic RAM must continually be charged by _____ circuits.

20. What are three general applications of RAM devices?

21. The data that appear at the output data lines of the 7489 IC are (opposite or the same) as that stored in the memory location being read.

22. To write data into a 7489 IC, the write enable line must be a (0, 1) and the memory enable line must be a (0, 1). To read data from a 7489 IC, the write enable line must be a (0, 1) and the memory enable line must be a (0, 1).

12.6 MEMORY CAPACITY

Memory capacity:

The total quantity of data that memory devices can store.

The total quantity of data that memory devices can store is sometimes expressed in bits. However, more often it is expressed in memory words, which can contain 4, 8, 16, 32, 64, or even 128 bits each. Memory devices are usually constructed into units of (K) word groupings. Contrary to what most people assume, the K does not amount to 1000. Instead, the number 1K used for memory devices is 1024, a number derived from the power-of-2 equation.

The capacity for memory devices is usually referred to in units of (K) groupings. For example, a 1K memory chip that stores 8-bit words stores 1024 bytes of information. Examples of memory capacities for some of the common computing devices that use 8-bit words are as follows:

ABBREVIATION	ACTUAL NUMBER OF MEMORY BYTES
1K	$1024 = 2^{10}$
2K	$2048 = 2^{11}$
4K	$4096 = 2^{12}$
8K	$8192 = 2^{13}$
16K	$16,384 = 2^{14}$
32K	$32,768 = 2^{15}$
64K	$64,536 = 2^{16}$

To find the total number of bits in a memory device, multiply the word size by the number of memory locations.

EXAMPLE 12.4

Find the total number of bits in a memory chip that uses 8-bit words with a capacity of 2K.

Solution

$$2K = 2048 \text{ byte capacity}$$

$$8\text{-bit word} = 1 \text{ byte}$$

$$\begin{array}{r} 2048 \\ \times \quad 8 \\ \hline \end{array}$$

16,348 bits or memory cells

12.7   MASS MEMORY STORAGE

Working memory:

The memory devices located in computing equipment from which data is continually stored and retrieved during its operation.

Mass memory:

External memory devices that are used by a computing device when the memory capacity requirements of the internal memory are exceeded.

One method used of rating a computer is its memory capacity. This refers to its internal **working memory,** which is where data are stored and retrieved very quickly. Ranging in size from several hundred to millions of bytes, the internal working storage of a computer provides enough data for most applications. However, there are many applications that require an amount of data that exceeds the internal memory capacity of a computer. For example, bank transactions, payroll, bookkeeping, and inventory records often require much more data than what the internal memory is capable of storing by itself. Therefore, a peripheral (external) auxiliary memory, which is also called **mass memory,** is required.

Despite being relatively slow in comparison to other types of memory devices, these memories are capable of storing a very large quantity of data inexpensively. For example, some types can store one bit of information for as low as 0.000000001 cent. These external memory devices are used to transfer data and information to a computer's internal working memory as needed. As the internal memory uses up the data, it periodically takes more data out of the external memory. Likewise, data and information are often transferred from the internal memory to the external memory for long-term storage.

Magnetic tape:

A mass storage memory device that uses tape to magnetically store binary information.

A common way of storing information is by using electromechanical memory devices such as **magnetic tape** and magnetic disks, as shown in Fig. 12-15. Both devices are made of a plastic substrate that has a thin layer of iron oxide or other ferromagnetic material. A coil, called a recording head, is located close to this surface. To write data onto a magnetic tape, the tape surface passes under the head, which magnetizes regions in either one direction or the other depending on which of two directions current passes through the coil. The iron particles magnetized in one direction represent a 1 state and those magnetized in the other direction represent a 0 state, as shown in Fig. 12-16(a).

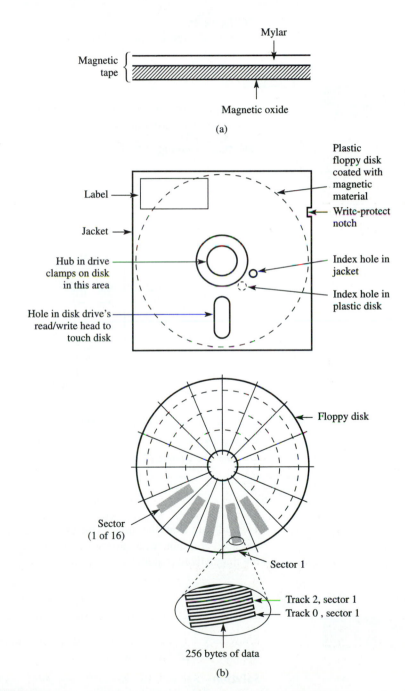

FIGURE 12.15
Electromechanical mass-storage devices: (a) Magnetic tape. (b) Magnetic disk.

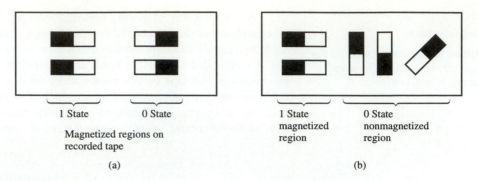

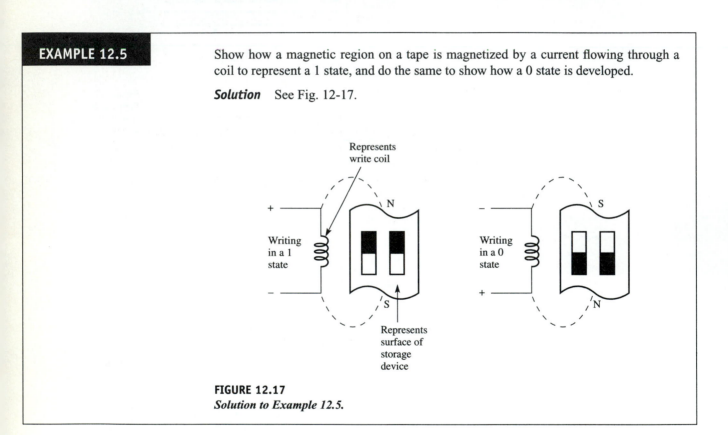

FIGURE 12.16
Magnetized surfaces representing both logic states: (a) Magnetic tape. (b) Magnetic disk.

EXAMPLE 12.5

Show how a magnetic region on a tape is magnetized by a current flowing through a coil to represent a 1 state, and do the same to show how a 0 state is developed.

Solution   See Fig. 12-17.

FIGURE 12.17
Solution to Example 12.5.

To read data from the already magnetized iron-oxide layer, the tape surface passes under the head and induces a current in one direction or the other, depending on the polarity of the magnetized region.

EXAMPLE 12.6

Show how a 1 and a 0 are induced into a magnetic head as the magnetized regions on the tape pass under the head.

Solution   See Fig. 12-18.

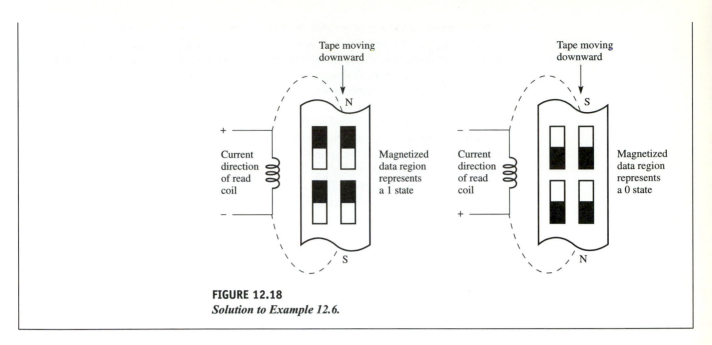

FIGURE 12.18
Solution to Example 12.6.

Figure 12-19 shows a recording head. It consists of a core made from high-permeability soft iron with a center-tapped coil wound around it. Half of the coil is used as a sense winding that is utilized in the read operation. The other half, made of a heavier-gauge wire, is used for supplying current during the write operation. The core has a small air gap (0.001 inch) that enables the read/write function to take place.

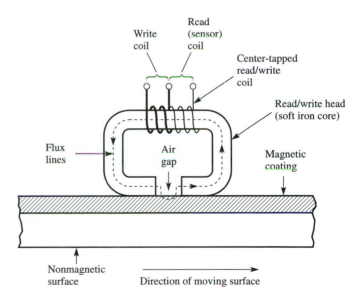

FIGURE 12.19
Recording head.

The magnetic disk stores bits in a different way than magnetic tape. The logic-1 state is stored by the presence of a magnetized region, and the logic-0 state is stored by the absence of a magnetic region, as shown in Fig. 12-16(b). A region is magnetized when current is passed through the coil of the recording head, and a region is not magnetized when current does not pass through the coil. The magnetic regions are in the form of concentric

Floppy disk:

A mass storage memory device that is in the shape of a flat disk 3 1/2 inches in diameter which can store or retrieve data.

Hard drive:

A mass storage memory device located inside a computer where data can be stored or retrieved.

tracks around the surface of the disk. The disk is spun at high speeds and a read/write head on a movable track permits easy access to each data track quickly. Magnetic disks are available in many forms and sizes. The most common is the 3.5-inch **floppy disk.** Also known as a *diskette,* its 3.5 designation refers to the outside dimension of its square protective jacket. It is a portable storage device that stores data into, or retrieves data from, a computer.

There are 135 tracks per inch on a conventional diskette. On each track, data is organized into *sectors,* each the shape of a slice of pie. To read a given block of data, its track and sector must be specified. Locating the proper sector is accomplished by using an index detector, a small coil of wire. A pulse is induced into the coil each time a small magnet placed on the side of the drive motor that spins the disk passes the coil. The pulse serves as a positional reference indicator and occurs when the first sector is located at the head. A standard double-sided, high-density diskette can store 1.44 megabytes of data.

Disks are also used for high-capacity storage inside a computer. However, unlike floppy disks, which are made of thin and flexible material, they are thick and rigid. These disks are called **hard drives.** A hard drive of 3.5 inches in diameter can store one gigabyte, which is a billion bytes. A simplified sketch of a hard drive is shown in Figure 12-20. By stacking several hard drive disks, the memory capacity can be increased even more. Because they are rigid, they can spin faster than a floppy disk and therefore information can be located more quickly. Hard drives are usually used to store programs and data that are used frequently by a computer operator. Memory is also stored on minicartridge tapes. Larger tapes can store quantities in the gigabyte range. Tapes are used for long-term archival or backup storage.

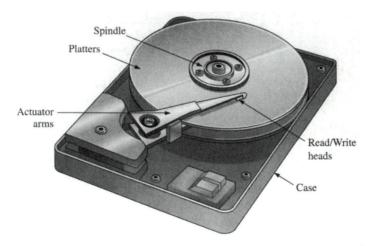

FIGURE 12.20
A simplified sketch of a hard drive.

The newest device to be developed for storing data is the *compact disc,* or *CD-ROM.* The ROM portion of the name indicates that it is a read-only device. It uses the same technology as music CDs. Unlike disks or tape storage, the compact disc is not magnetic. The data is represented by small pits pressed into the surface of the disc. Unlike the concentric paths of a disk, the pits are arranged in a long spiral that is several miles in length. Information about how data is read from a disc is explained in Section 13-9 of Chapter 13.

Magnetic bubble memory (MBM):

A mass storage memory device that uses individual microscopic bubbles to store binary bits of data.

Another type of mass storage device is **magnetic bubble memory (MBM).** Bubble memories were developed by Bell Laboratories in 1966 and are presently available as chips. For years, they have been used by Bell Telephone with their voice synthesizer that provides the message, "We're sorry, your call did not go through. Please check the number and dial again or ask your operator for assistance," when a telephone number is dialed incorrectly.

These memories use a nonmagnetic wafer-shaped substrate on which a very thin layer of magnetic material, called *epitaxial garnet,* is grown that forms magnetic domains, as shown in Fig. 12-21(a). When a magnetic field is placed perpendicular to the film surface, the domains having opposite polarity shrink into microscopic magnetic cylinders called bubbles, as shown in Fig. 12-21(b). The bubble shape of these domains remains as long as the magnetic field strength of an external magnet stays within a certain range.

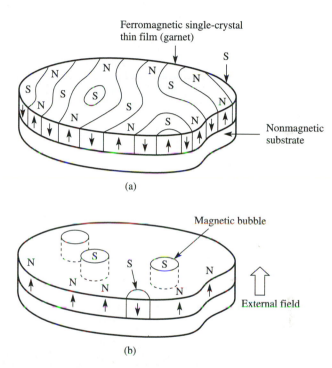

(a)

(b)

FIGURE 12.21
(a) Magnetic domains outside an external magnetic field. (b) Magnetic domains inside the field of an external magnet.

Data are entered into memory by a hairpin-shaped conductor called a *generator.* When a momentary pulse of current lasting during a clock period of 10 μs is sent through it, a magnetic field is developed that produces an individual bubble that represents a 1 state. If a pulse does not occur during a clock period, the absence of a bubble represents a 0 state. Permanent magnets are placed on each side of the wafer so that the bubble shapes are retained. Coils are wrapped around the wafer in a grid pattern, as shown in Fig. 12-22. As separate synchronized currents flow through individual coils, they distort the magnetic field of the permanent magnets every 10-μs clock period, which causes the bubbles to move from the generator to the conductor path.

Hall-effect device:

A conductor that changes resistance as the magnetic field it is within changes its strength.

Data are read from the memory by a bubble detector, which reads a 1 state with the presence of a bubble or a 0 state with the absence of a bubble. The detector consists of a **Hall-effect device,** which is a conductor that changes resistance as the surrounding magnetic field varies. When a bubble passes through the detector during a 10-μs clock period,

FIGURE 12.22
Orthogonal coils.

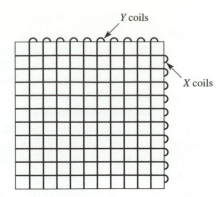

the magnetic field surrounding the Hall-effect device causes the current flowing through it to change 10 millivolts, which is enough to be converted to a standard logic pulse.

As the absence or presence of a bubble passes through the detector, it continues along the conductor and remains in memory. Data can be erased by a device called an **annihilator.** See Fig. 12-23. As bubbles pass through it, a pulse of current produces a magnetic field that is strong enough to cause the bubble to collapse.

Annihilator:

A bubble memory device that erases data from the memory.

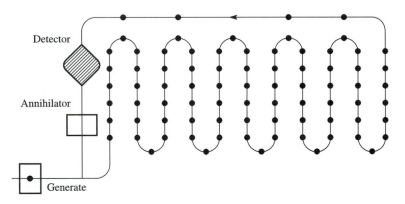

FIGURE 12.23
The basic bubble memory uses a serial-loop shift-register configuration. Access time to recover the data is long because for the data to be read, they must circulate through the entire loop.

The architecture of the MBM is available in two forms, the major loop and the minor loop, as shown in Fig. 12-24. In the major loop, data are stored along the single-conductor path and may contain up to 64K of memory. To obtain the desired memory, all the data must be shifted around in a serial loop. The memory capacity can be increased by connecting many minor loops to the major loop. A minor loop contains as much as 16K.

To explain in detail how these and other mass storage devices operate could easily occupy another book. Consequently, this subject will not be covered. The interested reader can consult popular electronic journals and computer books to obtain additional information.

■ REVIEW QUESTIONS

23. A 1K memory device stores _____ words of information.

24. To find the total number of bits in a memory device, multiply the _____ _____ by the _____ of memory locations.

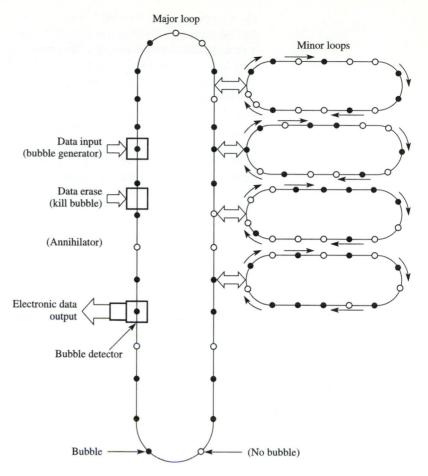

FIGURE 12.24
Minor loops are connected to one major loop to increase memory capacity.

25. When the amount of data a computer needs to work on exceeds the internal memory capacity, _____ _____ devices are used.
26. Four types of magnetic auxiliary memory devices used today include _____, _____, _____, and _____.

12.8 PROGRAMMABLE LOGIC DEVICES (PLDs)

Chapter 3 described the operation of basic logic gates. Chapter 5 illustrated combination logic circuits that use many logic gates to perform more complex logic functions. All of the circuits described in these two chapters are available as standard integrated circuit chips. Many more types of logic ICs than the ones covered in this book are available from manufacturers at a reasonably low cost to the customer. For many years, they have been used by circuit designers to perform an endless number of circuit functions. However, as digital equipment becomes more complex, there are some disadvantages to using them. For example, when standard ICs are utilized exclusively, large numbers of chips are often required. The result is that a considerable amount of circuit board space is used and manufacturing costs are high because of the time required for insertion, soldering, and testing.

An increasingly popular alternative to standard logic gates is the **programmable logic device (PLD).** The PLD is an IC that contains large numbers of logic gates that are all interconnected on the chip. Many of the interconnections are fusable links similar to the PROM memory cells shown in Fig. 12-7. The specific operation of the IC is determined by a process called *programming*. The programming function is performed by breaking the fuse links of some interconnections and leaving the other links intact to obtain a desired circuit pattern. The programming is usually performed by the manufacturer when a large quantity of one type of PLD is ordered. When smaller quantities of one circuit type are required, the programming is usually performed on a computer. Software packages prompt the programmer to write Boolean equations or fill in truth tables to specify the desired circuit operation. The software package performs several functions. First, if Boolean statements are used, it reduces the equation to its simplest form. Second, it determines if the input logic can be implemented on the PLD chip. Third, it determines which fuses are to be blown. Fourth, it produces a fuse map that conforms to an industry format called *JEDEC*. Fifth, it simulates the operation of the designed logic circuit configuration to verify that it functions properly. The final step is to use a hardware device called a *PLD programmer* to program the actual PLD IC. The software package of PLDs is available from integrated circuit manufacturers at a nominal cost. The PLD hardware devices range in cost from $100 to $2,500, depending on complexity and the features offered.

The most popular type of PLD is the Programmable Logic Array (PLA). Figure 12-25 shows the internal structure of a simplified PLA programmable logic device. Only a small portion of the IC is used. The configuration shown performs the operation of a full adder. It has three inputs, labeled *A, B,* and *C* (for carry-in), and two outputs, labeled SUM and CARRY-OUT. The inputs are connected to buffers, each of which has two output lines. The logic state applied to the input is passed straight through to the output without the bubble and is inverted at the output with the bubble. Each buffer output is initially

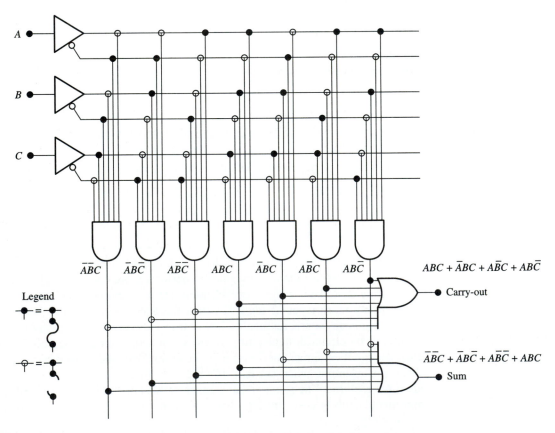

FIGURE 12.25
A PLD that operates as a full adder.

connected to a matrix of six input AND gates. The AND gate outputs are initially connected to each seven-input OR gate. During the programming process, some of the fusable interconnections are opened. A circle is used in the diagram to show each connection and they are shaded to show which connections are left intact. The unshaded circles indicate those connections that have been opened during programming. Unlike TTL input leads that recognize an open as a logic 1 state, the PAL gates observe open links as nonexistent inputs.

The circuit configuration of the PAL shown in the illustration performs the full-adder operation covered in Sections 4-8 and 5-9 of this textbook. The following sum-of-products (SOP) Boolean algebra equation describes the SUM output of the full-adder.

$$\text{SUM} = \overline{A}\,\overline{B}C + \overline{A}B\overline{C} + A\overline{B}\,\overline{C} + ABC$$

This means that output S (SUM) goes high if,

A and B are low and C is high, or

B is high and A and C are low, or

A is high and B and C are low, or

A and B and C are high.

Any other input combination applied to inputs A, B, and C causes the sum to be 0. The minterms of the SOP equation are listed at the AND gate that produces them.

The SOP Boolean algebra equation for the carry output is,

$$\text{CARRY–OUT} = \overline{A}BC + A\overline{B}C + AB\overline{C} + ABC$$

This means that output CARRY-OUT goes high if,

A is low and B and C are high, or

A and C are high and B is low, or

A and B are high and C is low, or

A and B and C are high.

Any other input combination applied to inputs A, B, and C causes the carry-out to be 0. The minterms of the SOP equation are listed at the AND gate that produces them.

The PLA example of the full adder is a simple illustration of how combination logic gate functions can be performed by a PLD. The actual internal configuration of PLDs is usually much more complex.

An example of a commercially available PLA is the PLS100 IC manufactured by the Signetics Corporation. It is a 28-pin device that contains 16 buffers, 144 AND gates, and 128 OR gates and has 16 inputs and 8 outputs. It is capable of replacing over 100 standard ICs. Another type, the Signetics PLS105, contains hundreds of AND gates, OR gates, and 14 flip-flops. It is capable of performing sequential circuit functions. As technological advancements are made, these programmable devices will become more versatile and provide an even larger variety of applications.

12.9   TROUBLESHOOTING A RAM DEVICE

Test 1   When an IC memory device appears to be defective, the first check made should be of the power supply and ground lines. This is because memory ICs are very sensitive to voltage levels that are a little too high, too low, or poorly filtered.

Test 2   If the V_{CC} and ground leads are all right, a dynamic test should be made on the data and address lines to see if a signal is present that alternates back and forth between a high and low. If any of the lines does not change, find out if this is supposed to happen. If an oscilloscope is used to observe the signals on the address and data lines, determine if they are all at the correct voltage levels. If they are not, the problem may be the result of another IC that is defective and that shares the same address or data lines as the ones being tested.

Test 3 Check the enable lines. Any ICs that are not enabled do not function. If the data, V_{CC}, and ground voltage levels are correct and the faults cannot be detected or isolated by using the dynamic test, the static test should be performed next by using the 16-bit 7489 RAM IC in Fig. 12-26(a). Test 4 describes how this should be accomplished.

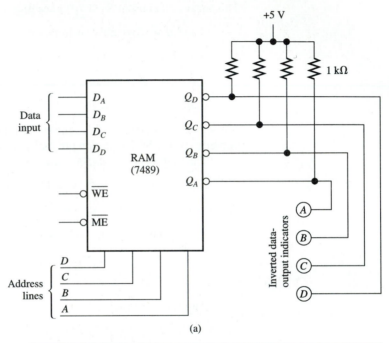

(a)

Addresses				Data inputs				Correct data outputs				Faulty data outputs			
D	C	B	A	D	C	B	A	D	C	B	A	D	C	B	A
0	0	0	0	1	0	1	0	0	1	0	1	0	1	0	1
0	0	0	1	0	1	0	1	1	0	1	0	1	1	1	0
0	0	1	0	1	0	1	0	0	1	0	1	0	1	0	1
0	0	1	1	0	1	0	1	1	0	1	0	1	1	1	0
0	1	0	0	1	0	1	0	0	1	0	1	0	1	0	1
0	1	0	1	0	1	0	1	1	0	1	0	1	1	1	0
0	1	1	0	1	0	1	0	0	1	0	1	0	1	0	1
0	1	1	1	0	1	0	1	1	0	1	0	1	1	1	0
1	0	0	0	1	0	1	0	0	1	0	1	0	1	0	1
1	0	0	1	0	1	0	1	1	0	1	0	1	1	1	0
1	0	1	0	1	0	1	0	0	1	0	1	0	1	0	1
1	0	1	1	0	1	0	1	1	0	1	0	1	1	1	0
1	1	0	0	1	0	1	0	0	1	0	1	0	1	0	1
1	1	0	1	0	1	0	1	1	0	1	0	1	1	1	0
1	1	1	0	1	0	1	0	0	1	0	1	0	1	0	1
1	1	1	1	0	1	0	1	1	0	1	0	1	1	1	0

(b)   (c)   (d)

FIGURE 12.26
Troubleshooting the 7489 RAM IC.

Test 4   To test if all of the address, data, and enable lines are operating properly, perform the *checkerboard* test. The checkerboard test involves storing 1s and 0s alternately in every odd address location and 0s and 1s in every even address location, as shown in Fig. 12-26(b).

1. *Action:*

   - Activate the memory enable line, $\overline{\text{ME}}$, by applying a low.
   - Apply a 0 at the write enable line, $\overline{\text{WE}}$, so that data can be stored into memory.
   - Load data into all 16 memory locations by storing 1010 into all even-numbered memory locations and 0101 into all odd-numbered memory locations.

   The checkerboard pattern is now stored in the memory.

2. *Action:*

   - Apply a low to the $\overline{\text{ME}}$ line.
   - Apply a 1 at the $\overline{\text{WE}}$ line so that data can be read from memory.
   - By sequencing numbers from 0000 to 1111 at the address inputs, read the data that were stored in all of the 16 different memory locations.

   Results: The contents of each memory location should appear at the outputs in its complementary form, as shown in Fig. 12-26(c). However, the contents appear as shown in Fig. 12-26(d).

   Conclusion: The pattern shows that 1s appear at bit *C* of each memory location output. Because each output bit is the complement of the data stored into it, a 0 must have been loaded into each address location.

3. *Action:* Connect a voltmeter to data-input line *C*.

   Results: A 0-volt potential is read.

   Conclusion: By using the checkerboard method, a data-input line *C* shorted to ground was easily detected.

■ SUMMARY

- One of the major advantages of digital equipment over analog equipment is its capability to receive, hold, and retrieve information.

- Four important characteristics to consider when selecting a type of memory device for a particular application are packing density, speed, power consumption, and cost.

- The workhorse of a computing device is the microprocessor, which is made of combination logic circuits that perform arithmetic functions and make various decisions. The microprocessor is provided with a list of instructions and data from memory devices that enable it to operate as it does.

- Addressing a memory device is done by either putting binary data in or reading data out of a certain memory location. Three common methods of addressing memories are one-dimensional, two-dimensional, and stacking.

- Read-only memory (ROM) is a type of memory that can only be read from. It is nonvolatile, and its different types are identified by the semiconductor device that makes up the individual memory cells. The three basic types of ROMs are diode matrix, bipolar transistor matrix, and MOSFET matrix.

- Some types of ROM memories that are also nonvolatile can be programmed by the user. These include PROMs, EPROMs, EAROMs, and EEPROMs.

- Random-access memory (RAM) is a type of memory from which data can be read from or written into. The two types of RAMs are static and dynamic.

- The capacity rating of a memory device is based on how many memory words it can store.

- Memory storage capacity can be expanded by using mass storage devices such as magnetic tape, magnetic bubbles, and magnetic disks.

■ PROBLEMS

1. Binary words can represent which of the following? (12-2)
   (a) Data
   (b) Memory addresses
   (c) Computer program instructions
   (d) All of the above

2. The higher the packing density of a memory device means (less, more) bits can be stored within a certain area. (12-1)

3. An 8-bit word is stored in how many memory cells? (12-2)

4. A 3-bit address decoder enables how many different words to be stored into or retrieved from a memory device? (12-2)

5. The (numerical, store, recall) key(s) of a calculator enters the data, the (numerical, store, recall) key(s) activates the write function of the memory, and the (numerical, store, recall) key(s) activates the read function of the memory. (12-1)

6. Memory elements constructed with (bipolar, MOS) devices generally consume less power and have greater packing density. (12-3)

7. Explain the difference between firmware and software. (12-3, 12-4)

8. Explain the difference between ROM and RAM devices. (12-3, 12-4)

9. Information that is frequently used by a computer is more likely stored in a (RAM, ROM) memory device. (12-3)

10. A (RAM, ROM) memory device that loses its contents if the power source is removed from it is called (volatile, non-volatile). (12-3)

11. Firmware is stored in a (RAM, ROM) memory device and software is stored in a (RAM, ROM) memory device. (12-3, 12-4)

12. During the evolution of computers, what computer devices have been used to make up the main working memory? (Working Memory)

13. What are three types of components used as memory cells for mask-programmable ROMs? (12-3)

14. What is addressing? (12-2)

15. The enable line (allows, prevents) data to be passed through to the output lines of a tristate buffer when it is in its active state. (12-3)

16. What does a computer program consist of? (Working Memory)

17. Explain the advantage of using a programmable type of ROM over a masked ROM device. (12-3)

18. What are three types of programmable ROM devices? Describe how data are programmed into them and how data are erased. (12-3)

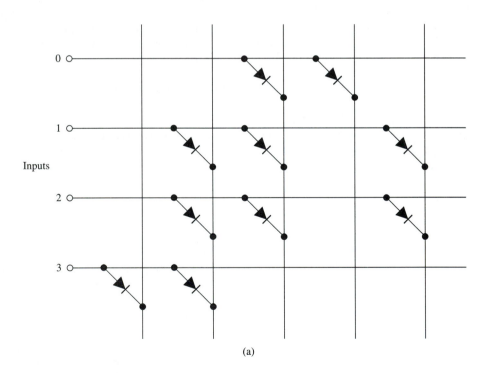

(a)

	Inputs		Outputs	
0	– – 0 0			
1	– – 0 1			
2	– – 1 0			
3	– – 1 1			

(b)

FIGURE 12.27
(a) ROM-matrix circuit and (b) truth table for Problem 30.

19. What are three functional applications of ROM devices? (12-3)

20. What are three ROM devices that can be reprogrammed? (12-3)

21. A _____ RAM memory cell uses a capacitive element for storing a data bit. (12-4)

22. RAM devices consist of cross-coupled circuits that form flip-flops; the dynamic RAM uses _____ _____ components, and the static RAM uses _____ _____ components. (12-4)

23. What are three functional applications of RAM devices? (12-4)

24. Why is a RAM device called "random"? (12-4)

25. How many bits of memory are stored by a computer that uses 8-bit words and has a capacity rating of 128K? (12-6)

26. What are mass storage devices used for? (12-7)

27. How are 1s and 0s placed on a magnetic tape for storage purposes? (12-7)

28. A one-dimensional memory device uses how many address decoders? (12-2)

29. To store 12 into memory location 13 of a 7489 IC, what logic states are applied to the following input lines? (12-5)

Data Inputs	Address Inputs		
$D\ C\ B\ A$	$D\ C\ B\ A$	$\overline{WE}$	$\overline{ME}$

List the logic states at each terminal when data is read from memory location 13.

Address Inputs			Data Outputs
$D\ C\ B\ A$	$\overline{WE}$	$\overline{ME}$	$D\ C\ B\ A$

30. Complete the output portion of the truth table of Fig. 12-27(b) for each vertical line of the ROM matrix circuit of Fig. 12-27(a). Also, indicate what type of logic gate each vertical line represents. (12-3)

Troubleshooting

31. To test a 7489 memory IC that appears to be faulty, a checkerboard pattern of bits is written into the chip. The incorrect pattern of Fig. 12-28 is then read from the chip. Which of the following causes this fault? (12-5, 12-9)
(a) Address line *B* is shorted to V_{CC}.
(b) Address line *A* is shorted to ground.
(c) Input data lines 1 and 4 are open.
(d) All of the above.

32. A pure 4-bit binary count is sequentially written into the 7489 RAM IC, as shown in column *A* of Fig. 12-29. Column *B* shows the data read from the chip. Which of the following is the most likely problem? (12-5, 12-9)
(a) Input line 8 is shorted to V_{CC}.
(b) Output line 8 is shorted to ground.
(c) Address line *D* is shorted to V_{CC}.
(d) All of the above.

Data written into memory				Data read out of memory			
(8)	(4)	(2)	(1)	(8)	(4)	(2)	(1)
1	0	1	0	0	1	0	1
0	1	0	1	0	1	0	1
1	0	1	0	0	1	0	1
0	1	0	1	0	1	0	1
1	0	1	0	0	1	0	1
0	1	0	1	0	1	0	1
1	0	1	0	0	1	0	1
0	1	0	1	0	1	0	1
1	0	1	0	0	1	0	1
0	1	0	1	0	1	0	1
1	0	1	0	0	1	0	1
0	1	0	1	0	1	0	1
1	0	1	0	0	1	0	1
0	1	0	1	0	1	0	1
1	0	1	0	0	1	0	1
0	1	0	1	0	1	0	1

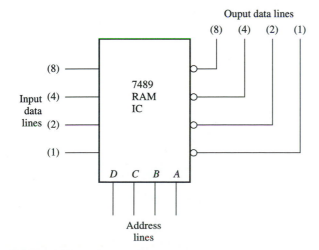

FIGURE 12.28
Bit patterns for Problem 31.

Column *A*				Column *B*			
(8)	(4)	(2)	(1)	(8)	(4)	(2)	(1)
0	0	0	0	0	1	1	1
0	0	0	1	0	1	1	0
0	0	1	0	0	1	0	1
0	0	1	1	0	1	0	0
0	1	0	0	0	0	1	1
0	1	0	1	0	0	1	0
0	1	1	0	0	0	0	1
0	1	1	1	0	0	0	0
1	0	0	0	0	1	1	1
1	0	0	1	0	1	1	0
1	0	1	0	0	1	0	1
1	0	1	1	0	1	0	0
1	1	0	0	0	0	1	1
1	1	0	1	0	0	1	0
1	1	1	0	0	0	0	1
1	1	1	1	0	0	0	0

FIGURE 12.29
Bit patterns for Problem 32.

■ ANSWERS TO REVIEW QUESTIONS

1. receive, hold, retrieve   2. storage
3. (1) Packing density, (2) power consumption,
   (3) speed, and (4) cost   4. memory cell
5. (1) The address decoder, (2) the memory storage elements,
   and (3) the output circuits   6. two
7. logic, arithemetic, decisions   8. ROM, RAM
9. nonvolatile   10. The manufacturer

11. diode, bipolar, MOSFET   12. 1, 0
13. EPROM, EAROM, EEPROM   14. True
15. repeatedly   16. RAM   17. volatile
18. flip-flop latch, bipolar transistor, MOSFET
19. capacitive, refresher   20. (1) Main working memory,
    (2) storing computer software programs, and
    (3) replacing logic gates   21. opposite
22. 0, 0, 1, 0   23. 1024   24. word size, number
25. mass memory   26. tapes, disks, CDs, bubbles

CHAPTER **13**

FUNCTIONAL DIGITAL CIRCUITS

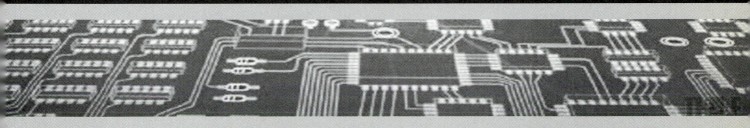

When you complete this chapter, you will be able to:

1. Explain the operation of a digital voltmeter.
2. Explain the operation of a digital frequency counter.
3. Describe the operation of a digital measuring device.
4. Describe how a digital device controls the velocity, physical position, and number of rotations of a DC motor.
5. Explain the operation of a digital clock.
6. Explain the operation of a digital locking device.
7. Describe the operation of the digital circuitry for a basketball shooting game.
8. Explain the operation of a sound synthesizer.
9. Explain the logic operation of a coffee vending machine.
10. Troubleshoot a digital clock.

13.1 DIGITAL VOLTMETER

Digital voltmeter:

A type of test equipment that uses internal digital circuitry to measure and display voltage levels.

The **digital voltmeter** illustrated in Fig. 13-1 is capable of measuring and displaying a positive potential ranging from 0.0 to 9.9 V. It is constructed primarily of two decade counters, two BCD decoder/drivers, two LED displays, an eight-input DAC, a comparator, and a voltage-regulator network. Two timing signals are provided, one by a single 20-kHz astable multivibrator and the other by a timing network that uses a one-shot at its output.

Functions

Timing Network   This network enables the astable multivibrator to produce a 1-ms negative reset pulse that causes both decade counters to clear 15 times per second.

Astable Multivibrator   This multivibrator produces a 20-kHz signal that is applied to the clock input of FF1. When a high is present at the J and K inputs of FF1, it toggles on every trailing edge of each clock pulse. This causes the cascaded tenths and ones digit decade counters to increment. However, if the J and K inputs are low, FF1 is in the hold mode; as a result, the decade counters are disabled and, therefore, are frozen.

D/A Converter   The Q outputs of both counters are applied to the digital inputs of the DAC. Because both counters are BCD instead of binary, the DAC is scaled for BCD-to-analog conversion. The analog output of the DAC, which ranges from 0 to 9.9 V, is proportional to the digital count applied to its input.

Decimal Display   The Q outputs of each decade counter cause LED segments to display the number of its count. A separate BCD decoder/driver interfaces each counter to its respective tenths and ones LED displays.

Comparator   The analog output voltage of the DAC is applied to the negative input of the comparator, and the voltage being measured with a probe is applied to the positive input. If the probe is connected to a voltage that is higher than the DAC output, the comparator output goes into saturation and produces an output of approximately 12.5 V. If the two voltages are the same, a 0-V low is produced at the output of the comparator. The output of the comparator is connected to the J and K inputs of FF1 through a voltage-regulator network.

Voltage-Regulator Network   The regulator network consists of a 5-V zener diode and a resistor. It is used to ensure that a voltage greater than 5 V is not applied to the J and K inputs of FF1 when the comparator output goes to a positive-voltage saturation.

Operation

- Assume that the timing circuit produces a 1-ms negative pulse that causes the decade counters to clear (reset). The count of 0000 0000$_\text{BCD}$ is applied to the input of the DAC. As a result, the analog output of the DAC is 0 V.
- The 20-kHz signal is applied to the clock input of the two cascaded decade counters. Suppose that the input probe of the voltmeter is connected to a 0-V potential. Because the output of the DAC is also 0 V, the comparator output generates a low and puts FF1 in the hold mode. Therefore, the cascaded up-counters do not allow the 20-kHz clock signal to cause them to increment.
- Suppose that the probe is connected to a +3.5-V potential. Also assume that the counters have just been reset by a pulse from the timing network. The output of the DAC is 0 V. Because the positive voltage at the noninverting input is greater than the inverting input of the comparator, its output goes to an approximate positive 12.5-V saturation potential. When this occurs, +5 V are dropped across the zener diode and the remaining 7.5 V are dropped across the resistor. Because the zener is in parallel with the J and K inputs of

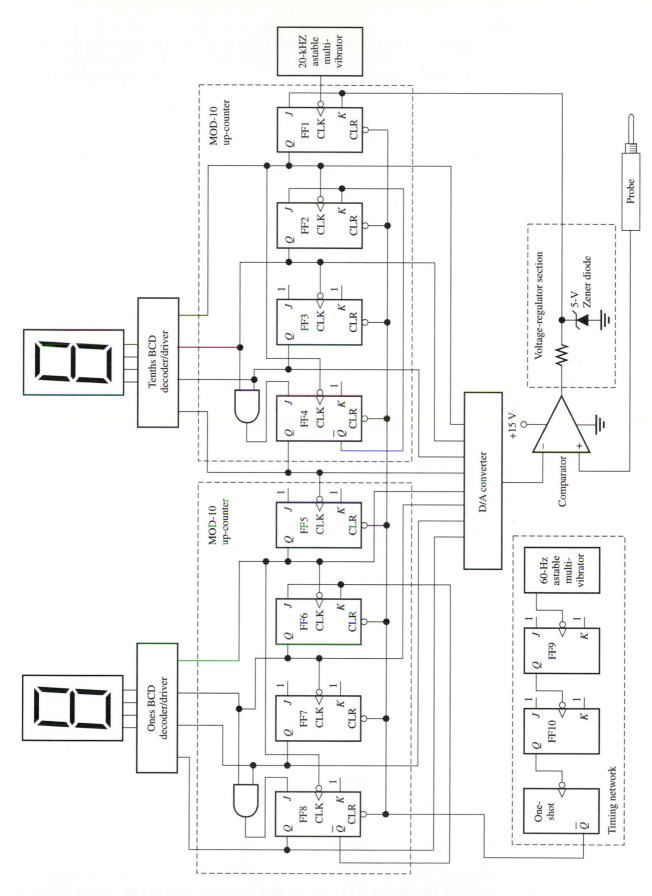

FIGURE 13.1
Digital voltmeter.

465

FF1, a +5-V high is applied to them. With a high at the *J* and *K* inputs of FF1, it is in the toggle mode, which enables the 20-kHz signal to increment the counters. They count up until the tenths counter reaches 5 and the ones counter reaches 3, which causes the output voltage of the DAC to be 3.5 V. Because both inputs of the comparator are 3.5 V, its output goes low and causes FF1 to go into the hold mode. The number 3.5 is temporarily frozen in the counters and is displayed on the LED readouts.

■ The function of the timing network is to reset the counters with the one-shot output pulse. Each pulse duration is 1 ms and occurs 15 times per second. The count-up time of the cascaded decade counters lasts up to only another 5 ms, until they are frozen. Therefore, the numbers are displayed for about 60 ms before the next reset pulse occurs. Because the display time is relatively long in comparison to the time duration of the reset pulse and the count-up period, 3.5 is read because the operator is unable to see the numbers of the LED readouts go to 0.0 and then count up.

■ Because the voltage the probe is connected to can change, the reset pulses clear the up-counters 15 times per second to enable the voltmeter to be continually updated.

■ REVIEW QUESTIONS

1. In a digital voltmeter, the output of the _____ enables and disables FF1 by alternately putting it in the hold or toggle mode.

2. When the positive voltage at the noninverting input of the comparator is (less, greater) than the voltage at the inverting input, the comparator output goes to a positive-voltage saturation potential.

3. The contents of the up-counter is updated _____ times per second so that any changed voltages can be immediately displayed.

13.2 DIGITAL FREQUENCY COUNTER

Digital frequency counter:

A type of test equipment that uses internal digital circuitry to measure and display the number of pulses, sine waves, or square waves that occur in a second.

A **digital frequency counter** is a standard type of testing instrument that is used for determining the frequency of sine waves, square waves, and pulses. The count of the signals being measured by these devices is usually displayed on a LED or LCD readout in hertz or cycles per second.

The circuit of Fig. 13-2 is a frequency counter with a four-digit display that is capable of measuring a frequency up to 9999 Hz. The functions of the components follow.

Functions

Astable Multivibrator   When power is applied to the circuit, the astable multivibrator generates a square wave output of 0.5 Hz. The square wave is used as the reference signal, and the accuracy of the frequency counter is determined by the stability of the multivibrator. The components that make up an astable multivibrator are subject to temperature changes and aging. Therefore, their values change, which in turn causes the output frequency to change and become inaccurate. To minimize frequency drift, the reference signal of frequency counters is often produced by crystal devices, which are extremely stable.

J–K Flip-Flop (FF1)   The square wave output of the astable multivibrator is applied to the negative-edge-triggered input of bistable multivibrator FF1. Both the *J* and *K* inputs are connected to 1 states, which causes the flip-flop to operate in the toggle mode. Therefore, on every negative edge of the reference-frequency square wave, the *Q* output of this flip-flop changes states. This occurs every 2 seconds.

One-Shot   The output of this monostable multivibrator is triggered by each positive-edge output of the flip-flop (FF1). As the one-shot is triggered, its output changes from its

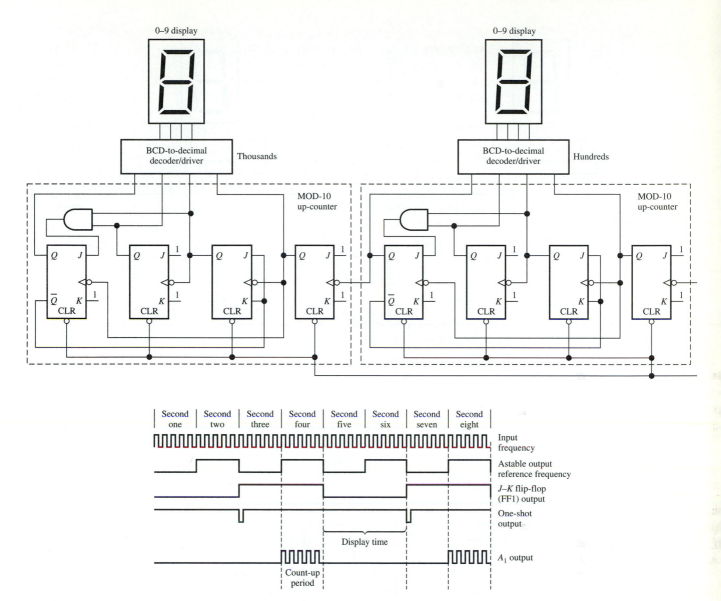

FIGURE 13.2
Digital frequency counter.

normal 1 state to a low for a 50-μs time duration. The output then changes back to a high until the next positive edge arrives at the input 4 seconds later. The purpose of the one-shot output pulse is to simultaneously reset the flip-flops of the counters every 4 seconds.

Schmitt Trigger The input signal applied to the input of the frequency counter can be a sine wave, a square wave, or pulses. Because digital devices accurately respond to rectangular waves at certain voltage levels, any nonrectangular signals such as sine waves can be transformed by a Schmitt trigger wave-shaping device into square waves.

Three-Input AND Gate Three separate signals are applied to the inputs of the AND gate. They are supplied from the following:

1. The Q output of the J–K flip-flop FF1.
2. The reference-frequency square wave of the astable multivibrator.
3. The test probe that makes the circuit connection at the location from which the frequency is being read.

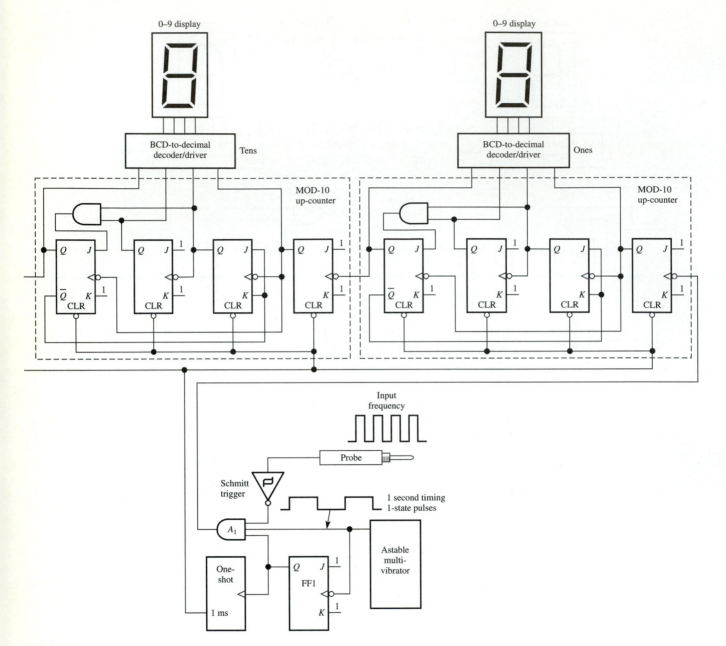

FIGURE 13.2
(Continued)

When the *Q* output of *J–K* flip-flop FF1 is high and the reference-frequency square wave is high, the frequency being read is sent through to the output of the AND gate.

Decade Counters Four separate decade counters are cascaded. The input frequency from the AND gate is applied to the clock input of the ones digit decade counter. After each cascaded counter increments from 0 to 9, it recycles on the following count, which causes the next counter to increment.

Decoder/Driver The output of each decade counter is applied to a decoder/driver that converts the BCD number to a code that causes the LED readout to display the equivalent decimal digit.

LED Readout Display Four separate readouts display the decimal ones, tens, hundreds, and thousands digits. Therefore, the maximum count that can be read by the frequency counter is 9999.

Operation

The functional description of how the frequency counter operates is divided into 1-second time intervals. A timing diagram is used as a reference by illustrating the appearance of waveforms at significant locations of the circuitry.

Second 1 The frequency signal being read is not passed on to the output of AND gate 1 because both the outputs of the astable multivibrator and *J–K* flip-flop (FF1) are not high.

Second 2 The astable output goes high while the (FF1) output remains low. The AND gate does not pass the input frequency to its output.

Second 3 As the reference frequency signal goes low, its trailing edge triggers *J–K* flip-flop FF1. With a low from the astable output and a high from FF1, the AND gate does not pass the input frequency to its output. Also, when the Q output of FF1 goes high, it provides a positive-edge signal that activates the one-shot multivibrator. The 1-ms output pulse of the one-shot clears all the counters.

Second 4 The astable output goes high. Because FF1 is already high, the input frequency is passed on through to the AND-gate output. This frequency is fed to the ones decade counter. During this 1-s time period, the four counters increment, and each of their counts is displayed on a LED readout after the count-up period ends.

Second 5 The count-up time stops as the reference frequency goes low. This creates a negative edge, which also causes the FF1 output to go low. The frequency is displayed during seconds 5 and 6.

The frequency-counter circuit operates on 4-s cycles. After the counter circuits clear, zeros are displayed for 1 s. The counters then increment for a period of 1 s and finally display the frequency for 2 s before starting over again. The cycles keep repeating to ensure that the frequencies being read are kept updated.

The frequency being read is counted in hertz. To display the count, the circuit of Fig. 13-2 counts the actual sine waves, square waves, or pulses over a 1-s period.

This frequency counter is a very simplified version of an actual commercial frequency-counting instrument. The operation of an actual frequency counter is such that the operator does not see the updating process of the decade counters. Instead, the numbers are viewed as a constant display. This can be accomplished by using circuits that are timed so that after being cleared, the counters are incremented by the incoming frequency for 0.1 s. The value representing the frequency is then displayed for 0.9 s before the next update cycle. Because the frequency that represents a count over a time period of 1 s is counted for only 0.1 s, the operator must multiply the number being displayed by a factor of 10.

■ REVIEW QUESTIONS

4. In a frequency counter, the frequency being read is sent through AND gate 1 to the cascaded decade counter when the Q output of FF1 is (low, high), and the astable multivibrator output is (low, high).

5. To change any nonrectangular waveform applied to the frequency counter into a square wave signal, a _____ _____ is used.

6. The frequency counter of Fig. 13-2 operates on _____ second cycles.

13.3   DIGITAL MEASURING INSTRUMENTS

Digital measuring instruments:

Equipment that uses internal digital circuitry to measure quantities of sound, temperature, humidity, etc.

Thermistor:

An electronic sensing component that varies its resistance when the temperature around it changes.

For many years, analog gauges have been used to measure and display information such as temperature, humidity, liquid level, pressure, and light intensity. Today, most gauges are being replaced by digital devices that display the value on LED or LCD digital displays.

One of the key elements that make up a **digital measuring instrument** is the sensor that is capable of detecting such values as degrees, pounds per square inch, etc. For example, a sensor that is capable of detecting differences in temperature is called a **thermistor.** This device gets it name from *thermo* and *resistor* because it operates like a variable resistor as its resistance changes when it senses ambient-temperature changes around it.

Figure 13-3 shows how a thermistor can be used with digital circuitry to operate as a thermometer that is capable of measuring temperatures from 0° to 99°.

Functions

Low-Frequency Clock Generator   This device consists of a 555 linear IC that is configured to operate as an astable multivibrator. It produces a square wave with a duty cycle that causes it to be high for most of the time and low for a brief moment, as shown in the timing diagram in Fig. 13-3.

One-Shot   This device consists of a 555 linear IC that is configured to operate as a monostable multivibrator. Its output, which is normally low, is activated to a temporary high by a positive-edged-triggered output signal from the low-frequency clock generator.

Variable-Frequency Clock Generator   This device consists of a 555 linear IC that is connected to operate as an astable multivibrator. Section 11-13 of Chapter 11, which describes the operation of this device, explains that the output frequency it produces is affected by the values of the external components connected to it. The thermistor, which is the primary element of the temperature probe, is one of these external components.

AND Gate 1   The output of this gate is connected to the clock input of the two cascaded BCD counters. The bottom input of the gate is connected to the output of the one-shot, and the top input is connected to the output of the variable-frequency clock generator. When the one-shot is high, it enables the AND gate to allow the output frequency from the variable-frequency clock generator to pass through to the clock input of the cascaded BCD counters.

BCD Counter   This device consists of two BCD up-counters that are cascaded and can count from 00 to 99.

Decoder/Drivers   There are two BCD-to-decimal decoder/drivers that convert BCD values from the counters into an output that causes the digital displays to show an equivalent decimal number.

Operation

Refer to the waveform diagram in Fig. 13-3.

Time Period 1   The low-frequency clock generator goes to a very brief low. This clears any count in the BCD counter.

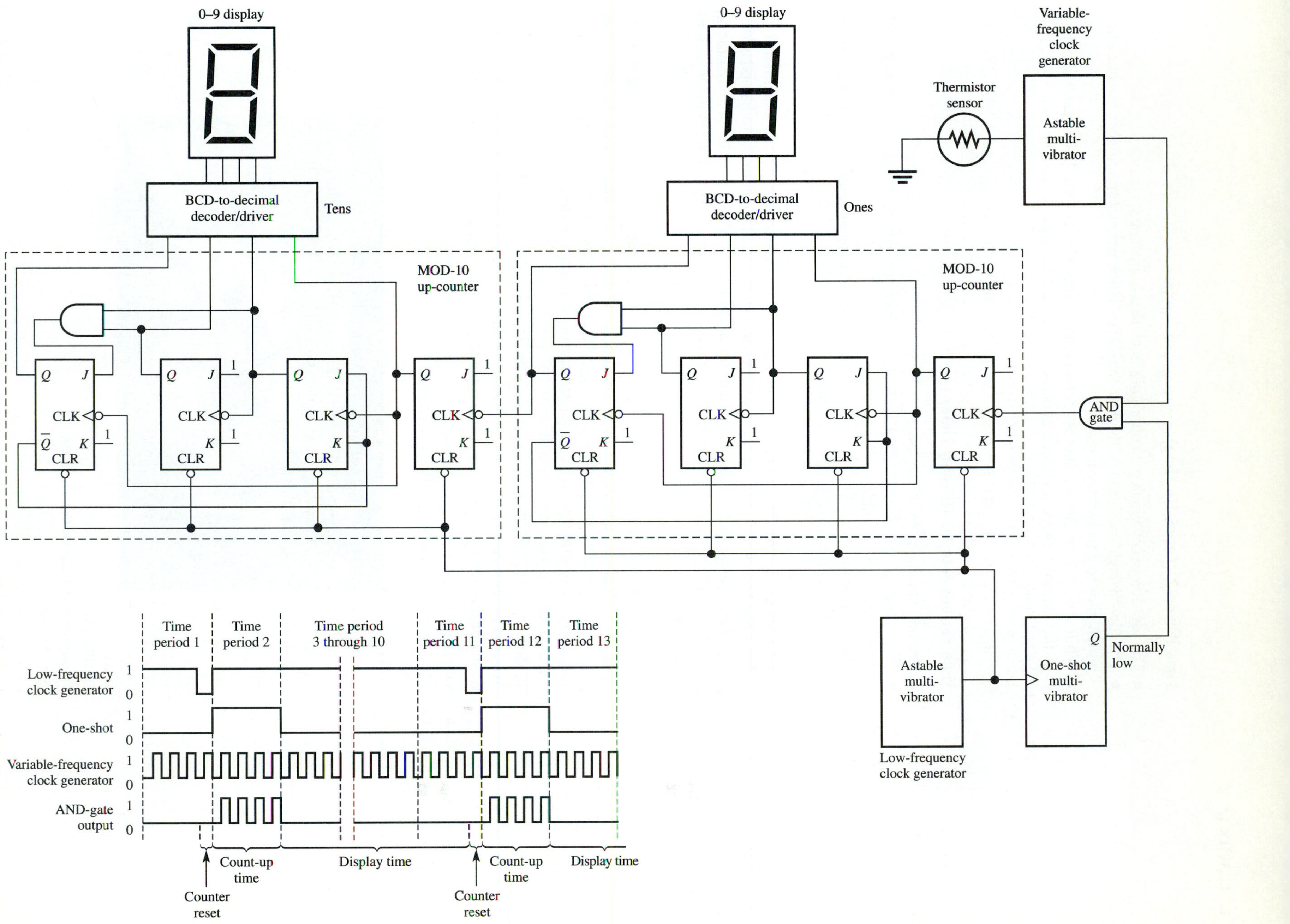

FIGURE 13.3 *Digital thermometer measuring instrument.*

471

Time Period 2 (Count-Up Time)

- The low-frequency clock generator goes back to a high state.
- The low-to-high transition of the low-frequency clock generator activates the one-shot, which goes to a temporary high.
- While a high from the one-shot is applied to the bottom lead of the AND gate, the output of the variable-frequency clock generator is passed through to the two cascaded BCD up-counters.

Time Periods 3 to 10 (Display Time)

- The one-shot goes back to a low.
- The output of the variable-frequency clock generator can no longer pass through the AND gate to the counter.
- The number of pulses from the variable-frequency clock generator that incremented the up-counter during time period 2 is shown on the display.

Time Period 11   At the end of this time period, the low-frequency clock generator produces a temporary low that clears the counter and allows the thermometer to be updated.

The duration of the reset and count-up times is very short, and the display time is present most of the time. Therefore, the user is only capable of viewing the number present during the display time.

The thermistor is a device that has a negative temperature coefficient, meaning that as the temperature increases around it, the resistance decreases. Suppose that the temperature is 20 degrees. The resistance of the thermistor is a value that causes the 555 timer to produce 20 square waves during time period 2, which enables the counter to increment and display 20. Time period 2 is a predetermined reference time controlled by the unstable output of the one-shot and must be very precise. If the temperature increases to 40 degrees, the resistance of the thermistor decreases by half and causes the variable-frequency clock generator to operate twice as fast and to produce an output of 40 square waves during the next count-up time. A photograph of a digital thermometer is shown in Figure 13-4.

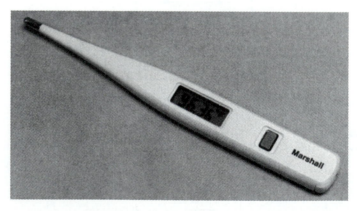

FIGURE 13.4
Photograph of a digital thermometer.

Digital instruments that measure other variables such as humidity and light are constructed with whatever sensor is needed to detect the variable change. The remainder of the circuitry is fundamentally the same as that of the thermometer.

■ **REVIEW QUESTIONS**

7. The primary difference between different digital measuring devices is the _____ that is used to detect variable changes.

8. On which of the following is the accuracy of the digital thermometer dependent?
   (a) Thermistor
   (b) One-shot
   (c) Variable-frequency clock generator
   (d) All of the above

9. As the temperature increases, the thermistor resistance (increases, decreases) and the variable-frequency clock generator (increases, decreases).

10. The count-up time occurs when the one-shot output is (low, high).

11. The user is unable to see the thermometer reset and count up because it happens very quickly and because the display period is about (1, 4, 9, 13) times greater.

13.4 DIGITAL DEVICE CONTROLLING THE VELOCITY OF A MOTOR

The circuit of Fig. 13-5 accepts a digital command signal that represents a numerical value. After receiving the command input, the circuit causes a DC motor to rotate at a speed that is proportional to the numerical value entered on the keypad. The number of different speeds is limited to a range of 1 to 9.

Operation

Suppose that the machine operator wants to have the motor rotate at about half speed. A description of how the circuit performs this function follows:

1. The operator presses key number 5.
2. The push-button closure activates a one-shot multivibrator inside the keypad circuitry that generates a momentary 10-ms 0-state pulse at the $\overline{R}$ output. This negative pulse is sent to the clear inputs of flip-flops FF1-FF4, which causes them to reset simultaneously.
3. The completion of the one-shot pulse during the closure of the number 5 key causes outputs $\overline{1}$ and $\overline{4}$ to go low and outputs $\overline{2}$ and $\overline{8}$ to remain high. Therefore, the parallel-loaded shift register consisting of FF1-FF4 is preset to a BCD 5.
4. The LED readout displays a 5 as the Q outputs of each flip-flop are applied to the BCD-to-decimal decoder/driver.
5. The Q outputs of each flip-flop are also applied to a digital-to-analog converter. The analog output voltage is proportional to the BCD input. The output is connected to the amplifier.
6. The output voltage of the amplifier is proportional to the applied input voltage from the DAC. The amplifier is needed to provide sufficient power to drive the DC motor.
7. The speed of the motor is proportional to the amplifier output voltage. Also, depressing the 0 key stops the rotation of the motor.

   The speed of the motor changes every time another key is depressed. Each key closure creates an $\overline{R}$ pulse that clears the register before the new BCD number is parallel-loaded. As the new number generates a different DAC analog output voltage, the motor speed increases or decreases.

■ **REVIEW QUESTIONS**

12. The motor rotates at its highest velocity when key _____ is pressed and stops when key _____ is pressed.

13. The numerical outputs of the keypad will not be activated unless the reset pulse is completed and the key closure is still being made. True or false.

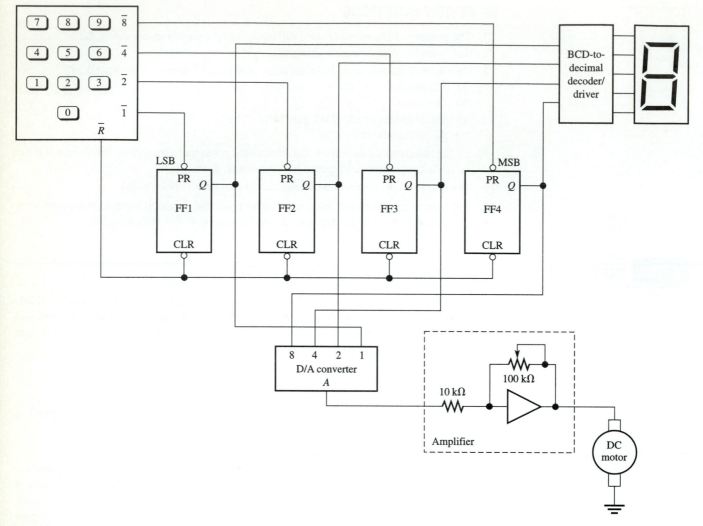

FIGURE 13.5
Digital device controlling the velocity of a motor.

14. If key 7 is pressed, keypad outputs _____, _____, and _____ go low, and output(s) _____ remains high.

13.5 DIGITAL DEVICE CONTROLLING THE NUMBER OF MOTOR REVOLUTIONS

The circuit of Fig. 13-6 accepts a digital command signal that represents a numerical value. After receiving the command input, the circuit causes a DC motor to rotate at the same number of revolutions as the numerical value that was entered on the keypad. The number of rotations is limited to a range of 1 to 9.

Operation

Suppose that the machine operator wants to have the motor make five revolutions. A description of how the circuit performs this function follows:

Assumptions FF5 is cleared.

1. The operator presses key number 5.
2. The push-button closure activates a 10-ms one-shot multivibrator inside the keypad circuitry, which generates a momentary 0-state pulse at the $\bar{R}$ output of the keypad. This

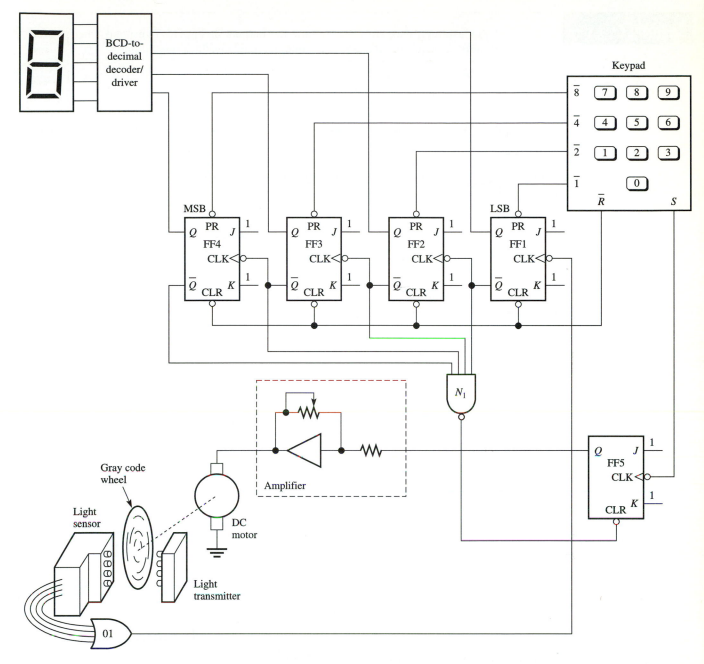

FIGURE 13.6
Digital device controlling the number of motor revolutions.

momentary low is applied to all of the clear inputs of FF1–FF4; therefore, all of the flip-flops that make up a down-counter are cleared.

3. As soon as the 10-ms reset pulse is finished, it enables the keypad's numerical outputs to be activated, which causes the normally high 4 and 1 outputs to go low and outputs 2 and 8 to remain high. Because FF1 and FF3 are preset, a BCD 5 is parallel-loaded into the down-counter.

4. The BCD 5 is fed to the BCD-to-decimal decoder/driver and causes the LED readout to display a 5.

5. Before any keys are pressed, the strobe (*S*) output of the keypad circuitry is low. While one of the keys is activated, the *S* output goes high. When the key is released, the *S* output returns to a low.

EXAMPLE 13.1

Draw a timing diagram to show how the outputs in Fig. 13-7 respond when a 3 is inserted into the circuit of Fig. 13-6 until it completes three revolutions.

Solution   See Fig. 13-7.

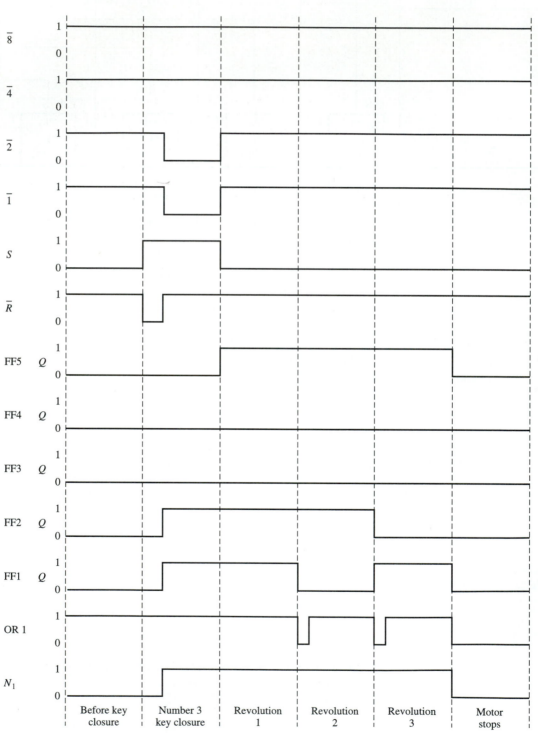

FIGURE 13.7
Timing diagram for Example 13.1.

6. The high-to-low transition of the *S* output caused by releasing a key enables FF5 to toggle, so that the *Q* output goes to a 1 state. When this happens, the 1 is applied to the input of the amplifier, which starts the DC motor rotating.

7. The motor is mechanically coupled to a Gray code wheel. Therefore, it rotates at the same rpm as the motor. As the Gray code wheel rotates, 4-bit number combinations consisting of 1s and 0s are sensed by the light sensor and are applied to OR gate 1. The only time all zeros are applied to the four inputs of OR gate 1 is when the Gray code wheel is in the 0-degree position. When this happens, the output of 01 makes a high-to-low transition. This signal is received by the clock input of FF1 and is read as a trailing edge of a clock pulse, which causes FF1 to toggle. The first time this happens, the down-counter decreases to 4. As soon as the Gray code wheel rotates to the 0001 position, the output of OR gate 1 goes back to a high.

8. Because the high-to-low transition generated at OR Gate 1 occurs only once during each 360-degree revolution, the down-counter decrements only once during each rotation. The motor rotates four more times as the down-counter decrements to 0000.

9. When the down-counter decrements to 0000, each *Q* output of FF1–FF4 is at a low and each $\overline{Q}$ output is at 1. With all 1 states being applied to NAND gate 1, a low is generated at the output that causes FF5 to clear. Therefore, the *Q* output of FF5 goes to a low (0-V potential) and causes the amplifier to lose its output voltage. Therefore, the motor stops after five revolutions.

■ REVIEW QUESTIONS

15. When the motor is running, the *Q* output of FF5 is (low, high), and when it is stopped, the *Q* output is (low, high).

16. The high-to-low transition that causes the down-counter to decrement occurs when the Gray code output goes from its highest number to _____ .

17. The motor stops when the $\overline{Q}$ outputs of FF1-FF4 are _____ . These outputs are applied to _____ , which produces a (low, high) output that (presets, clears) FF5.

13.6   DIGITAL CONTROL OF A PHYSICAL POSITION

Rack and pinion:

A mechanical device that is made up of a gear and a bar with gear teeth that converts rotary motion into linear movement.

Figure 13-8 illustrates how the positioning of a linear-motion device can be controlled by a digital command that is entered by a keypad input with numbers 0–9. A bar mechanism positioned by a gear assembly (known as a **rack and pinion**) has 10 position regions, with mechanical stops at both ends of travel. The bar is shown at position 0, and its location is detected by a Gray code wheel mounted on the same motor shaft as the gear. When a number greater than 0 is entered into the keypad, the gear turns in a clockwise direction, causing the bar to travel to the left. When the bar reaches the region that is the same as the number entered into the keypad, the gear stops and a light turns on. If a second number greater than the first entry is registered, the gear will again rotate clockwise until the bar reaches the new location. However, if the second entry was less than the first number entered, the gear would turn counterclockwise, pushing the bar to the right until its position equaled the number entry, at which time the gear would stop and the light would turn on.

Functions

To better explain the operation of the positioner, functional descriptions of the various sections are provided.

Input Section (Keypad)   To move the bar mechanism to any one of 10 locations, a command is entered by a keypad that has 10 different keys, numbered 0–9. As the desired value is inserted by pressing one of the keys, a 10-ms negative reset pulse ($\overline{R}$) first clears the flip-flops. Then the active-low numerical outputs are activated as an equivalent binary number is parallel-loaded into a 4-bit register.

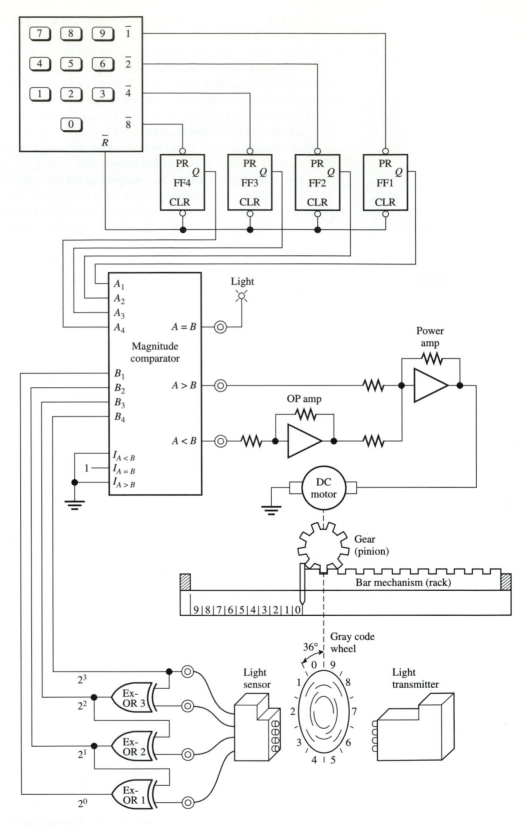

FIGURE 13.8
Digital control of physical position.

Decision Section (Magnitude Comparator) The control portion of the positioner is made up of a combination circuit called a *magnitude comparator*. As described in Chapter 5, it is made up of two separate 4-bit inputs and three single-lead outputs. The function of the circuit is to compare 4-bit input A to 4-bit input B. If the binary numerical value applied to input A is larger than B, the $A > B$ output line will go high while the other two outputs go low. If the numerical value of input B is larger than A, the $A < B$ line will go high while lows are generated at the other two outputs. If the binary numerical values of inputs A and B are the same, the $A = B$ line will go high while the other two outputs go low.

Power Section The power section consists of two separate inputs and one output line. One of the inputs is connected to the $A > B$ output of the magnitude comparator and the other input is connected to the $A < B$ output. The function of the output line of the power section is to provide an amplified potential that is sufficient to drive the DC positioning motor of the circuit.

When output $A < B$ of the comparator goes high, it is applied to the input of the operational amplifier. Because the op-amp is set at a gain of 1, its only function is to invert the +5-V signal to a −5 V. As the negative 5 V are applied to the power-amplifier input, it is amplified and inverted to a positive potential.

When output $A > B$ of the comparator goes to a +5-V high, it is applied directly to the power amplifier, which amplifies and inverts the signal to a negative potential.

Prime-Mover Section The device that causes the actual physical positioning in the system is the DC motor. The potential used to activate the motor is provided by the power amplifier of the power section. When the power amplifier section output is positive, electron current flows from ground through the motor to the output terminal. When the power amplifier output goes negative, electron current flows from its output through the motor and then to ground. Because current flows through the motor in two different directions, it can rotate clockwise or counterclockwise.

Sensing Section Attached to the shaft of the motor is a sensing device that detects the positioning of the system. Figure 13-8 shows that the sensor is comprised of a disc, four lights, and four optical light sensors. A series of slots is arranged in four rows in a circular fashion over the entire surface of the disc. As the disc rotates, each light is detected by its corresponding sensor when a slot is positioned to allow light to pass through. A sensor that detects light generates a 1 state. Likewise, whenever the disc is positioned so that the absence of a slot does not allow light to pass through to its corresponding sensor, a 0 state is generated.

The slots on the disc are arranged in a pattern of 1s and 0s that are in a Gray code format. A close examination of the disc reveals that the Gray code is divided into 10 pie-shaped sectors of equal size, each representing a 36-degree region.

Rack and Pinion Attached to the end of the motor shaft is a (pinion) gear that meshes with the teeth on the bar mechanism. When the gear rotates clockwise, the bar moves to the left. Likewise, when the gear rotates counterclockwise, the bar moves to the right.

Conversion Circuit As the Gray code numbers are being generated by the four sensors, they represent the 10 different physical locations of the digital positioner. However, before the four Gray code numbers are applied to input B of the magnitude comparator, they are converted into their equivalent pure binary values. This conversion process is necessary because a binary number is applied to input A of the magnitude comparator; the comparator must compare values from the same number system to operate properly. The conversion process is accomplished by the three exclusive-OR gates (1, 2, and 3) that act as a Gray-code-to-binary encoder.

Operation

1. Suppose that the bar mechanism is located at position 0. The Gray code wheel mounted on the motor shaft generates a 4-bit output of 0000 that is applied to the encoder, which also generates an output of 0000. Also assume that the number stored in the register is 0000.

2. Because the register output applied to input A of the comparator is the same as the encoder output applied to input B, the $A = B$ output is high and turns on the light. Meanwhile, the $A < B$ and $A > B$ outputs are low.

3. If key 6 is pressed on the keypad, number 0110 is loaded into the register. With the bar mechanism still at location 0, the A input is greater than the B input. Therefore, the $A > B$ output of the magnitude comparator goes high.

4. The $+5$-V signal from the $A > B$ output is applied to the power amplifier, becomes inverted, and is amplified to produce an adequate potential to drive the motor.

5. As electron current flows through the motor from the negative output of the power amplifier to ground, the motor rotates in a clockwise direction. Because the disc and gear are mounted on the same motor shaft, they rotate together. The length of each region at the outer portion of the gear is the same size as each region on the bar mechanism. Therefore, as the gear (and disc) turn 36 degrees, the bar mechanism moves the distance of one region.

6. When the bar mechanism reaches region 6, the Gray code wheel generates a 0101 that is converted into a binary 0110 by the encoder.

7. As the register supplies a 0110 to input A of the magnitude comparator and the encoder output of 0110 is applied to input B, the $A > B$ output goes low and stops the motor. Because the gear-and-bar mechanism also stops, the $A = B$ output goes high and turns on the light.

To modify the positioning device so that it stops within an area less than 36 degrees, it is necessary to make the following changes:

- Use a keypad with more than 10 number entries.
- Use a register capable of storing a number greater than 9.
- Use a Gray code wheel with more than 10 number regions.

■ REVIEW QUESTIONS

18. The purpose of the three _____ gates is to convert the Gray code number into its equivalent binary number.

19. How many different Gray code numbers are on the Gray code wheel in Fig. 13-8?

20. When input A of the magnitude comparator is less then input B, the motor turns (clockwise, counterclockwise, nowhere).

21. The motor stops rotating when input A of the magnitude comparator is (greater than, less than, equal to) input B, at which time the light turns (on, off).

13.7 DIGITAL CLOCK

The digital clock illustrated in Fig. 13-9 is capable of displaying hours, minutes, and seconds. There is a *fast clock-set* push button and a *slow clock-set* push button that are used to set the time of the clock. Two separate LEDs are used to indicate whether the time displayed represents AM or PM.

Operation

1. The input of the circuit is controlled by a timing network that generates an output originating at the 60-Hz 120-V AC power source used throughout the United States. Because the power company is required to generate an accurate 60-Hz AC voltage, this signal is reliable when used as a reference.

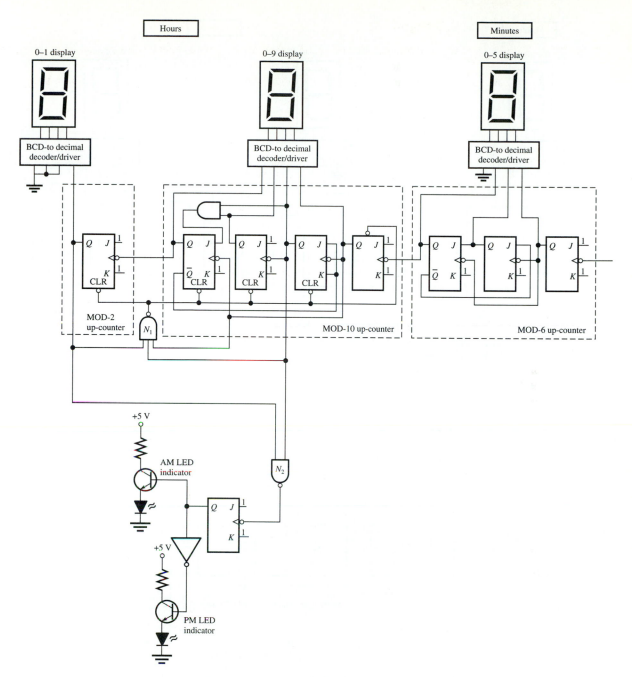

FIGURE 13.9
Digital clock.

2. The 120-V source is applied to a step-down transformer that reduces the potential to 12.6 V. After being sent through a rectifier and filter, the 12.6 V are transformed into a +5-V DC potential that is used to power the IC chips used in the circuit.

3. Another secondary AC voltage is connected to a 1N749 zener diode through two current-limiting 1-kΩ resistors. During one alternation, the zener "breaks down" at 4.3 V, and during the other alternation, the zener breaks down at −0.7 V. The Schmitt trigger conditions the signal so that it is an acceptable square wave for TTL circuitry.

4. The 60-Hz square wave from the Schmitt trigger is applied to the input of a mod-10 up-counter that is cascaded to a mod-6 up-counter. Connected together, they create a

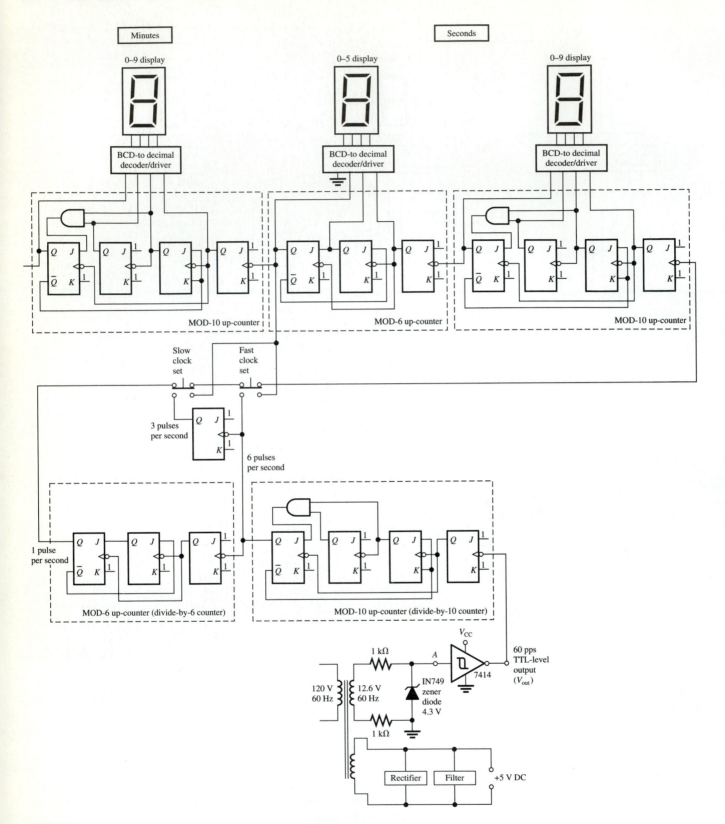

FIGURE 13.9
(Continued)

divide-by-60 circuit that provides a 1-pps clock that is fed into the least significant digit of the seconds counter.

5. The seconds network is made up of two cascaded counters that count up from 00 to 59. Therefore, a mod-10 up-counter is used to count up the least significant seconds digit, and a mod-6 up-counter is used to count up the most significant seconds digit.

6. When the seconds recycle from 59 to 00, the trailing-edge signal from the MSD output of the mod-6 seconds counter triggers the input of the LSD minutes counter. The minutes network is also made up of two cascaded counters that increment from 00 to 59.

7. Every time the minutes recycle from 59 to 00, they trigger the hour counters of the clock. The hours network is made up of two cascaded counters that increment from 01 to 12. A modified mod-10 counter is used for the LSD of the hours network. A mod-2 counter that consists of only one flip-flop alternates between 0 and 1 and is used for the MSD of the hours network.

8. After the hours network counts up from 01 to 09, the mod-10 counter recycles on the next incoming trailing-edge signal and increments the mod-2 counter so that a 10 is displayed.

9. When the hours network increments to 13, the Q output of the mod-2 counter flip-flop and the first two flip-flops of the mod-10 counter are all high. These three 1-state outputs are applied to the inputs of NAND gate N_1, which generates a low. At this moment, the mod-2 flip-flop is reset along with the three MSB flip-flops of the mod-10 counter. At the same time, the LSB flip-flop of the mod-10 counter is preset. The count of 13 is too brief to be seen. As a result, the hours network appears to recycle from 12 to 01.

10. BCD decoder/drivers are used to convert the binary numbers from the counters to the code used by the LED readouts. Note that the inputs not used by the decoder/driver of the mod-6 and mod-2 counters are tied to grounded lows. This ensures that the readout never displays a number greater than it is supposed to.

The digital clock also has a feature that uses two separate LEDs to display whether the time represents AM or PM.

11. The inputs of NAND gate N_2 are connected to the Q output of the mod-2 counter and the Q output of the second flip-flop of the mod-10 counter in the hours network. Normally, the NAND gate generates a high. However, every time the hours network changes from 11 to 12, the gate output goes low and triggers a toggle flip-flop that turns on one of the LEDs. When the counter changes from 12, the gate output goes high again until the next time the hours network increments from 11 to 12, at which time the flip-flop toggles and turns on the other LED.

The digital clock also has fast and slow clock-set buttons to provide a way for presetting a desired time.

12. To rapidly change the clock time, the fast clock-set button is pressed, which injects a 6-Hz clock pulse into the minutes network.

13. To change the clock at a more manageable speed, the slow clock-set button is pressed, which injects a 3-Hz clock pulse into the minutes network input. Note that the 3-Hz signal is obtained by using a single flip-flop to divide the 6-Hz signal by 2. The reason the clock-set inputs are not applied to the seconds network is that it is unlikely that the user would want to preset the clock to the nearest second.

■ REVIEW QUESTIONS

22. The reference clock signal is obtained from where?

23. The purpose of the _____ _____ is to condition a sine wave signal into an acceptable square wave.

24. How often does a negative pulse occur at the input of the mod-6 *minutes* up-counter?

25. The AM/PM LEDs toggle every time the hours counters increment from _____ to _____ .

13.8 BASKETBALL SHOOTING GAME

The pictorial diagram of Fig. 13-10 shows a basketball shooting game found in arcades. The objective of the game is to score as many shots as possible within 59 seconds using a miniature basketball. Every time a shot is scored, the ball goes through a collar located at the bottom of the net. A sensor on the collar then sends a signal to a counter that records the score.

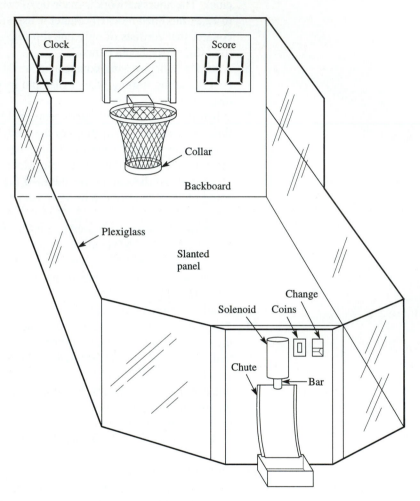

FIGURE 13.10
Pictorial of a basketball shooting game.

Whenever a shot is taken, the ball either bounces off the backboard, rim, or goes through the hoop. It always returns to the player because a bottom panel with plexiglass sides is slanted downward, forcing the ball to roll through a chute to where the player is standing. When the decrementing clock that displays the time goes to 00, a solenoid is deactivated, which causes a bar to block the chute, preventing the ball from returning to the player. The score remains displayed until the next player inserts coins into the machine. When the game starts, the solenoid is activated, which pulls the bar back, opening the chute, and allows the ball to roll through to the player. The clock timer is preset to 59 seconds, and the counter/display used to show the score is reset to 00.

Operation

Refer to Fig. 13-11. The game is activated when OR gate 03 goes high as a result of 40 cents being inserted into the machine. A series of combination logic circuits is used to detect whether one of several possible payment requirements are met. When one is met, a 1 state is applied to one of the inputs of OR gate 03. A high is applied to one of the inputs of OR gate 3 in one of the following ways:

Eight nickels (N_1–N_8) are inserted,

or one dime (D_1) and six nickels (N_1–N_6) are inserted,

or two dimes (D_1 and D_2) and four nickels (N_1–N_4) are inserted,

or three dimes (D_1, D_2, and D_3) and two nickels (N_1 and N_2) are inserted,

or four dimes (D_1–D_4) are inserted,

or one quarter (Q_1) and three nickels (N_1–N_3) are inserted,

or one quarter (Q_1) and one dime (D_1) and one nickel (N_1) are inserted,

or one quarter (Q_1) and two dimes (D_1 and D_2) are inserted,

or two quarters (Q_1 and Q_2) are inserted.

Three different possibilities of change are provided under the following circumstances:

Chg 1: One dime that is produced if two quarters are inserted (Q_1 and Q_2) *and* a dime is detected by the change sensor S_{D1} (Chg 1 output goes high).

Chg 2: Two nickels are produced if two quarters are inserted (Q_1 and Q_2), *and* a dime is not detected at the change sensor S_{D1}, *and* two nickels are detected by the change sensors S_{N1} and S_{N2} (Chg 2 output goes high).

Chg 3: One nickel is produced if one quarter (Q_1) and two dimes (D_1 and D_2) are inserted *and* a nickel is detected by the change sensor (S_{N1}) (Chg 3 output goes high).

The coins will be returned if:

Two quarters (Q_1 and Q_2) are inserted and one dime is not detected by sensor (S_{D1}) *and* two nickels are not detected by sensors S_{N1} and S_{N2}.

or one quarter (Q_1) and two dimes (D_1 and D_2) are inserted *and* one nickel is not sensed by detector S_{N1}.

When referring to the inputs, the presence of a coin represents a 1 state and the absence of a coin represents a 0 state.

- When one of the combinations of coins amounting to 40 cents (along with any necessary change) is provided, the output of OR gate 3 goes high. When the high is produced, a mechanism is triggered that causes the inserted coins to be disposed into a bucket inside the machine. The 1 state is also applied to the input of a one-shot.
- When the coins are sent into the bucket, the 1-state signal applied to one of the inputs of OR gate 3 goes to a 0 state. Therefore, the OR gate 3 output also goes to a 0 state. This high-to-low transition triggers the one-shot, causing the $\overline{Q}$ output to go to a temporary 100-ms low before returning to its normal 1 state. The 100-ms signal sets the clock to 59 and clears the scoring counter to 00.
- When FF1–FF7 are set to a BCD 59, the $\overline{Q}$ outputs of these flip-flops connected to the inputs of N_1 are a combination of 1 and 0 states. Recall that when at least one input to a NAND gate is low, it generates a high at its output. Therefore, a 1-state output is applied to a solenoid, which activates and retracts the bar to allow the ball to pass through the chute. The 1 state from N_1 also enables AND gates 20 and 21. By doing so, it allows the 1-Hz clock pulse to pass through to the timer and the pulses from the sensor to pass through to the Score Display counter every time a basket is made. At this time, the game begins.

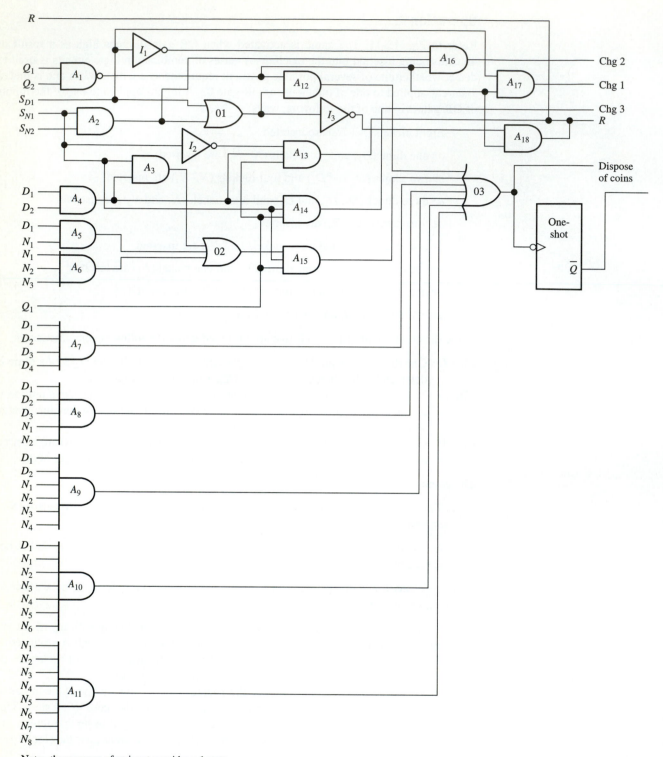

Note: the presence of an input provides a 1 state

FIGURE 13.11
Schematic diagram of a basketball shooting game.

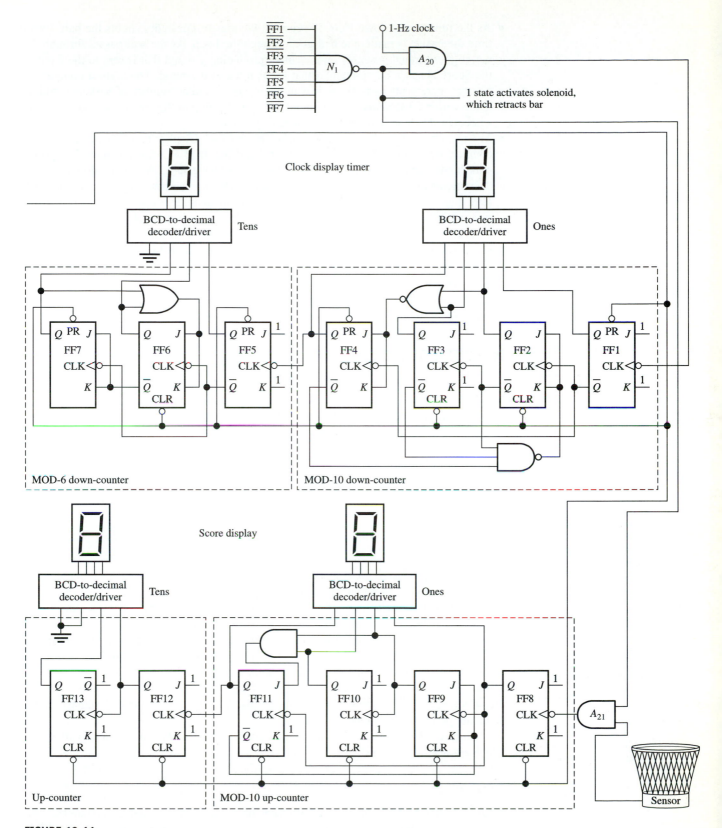

FIGURE 13.11
(Continued)

■ As the time counts down from 59 seconds, the player repeatedly shoots the ball. Every time the basket is made, the ball goes through the hoop. As the ball passes through the net, it goes through a sensor on the collar, producing a signal that is sent to the input of the Score Display, an up-counter that keeps track of the score. The highest number the counter increments to is 39, as it is unlikely that a greater number of baskets could be made within 59 seconds. The Q outputs of each flip-flop in the counter are connected to a decoder/driver so that the score can be displayed.

■ The Q outputs of the flip-flops that make up the Clock Display Timer, which is a down-counter, are connected to a decoder/driver that drives the LED clock display. The $\overline{Q}$ outputs of the same flip-flops are connected to NAND gate 1. When the down-counter decrements to 00, all the inputs of N_1 are high, which generates a low at its output. A 0 state deactivates the solenoid, causing the spring-loaded bar to return to its normal position, which results in blocking the chute. The low also disables A_{20}, which blocks the 1-Hz clock pulse from passing through to the clock input of FF1. It is possible that the player has possession of the ball after the chute closes. Therefore, the low also disables AND gate 21, which prevents the score from being increased if the player shoots the ball and makes the basket after the time expires.

■ At this time, the game is over and the score is displayed until another 40 cents is deposited into the machine.

■ REVIEW QUESTIONS

26. The presence of a coin provides a (low, high) at the inputs of logic gates A_1 to A_{11}.

27. A high-to-low transition at the output of OR gate 3 results in which of the following?
    (a) Empties the inserted coins into a bucket.
    (b) Activates a one-shot to preset the clock to 59.
    (c) Activates a one-shot to clear the score display.
    (d) All of the above.

28. When the clock counts down to 00, the _____ outputs of FF1–FF7 apply all (0s, 1s) to the input of NAND gate 1, which causes what three operations to take place?

13.9 ■ DIGITAL SOUND SYNTHESIZER

Sound synthesizer:

A digital device that electronically converts digital data into almost any sound.

Voice synthesizer:

A digital device that electronically converts digital data into human speech.

Sounds that reproduce human speech or music by a speaker are the result of analog electrical signals applied to it that vary both in frequency and amplitude. Fig. 13-12 shows examples of such electrical waveforms that produce various sounds. Because an analog waveform can be varied in any way, any type of sound can be created.

Digital systems called **sound synthesizers** can produce electrical signals similar to the analog signals shown in Fig. 13-12. A device called a **voice synthesizer** is used to replicate the sound of a human voice. For example, the recording that a telephone user receives, "I'm sorry, the number you have dialed has been changed, . . . ," is the result of digital signals from a bubble memory device that are converted to analog signals before being applied to the earphone speaker.

Figure 13-13 shows the fundamental sections that make up a voice synthesizer, which are described below.

Functions

Astable Multivibrator   This device produces no output until a signal to activate the synthesizer arrives. When the signal does arrive, a square wave with a predetermined frequency is applied to the counter.

Counter   This device is a binary up-counter that begins its count at 0000. The output of the counter is applied to the address input of the ROM device.

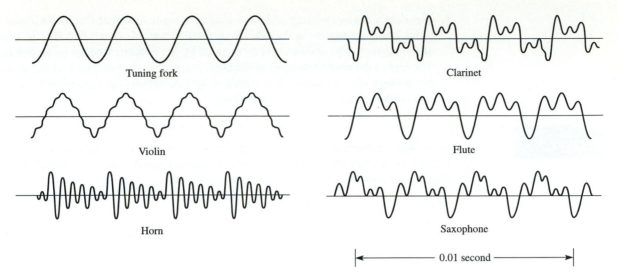

FIGURE 13.12
Typical audio waveforms.

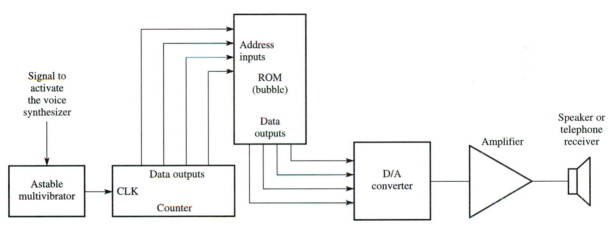

FIGURE 13.13
Sections of a voice synthesizer.

ROM This device is a read-only memory device that uses bubbles as memory elements to store 1s and 0s. Which data word stored in memory is to be read from the memory data output lines is determined by the binary number applied to the memory address input lines.

D/A Converter The D/A converter produces an analog output voltage that is proportional to the binary number applied to its digital input lines. Each digital input signal is a data word from the ROM device.

Amplifier The amplifier increases the analog output voltage of the D/A converter to a higher amplitude level necessary to drive a speaker.

Speaker The speaker can be an earphone, telephone receiver, or conventional loud-speaker device.

Operation

1. When the voice synthesizer is activated, the astable multivibrator begins to run.
2. The clock signal to the counter causes it to increment from 0 to its maximum count.

3. The data words consisting of binary numbers are read out of the ROM in sequential order. They are applied to the DAC to produce electrical signals that vary in amplitude and frequency. The amplitude is determined by the numerical value of each data word. The higher the number applied to the DAC, the higher the amplitude of the analog output voltage. The frequency is determined by how rapidly the numerical values in several consecutive memory locations change.

EXAMPLE 13.2

Show how numerical values in several consecutive memory locations, as shown in Fig. 13-14(A), first create a high-amplitude high-frequency waveform signal and then change to a low-amplitude low-frequency waveform.

Solution See Fig. 13-14(B).

FIGURE 13.14
Memory locations and waveform for Example 13.2.

Address	Binary data words				Address	Binary data words			
1	0	0	0	0	14	0	0	1	1
2	0	0	0	1	15	0	1	0	0
3	0	0	1	0	16	0	1	0	1
4	0	1	0	0	17	0	1	1	0
5	1	0	0	0	18	0	1	1	1
6	1	1	1	1	19	0	1	1	1
7	1	0	0	0	20	0	1	1	0
8	0	1	0	0	21	0	1	0	1
9	0	0	1	0	22	0	1	0	0
10	0	0	0	1	23	0	0	1	1
11	0	0	0	0	24	0	0	1	0
12	0	0	0	1	25	0	0	1	0
13	0	0	1	0	26	0	0	1	1

(a)

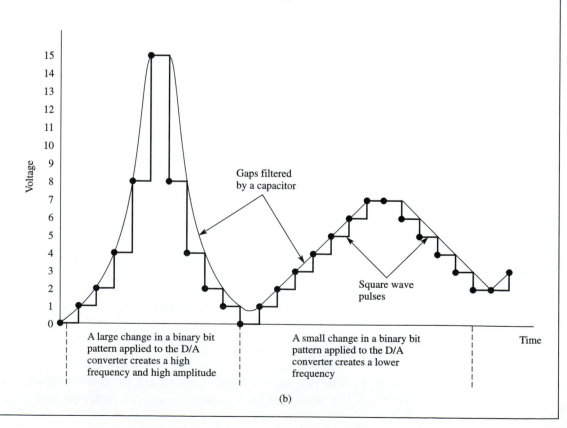

(b)

The changing output voltage of the DAC shown in Fig. 13-14(B), which consists of square wave pulses, can be smoothed out by using a filtering device, such as a capacitor circuit.

The particular bit pattern of several consecutive memory locations in a voice synthesizer can produce a basic sound of speech called a **phoneme.** The bit pattern and the analog voltage waveform it produces to create the vowel *o* is shown in Fig. 13-15. The English language is made up of 64 different phonemes. By putting different groups of phonemes together properly, human speech can be simulated.

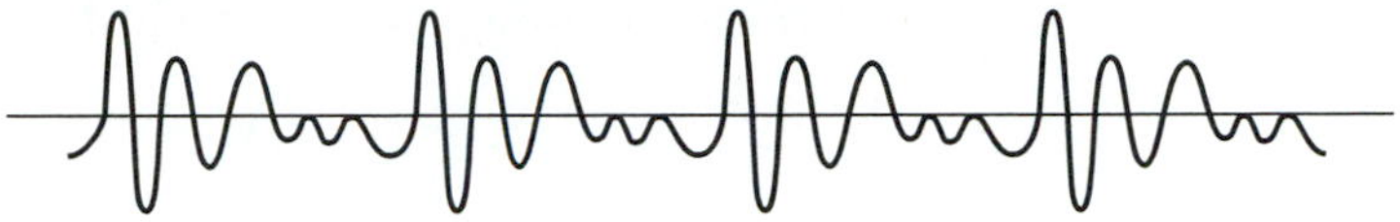

Phoneme:

One of 64 basic sounds that make up a speech pattern.

FIGURE 13.15
Voice vowel o.

The principle of producing sound from a series of 1- and 0-state stages is also used in digital audio technology. For example, one type of digital audio equipment is the compact disc. Instead of using ROM chips to store music in the form of 1s and 0s, binary bits are stored on the disc. A series of microscopic pits are etched into the disc surface with a laser by the recording manufacturer. When the disc is played, the CD player focuses a beam through lenses and mirrors from a low-powered solid state laser onto the CD surface. As the CD rotates, the lenses and beam follow the track that contains the pits and the "flats" in between. The light from the beam varies in brightness as it reflects off the disc before being received by a photodiode light sensor. When the laser hits the flat area between the pits, its light is dispersed. When it hits a pit, the light is reflected directly back to the sensor. Therefore, pit reflections are interpreted as one state and flat-surface reflections become the other state. How many pits there are and how they are spread apart determine the pattern of 1s and 0s. Figure 13-16 shows a basic diagram to illustrate this action.

The information on the disc is arranged on a spiral track similar to that of a vinyl record. The pits and flat areas on the track that contains the music are divided into 16-bit words. The resultant information is then fed serially into a serial-in parallel-out shift register before being parallel-loaded into a DAC to emulate an analog audio signal.

The information is read from the disc at a rate of 44,100 16-bit words per second. This includes both the right and left stereo channels, which are processed sequentially. To accomplish this function, 1.25 meters of information must be encoded per second. The information is read from the center out. To maintain the constant rate of information flow as the diameter changes, the CD spins at about 500 rpm at the center and about 200 rpm near the outer edge. To increase the accuracy of the information being read, newer CD player designs use low-cost high-quality computer chip circuits to reread each 16-bit data word two, three, four, or as much as eight times, and they then average the results to improve the sound quality. This is called **oversampling.**

Oversampling:

A method used by electronic circuitry that involves reading a digital signal several times and taking the average of the readings before using it.

■ REVIEW QUESTIONS

29. Analog electrical waveforms that are applied to a speaker system to produce various sounds vary both in _____ and _____.

30. A device that converts digital information consisting of binary numbers into analog waveforms to replicate the sound of a human voice is called a _____ _____.

31. The data word from the ROM chip applied to the DAC with a high binary numerical value produces an analog output voltage that has a (low, high) amplitude.

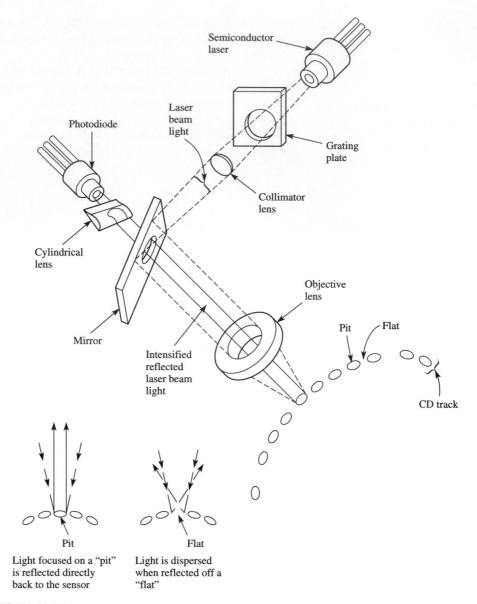

FIGURE 13.16
The laser beam passes through a series of lenses. It then focuses on and reads the information on the CD track. A beam reflected off a "pit" passes directly to a light-sensing photodiode. A beam reflected off a "flat" is dispersed and does not pass to the photodiode.

32. A low-frequency waveform is produced by the synthesizer when the numerical binary value of several consecutive data words from memory change (slowly, rapidly).

33. The English language is made up of 64 different basic sounds called _____.

13.10 DIGITAL LOCK

Figure 13-17 illustrates the circuitry that is used to operate an electronic digital lock. To activate, a combination of four numbers must be entered into a touch keypad consisting of nine keys numbered 1–9. When activated, a 1-state output causes a solenoid to energize, which retracts a bolt and allows a door to be opened. After a 10-s period, the solenoid deenergizes and the spring-return bolt extends and locks the door until the correct combination of four numbers is again entered into the keypad.

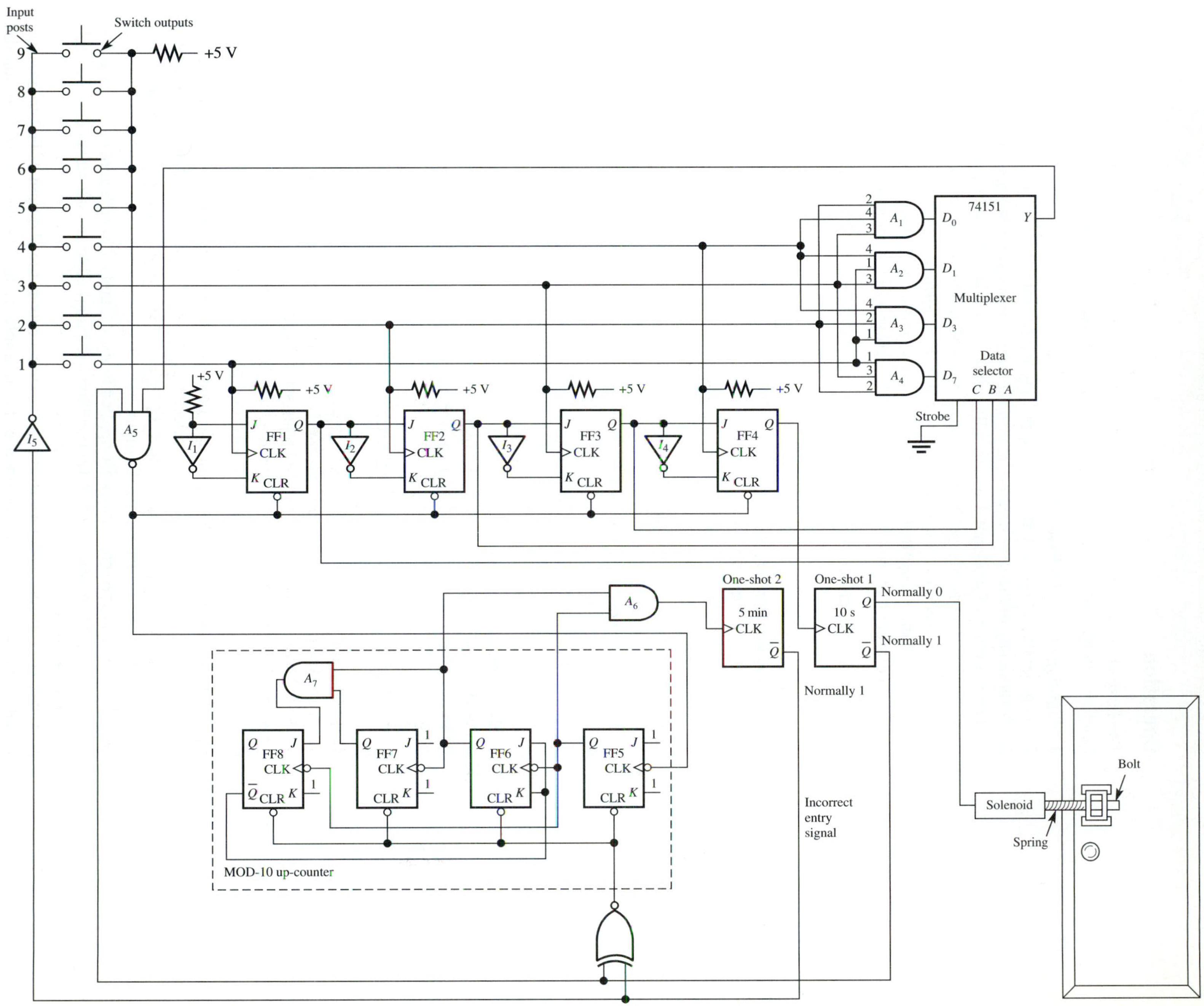

FIGURE 13.17
Digital lock.

This electronic lock includes a feature that disables the keypad for 5 minutes in the event that three consecutive incorrect combinations are entered.

Operation

The keypad in Fig. 13-17 is designed so that the correct combination is 1-2-3-4. The combination can be changed by rewiring the connections to the keypad.

1. Flip-flops FF1–FF4 have been cleared.
2. A_5 generates a high as 1 states are applied to its three inputs.
3. Without any of the keys pressed, their outputs are tied to a +5-V high through resistors. The $\overline{Q}$ output of one-shot 2 is normally a 1 state and is complemented to a low by inverter I_5 before being applied to the input posts of all the push buttons. Therefore, when a key is pressed, its switch output is pulled from a normal high to a low.
4. A +5-V 1 state is applied to the J input of FF1, and its complement is connected to the K input through inverter I_1. Therefore, FF1 is always in the set mode.

When the correct combination is entered, the following sequence of events takes place.

1. When key 1 is pressed, its output applied to the FF1 clock input goes low. At the moment the key is released, it generates a low-to-high transition and triggers FF1, which causes the Q output to go to a 1 state.
2. FF2 goes into the set mode when its J input receives a high from the Q output of FF1. When key 2 is pressed, its output goes from a high to a low. As the key is released, the low-to-high transition causes the Q output of FF2 to go high.
3. FF3 goes into the set mode when its J input receives a high from the Q output of FF2. Key 3 is then pressed and released, which causes the Q output of FF3 to go high.
4. FF4 goes into the set mode as its J input receives a high from the Q output of FF3. When key 4 is pressed and then released, the Q output of FF4 goes high.
5. When the Q output of FF4 goes high, the positive-edge-triggered signal activates one-shot 1, causing the Q output to go high for a 10-s period, which activates the solenoid that retracts the locking bolt from the door.
6. When the door is unlocked, the $\overline{Q}$ output of one-shot 1 goes to a 0 state and is applied to A_5, which resets FF1–FF4, preparing them for the next series of combination key code entries.

The incorrect entry of the combination code is detected two ways.

1. When the Incorrect Sequence of Numbers 1, 2, 3, and 4 Is Entered

The proper sequence of the four key closures is ensured by the 74151 multiplexer. When the numbers are entered, it is necessary for output Y connected to AND gate 5 to remain high. When the keys are inserted in an improper sequence, output Y goes low, which also causes the output of A_5 to go low and clears FF1–FF4. The multiplexer operates as follows:

1. Before any keys are pressed, all flip-flops are cleared. Therefore, the Q outputs of FF1–FF3, which are connected to selector inputs A, B, and C, are 000. This activates input D_0, which is at a high because all inputs of A_1 are at 1 states. As a result, the 1 state at D_0 is transferred to the Y output.
2. When key 1 is pressed, 0 states are applied to A_2, A_3, and A_4, which generate lows at D_1, D_3, and D_7. Meanwhile all 1 states continue to be applied to A_1, which generates a high at D_0, causing the Y output to remain high.
3. When key 1 is released, its positive-going signal causes the Q output of FF1 to go high. This causes the data selector input of the multiplexer to increment to 001 so that the input of D_1 can be passed to output Y.
4. When key 2 is pressed, 0 states are applied to A_1, A_3, and A_4, which generate lows at D_0, D_3, and D_7. Meanwhile, all 1 states are applied to A_2, which generates a high at D_1 that is passed to the Y output.

5. When key 2 is released, its positive-going signal causes the Q output of FF2 to go high. With the Q outputs of FF1 and FF2 high, the selector input of the multiplexer increments to 011 so that the input of D_3 can be passed to the Y output.

6. When key 3 is pressed, 0 states are applied to A_1, A_2, and A_4, which generate lows at D_0, D_1, and D_7. Meanwhile, all 1 states are applied to A_3, which generates a high at D_3 that is passed to the Y output.

7. When key 3 is released, its positive-going signal causes the Q output of FF3 to go high. With the Q outputs of FF1, FF2, and FF3 high, the selector input of the multiplexer increments to 111 so that the signal at the input of D_7 can be passed to the Y output.

8. When key 4 is pressed, 0 states are applied to A_1, A_2, and A_3, which generate lows at D_0, D_1, and D_3. Meanwhile all 1 states are applied to A_4, which generates a high at D_7 that is passed to the Y output.

9. When key 4 is released, its positive-going signal causes the Q output of FF4 to go high, which activates the solenoid that retracts the locking bolt from the door.

The nine steps just described explain how the multiplexer operates correctly when the required combination sequence of buttons is pressed. A 1 state is applied to D_0 when the data selector input is 000, D_1 when the data selector input is 001, D_3 when the data selector input is 011, and, finally, D_7 when the data selector input is 111. Therefore, a high is always passed to the Y output of the multiplexer.

However, if one of the 1–4 keys is pressed out of the proper sequence, the Y output will go low, which causes the A_5 output to go low, clearing FF1–FF4. For example,

1. Suppose that after key 1 is properly pressed on the first entry, the Q output of FF1 correctly goes to a high.
2. The data selector input of the multiplexer becomes 001.
3. If key 3 were incorrectly pressed on the second entry instead of key 2, a low would be applied to the input of A_2.
4. This 0 state would then be passed to the Y output of the multiplexer and then the input of A_5. With the low applied to A_5, a 0 state would be generated at its output, causing FF1–FF4 to be cleared.

2. When Keys 5–9 Are Pressed   In the event that any of keys 5–9 is pressed, its outputs are pulled low. From Fig. 13-17, note that all these keys are tied together and are connected to AND gate 5. Therefore, a 0 state generated by one of these key closures causes the output of A_5 to go low. As a result, FF1–FF4 are all reset.

1. When an incorrect key entry is made, the mod-10 (decade) up-counter is incremented. After three consecutive incorrect entries are made, the counter increments to 0011_3. This causes the Q outputs of FF5 and FF6 to go high.
2. When two highs are applied to both inputs of AND gate 6, a 1 is generated, which activates one-shot 2. Its $\bar{Q}$ output goes to a 0 state for 5 minutes.
3. By inverting the 5-minute low through I_5, a 1 is generated and applied to the input posts of the push button keys. Therefore, the circuit is disabled during this time because any key entries are unable to generate a low at the switch output.

An exclusive-NOR gate is used to reset the decade counter by generating a low when either one-shot 1 or one-shot 2 is triggered. This occurs when either the solenoid that retracts the locking bolt is activated or when the circuit is disabled for 5 minutes after three consecutive entries are made.

■ REVIEW QUESTIONS

34. The incorrect combination code is detected in what two ways?
35. Which of the following occurs when the incorrect combination is entered?
    (a) A_5 goes low.
    (b) FF1–FF4 are cleared.

    (c) Mod-10 up-counter increments.

    (d) All of the above.

36. During the proper sequence of four entries of the keys, output Y of the multiplexer remains (low, high), and an incorrect entry causes Y to go (low, high).

37. When _____ incorrect consecutive entries are made, the $\overline{Q}$ output signal of one-shot 2 goes (low, high) for 5 minutes and is inverted by I_5, which creates a high at its output. Therefore, FF1–FF4 are deactivated because the closure of keys 1–4 does not allow the necessary (low-to-high, high-to-low) transition at their clock inputs.

13.11   DIGITAL MUSIC SYNTHESIZER

Piano music that has been recorded can be reproduced by sending electrical signals through a speaker. A song is produced by varying the frequency of the signal. Each note is created by a specific frequency.

By generating frequencies corresponding to musical notes, digital music synthesizers are capable of simulating piano music. Each desired frequency is developed by the process called *frequency division*. A six-bit binary counter, shown in Fig. 13-18, illustrates the concept of frequency division. Each flip-flop in the counter divides the frequency applied to its input by 2. Therefore, six flip-flops cascaded divide the frequency applied to its first flip-flop by 64 (2^6). However, the divisor value of 64 can be altered by presetting the counter with a number (other than zero) before applying the clock pulses. For example, if the counter is preset to start at a count of 16, then the division is 48(64 − 16). Therefore, if a frequency of 14,000 is applied to the counter's input, its output frequency will be 291.67 Hz (14,000/48). Thus, by presetting the counter to different divisors, the resultant output frequency can be controlled.

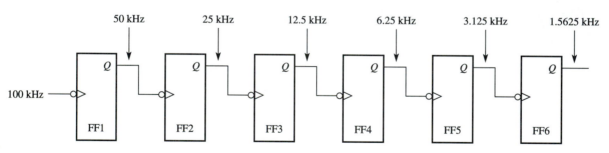

FIGURE 13.18
Frequency division in a 6-bit binary counter.

The digital circuitry that produces tones for a partial piano keyboard is shown in Fig. 13-19. It shows a 6-bit binary up-counter, a clock generator, a partial keyboard that produces a high and a low octave, an encoder, an octave frequency divider, and an amplifier to drive a speaker.

The clock generator produces pulses at a frequency of 16,755 Hz. These pulses are fed to the chain of flip-flops that form the counter and a frequency divider that creates the low and high octave for the keyboard. As a key is pressed, a binary pattern from the keyboard is encoded to preset a corresponding number into the counter. The clock pulses from the generator then cause the counter to increment from the preset value. When the counter reaches its highest count (64), it recycles to 0. If the key remains pressed, the same preset value is reloaded into the counter, and then the counting sequence is repeated.

The process of reloading a preset value into the counter occurs when it recycles from its maximum count to zero. When this transition takes place, the $\overline{Q}$ output at FF6 goes high. Inverter I_1 responds by producing a low at its output, which is applied to the active-low *set* input of FF6. As flip-flop 6 is preset, its Q output goes high and the $\overline{Q}$ output goes

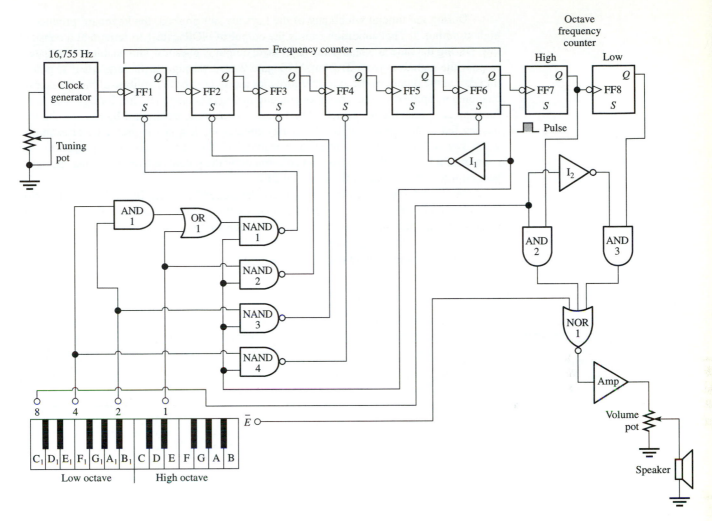

FIGURE 13.19
Music synthesizer circuitry.

back to a low. This monostable-type preset pulse is also applied to one of the inputs of each NAND gate in the encoder. Bits from the keyboard are connected to the other input of each encoder's NAND gate. Any NAND gate that has a high applied to its input from the keyboard will produce a momentary low when it receives the one-shot pulse from FF6. Therefore, any flip-flop in the counter that receives a momentary low will be set. The remaining flip-flops in the counter are cleared. If a different key is pressed, a new number is preset into the counter when the next monostable pulse from FF6 is fed into the encoder. This action alters the division of the string of flip-flops, which results in a change in frequency.

Each time the counter recycles from its maximum count to 0, a negative-going pulse is created at the Q output of FF6. The number of these pulses each second develops the frequency used for producing the sound. Each of these pulses cause FF7 to toggle. The FF7 output frequency, which forms the high octave tones, are also fed to FF8, which divides the signal by 2 to produce a frequency lower by half (an octave).

When the high octave keys are pressed, output 8 of the keyboard goes high and enables AND gate 2 to pass the pulses from FF7 to NOR gate 1. The frequency produced by FF7 divides the frequency from FF6 by half. Whenever any of the low octave keys are pressed, a logic 0 state is produced by output 8 of the keyboard. Inverter 2 complements this signal and enables AND gate 3 to pass the pulses from FF8 to NOR gate 1. FF8 produces the low octave because it reduces the frequency from FF7 by half.

During the time at which any of the keys are not pressed, the keyboard produces a high at output $\overline{E}$. This condition causes the output of NOR gate 1 to remain at a constant low. During the time at which a key is pressed, output $\overline{E}$ goes low and enables NOR gate 1 to pass the signals (inverted) from AND gate 2 or 3 through to the amplifier. A potentiometer at the output of the amplifier provides a way to vary the volume of the speaker by changing its applied voltage. Another potentiometer is connected to the clock generator. Its function is to tune the instrument. The tuning procedure is performed by reading the frequency fed to the speaker with a frequency counter while a key is pressed. For example, when key A (low octave) is pressed, a reading of 220 Hz should be displayed by the counter. If it is too high or too low, an adjustment by the potentiometer will cause the clock generator to alter the output frequency of FF7.

Table 13-1 shows the binary bit pattern developed by the keyboard for each key closure, their corresponding binary numbers that are preset into the counter, the resultant divisor of the counter, the value at which the counter divides the frequency, and the output frequency produced during the low and high octave modes. Table 13-2 shows which keys in each octave are pressed to play the notes for a familiar song.

TABLE 13.1
Function table of the music synthesizer

Note	Keyboard Output 4	2	1	(1) FF1	(2) FF2	(4) FF3	(8) FF4	(16) FF5	(32) FF6	Divisor Value (64-Preset Count)	FF6 Output (16,755÷ Divisor)	High Octave FF7 Output (FF6 Output ÷2)	Low Octave FF8 Output (FF7 Output ÷2)
C	0	0	0	0	0	0	0	0	1	32	523.59	261.79	130.89
D	0	1	0	0	0	1	0	0	1	28	598.39	299.20	149.60
E	0	1	1	1	1	1	0	0	1	25	670.20	335.10	167.55
F	1	0	0	0	0	0	1	0	1	24	698.13	349.06	174.53
G	1	0	1	1	1	0	1	0	1	21	797.86	398.93	199.46
A	1	1	0	1	0	1	1	0	1	19	881.84	440.92	220.46
B	1	1	1	1	1	1	1	0	1	17	985.59	492.79	246.40

TABLE 13.2

ON C_1	TOP C_1	OF E_1	OLD G_1	SMO- C	KEY A_1
ALL A_1	COV- F_1	ERED G_1	WITH A_1	SNOW G_1	
I G_1	LOST C_1	MY E_1	TRUE G_1	LOV- G_1	ER D_1
FROM E_1	A E_1	COURT- F_1	IN E_1	TOO D_1	SLOW C_1

■ **REVIEW QUESTIONS**

38. A five-bit binary up-counter will divide the applied frequency by _____.
    (a) 2
    (b) 5
    (c) 32
    (d) 64

39. A five-bit binary up-counter will divide the applied frequency by _____ if a count of 6 is preset into its flip-flop.

(a) 2

(b) 5

(c) 26

(d) 32

Refer to Fig. 13-19 to answer the following problems:

40. Suppose a closure of the D key produces an output 0010 at the keyboard. Write the binary bits that are preset into each of the counter flip-flops.

FF1   FF2   FF3   FF4   FF5   FF6

41. What is the high octave frequency applied to the speaker when key G is pressed. Assumed that the clock generator's frequency is 16,755 Hz.

42. While a high octave key is pressed, output 8 of the keyboard is _____ (low, high).

13.12 COFFEE VENDING MACHINE

A practical application of a combination logic circuit is shown in Fig. 13-20. Forty logic devices are used in the control section of a coffee vending machine. The purchase price is 50 cents, and the customer has a choice of four coffee selections to choose from, which are black, cream, sugar, and cream and sugar.

To describe the operation of the combination circuitry, several external input variables must first be defined.

Input Variable	Symbol
Quarter 1, 2	Q_1, Q_2
Dime 1, 2, 3, 4, 5	D_1, D_2, D_3, D_4, D_5
Nickel 1, 2, 3, 4, 5, 6, 7, 8, 9, 10	$N_1, N_2, N_3, N_4, N_5, N_6, N_7, N_8, N_9, N_{10}$
Black button	B_B
Cream button	C_B
Sugar button	S_B
Cream and sugar button	A_B
Coin return	C_R

The price of each coffee is 50 cents, so the following coin combinations of change will provide the required money when inserted into the machine:

Ten nickels ($N_1, N_2, N_3, N_4, N_5, N_6, N_7, N_8, N_9, N_{10}$), or

Eight nickels and one dime ($N_1, N_2, N_3, N_4, N_5, N_6, N_7, N_8, D_1$), or

Six nickels and two dimes ($N_1, N_2, N_3, N_4, N_5, N_6, D_1, D_2$), or

Four nickels and three dimes ($N_1, N_2, N_3, N_4, D_1, D_2, D_3$), or

Two nickels and four dimes ($N_1, N_2, D_1, D_2, D_3, D_4$), or

Five dimes (D_1, D_2, D_3, D_4, D_5), or

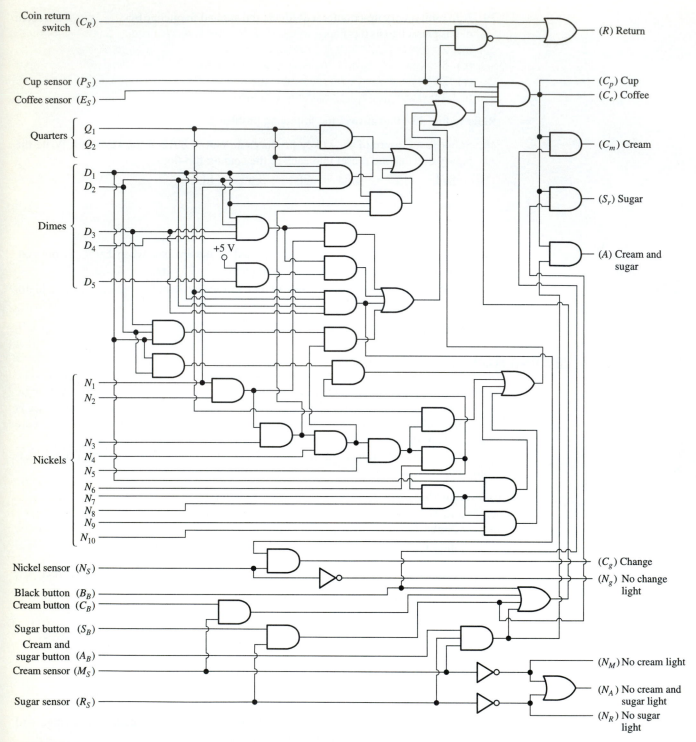

FIGURE 13.20
Logic circuitry of the control section for the coffee vending machine.

Two quarters (Q_1, Q_2), or

One quarter and five nickels (Q_1, N_1, N_2, N_3, N_4, N_5), or

One quarter, one dime, and three nickels (Q_1, D_1, N_1, N_2, N_3), or

One quarter, two dimes, and one nickel (Q_1, D_1, D_2, N_1), or

One quarter and three dimes (Q_1, D_1, D_2, D_3).

One of four separate buttons is used by the customer to choose from the four coffee selections, black (B_B), cream (C_B), sugar (S_B), or cream and sugar (A_B). When one of these buttons is pressed, it activates the coffee dispensing operation, which causes the cup to be lowered and the coffee to be dispensed. A coin return switch (C_R) is also provided to return the inserted coins if activated by the customer.

The external output variables are identified as:

External Output Variable	Variable Symbol
Cup	C_P
Coffee	C_e
Cream	C_m
Sugar	S_r
Cream and sugar	A
Return	R
Change	C_g
No change light	N_g
No cream light	N_M
No sugar light	N_R
No cream and sugar light	N_A

In addition to being activated by the coin return (C_R) switch, the return output (R) is also generated by the absence of the cups or coffee supplies.

A nickel is available as change when a quarter and three dimes are inserted. If no change is available, a *no change* (N_g) light on the vending machine front panel illuminates. The customer who inserts one quarter and three dimes and presses one of the four selection buttons with this light on forfeits a nickel. When there is no cream available, a *no cream* (N_M) light illuminates on the panel. Likewise, if no sugar is available, a *no sugar* (N_R) illuminates. When there is no cream or sugar available, a *no cream and sugar* (N_A) light illuminates on the panel. If a corresponding selection button to one of these illuminated lights is pressed, nothing will happen. The customer can, therefore, press another selection button, or activate the coin return switch.

There are several internal input variables that are used to determine the conditions just described. These are activated by sensors that provide data about the presence or absence of certain supplies. These internal input variables are:

Internal Input Variable	Variable Symbol
Nickel sense	N_S
Cup sense	P_S
Coffee sense	E_S
Cream sense	M_S
Sugar sense	R_S

The *nickel sense* (N_S) is activated if at least one nickel is available for change. Sensors are also used to detect if the cup (P_S), coffee (E_S), cream (M_S), and sugar (R_S) supplies are present.

Design Steps

There are three steps required to design the vending machine control section:

Step 1. The first step is to write expressions for each output variable. The process begins by making three assumptions:

1. The absence of any input equals a 0 logic state.
2. The presence or activation of any input equals a logic 1 state.
3. An output is generated by a logic 1 state.

Cup and Coffee Output ($C_P C_e$)

The expression for the cup, C_P, is the same as for the coffee, C_e, since one should not be dispensed without the other. Therefore, the logic expression for the cup and coffee is:

$$C_e \text{ and } C_p = P_S E_S \ (N_1 N_2 N_3 N_4 N_5 N_6 N_7 N_8 N_9 N_{10} + N_1 N_2 N_3 N_4 N_5 N_6 N_7 N_8 D_1 + N_1 N_2 N_3 N_4 N_5 N_6 D_1 D_2 + N_1 N_2 N_3 N_4 D_1 D_2 D_3 + N_1 N_2 D_1 D_2 D_3 D_4 + D_1 D_2 D_3 D_4 D_5 + Q_1 Q_2 + Q_1 N_1 N_2 N_3 N_4 N_5 + Q_1 D_1 N_1 N_2 N_3 + Q_1 D_1 D_2 N_1 + Q_1 D_1 D_2 D_3) (B_B + C_B M_S + S_B R_S + A_B M_S R_S)$$

which states that the coffee and cup will be dispensed if:

A cup is available,

AND coffee is available,

AND ten nickels have been inserted,

OR eight nickels and one dime have been inserted,

OR six nickels and two dimes have been inserted,

OR four nickels and three dimes have been inserted,

OR two nickels and four dimes have been inserted,

OR five dimes have been inserted,

OR two quarters have been inserted,

OR one quarter and five nickels have been inserted,

OR one quarter and one dime and three nickels have been inserted,

OR one quarter and two dimes and one nickel have been inserted,

OR one quarter and three dimes have been inserted,

AND the *black* select button has been pressed,

OR the *cream* button is pressed and the *cream sensor* is activated,

OR the *sugar* button is pressed and the *sugar sensor* is activated,

OR the *cream and sugar* button is pressed and the *cream sensor* and the *sugar sensor* are activated.

Coin Return Output

The expression for the *return* output is as follows:

$$R = C_R + \overline{P_S E_S}$$

which states that the *return* output goes to a logic high if:

The C_R switch is activated,

OR the cup, the coffee, or both are not present.

Change Output

The expression for the *change* output is as follows:

$$C_g = (Q_1\, D_1\, D_2\, D_3)\, N_S$$

which states that the *change* output goes to a logic high if:

The quarter, dime 1, dime 2, and dime 3 inputs are activated,

AND a nickel is present.

No Change Light

The expression for this indicator light is:

$$N_g = \overline{N_S}$$

which states that the *no change* indicator light will illuminate if:

No nickel is available.

No Cream, No Sugar, No Cream and Sugar Output Lights

The expression for these light indicators are:

$$N_M = \overline{M_S}$$

which state that the *no cream* light illuminates if:

No cream is available.

$$N_R = \overline{R_S}$$

which states that the *no sugar* light illuminates if:

No sugar is available.

$$N_A = \overline{M_S} + \overline{R_S}$$

which states that the *no cream* and *sugar* light illuminates if:

No cream is available,

OR no sugar is available.

Cream, Sugar, Cream and Sugar Is Dispensed with Coffee and the Cup

$$C_M = (C_P\, C_e)(C_B\, M_S)$$

which states that the cream will be dispensed with the coffee and cup if:

The cup and coffee are present,

AND the *cream selector button* is pressed and the *cream* is available.

$$S_r = (C_p\, C_e)(S_B\, R_S)$$

which states that the sugar will be dispensed with the coffee and cup if:

The cup and coffee are present,

AND the *sugar selector button* is pressed and the *sugar* is available.

$$A = (C_p\, C_e)(A_B\, M_S\, R_S)$$

which states that the cream and sugar will be dispensed with the coffee and cup if:

The cup and coffee are present,

AND the *cream and sugar* selector button is pressed and the *cream and sugar* is available.

Step 2. The second step in the implementation process is to develop a logic circuit diagram from the expressions. Figure 13-20 shows the completed logic diagram of the control section of the coffee vending machine.

Step 3. The final step of the implementation process in designing this circuit is to convert the diagram into a functional unit using the TTL 7400 family integrated circuit chips.

■ REVIEW QUESTIONS

43. Coffee will be dispensed if the *cream and sugar* button is pressed when the *no sugar* light illuminates. True or false.

44. Write the Boolean statement that causes the *change* output to go to a logic high.

45. Assuming that the money requirements are met, which of the remaining sensors must be a logic 1 to dispense coffee with sugar?

46. What denomination of money does the customer receive when the *return* output is activated?

13.13   TROUBLESHOOTING A DIGITAL CLOCK

The digital clock shown in Fig. 13-9 is made up of eight cascaded up-counters. Because LED segments visually display the operation of each counter, the technician can use them to troubleshoot the clock when it malfunctions.

Suppose the clock begins to count from 00 to 19 o'clock. To determine what may be the cause, the following troubleshooting steps are used to find the problem.

1. *Action:* Press the slow clock-set button and observe the count of the minutes and hours displays.

   Results: Both the 0–9 and 0–5 minutes displays increment and recycle in the proper sequence. The hours display counts from 01 to 19 before recycling instead of from 01 to 12.

   Conclusion 1: By pressing the slow clock-set button, the cascaded seconds counters are bypassed. If a fault did not show while the button was pressed, then the seconds counter would be suspected of being bad. However, because the problem shows up on the hours counter, it is unlikely that the seconds counters are faulty, so they do not have to be tested.

   Conclusion 2: Because the hours counter does not return to 01 after the twelfth hour ends, N_1, which is the device that causes this function to occur, should be tested to determine if a negative pulse is generated.

2. *Action:* Use the slow clock-set button to preset the hours counters to display the thirteenth hour. Use a logic probe to find the input and output logic states of N_1.

   Results:
   - Inputs: All high.
   - Output: Bad voltage level. Further testing reveals an open at the output of N_1 due to an IC pin being bent and not making contact in the IC socket.

   Conclusion: Because of the open output of N_1, the flip-flops in the cascaded hours counters never receive a negative pulse when the count increments to 13.

■ PROBLEMS

Refer to Fig. 13-1 for Problems 1 to 5.

1. Why is it necessary for the voltmeter to continually be reset? (13-1)

2. The voltmeter reset pulse has a duration of _____ milliseconds, the count-up time lasts for up to _____ milliseconds, and the numbers are displayed for about _____ milliseconds, which allows the display to be seen. (13-1)

3. What is the function of the comparator? (13-1)

4. What is the output frequency of FF9, FF10, and the one-shot? (13-1)

Troubleshooting Problem

5. Which of the following faults causes the voltmeter to be frozen at a count of 00? (13-1)
   (a) The one-shot is frozen in its unstable state.
   (b) The 20-kHz astable multivibrator is not working.
   (c) The positive power supply lead at the comparator is open.
   (d) All of the above.

Refer to Fig. 13-2 for Problems 6 to 9.

6. What is the purpose of the Schmitt trigger? (13-2)

7. If the input frequency is 2 kHz, how many pulses are counted during the count-up period if it were 0.1 s? (13-2)

8. What is the likely count-up time of frequency counters that require that the displayed number be multiplied by a factor of 10? (13-2)

Troubleshooting Problem

9. A technician tests the frequency counter and finds that the displayed count is exactly twice the actual frequency. Which of the following is the most likely cause of the fault and why? (13-2)
   (a) The astable multivibrator frequency is twice the amount it should be.
   (b) The one-shot pulses are twice the amount they should be.
   (c) The middle input line at A_1 is open.
   (d) All of the above.

Refer to Fig. 13-3 for Problems 10 to 12.

10. The frequency of the digital thermometer's variable-frequency clock generator is primarily determined by what component? (13-3)

11. How many clock pulses are applied to the up-counter during the count-up period when 50 degrees is measured? (13-3)

Troubleshooting Problem

12. If the thermometer always displays a reading that is too high, which of the following devices will not cause the defect? (13-3)
   (a) Low-frequency clock generator
   (b) Thermistor sensor
   (c) Variable-frequency clock generator
   (d) One-shot multivibrator

Refer to Fig. 13-5 for Problems 13 to 15.

13. The analog voltage at the output of the _____ that is proportional to the digital input number is applied to an _____, which provides sufficient power to drive the DC motor. (13-4)

14. What kind of device inside the keypad causes the reset output to generate a momentary low? (13-4)

Troubleshooting Problem

15. The following faulty symptoms develop with the circuit. (13-4)

ACTION	DISPLAY
Key 7 pressed	5
Key 9 pressed	9
Key 3 pressed	1
Key 6 pressed	4

Which of the following is the most likely cause of the defective circuit?
   (a) No reset pulse.
   (b) Input 2 of the DAC is open.
   (c) FF3 is faulty.
   (d) A bad LED display.

Refer to Fig. 13-6 for Problems 16 to 18.

16. Describe the function of the strobe signal. (13-5)

17. From the circuit, how could the speed of each rotation be increased? (13-5)

Troubleshooting Problem

18. The following faulty symptoms develop with the circuit.

CONDITION	ACTION	RESULTS
1	Key 8 pressed	■ Motor runs and does not stop. ■ Display alternates between 8 and 9.
2	Key 4 pressed	■ Motor runs and does not stop. ■ Display alternates between 4 and 5.
3	Key 3 pressed	■ Motor runs and does not stop. Display alternates between 2 and 3.
4	Key 2 pressed	■ Motor runs and does not stop. Display alternates between 2 and 3.

Which of the following is the cause of the malfunctions?
(a) K input of FF2 is shorted to ground.
(b) $\overline{Q}$ output of FF1 is open.
(c) Q output of FF2 is shorted to ground.
(d) Q output of FF2 is open.
(e) All of the above.

Refer to Fig. 13-8 for Problems 19 and 20.

19. What necessary circuit modifications would have to be made to enable the positioning device to rotate to 1-degree sectors? (13-6)

20. Suppose that a 6 on the keypad is pressed and the motor turns until the device reaches region 6. Why would the motor make a slight movement instead of rotating 36 degrees if key 5 were pressed? (13-6)

Refer to Fig. 13-9 for Problems 21 and 22.

21. When the cascaded hour counters increment to 13, why isn't that number displayed? (13-7)

22. How could the circuit be modified to display tenths of a second? (13-7)

Refer to Fig. 13-11 for Problems 23 to 26.

23. What do the labels $\overline{FF1}$ to $\overline{FF7}$ at the inputs of N_1 represent? (13-8)

24. Why is the tens MSD BCD-to-decimal decoder/driver input of the clock display grounded? (13-8)

25. If a quarter and two dimes are inserted into the panel of the basketball game, and there are no nickels present, explain how the logic gates respond. (13-8)

Troubleshooting Problem

26. The following faulty symptoms develop in the circuit:

   ■ The tens clock display alternates between 5 and 4 every 10 seconds.
   ■ The timer never decrements to 00.
   ■ The score display on the scoreboard is never disabled.
   ■ The bar never blocks the chute.

   Which of the following is the cause of the malfunction? (13-8)
   (a) The clock input to FF6 is shorted to ground.
   (b) The $\overline{Q}$ output of FF5 is open.
   (c) The clock input to FF7 is shorted to ground.
   (d) All of the above.

Refer to Fig. 13-13 for Problems 27 and 28.

27. The frequency of the analog waveform applied to the speaker device is determined by which of the following? (13-9)
   (a) How fast the counter output changes.
   (b) The amplifier gain.
   (c) The data words stored in ROM.
   (d) The DAC.
   (e) All of the above.

Troubleshooting Problem

28. What section of the voice synthesizer would create a malfunction that would make the synthesizer sound like the "chipmunk" cartoon characters? (13-9)

Refer to Fig. 13-17 for Problems 29 and 30.

29. What components or circuits detect that any of the improper keys 5 to 9 are pressed? (13-10)

30. What components or circuits detect that any of the required keys 1 to 4 are pressed out of sequence? (13-10)

Refer to Fig. 13-19 to answer Problems 31 through 34.

31. The preset pulse clears the counter. True or false.

32. The logic state produced by output 8 of the keyboard determines which octave is produced by the synthesizer. True or false.

33. When a lower-octave key G is pressed, the output frequency of FF8 is _____ frequency value as when the higher-octave key G is pressed.
   (a) half the
   (b) twice the
   (c) the same

34. A six-bit binary up-counter will divide the applied frequency by _____.
   (a) 2
   (b) 6
   (c) 32
   (d) 64

Refer to Fig. 13-20 to answer Problems 35 through 37.

35. A logic _____ (0,1) activates the inputs and a logic _____ (0,1) activates the outputs of the combination circuitry for the coffee vending machine.

36. If one of the four selector buttons is pressed when the *no change* light illuminates, the change is forfeited. True or false.

37. List three sensors that cause the coins to return if they are activated by the appropriate logic state.

■ ANSWERS TO REVIEW QUESTIONS

1. comparator   2. greater   3. 15   4. high, high
5. Schmitt trigger   6. 4   7. Sensor   8. (d)
9. decreases, increases   10. high   11. 9
12. 9, 0   13. True   14. $\overline{4}, \overline{2}, \overline{1}, \overline{8}$ 15. high, low
16. 0000   17. 1111, N_1, low, clears   18. exclusive-OR
19. 10   20. counterclockwise   21. equal to, on
22. 60-Hz AC power source   23. Schmitt trigger
24. Once every 10 minutes   25. 11, 12   26. high
27. (a)   28. Q, 1s; (1) extends the bar, (2) disables the clock, and (3) disables the scoring counter
29. amplitude, frequency   30. voice synthesizer
31. high   32. slowly   33. phonemes
34. (1) Wrong keys 5–9, and (2) incorrect sequence
35. (d)   36. high, low   37. 3, low, low-to-high
38. (c)   39. (c)   40. FF1 = 0   FF2 = 0
   FF3 = 1   FF4 = 0   FF5 = 0   FF6 = 1
41. $\begin{array}{r} 64 \\ -43 \\ \hline 21 \end{array}$ 16,755/21 = 797.85
   797.85/2 = 398.93   42. high
43. false   44. $C_g = (Q_1 D_1 D_2 D_3) N_S$
45. R_S, P_S, E_S 46. Nickel

TROUBLESHOOTING DIGITAL SYSTEMS

Applications of digital electronics are increasing daily. However, like all other types of electronic equipment, digital systems also malfunction. These problems can develop in the manufacturing process as a result of voltage surges and spikes, exposure to extreme temperatures and vibrations, and by just plain wear. Therefore, it is extremely important for maintenance technicians to learn how to troubleshoot these devices. This appendix and troubleshooting sections throughout the book provide information about analytical techniques and tools used to troubleshoot digital circuitry. Practical examples are included to develop basic skills on this subject.

Before troubleshooting any digital circuitry, it is essential that the technician should have some knowledge of how the circuit is supposed to operate. Familiarization can be obtained by receiving formal training on the equipment. For those not trained on the device, some equipment manuals provide information about the operation along with a troubleshooting flow chart that describes a sequence of suggested troubleshooting steps.

A.1 TROUBLESHOOTING STEPS

The procedure necessary to effectively troubleshoot involves a variation of the following steps:

Step 1. Gather Information   Observe the system's operation and compare it to the correct operation. Then analyze what the symptoms are and try to determine what could be causing the problem.

If the reason for the fault is not obvious at first, the next step may be to obtain historical information on the system to find out if the problem has occurred before. This can be in the form of the equipment repair log provided either by the person who operates the system or by the service personnel who have worked on this or similar equipment failures in the past.

Step 2. Isolate the Fault   Based on the symptom of the problem and knowledge of the operation, perform necessary tests and take measurements to localize the fault to a functional section.

Some types of equipment have indicator lights and test switches that provide information to make the diagnosis of the malfunction easier. If these features are not available, a troubleshooting flow chart in the form of a block diagram is sometimes provided in the equipment manual. Test points at certain key locations are used in the flow-chart procedure to guide the troubleshooter to the faulty area in the fewest steps. The troubleshooter looks for desired logic signals. If a signal at a test point is correct, the section being tested and the sections that feed signals into it are operating properly. Therefore, the sections that are driven from the test-point location need to be checked.

Digital test instruments check digital components and circuits. They are used to detect the absence or presence of a pulse, a certain waveform pattern consisting of a train of pulses, or stationary logic states of a digital circuit. These tests are performed when the system is either in the dynamic or static mode of operation.

Dynamic Tests   Dynamic tests are performed when the circuit is in its normal mode of operation. The circuit can be in a stable state or operate at a very fast rate of speed. When the circuit is operating at a fast rate of speed, it is often necessary to use an oscilloscope to observe the very rapidly changing square waves or pulses.

Static Tests   Static tests are performed by taking readings of logic signals in their stable states. This is accomplished by sequencing the circuit through the conditions listed in its truth table. The results of the test are then compared with those indicated in the truth table output.

Dynamic tests are usually performed first. The troubleshooter looks for signal activity (changing states) until the absence of a signal is found (not changing states). When a steady logic state is found, the circuit diagram should be consulted to verify that the signal should indeed be there. If not, a static test, which is the easiest to perform, should be used to find the cause of the improper signal. However, dynamic tests are used exclusively when timing faults or noise problems are the suspected causes of the problem.

DC Voltmeter   Voltmeters are primarily used to read the very critical voltages found in all logic circuits. Out-of-spec voltages usually cause erratic circuit performance. Voltmeters can also be used for reading static (unchanged) logic signals. However, they cannot be used to measure logic pulses that are rapidly changing states or pulses that occur very quickly.

Oscilloscope   The intrument that is primarily used for dynamic testing is the oscilloscope. It is capable of measuring AC and DC voltages and viewing pulse waveforms and time

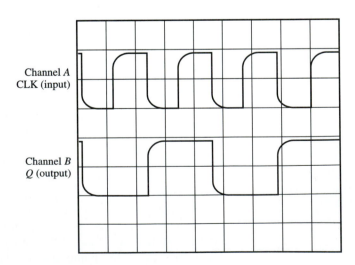

FIGURE A.1
Dual-trace oscilloscope screen showing the input and output waveform relationship of a **J–K** *negative-edge-triggered flip-flop.*

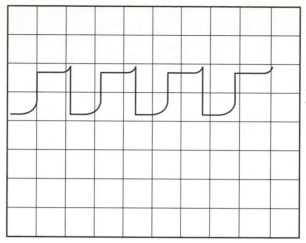

Test point 32
Volts/division = 2 V or 3.5 V p-p
Time/division = 2 ms or 200 Hz

FIGURE A.2
Proper waveform at a test point that should be observed on a scope when switches are at their proper settings.

relationships. Waveforms that cause circuit problems are those that are sufficiently distorted so that they do not have fast enough rise or fall times and those that have voltages too low or too high. Dual-trace scopes are useful because they show the true relationship between several logic signals, as shown in Fig. A-1. Observing a waveform, as shown in Fig. A-2, is another function of the oscilloscope. Equipment manuals provide actual pictures of what proper signals should look like on a scope at certain test points. Information about the frequency and voltage settings on the scope is also included.

Logic Analyzers   A specially designed test instrument for digital equipment is the logic analyzer. It is used for testing applications that require several logic levels to be tested simultaneously. As shown in Fig. A-3, a logic analyzer can display up to 16 test signals.

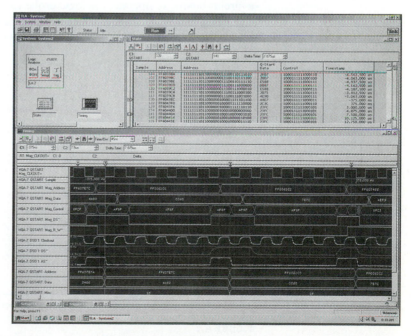

FIGURE A.3
A logic analyzer display. Courtesy of Tektronix, Inc.

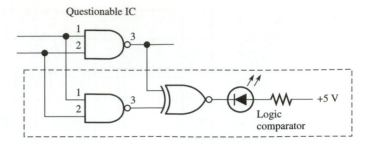

FIGURE A.4
Comparing the operation of two identical ICs.

When triggered, it captures in memory the tested states and displays a fixed pattern of 1 and 0 states on the scope display. Then the technician easily can analyze events that took place in nanoseconds. These devices are especially useful in finding transient faults that are difficult to catch and narrow pulses with low repetition rates that are difficult to see.

Logic Comparator   A logic comparator is a hand-held device that compares a questionable IC with an identical reference IC that is known to be good. Figure A-4 illustrates the basic idea of how the comparator tests the operation of an IC that consists of a questionable NAND gate. The reference IC is inserted into a socket on the comparator. Two input leads from the reference are connected in parallel to two input pins of the IC in question. The input signals are provided by the system of which the IC under test is part. The output of each IC is connected to one of the two input leads of the exclusive NOR gate. If any difference in the two outputs exists, an LED on the comparator turns on, indicating a fault.

Logic Probe   A logic probe is a compact test and troubleshooting tool for all types of digital applications. Figure A-5 shows a logic probe used to troubleshoot components on a circuit board. To operate, the logic probe requires power, which it receives from the circuit being tested. By simply connecting the probe to a circuit node under test, an indication of the circuit conditions is provided. There is a variety of different logic-probe de-

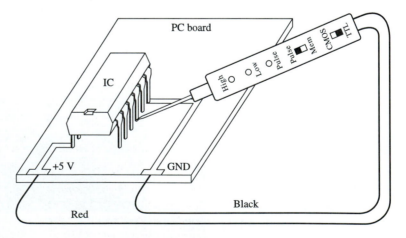

FIGURE A.5
A logic probe.

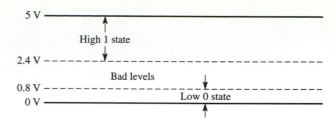

FIGURE A.6
TTL threshold voltages.

signs available. The type described in Section A-3 has three LEDs. One of them indicates highs, another indicates lows, and the third indicates a single pulse or pulse trains. If none of the LEDs light, this indicates bad levels (between high and low). The voltages between 0.8 and 2.4 V are considered "bad levels" because logic devices do not respond to them. See Fig. A-6.

A.3　HOW TO USE THE LOGIC PROBE

Applying Power to the Probe

The logic probe is protected against overvoltage and reverse voltage on its power leads. Connect the black clip lead to the common ($-$) and the red clip lead to plus (V_{CC}).

Multifamily Use

The probe is compatible with TTL IC family chips at one switch setting and CMOS at the other setting.

Interpreting the LEDs

Refer to Fig. A-7.

Steady States

- When reading logic voltage levels that are not changing states, place the memory/pulse switch in the pulse position.
- The low LED will turn on if the logic state is at the 0 level. See (A) in Fig. A-7.
- The high LED will turn on if the logic state is at the 1 level. See (B) in Fig. A-7.

All LEDs Off The absence of any indication by the LEDs is the result of the following circuit conditions.

1. The test point is an open circuit.
2. The logic signal is in the "bad level" range.
3. The probe is not connected to power.
4. The node or circuit is not powered.

See (C) in Fig. A-7.

LED states			Input signal	Operational description	
High	Low	Pulse			
○	●	○	○——————	Logic 0 no pulse activity	(A)
●	○	○	○—————	Logic 1 no pulse activity	(B)
○	○	○	————	All LEDs off 1. Test point is an open circuit 2. Out-of-tolerance signal 3. Probe not connected to power 4. Node or circuit not powered	(C)
●	●	*	○⊓⊓⊓	The shared brightness of the high and low LEDs indicates a 50% duty cycle at the test point	(D)
○	○	*	○⊓⊓⊓⊓⊓	High-frequency square wave (>MHz) at test node. As the high frequency-signal duty cycle shifts from a square wave to either a high or low duty cycle pulse train, either the low or high LED becomes activated	
○	●	*	○⊓⊓⊓⊓	When logic pulses are low more than 70% of the time	
●	○	*	○⊓⊓⊓	When logic pulses are high more than 70% of the time	

● LED on
○ LED off
* Blinking LED

FIGURE A.7
Interpreting the LEDs.

Pulse Indicator For continuous trains of pulses up to 10 MHz, the pulse LED blinks at a steady rate of 3 Hz. See (D) of Fig. A-7.

Pulse/Memory Switch

Pulse Position: For displaying short single-shot pulses, the probe has a pulse stretcher that detects pulses as short as 50 ns and stretches them to 1\3 s so that they can be seen on the pulse LED. These pulses include glitches and spikes, which a scope or voltmeter cannot detect.

Memory Position: A memory circuit is capable of catching and holding level transitions or pulses as narrow as 10 ns by keeping the pulse LED lighted. This

FIGURE A.8
A logic probe. Courtesy of Global Specialties.

feature provides a means of preventing failing to detect a short pulse because the operator blinked or turned while the LED flashed. It also allows the technician to clip the probe in place and wait as long as necessary to trap a troublesome glitch.

Logic Pulser A device frequently used with a logic probe is a logic pulser. See Fig. A-8. It is an input stimulus device that provides a signal to a static (unchanged) IC input. The technician using a pulser manually injects a logic input at will even if it is connected to another circuit output. Like the logic probe, the pulser obtains its power from the same power supply of the circuit that is being tested.

When the pulser tip is connected to the circuit node to be tested, the pulser automatically senses the logic level at the test point. The pulser also self-adjusts so that the output pulse it generates is within the voltage-level requirements of the circuit under test.

Before activating the pulser, the TTL/CMOS selector switch should be placed in the correct position. When the trigger switch is depressed, it overrides any logic level and automatically supplies a pulse of the correct polarity needed to change the state of the circuit node to which it is connected.

When used together, the logic pulser and logic probe are capable of effectively troubleshooting most logic-circuit malfunctions. Figure A-9 illustrates how to troubleshoot an open failure of a gate input by using a logic pulser and a logic probe. Figure A-9(a) shows an open at the bottom lead of an AND gate. Figure A-9(b) shows that when a string of pulses is applied to the top lead, the pulse LED on the probe flashes. Figure A-9(c) shows that when a string of pulses is applied to the bottom lead, the high LED stays constantly lighted (the gate likely observes the open as a high) because the pulses do not reach the internal circuit of the gate.

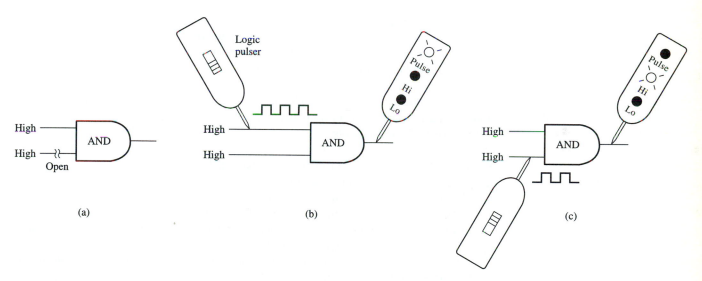

FIGURE A.9
Troubleshooting an open input.

Figure A-10 illustrates how to troubleshoot an open failure of an OR-gate output. Figure A-10(a) shows the absence of a pulse train at the output when the top lead is pulsed. Figure A-10(b) also shows that no LEDs light when the bottom lead is pulsed. The LEDs on the probe do not light because an open produces a 1.5-volt "bad level" potential that the probe cannot display.

If the LO probe light were constantly on, it would indicate a probable short of the output to ground. If the HI LED were constantly on, it would indicate a probable short of the output to V_{CC}.

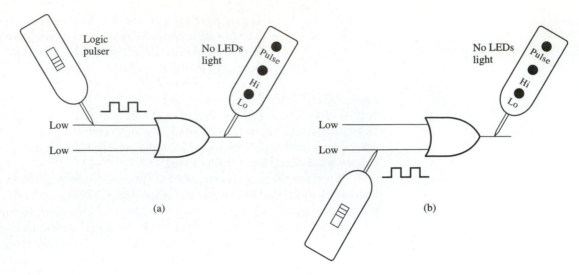

FIGURE A.10
Troubleshooting an open output.

LOGIC CLIP AND CURRENT TRACER

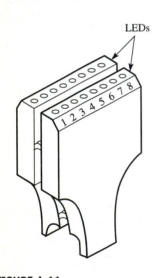

FIGURE A.11
Logic clip.

Logic Clip   A logic clip is another easy-to-use digital testing device. Unlike the logic probe, it checks a number of points simultaneously.

A logic clip is shown in Fig. A-11. By clamping it over an IC DIP package, the two rows of LEDs instantly indicate the logic states of all pins. When any IC pin is 2.4 V or above, the LED that corresponds to it turns on. The pins that have a voltage below 2.4 V do not activate their corresponding LEDs. There are no controls to set and no power leads to connect. A built-in power-seeking gating network automatically locates the most positive and negative voltages applied to the IC under test, which allows the clip to be connected either way.

The shortcomings of a logic clip are that it can only detect a changing signal up to about 10 Hz and that it can draw up to 200 mA when all the LEDs are on, which can affect the operation of some circuits.

Current Tracer   A current tracer is another hand-held troubleshooting device that detects a changing current in a conductor. It has an insulated tip that contains a magnetic pick-up coil. By placing the tip next to a conductor, it senses any changing magnetic field produced by a changing current between 1 mA to 1 A, which causes a small LED to flash. If no current is changing in a conductor, the light will not flash. The current tracer also does not respond to any static current logic levels.

Unlike the logic probe, the current tracer is able to locate where a short exists in a defective circuit.

THE ISOLATED FAULT

Once the problem has been isolated to a particular area, say, a printed circuit (PC) board, it is sometimes advisable to interchange the board with one that is known to be operating properly. If the problem is corrected, the malfunction is somewhere on the board.

Once the problem has been traced to a specific PC board, the technician should attempt to isolate the problem to the defect itself. The next step for the technician is to use his or her physical senses.

■ Touch the flat top of the IC to determine if it is hot. This would result from excessive current flow due to an internal short.

- Visually scan the board and look for obvious faults such as broken connections, bad soldering joints, discolored components, or any charred spots that result from excessive heat.
- Smell for possible overheating.

    If these procedure steps are not successful, then the use of test instruments is necessary to locate the specific problem, which is either an internal IC fault or one that is external to the IC.

A.6   INTERNAL DIGITAL IC FAULTS

The most common internal IC failures are as follows:

- The inputs are internally shorted to ground or to the +5-V supply. If it is shorted to ground, a low will be permanently present, as shown in Fig. A-12(a). Therefore, the output will always be high. Similarly, the pin could be shorted to a high, as shown in Fig. A-12(b).

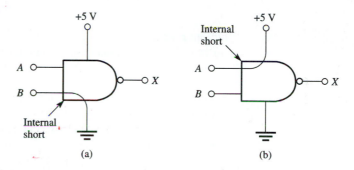

FIGURE A.12
(a) The IC input is internally shorted to ground. (b) The IC input is internally shorted to the supply voltage.

- The output is internally shorted to ground or to the +5-V supply. If it is shorted to ground, a low will be permanently present at the output, as shown in Fig. A-13(a). If it is shorted to the power supply, a high will always be present at the output, as shown in Fig. A-13(b). Both conditions result in no signal changes beyond the point where the short exists. Shorted inputs or outputs are called *stuck-low* or *stuck-high,* respectively.

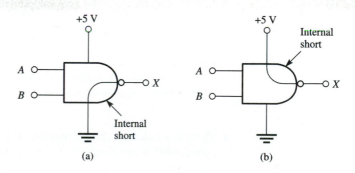

FIGURE A.13
(a) The IC output is internally shorted to ground. (b) The IC output is internally shorted to the supply voltage.

- There is an open-circuited input or output pin. This condition occurs when the very fine wire that connects the external IC pin to the internal circuitry of the IC breaks. If an input were open, as shown in Fig. A-14(a), the logic signal applied to that input will have no effect on the output. Instead, the opens may act as stuck-low or stuck-high inputs or may oscillate between lows and highs due to noise voltages. When an output is open, as shown in Fig. A-14(b), there will be no voltage present regardless of whatever is applied to the inputs. Open inputs or outputs are sometimes referred to as "floating" terminals.

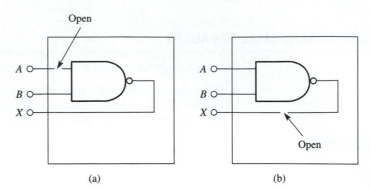

FIGURE A.14
(a) The IC with an internally open input. (b) The IC with an internally open output pin.

■ There is a short between two input pins. As shown in Fig. A-15, this condition causes these two terminals always to be identical and never permits the two signals applied to the inputs to be opposite. This condition usually causes the gate to produce an output with three distinct levels.

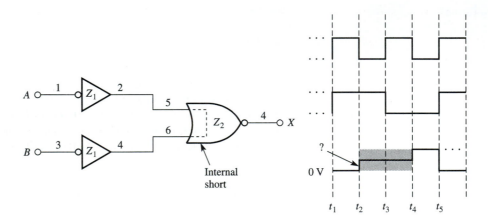

FIGURE A.15
When two input pins are internally shorted, they force the signals driving these pins to be identical and usually produce a signal with three distinct levels.

■ There is a failure of the internal circuitry. If a failure occurs inside the miniature chip circuitry itself, it can cause stuck-low or stuck-high inputs or outputs.

A.7   EXTERNAL DIGITAL IC FAULTS

Most digital electronic circuit problems are the result of things that go wrong external to the ICs. The most common external IC failures are as follows:

- Open input or output lines. This problem is the result of a break that occurs in a conducting path, which prevents a proper logic signal from going from one location to another. Some of the more common causes of this problem are:

  –A broken wire.

  –A crack on a circuit path of a printed-circuit board.

  –A poor solder connection.

  –A bent or broken IC pin.

  –A faulty IC socket.

- Shorted input or output lines. This problem usually causes an improper logic signal to be present at a circuit location. Some of the more common causes of this problem are:

  –Solder bridges that are splashes of solder that short two conductors together, such as printed-wiring conductor paths or IC pins.

  –A conducting lead of a component touching another conducting path.

  –Foreign particles made of conducting materials that short out two circuit paths.

- A faulty power supply. Each TTL family IC requires $+5$ V to power the internal circuity of the chip. Any variation from the $+5$ V causes the chip to operate improperly. The types of IC power supply malfunctions that occur are:

  –No $+5$-V potential. This can result from an open fuse that is blown because of faulty components in the power supply circuitry.

  –A power supply that does not provide a voltage that is within the tolerance a chip can handle. It can be a voltage that is too low or one that changes its level erratically. These two problems usually are caused by a faulty chip that is operating erratically or drawing too much current. It can also be a leaky filter capacitor that causes poor regulation.

  –A ground lead that is not providing a 0-V potential. This sometimes results in an open ground lead or a short that pulls the ground higher than 0 V.

- Failure of a discrete component. In addition to using ICs, most digital circuits use discrete components such as resistors, capacitors, diodes, and transistors. If one of these devices is found to be defective, it is a good idea to check if another component caused the failure.

A.8 INTERMITTENT PROBLEMS

Sometimes a circuit malfunction occurs intermittently. Finding such a problem can be an extremely frustrating situation for the troubleshooter who tries to isolate a fault that develops unpredictably or for only a brief moment. Methods used to troubleshoot such a malfunction are as follows:

Stress Test Intermittent problems usually occur because of a marginal discrete component or chip, a hairline fracture of a lead or conducting path, or a poor solder joint. Using a stress test can cause these types of faults to temporarily improve or become worse, which helps in pinpointing the trouble. Stress tests are performed in the following ways:

Physical Stress Test: Circuit boards are physically stressed by tapping or twisting them.

Thermal Stress Test: By applying extreme temperatures to a circuit board or components, it can cause the conductors internal or external to the IC to bend slightly and thereby make or break a connection. A thermal test can be done by using either a heat gun to make it hot or a spray can of freon to make it cold. The advantages of either thermal procedure are that the test can be applied to a single suspected component and can cause the intermittent problem to occur for a longer and more predictable time.

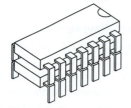

FIGURE A.16
Piggybacking an IC.

Piggyback A very common reason for an intermittent IC problem is an internal broken bond between the external pins and the chip mounted inside. When a chip does not work for most of the time, a technique to find such a problem is to use the piggyback method. A good IC is placed in parallel on top of the suspected IC, as shown in Fig. A-16. The new IC is fed the input data and provides the desired output signals. If the intermittent problem disappears, the defective IC has been found.

A.9 MORE TROUBLESHOOTING STEPS

Step 3. Correct the Fault Once the cause of the fault is found, it should be corrected. This can be accomplished by repairing or replacing a faulty component or repairing a faulty connection. Once the correction is made, the equipment should be tested for proper operation.

When a malfunction occurs, the amount of time required to correct it is often very important. For example, a faulty programmable controller that causes a paper machine to shut down can cost the company many thousands of dollars per hour. A broken bank computer can cause most of the bank operations to discontinue until the problem is corrected. In either situation, it is vital to troubleshoot and repair the equipment as quickly as possible.

The time in which a technician can get the equipment running again often depends on efficient troubleshooting techniques that help to find the problem quickly. Some of these techniques are as follows:

TABLE A.1
Relative Failure Rates of Some Common Components

FAILURE RATE	*COMPONENT*
High	Fuses and lamps
↑	Switches and relays
	Mechanical I/O devices
	Power supplies
	Connectors and cables
	Capacitors and resistors
	Transistors and diodes
	LEDs
↓	Integrated circuits
Low	Printed circuit boards

Technique 1. **Make the Easy Test First:** First check the circuit locations that are relatively easy to test. If the equipment is designed properly, test points will be available at locations where the equipment is most likely to fail.

Technique 2. **Check the Components with High Failure Rates First:** There are some components that are likely to fail at a higher rate than others. These should be checked first. Table A-1 provides a list of common components in the order of their relative failure rates.

Technique 3. **Use Information Provided by Digital Circuit Diagrams:** Some manufacturers use logic symbols with small circles (bubbles) to help the troubleshooter identify the desired logic-state locations throughout the circuitry. A bubble means that an input or output is low when the circuit is activated. An input or output without the bubble means that the signal is high when the circuit is activated. Figure A-17 shows a push button with two inverters and a LED that represents the output. When the push button is not pressed, test points *A* and *C* are high and test point *B* is low. Therefore, the output LED will not be activated by turning on. When the push

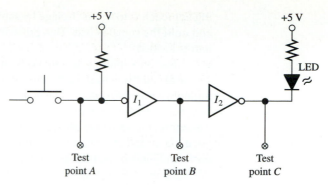

FIGURE A.17
Using logic-state indicators in troubleshooting.

button is pressed, test points A and C are low and test point B is high. Therefore, the output LED will be activated by turning on.

By using the bubbles, the troubleshooter can go to anywhere in the circuit and determine by the bubble locations what the signal should be when the light is supposed to be activated or deactivated. For example, because there is no bubble at the output of I_1 and the input of I_2, the troubleshooter knows that the signal is low when the LED output is off and high when the LED output is on. The reason this information is valuable is that the troubleshooter does not have to analyze the entire circuit to determine what signals should be present at certain locations, and this saves time.

Technique 4.   The Half-Splitting Technique: A common type of digital troubleshooting approach is the half-splitting technique. This method works best in circuits that have unidirectional signal paths without large feedback loops. For example, Fig. A-18 shows such a circuit with several stages connected in series.

Suppose that a malfunction develops in one of the stages and the known condition is that the input signal is good, but the signal at the output is not present. One troubleshoot-

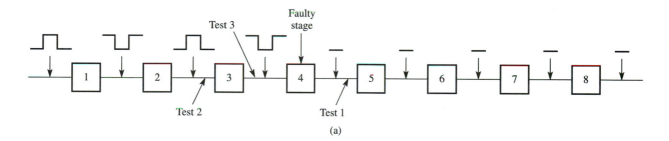

(a)

Number of stages	Number of readings
8	3
16	4
32	5
64	6
128	7
256	8

(b)

FIGURE A.18
Half-splitting technique.

ing approach is to test each stage by starting at the beginning stage and working toward the end until the signal is lost. This procedure could take up to nine readings before the problem is located.

The half-splitting technique is more efficient. It begins by taking a reading at the center of the string (test 1), because it is just as likely that a fault exists before as after this midpoint. A good reading indicates that the fault is located in the second half of the string. Suppose the reading is bad. This means the fault is located in the first half of the string. Therefore, the second reading (test 2) is made in the middle of the first half of the string. A good signal indicates that the fault is located between where tests 1 and 2 were taken. A third reading is made (test 3). This test is good, so the problem is located in stage 4. Instead of taking up to 9 readings, this technique requires 3 to find a fault at any location in a serial circuit with eight stages. Figure A-18(b) provides serial-stage samples of different sizes and shows how many readings using the half-splitting technique are needed to find a fault in each.

IEEE STANDARD LOGIC SYMBOLS

In 1984, a new set of standard symbols for digital devices was introduced in the United States by the Institute of Electrical and Electronics Engineers (IEEE) Committee SCC 11.9. The new standard, numbered simply IEEE Std. 91–1984, includes all of the approved work that has been developed by the International Electrotechnical Commission (IEC). The advantage of this system is that it provides a method of determining the complete logical operation of a given device by interpreting the notations on the symbol for the device.

Because U.S. military contracts now require the use of these symbols, and the symbols are gradually being accepted by electronic companies and IC manufacturers, it is important to become familiar with them.

The following information about the new symbology is condensed from the 147-page *IEEE Standard Graphic Symbols for Logic Functions.*

B.1   SYMBOL COMPOSITION

The IEEE Std. 91–1984 system uses a rectangular-shaped symbol called an *outline* for all devices, as shown in Fig. B-1. To indicate what operation is performed by the device, a general qualifying symbol is placed at the top center of the outline. Unless otherwise indicated, input lines are placed on the left and output lines are placed on the right.

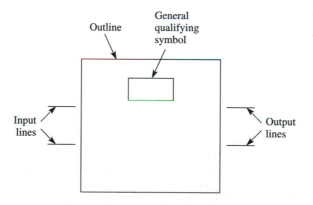

FIGURE B.1
IEEE symbol composition.

B.2   LOGIC GATES

Table B-1 shows the traditional symbols with the IEEE symbols for the basic logic gates.

The traditional symbols shown throughout this text use the bubble to either indicate an inversion of a logic level or an active-low input or output terminal. The IEEE system uses a small right triangle (◁) in place of the bubble. See Table B-2.

TABLE B.1

Comparing Tradition Logic Symbols with IEEE Logic Symbols

	Traditional logic symbol	IEEE logic symbol	Explanation
AND	A, B → X	A, B → & → X	The general qualifying symbol & means that the output goes to its active-high state when *all* inputs are in their active-high state.
OR	A, B → X	A, B → ≥1 → X	The general qualifying symbol ≥ means that the output goes to its active-high state whenever one or more inputs are in their active-high state.
Exclusive-OR	A, B → X	A, B → = 1 → X	The general qualifying symbol = 1 means that the output goes active high only when one input is high.

TABLE B.2

Active-Low Output Logic Symbols

	Traditional logic symbol	IEEE logic symbol	Description
Inverter	A → X	A, B → 1 → X	The general qualifying symbol 1 indicates that a device has only one input. The triangle on the output indicates that the output goes to its active-low state when the input is in its active-high state.
NAND	A, B → X	A, B → & → X	The general qualifying symbol & and the inversion triangle at the output lead mean that the output goes to its active-low state when all inputs are in their active-high state.
NOR	A, B → X	A, B → ≥1 → X	The general qualifying symbol ≥ and the inversion triangle at the output lead mean that the output goes to its active-low state whenever one or more inputs are in their active-high state.
Exclusive-NOR	A, B → X	A, B → = 1 → X	The general qualifying symbol = 1 and the inversion triangle at the output mean that the output goes active-low only when one input is high.

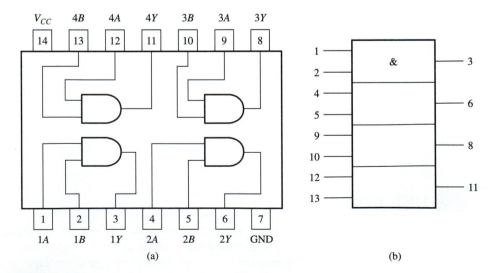

FIGURE B.2

7408 quad two-input AND gates.

The IEEE symbols can also be used to represent a complete IC chip that contains several independent logic gates. For example, Fig. B-2(a) shows the traditional pin diagram of a 7408 quad two-input AND gate package and Fig. B-2(b) shows the equivalent IEEE symbol. Each rectangular box represents a separate logic gate. The general qualifying AND gate symbol is located only in the top block and indicates that all of the other blocks are also AND gates.

B.3 DEPENDENCY NOTATION

TABLE B.3
Dependency Notation

G, AND
V, OR
N, Negate (Exclusive-OR)
Z, Interconnection
X, Transmission
C, Control
S, Set and R, Reset
EN, Enable
M, Mode
A, Address

If the IEEE system were used only for basic logic gates, there would be no advantage over the traditional symbols that use distinctive shapes for each type of gate. However, for digital devices that are more complex, *dependency notation* with IEEE qualifying symbols is capable of specifying the complete operation of the device. This information provides a compact, meaningful method of indicating the relationship between the inputs and outputs without actually showing all of the internal circuitry or function tables.

There are 11 types of dependency notation, as shown in Table B-3. Each notation is placed just inside the symbol outlines where an input or output terminal is connected. Dependency notation is accomplished by the following:

1. Labeling the input or output *affecting* other inputs or outputs with one of the letters in Table B-3 that indicates the function involved (e.g., V for OR) followed by an identification number.
2. Labeling each input or output that is *affected* by the affecting input or output with that same number.

Detailed information on how to use dependency notation can be obtained from the manual *IEEE Standard Graphic Symbols for Logic Functions*. Simplified examples of how dependency notation is used with brief explanations are provided with some of the remaining IEEE logic symbols that are shown in this appendix.

In the decoder of Fig. B-3, the label BIN/OCT indicates that the circuit is a binary-to-octal decoder. The three inputs with the prefix $\overline{E}$ are applied to an AND block to produce an internal signal, EN.

In the encoder of Fig. B-4, the label HPR1/BCD indicates that the circuit converts the active-low input with the highest priority to its BCD code.

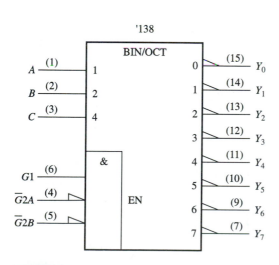

FIGURE B.3
The 74138 decoder IC.

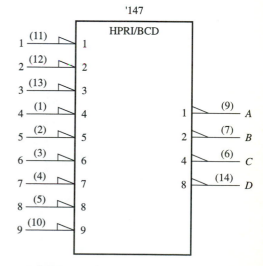

FIGURE B.4
The 74147 encoder IC.

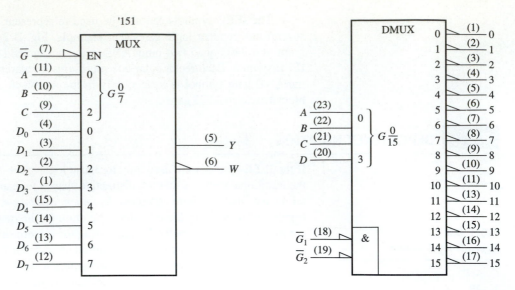

'151

FIGURE B.5
The 74151 multiplexer IC.

FIGURE B.6
The 74138 demultiplexer IC.

In the multiplexer of Fig. B-5, the label MUX indicates that the circuit is a multiplexer. The $G\frac{0}{7}$ label is a dependency notation that indicates each data input line 0 through 7 is ANDed with one specific combination of selected inputs, and when that combination occurs, that data input is routed to the output line.

In the demultiplexer of Fig. B-6, the label DEMUX indicates that the circuit is a demultiplexer. The $G\frac{0}{15}$ label is a dependency notation that indicates each data output line 0 through 15 is ANDed with one specific combination of selected inputs, and when that combination occurs, that data input is routed to the output line.

In the comparator of Fig. B-7, the label COMP indicates that the circuit is a comparator. The letters P and Q are the preferred designation used by the IEEE standard to represent input variables.

Figure B-8 shows a IEEE logic symbol for a *J–K* flip-flop. The right triangle at the C (CLK) with the diamond inside the symbol outline indicates that the flip-flop is negative-

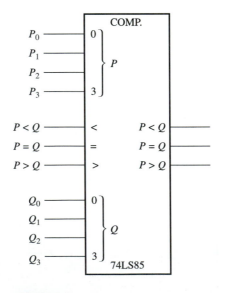

FIGURE B.7
The 7485 comparator IC.

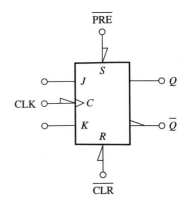

FIGURE B.8
The J–K flip-flop.

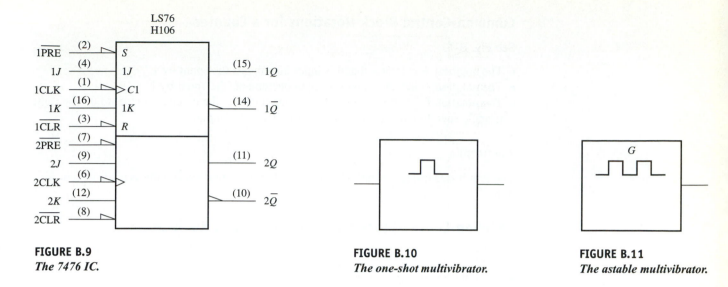

FIGURE B.9
The 7476 IC.

FIGURE B.10
The one-shot multivibrator.

FIGURE B.11
The astable multivibrator.

edge-triggered. The right triangles at the S (set) *and* R (reset) leads indicate that the inputs are active low and the $\overline{Q}$ output is an inverted signal.

Figure B-9 shows the IEEE logic symbol for a 7476 IC. Each block inside the outline represents an individual *J–K* flip-flop.

Figure B-10 shows the general qualifying symbol indicating a one-shot monostable multivibrator.

Figure B-11 shows the general qualifying symbol indicating an astable multivibrator.

B.4   COMMON-CONTROL BLOCK

Figure B-12(a) shows the IEEE standard symbol called a common-control block. It is used when a circuit has one or more inputs that are common to more than one element of the circuit, such as the one shown in Fig. B-12(b).

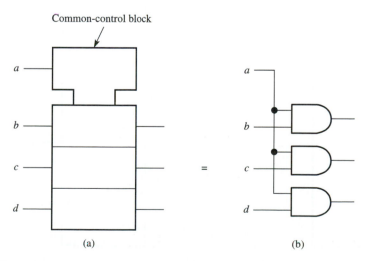

(a)

(b)

FIGURE B.12
Common-control block.

Common-Control Block Notations for a Counter

See Fig. B-13.

- The notation $+$ indicates that this input increments the count by 1.
- The notation $-$ indicates that this input decrements the count by 1.
- The notation $CT = 0$ indicates that this input causes the counter to reset to 0. The right triangle specifies that a low is needed to activate the input.

The Outline

The four boxes inside the outline of Fig. B-13 indicate that the counter is made of four flip-flops.

Common-Control Block Notations for a Shift Register

See Fig. B-14.

- The notation SR signifies that the device is a shift register.
- The notation $\rightarrow$ indicates that this input causes the data to shift to the right.
- The notation $\leftarrow$ indicates that this input causes the data to shift to the left.

The Outline

- The four boxes inside the outline indicate that the register is made up of four flip-flops.
- The S represents the set input.
- The R represents the reset input.
- The leads on the right with and without the right triangle represent complementing outputs.

Common-Control Block Notation for an 8 × 4 RAM

See Fig. B-15.

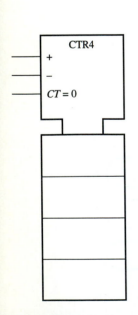

FIGURE B.13
The counter.

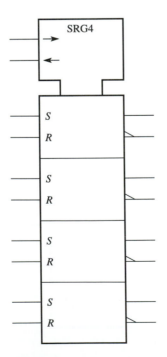

FIGURE B.14
The shift register.

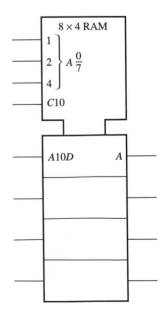

FIGURE B.15
The 8 × 4 RAM.

- The notation 8×4 RAM specifies that the device is a random-access memory device capable of storing eight 4-bit words.
- The 1, 2, and 4 represent the binary-weighted values of each input. The A specifies that the leads are address lines, and the $\frac{0}{7}$ indicates that address locations 0 to 7 are used.
- The C indicates a control (enable) line. The 1 and 0 after the C specify that when one state is applied to the line, information can be written into the device, and when the other state is applied, data can be read from it.

ANSWERS TO SELECTED PROBLEMS

CHAPTER 1

1. gradually, discretely   3. proportional
5. 1   0
   Yes   No
   True   False
   On   Off
   High   Low
6. memory, decision or logic   8. (d)
10. Electromagnetic radiation from lightning storms, sun spots, high-wattage devices, and electrical motors
13. (e)   15. (a)

CHAPTER 2

1. Type of symbol used, position where a symbol is located, and use of zero
3. Binary. Because they operate on signals that are at two voltage levels, which represent 1s and 0s
5. 16
7. modulo division:

- Repeatedly divide the decimal number by 2.
- Write down the remainder after each division until a quotient of 0 is obtained.
- The first remainder is the LSB and the last remainder is the MSB.

9. addition

11. **0011**   13.   **0110**
    0100   0111
    0101   1000
    0110   1001
    0111   0001   0000
    1000   0001   0001
    1001   0001   0010
    1010   0001   0011
    1011   0001   0100
    1100   0001   0101
    1101   0001   0110
    1110   0001   0111
    1111   0001   1000
    10000   **0001**   **1001**

15.

DECIMAL	BINARY	OCTAL	BCD	HEXA-DECIMAL
33	100001	41	0011 0011	2I
17	10001	21	0001 0111	11
25	11001	31	0010 0101	19
44	101100	54	0100 0100	2C
125	1111101	175	0001 0010 0101	7D

16. 10111, 11110, 111110, 110110, and 11001000
17. 1000, 0101, 00101, 01011, and 1001000
18. FOX

CHAPTER 3

1. See Fig. 3-68.
3. (a) $\overline{A + B}$
   (b) $A \oplus B$
   (c) $A = \overline{A}$
   (d) $\overline{A \cdot B}$
   (e) $A + B$
   (f) $A \cdot B$
   (g) $\overline{A \oplus B}$
6. Connecting ground, V_{CC}, gate inputs, and gate outputs.
7. AND–NAND, OR–NOR, exclusive-OR–exclusive-NOR
9. The word is derived from the inverter producing an output such as "Not A" ($\overline{A}$) when an A is applied to its input.
10. False   11. counterclockwise   16. Two   17. (d)
19. vertical, horizontal, left, right
22. Any gate producing a high will be pulled low by any of the other gates producing a 0 state.
24. NAND, OR, AND, NOR
26. They show which logic state input and output leads are in during their active and inactive (resting) circuit conditions. This information is especially useful in troubleshooting.
27. See Fig. 3.69.
29. See Fig. 3.70.
31. High
34. (a)   (i)
    (b)   (iii)

Gate rule	Truth table	Switch analogy	Symbol

Row 1:

Output = 1 if inputs are the same

	Inputs		Output
	Switch A	Switch B	Y
	0	0	1
	0	1	0
	1	0	0
	1	1	1

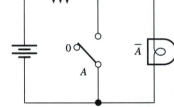

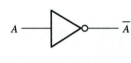

Row 2:

Output = 0 if and only if all inputs = 1

A	B	Y
0	0	1
0	1	1
1	0	1
1	1	0

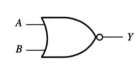

Row 3:

Output = 1 if input = 0 and vice versa

A	$\overline{A}$
0	1
1	0

Row 4:

Output = 0 if one or more inputs = 1

A	B	Y
0	0	1
0	1	0
1	0	0
1	1	0

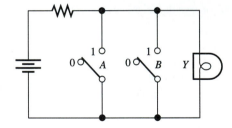

FIGURE 3.68

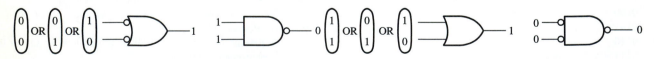

FIGURE 3.69

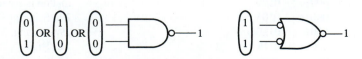

FIGURE 3.70

36. (a) (iv)
    (b) (iv)

CHAPTER 4

1.

SYMBOLS	OPERATORS	FUNCTIONS
⊃D—	$(\cdot)$	Boolean multiplication
⊃D—	$(+)$	Boolean addition
—▷—	$\overline{x}$	Complementation

3. inputs, output
5. (a) product-of-sums, (b) sum-of-products
7. (a) AB (commutative);
    (b) ABC (associative);
    (c) $A + B + C$ (associative)
9. input
11. It describes what state each input line must be in to cause the output (test point) to go high. A letter representing an input with an overbar must be low, and a letter without an overbar must be high.

CHAPTER 5

1. three, two   **3.** encoding, decoding   **5.** one
7. demultiplexer   **9.** nibble   **10.** highest
11. It determines if one binary number is greater than, less than, or equal to the other binary number.
13. serial, parallel
15. (a) a, b, g, e, d
    (b) b, c, f, g
    (c) a, b, c, d, e, f, g
17. input, output   **19.** address   **20.** $a, b, c, f,$ and g
23. low
25. Don't-care conditions. These indicators specify that these inputs do not influence the output of the circuit because they are overridden by another input condition.
27. The small circles at the outputs of the gates inform the technician that an output must be low when it is activated.
29. (1) Apply a low to pin $\overline{LT}$. (2) Apply the following input bit pattern to the 7447 IC: $A = 0, B = 0, C = 0,$ and $D = 1$.
33. Lamp 3   **35.** (a)
36. The strobe input (pin 9) is high, which deactivates the multiplexer. Therefore, output W (pin 10) should be high instead of low.

CHAPTER 6

1. (a) SSI: 20 to 50, logic gates
    (b) MSI: 500, decoders and multiplexers
    (c) LSI: 20,000, microprocessor
    (d) VLSI: 20,000 to 100,000, memory
    (e) SLSI: over 100,000, memory
3. Size, cost, low power consumption, better reliability
5. families
7. The difference between the two is based on the type of transistor formed on the substrate. Bipolar ICs use common NPN or PNP transistors, and MOS ICs use field-effect transistors.
9. slower   **12.** PMOS, NMOS   **14.** (c)
15. DM: prefix
    54: temperature

LS: subfamily
04: functional type
N: package
16. temperature range, plastic, ceramic
18. (a) 7400: standard TTL
    (b) 74H00: high-power TTL
    (c) 74S00: Schottky-clamped TTL
    (d) 74L00: low-power TTL
    (e) 74LS00: low-power Schottky TTL
21. invalid region, unpredictable
23. 12 mA/-2 mA = 6 (a worst-case situation), fan-out, sinking
24. It indicates that the direction of conventional current is flowing out of the device.
26. sinking
28. Family $A = 0.7$ V and family $B = 0.4$ V; therefore family A.
30. (d)   **32.** switch time
33.

INPUT	OUTPUT
Low: 0.0–0.8 V	Low: 0.0–0.4 V
High: 2.0–5.0 V	High: 2.4–5.0 V

34. 12 mA is the result of one gate output being low. The current value for two gates going low is calculated as follows:

$$2 \times 12 \text{ mA} = 24 \text{ mA}$$

36. Position 2 to 1:

I_1 = high to low	7 ns
I_2 = low to high	12 ns
I_3 = high to low	7 ns
	26 ns

Position 1 to 2:

I_1 = low to high	12 ns
I_2 = high to low	7 ns
I_3 = low to high	12 ns
	31 ns

Therefore, position 2 to 1.
38. Recommended Operating Conditions
40. Different logic-level voltages and different power supply voltages
42. (c) or (d). All conditions could cause this problem. The most likely answer is (c) because it is impossible to obtain a high output for an open-collector IC without a pullup resistor.
43. (a)

CHAPTER 7

1. Sequential circuits have the ability to store binary information. Combination circuits make logic decisions.
3. memory   **5.** one   **6.** (b)
8. asynchronous, synchronous
10. During the transition from low to high or high to low of the enable signal.
12. (b)   **13.** (d)
15. Through (1) data and clock inputs, (2) preset input, and (3) clear input

17. low   **19.** inverter   **20.** edge, low
23. (1) Hold, (2) reset, (3) set, (4) toggle   **27.** 8 kHz
30. (b)   **31.** (d)
32. In Steps 3 and 6, the flip-flop is in the toggle mode and should change states. Instead, it appears as if it is in the reset mode. This would occur if the J input were shorted to ground.

CHAPTER 8

1. sequential
3. From the output of the preceding flip-flop.

5. modulo, 5 (since 0 occupies one count)
7. (c)   **8.** LSB   **9.** asynchronous
11. parallel   **13.** preset   **15.** most
17. 6.25 kHz   **18.** 500 Hz   **21.** high
22. 2, 12, 1   **24.** 6
27. (a) mod-2, mod-8, mod-16
(b) up
(c) asynchronous
(d) both high
29. (d)
30. (d)

TABLE 8.2
Answer to Problem 28

EVENT NUMBER	INPUTS				OUTPUTS					
	$(CP)_U$	$(CP)_D$	$\overline{PL}$	MR	Q_0	Q_1	Q_2	Q_3	$(TC)_U$	$(TC)_D$
1	1	1	0	0	1	0	1	1	1	1
2	0	1	1	0	1	0	1	1	1	1
3	1	1	1	0	0	1	1	1	1	1
4	0	1	1	0	0	1	1	1	1	1
5	1	1	1	0	1	1	1	1	1	1
6	0	1	1	0	1	1	1	1	0	1
7	1	1	1	0	0	0	0	0	1	1
8	0	1	1	0	0	0	0	0	1	1
9	1	1	1	0	1	0	0	0	1	1
10	0	1	1	0	1	0	0	0	1	1
11	1	1	1	0	0	1	0	0	1	1
12	1	0	1	0	0	1	0	0	1	1
13	1	1	1	0	1	0	0	0	1	1
14	1	0	1	0	1	0	0	0	1	1
15	1	1	1	0	0	0	0	0	1	1
16	1	0	1	0	0	0	0	0	1	0
17	1	1	1	0	1	1	1	1	1	1
18	1	1	1	1	0	0	0	0	1	1
19	1	0	1	0	0	0	0	0	1	0
20	1	1	1	0	1	1	1	1	1	1

CHAPTER 9

1. (c)   **3.** shift   **5.** SISO   **7.** (d)
10. (a) SISO: 8
(b) PISO: 4
(c) PIPO: 0
(d) SIPO: 4
12. 0011, 1001   **13.** 32   **14.** preset, clear   **16.** 1100

18. 1011 when loaded; 1101 after the first clock pulse; 1110 after the second clock pulse; and 0111 after the third clock pulse.
19. one   **21.** 1000
22. (c). An open at the K input is recognized as a high, which puts FFD in the toggle mode.
23. (c)
24. 1

TABLE 9.3
Answer for Problem 20

EVENT NUMBER	CLK	S_0	S_1	$\overline{CLR}$	D_{SR}	D_{SL}	A	B	C	D	Q_A	Q_B	Q_C	Q_D
							INPUTS PARALLEL DATA				OUTPUTS			
1	—	1	1	1	0	0	0	1	0	1	0	0	0	0
2	⎍	1	1	1	0	0	0	1	0	1	0	1	0	1
3	—	1	0	1	0	0	0	1	0	1	0	1	0	1
4	⎍	1	0	1	0	0	0	1	0	1	0	0	1	0
5	—	1	0	1	1	0	0	1	0	1	0	0	1	0
6	⎍	1	0	1	1	0	0	1	0	1	1	0	0	1
7	—	1	0	0	1	0	0	1	0	1	0	0	0	0
8	⎍	1	0	1	1	0	0	1	0	1	1	0	0	0
9	—	0	1	1	0	1	0	1	0	1	1	0	0	0
10	⎍	0	1	1	0	1	0	1	0	1	0	0	0	1
11	—	0	1	1	0	1	0	1	0	1	0	0	0	1
12	⎍	0	1	1	0	1	0	1	0	1	0	0	1	1
13	—	0	1	1	0	0	0	1	0	1	0	0	1	1
14	⎍	0	1	1	0	0	0	1	0	1	0	1	1	0
15	—	0	0	1	0	0	0	1	0	1	0	1	1	0
16	⎍	0	0	1	0	0	0	1	0	1	0	1	1	0
17	—	1	1	1	0	0	1	1	0	0	0	1	1	0
18	⎍	1	1	1	0	0	1	1	0	0	1	1	0	0

CHAPTER 10

1. It causes the exclusive-OR gate to operate as either a straight wire or an inverter to the input not connected to the switch. When the switch is in the add (0 state) position, it causes the data from register *B* to pass straight through the exclusive-OR gate to input *B* of the full adder. When the switch is in the subtract (1 state) position, it causes the data from register *B* to be inverted as it passes through to input *B* of the full adder.

3. True   **5.** It is stored in the carry flip-flop.   **6.** (A, C, D)

8. When the outputs of each full adder are applied to the *J* input (and *K* input through the inverter), each flip-flop in the accumulator is put in either the set or reset mode. When a clock pulse arrives, the appropriate 1s or 0s are then stored (parallel-loaded) at the *Q* output of each flip-flop.

9. Five add cycles. Every time an add cycle occurs, the multiplier, which is a down-counter, decrements one count. When the multiplier is at 0000, N_1 goes low, which disables AND gate 2 and prevents the clock pulse from passing through to the rest of the circuit. The number in the accumulator displays the product answer.

12. (a) 16
      (b) 16
      (c) 24

13. able

14. On every high-to-low transition, it increments the up-counter. This occurs at the end of each add cycle.

16. As the numbers in registers *A* and *B* are shifted around during each add cycle, there are times that the contents of register *B* are greater than *A*. Therefore, the only time a valid comparison should be made is at the end of each add cycle, which occurs when the clock timer is 00.

17. Fifteen, or 1111_2.

18. (b). During the addition, exclusive-OR 2 should pass the 1 in the subtrahend straight through to its output. During the subtraction, exclusive-OR 2 should invert the 1 in the subtrahend. The open at the exclusive-OR 2 output causes a 1 to be always present.

19. (a). A short to ground causes the logic state at input 2 of each exclusive-OR gate and C_{IN} of FA1 to be pulled low because they are on a common line. Therefore, the circuit operates as an adder instead of a subtractor.

CHAPTER 11

1. The power supplies are at voltages of opposite polarities. This enables the operational amplifier to generate an output voltage that goes positive and negative.

3. (a) (<)
      (b) (=)
      (c) (>)

5. −10 V   **7.** −18 V   **9.** 3   **11.** twice

12. $2^5 = 32$. Therefore, 31 steps and 15/31 = 0.48 V.

15. (1) Reduce V_{REF} to 5 V. (2) Increase the resistor at pin 14 to 10 kΩ. (3) Reduce resistor R_f across the op-amp to 2.5 kΩ.

17. Counter   **19.** They are equal.

21. 255, 8   **22.** 2.56   **23.** SPDT   **24.** Schmitt trigger

26. The resistor and capacitor values of the external components that make up the *RC* network.

27. 0.1 μF   **28.** high

30. Place a diode across R_b with the cathode lead connected to discharge pin 7 and the anode lead to threshold pin 6.

32. $T = 1.1RC = 1.1 \times 5 \text{ k}\Omega \times 5 \text{ μF} = 27.5 \text{ ms}$

35. 32, 11.25°   **37.** 0111

38. Pin B(2) is shorted to ground.   **39.** (c)

CHAPTER 12

1. (d)   **3.** 8   **5.** numerical, store, recall

7. Firmware is nonvolatile data permanently stored in ROM devices, whereas software is volatile data that can be changed at any time in RAM devices.

8. ROM devices only can have data read out of them, whereas data can both be easily written into and read out of RAM devices.

10. RAM, volatile   **11.** ROM, RAM

13. (1) Diodes, (2) bipolar transistors, and (3) MOSFETs

16. A program is a list of binary words that contain numerical data, memory locations, and computer instructions.

19. (1) Computer programs, (2) lookup tables, and (3) logic-circuit functions

21. dynamic

23. (1) The main working memory of a computer, (2) storing computer software, and (3) logic-circuit functions

25. $1024 \times 128 \times 8 = 1,048,576$ bits

26. Mass storage devices provide extra memory storage capacity for a computing device beyond its internal storage capacity. This extra capacity is needed by a bank, for example, that must record a large number of money transactions.

29.

Data Inputs				Address Inputs				$\overline{\text{WE}}$	$\overline{\text{ME}}$
D	C	B	A	D	C	B	A		
1	1	0	0	1	1	0	1	0	0

Address inputs				$\overline{\text{WE}}$	$\overline{\text{ME}}$	Data Outputs			
D	C	B	A			D	C	B	A
1	1	0	1	1	0	0	0	1	1

31. (b)   **32.** (d)

CHAPTER 13

1. Because the voltmeter consists of an up-counter, it is not capable of decrementing to a lower number if the voltage that it is reading decreases. Therefore, by clearing, it then counts up to a number that represents any new voltage to which the probe is connected. The voltmeter updates itself 15 times per second.

4. FF9 = 30 Hz, FF10 = 15 Hz, one-shot = 15 Hz (pulses)

5. (d)   **7.** 200 Hz

9. (c). The open at the middle of A_1 acts like a high, which allows the unknown frequency pulses through the gate while the Q of FFI output is high. This is twice the normal time period, which allows the frequency counter to count twice the normal number of pulses.

11. 50   **12.** (a)   **15.** (b)   **18.** (b)

20. Because the motor stopped as soon as it left sector 5 to go into sector 6. When it is required to move back to sector 5 from 6, it only has to travel less than 1 degree instead of 36 degrees in the direction opposite to that of the original rotation.

21. Because once 13 is in the counter, all of the inputs to N_1 are high, which produces a low at its output so that the hour counters are preset to 01. The duration of 13 is so brief that it cannot be seen.

23. The labels represent the $\overline{Q}$ outputs of flip-flops FF1 through FF7.

26. (d)   **27.** (c)

29. AND gate 5, which goes low because the middle input is pulled low.

31. False

33. (c)

35. 1, 1

37. CR, $\overline{P}_S$, $\overline{E}_S$

Accumulator: A shift register that receives the answer from a full adder at the completion of an arithmetic operation.

Adder: A circuit that performs the addition operation.

Address: A numerical value that designates a specific location in a memory device, or a circuit terminal to be enabled.

Analog signal: A voltage or current value that is proportional to the quantity it represents.

Analog-to-digital converter (ADC): A circuit that converts an analog voltage applied to its input into a proportional digital output.

AND gate: A basic logic device that produces a high at its output when all inputs are high.

Annihilator: A bubble memory device that erases data from the memory.

ASCII (American Standard Code for Information Interchange): An alphanumeric code used for word processing software programming.

Associative law: Allows parentheses to be removed when an enclosed expression is ANDed and that expression is also ANDed with an outside expression; the same rule applies to an ORed expression ORed to an external expression.

Astable multivibrator: A circuit that generates a continuous square wave output.

Asynchronous: Latches that respond to signals as soon as they are applied to the inputs.

Asynchronous counter: A basic counter consisting of several cascading *J–K* flip-flops that receive their clock signal from the output of the preceding flip-flop.

BCD counter: Also known as a decade counter, it counts in BCD from 0000 to 1001.

BCD-to-decimal decoder: A type of decoder that converts binary-coded decimal numbers into equivalent decimal values.

BCD number system: A number system that provides a way of encoding the digits of a decimal number into groups of binary digits.

BCD-to-seven-segment decoder/driver: A type of decoder that lights up an electronic readout to display one of the same numbers 0–9 as the BCD value applied to its input.

Bidirectional shift register: A serial shift register capable of moving data to the right or to the left.

Binary adder: A logic circuit that is capable of adding two bits and a carry and produces a sum and carry-out.

Binary counter: An up-counter capable of counting in pure binary.

Binary number system: Also called the base 2 system, it contains the two functional characters 0 and 1.

Bipolar: The type of integrated circuit that consists of miniature internal NPN or PNP transistors.

Bipolar transistor ROM: A type of read-only memory device that uses bipolar transistors to store each data bit.

Bistable multivibrator: Also known as a flip-flop, the circuit produces one logic level signal until an input signal causes it to produce the opposite logic level output signal.

Boolean addition: When binary numbers are added by an OR function.

Boolean multiplication: When binary numbers are multiplied by an AND function.

Byte: A binary word that is divided into eight bits.

Carry flip-flop: A flip-flop that accepts a carry-out bit from the full adder, and then places it back in the full adder for when the next column is added.

Chip: Another name for an integrated circuit.

Clock pulses: Pulses that are applied to digital circuits usually at precise fixed intervals.

Clock pulse triggering: When latches or flip-flops respond to a clock signal.

Combination circuits: Digital logic circuits made up of a combination of gates and inverters. Also referred to as *combination logic circuits.*

Commutative law: When two variables are ANDed or ORed, they will yield the same result.

Comparator: A type of operational amplifier circuit that produces an output voltage when both inputs are not at the same potential.

Complementary law (complementation law): If 0 and 1 are ANDed together, the result is 0; if 0 and 1 are ORed together, the result is 1.

Complementary operation: When a binary number is changed to the opposite binary number by an inverter function.

Complementation: The process of changing a logic level to the opposite logic level.

Computer word: Binary bits divided into organized groups that are stored into registers.

Counter: A common digital circuit made up primarily of flip-flops that tally the number of pulses arriving at its input.

Counter ramp: A type of analog-to-digital converter that uses a digital-to-analog converter to produce an analog waveform that resembles a staircase when digital numbers applied to its input increment.

Current sinking: Whenever an output lead of a digital device has conventional current flowing into it.

Current sourcing: Whenever an output lead of a digital device has conventional current flowing out of it.

Cutoff: When a transistor is in a state similar to an open switch because current does not flow through it.

Data conversion: The process of converting the movement of data from parallel to serial or from serial to parallel.

Data latch: A storage device capable of holding data one bit in length; several connected together form a storage register.

Data manipulation: The process of moving binary data by a shift register within itself.

Data sheets: Literature written by the IC manufacturer that provides information about the minimum and maximum operation conditions of each type of IC.

Data word: Binary bits divided into organized groups that are usually 4, 8, 16, or 32 bits in length.

Decade counter: An up-counter that can count 10 different numbers, 0 through 9.

Decimal number system: Also referred to as the base 10 number system, it contains the 10 characters 0 to 9.

Decimal-to-binary encoder: A type of encoding device that converts decimal inputs into equivalent binary values at the output leads.

Decoder: Any logic device that converts a binary code into an equivalent nonbinary code.

Demultiplexer: Circuit capable of transmitting data from an input line to one of several output terminals.

Digital-to-analog converter (DAC): A circuit that converts a digital value into a proportional analog output voltage.

Digital frequency counter: A type of test equipment that uses internal digital circuitry to measure and display the number of pulses, sine waves, or square waves that occur in a second.

Digital measuring instruments: Equipment that uses internal digital circuitry to measure quantities of sound, temperature, humidity, etc.

Digital signal: A voltage or current value that abruptly alternates between two different levels.

Digital voltmeter: A type of test equipment that uses internal digital circuitry to measure and display voltage levels.

Diode matrix: A type of read-only memory device that uses diodes to store each data bit.

Divider circuit: A network of logic gates and flip-flops capable of dividing binary numbers.

D latch: A storage device that has an inverter connected between its R and S inputs to ensure that data applied to them is always opposite.

Don't-care conditions: Xs used in truth tables to specify that the inputs where they are placed are overridden by another input condition.

Double negation law: A mathematical method which states that if a binary bit is complemented twice, the result is the binary bit itself.

Down-counter: A counter consisting of several cascading flip-flops that decreases its count by one on every clock pulse applied to its input.

Dual-in-line package (DIP): A type of integrated circuit package used in the construction of most digital electronic circuits.

Duty cycle: The ratio of time a square wave signal is high to the total time period of one cycle.

Dynamic RAM: A type of random-access memory device that uses a capacitive element to store a data bit.

Edge-triggering: When data applied to the input leads of a flip-flop is transferred to its output during the low-to-high or high-to-low transition of a clock pulse.

Electronically alterable read-only memory (EAROM): A type of memory device that can be reprogrammed after its previous contents are erased by a temporary applied voltage.

Electronically erasable programmable read-only memory (EEPROM): A type of memory device that can be reprogrammed through the use of a computer or special programming device.

Encoder: Any logic device that converts a nonbinary number code into an equivalent binary bode.

Erasable programmable read-only memory (EPROM): A type of memory device that can be reprogrammed after its previous contents are erased by shining ultraviolet light through a window located on the top of the IC package.

Exclusive-NOR gate: A basic logic device that produces a high at its output when the logic states applied to its inputs are the same.

Exclusive-OR gate: A basic logic device that produces a high at its output when the logic states applied to its inputs are different.

Exponent: A symbol written above and to the right of a larger number to indicate how many times the number is multiplied by itself.

Fan-in: The number of devices that can be connected to the input on a digital device.

Fan-out: The number of digital inputs that one output of a digital device can reliably drive.

Feedback: A technique used to control the gain of an operational amplifier.

Field-effect transistor (FET): A type of transistor used in the internal circuitry of CMOS ICs.

Firmware: Programs that are permanently stored in a ROM memory device.

Flip-flop: A memory device that is capable of storing one digital level during the transition of the enable signal applied to the clock input.

Floppy disk: A mass-storage memory device that is in the shape of a flat disk $3\frac{1}{2}$ inches in diameter on which data can be stored or from which data can be retrieved.

Frequency divider: A circuit that after receiving a certain number of input pulses produces an output pulse.

Full adder: A combination logic circuit capable of adding two binary bits and a carry-in bit, which generates a sum and carry-out.

Functional logic circuits: Standard combination logic circuits such as encoders, decoders, multiplexers, demultiplexers, parity circuits, adders, and comparators.

Gain: The amount an analog input voltage is increased by an amplifier at the output terminal.

Gate: A device with at least two inputs and one output that is capable of making logic decisions.

Glitch: An unwanted logic pulse that often results when two or more logic components operate at different speeds rather than simultaneously.

Gray code: A number system that uses multibit 0s and 1s; only one of the bits changes when incrementing or decrementing the count.

Hall-effect device: A conductor that changes resistance as the magnetic field it is within changes its strength.

Hard disk: A mass-storage memory device located inside a computer where data can be stored or retrieved.

Hexadecimal number system: Also referred to as the base 16 number system, it uses 16 different characters, which are both numbers and letters of the alphabet.

Idempotent law: A mathematical method that states if a variable is ANDed or ORed with itself, it will give the variable itself as an output.

Integrated circuit (IC): A miniature chip of silicon and germanium on which an entire electronic circuit, consisting of resistors, capacitors, diodes, and transistors, is built.

Interfacing: Electronically making different circuits compatible with one another in terms of speed, power dissipation, and logic-level voltages.

Invalid region: The voltage range between the minimum logic voltage level for a high state and the maximum logic voltage level for a low state.

Inversion: The process of changing a logic level to the opposite logic level.

Inverter: A circuit that changes the logic level applied to its input to the opposite level at the output.

J–K flip-flop: A universal type of flip-flop that can function as most other types of flip-flops.

Large-scale integration (LSI): ICs that contain 20,000 components on a single silicon chip.

Latch: A basic type of storage element capable of storing one bit of information when its clock input is at the required logic level.

Law of intersection: A mathematical method that states that if a 0 is ANDed with a 1, the result is 0, or if a 1 is ANDed with a 1, the result is 1.

Law of union: A mathematical method that states that if an ORed input is 0, the output depends on the other inputs; if an input is 1, the output is 1.

Level-triggered latch: A latch that allows data applied to its input to be transferred to its output when the signal applied to the clock input is at a high or a low.

Light-emitting diode (LED): A device that allows current to flow through it in one direction and gives off light when current passes through it.

Logic gates: The most basic type of digital circuit, which consists of two or more inputs and one output.

Magnetic bubble memory (MBM): A mass-storage memory device that uses individual microscopic bubbles to store binary bits of data.

Magnetic tape: A mass-storage memory device that uses tape to magnetically store binary information.

Magnitude comparator: A circuit that compares two binary numbers and indicates whether one number is larger than, less than, or equal to the other.

Mask-programmable ROM: A type of read-only memory device that can only be programmed by the IC manufacturer.

Mass memory: External memory devices that are used by a computing device when the memory capacity requirements of the internal memory are exceeded.

Medium-scale integration (MSI): ICs that contain 500 components on a single chip.

Memory capacity: The total quantity of data that memory devices can store.

Memory cell: A basic element of a memory that stores one bit of information.

Memory circuits: Circuits found in digital equipment that are used to store data.

Memory device: A device capable of storing both types of digital signals.

Metal-oxide semiconductor (MOS): The type of integrated circuit that consists of miniature internal FETs.

Microprocessor: The central processing unit of a computer, which is in the form of an integrated circuit.

Mod-6 down-counter: A counter that makes six counts backwards from 5 to 0 before recycling back to 5.

Mod-6 up-counter: An up-counter capable of counting six counts from 0 to 5 before recycling back to 0.

Mod-10 down-counter: A counter that makes 10 counts backwards from 9 to 0 before recycling back to 9.

Modulo: The maximum number of counts a counter is capable of making.

Modulo-division method: A mathematical method of converting decimal numbers into other number systems.

Monostable multivibrator: Also known as a one-shot, a circuit that produces a temporary logic-level voltage after an activating signal is applied to its input.

MOS ROM: A read-only memory device that uses miniature field-effect transistors to store binary data.

Multiplexer: A logic circuit that directs data from one of several inputs to a single output.

Multiplication circuit: A network of logic gates and flip-flops capable of multiplying binary numbers.

Multivibrator: A device that has two complementary outputs that produce rectangular output pulses.

NAND gate: A basic logic device that produces a low at its output when all inputs are high.

Negative-edge-triggered flip-flop: A flip-flop that allows data applied to its input to be transferred to its output when the high-to-low transition of a signal occurs.

Nibbles: Binary bits split into groups of four.

Node: A circuit junction point that is common to two or more gates or other elements.

Noise: An electromagnetic field produced by various environmental effects and artificial devices that can induce an unwanted signal into a circuit conductor path.

NOR gate: A basic logic device that produces a high when all inputs are low.

NOT gate: Another word used in place of an inverter, the term is derived from its output that produces "not A" when an A is applied to its input.

Octal-to-binary encoder: A type of encoder that converts octal inputs into equivalent binary output values.

One-dimensional memory: A type of memory configuration that uses one decoder to address a memory device.

Operational amplifier: An analog electronic circuit that has a high input impedance, low output impedance, and a high voltage gain.

Optical encoder: A circuit that uses opto couplers to sense the positional location of a linear or rotary device.

OR gate: A basic logic device that produces a high at its output when at least one of its inputs is high.

Oversampling: A method used by electronic circuitry to read a digital signal several times and take the average of the readings before using it.

Parallel adder/subtractor circuit: A network of logic gates and flip-flops capable of adding or subtracting two multibit binary numbers quickly by using one clock pulse.

Parallel addition: The process of adding multibit binary numbers simultaneously.

Parity circuit: A device that detects an unwanted change in a logic signal while it's being transmitted from one location to another.

Phoneme: One of 64 basic sounds that make up a speech pattern.

PIPO (parallel-in parallel-out): A type of shift register that enters and removes data in parallel form.

PISO (parallel-in serial-out): A type of shift register that enters data in parallel form and removes data in serial form.

Positive-edge-triggered flip-flops: A flip-flop that allows data applied to its input to be transferred to its output when the low-to-high transition of a signal occurs.

Product-of-sums equation: A logic expression that represents inputs that feed OR gates and outputs that feed an AND gate.

Program: A list of instructions stored in a memory device that tells a computing device what operations to perform.

Programmable controller: An industrial computer into which ladder diagrams are programmed.

Programmable logic device (PLD): An IC that contains a large array of logic gates that can be programmed to achieve specific logic functions.

Programmable read-only memory (PROM): A read-only memory device into which data is permanently programmed by a special programming device.

Propagation delay: The delay time from when an input transition of an IC takes place until the output changes states.

Rack and pinion: A mechanical device that is made up of a gear and a bar with gear teeth that converts rotary motion into linear movement.

Radix: The sum of the different symbols used in a given number system.

Read-only memory (ROM): A semiconductor memory device from which data can only be read and in which data is permanently stored.

Recirculating shift register: The process of preserving data in a SISO shift register by reloading the data that leaves the output back into the input.

Register: A circuit consisting of several flip-flops that performs storage, counting, data manipulation, or arithmetic operations.

Reprogrammable read-only memory: A type of ROM device that can be reprogrammed through the use of special equipment.

Resolution: The number of equal divisions a digital-to-analog converter divides the reference voltage into.

Ripple counter: A counter consisting of several cascading J–K flip-flops that receive their clock signal from the output of the preceding flip-flop.

Saturation: When a transistor is in a state similar to a closed switch because almost all the current flows through it.

Schmitt trigger: A device that produces sharply defined square waves from distorted signals or sine waves.

Sequential circuit: A circuit made up of several flip-flops that are classified into two categories, counters and registers.

Serial adder circuit: A network of logic gates and flip-flops capable of adding two multibit binary numbers one column at a time.

Serial addition: The process of adding multibit binary numbers one column at a time.

Serial subtractor circuit: A network of logic gates and flip-flops that subtract two multibit binary numbers one column at a time using the 2's complement method.

Shift register: Circuits consisting primarily of flip-flops that are used to store and move data within itself.

SIPO (serial-in parallel-out): A type of shift register that enters data in serial form and removes data in parallel form.

SISO (serial-in serial-out): A type of shift register that enters and removes data in serial form.

Small-scale integration (SSI): ICs that contain 20 to 50 components on a single silicon chip.

Software: Programs that are temporarily stored in a RAM memory device.

Sound synthesizer: A digital device that electronically converts digital data into almost any sound.

Stacked memory: Memory architecture where memory cells are placed on planes that are stacked on top of each other.

Static RAM: A memory device that uses flip-flops for its basic storage cell element.

Storage device: Memory circuits used by digital equipment to store data.

Storage element: A basic digital circuit capable of the memory function.

Storage registers: Circuits consisting primarily of flip-flops that are used to temporarily store digital information.

Subtract and count method: A procedure that obtains an answer for division by subtracting numbers.

Successive-approximation register (SAR): A type of analog-to-digital converter that requires the same number of clock pulses to operate as the number of digital bits being converted.

Sum-of-products equation: A logic expression that represents inputs that feed AND gates and outputs that feed an OR gate.

Sum of weighted coefficient method: A mathematical method of converting numbers from most number systems into decimal numbers.

Super large-scale integration (SLSI): ICs that contain over 100,000 components on a single silicon chip.

Switch bounce: When a mechanical switch is closed, it bounces several times before its contacts come to rest.

Synchronous: Latches that respond to data signals applied to their inputs only when they are enabled by a clock signal.

Synchronous counter: A counter in which all of its flip-flops are triggered simultaneously by the same clock pulse.

T flip-flop: Also known as a toggle flip-flop, it changes states on every triggering portion of a clock signal.

Thermistor: An electronic sensing component that varies its resistance when the temperature around it changes.

Timing circuit: A device that produces a continuous square wave output in a digital-type circuit.

Totem pole: A pair of miniature transistors in an IC with their collector-to-emitter circuits in series to provide fast switching capabilities.

Transient: An unwanted electromagnetic signal picked up on a transmission line that carries digital signals.

Transistor-transistor logic (TTL): A popular IC that uses transistors at both the inputs and outputs.

Truth table: A graphical method of showing the output behavior of a logic device for every possible set of input conditions.

Two-dimensional memory: A type of memory configuration that uses two decoders to address a memory device.

2's complement notation: A method of subtracting binary numbers through addition.

Two-state logic: Logic devices that produce either a 0 or 1 at their output.

Universal shift register: A register capable of functioning in any of the SISO, SIPO, PISO, and PIPO modes of operation.

Up-counter: A counter consisting of several cascading flip-flops that increases its count by one on every clock pulse applied to its input.

Very large-scale integration (VLSI): ICs that contain between 20,000 and 100,000 components on a single silicon chip.

Voice synthesizer: A digital device that electronically converts digital data into human speech.

Volatile: Stored information in a memory device that is lost when power is removed from the circuit.

Voltage-summing amplifier: A type of operational amplifier configuration capable of adding the algebraic sum of several voltages applied to one of its input lines.

Wire anding: When multiple open collector outputs are connected together.

Word: Binary bits that are divided into organized groups.

Working memory: The memory devices located in computing equipment that continually store and retrieve data during its operation.

INDEX